Lecture Notes in Networks and Systems

Volume 607

The series "Lecture Notes in Networks and Systems" publishes the latest developments in Networks and Systems—quickly, informally and with high quality. Original research reported in proceedings and post-proceedings represents the core of LNNS.

Volumes published in LNNS embrace all aspects and subfields of, as well as new challenges in, Networks and Systems.

The series contains proceedings and edited volumes in systems and networks, spanning the areas of Cyber-Physical Systems, Autonomous Systems, Sensor Networks, Control Systems, Energy Systems, Automotive Systems, Biological Systems, Vehicular Networking and Connected Vehicles, Aerospace Systems, Automation, Manufacturing, Smart Grids, Nonlinear Systems, Power Systems, Robotics, Social Systems, Economic Systems and other. Of particular value to both the contributors and the readership are the short publication timeframe and the world-wide distribution and exposure which enable both a wide and rapid dissemination of research output.

The series covers the theory, applications, and perspectives on the state of the art and future developments relevant to systems and networks, decision making, control, complex processes and related areas, as embedded in the fields of interdisciplinary and applied sciences, engineering, computer science, physics, economics, social, and life sciences, as well as the paradigms and methodologies behind them.

Indexed by SCOPUS, INSPEC, WTI Frankfurt eG, zbMATH, SCImago.

All books published in the series are submitted for consideration in Web of Science.

For proposals from Asia please contact Aninda Bose (aninda.bose@springer.com).

More information about this series at https://link.springer.com/bookseries/15179

Vladimir Robles-Bykbaev ·
Josefa Mula · Gilberto Reynoso-Meza
Editors

Intelligent Technologies: Design and Applications for Society

Proceedings of CITIS 2022

Springer

Editors
Vladimir Robles-Bykbaev
Campus El Vecino
Universidad Politécnica Salesiana
Cuenca, Ecuador

Gilberto Reynoso-Meza
Rua Imaculada Conceiçao
Pontifícia Universidade Católica do
Paraná – PUCPR
Curitiba, Brazil

Josefa Mula
Research Centre on Production Management
and Engineering - CIGIP
Universitat Politècnica de València
Alcoy, Spain

ISSN 2367-3370 ISSN 2367-3389 (electronic)
Lecture Notes in Networks and Systems
ISBN 978-3-031-24326-4 ISBN 978-3-031-24327-1 (eBook)
https://doi.org/10.1007/978-3-031-24327-1

This Springer imprint is published by the registered company Springer Nature Switzerland AG
The registered company address is: Gewerbestrasse 11, 6330 Cham, Switzerland

Preface

This book comprises the papers accepted for presentation and discussion at the 8th International Conference on Science, Technology and Innovation for Society (CITIS 2022). CITIS is an international conference on engineering at the service of social transformation that has been organized since 2014 by Universidad Politécnica Salesiana, Ecuador. The main goal of the conference is to propose a platform so that engineering professionals, academics, and students can critically and objectively rethink the theoretical and practical problems of the most significant impact on modern society through an engineering perspective. Due to the pandemic affecting humanity since the beginning of 2020 and the social conflicts that took place in Ecuador during the event days, the eighth edition of CITIS took place as a hybrid edition hosted from Guayaquil during June 22–24, 2022.

The Program Committee of CITIS 2022 was composed of a multidisciplinary group of 95 experts from 19 countries who were responsible for evaluating the papers received at the conference in a 'double-blind review' process. The main topics covered in this edition focused on six thematic axes: science, technology, and innovation for health, engineering education, industry, mobility, environmental sustainability, information, and telecommunications. The accepted contributions recreate recent research trends in applications regarding several topics, like biomechanics, augmented reality, cloud communication, software engineering, nanotechnology, machine learning, Internet of Things, virtual learning environments, processes control, power systems, artificial vision, simulation, water pollution, photovoltaic systems, mechatronics, and production engineering.

CITIS 2022 also included nine workshop sessions, and nine keynote speeches were given on cutting-edge topics related to the thematic axes of the conference. In addition, parallel working sessions were held to strengthen entrepreneurship and innovation skills for engineering students.

We acknowledge all of those that contributed to the staging of CITIS 2022 (authors, committees, workshop instructors, and our attendees). We deeply appreciate their involvement and support that was crucial for the success of CITIS 2022.

June 2022

Vladimir Robles-Bykbaev
Josefa Mula
Gilberto Reynoso-Meza

Contents

Mobility

Education

Health

Evaluation of Biomechanical Conditions in Infants from 20 to 48 Months of Age in Gait

Yessenia Maribel Armas, Victor Hugo Cabrera(✉), and Adriana López

Universidad Politécnica Salesiana, Quito, Ecuador
vcabrera@ups.edu.ec

Abstract. The aim was to analyze independent child gait in children 20 to 48 months with typical motor development to establish a valid set of angular position data. A total of five children were recruited for this research. Five Xsens MTw Awinda wearable inertial movement sensors located at the pelvis, between L4/L5 OR S1/S2, the quadriceps and anterior tibialis were used to quantify the gait cycle on rigid floors. Five measurements were taken from each child and only the four bests were taken. Data analysis was performed using polynomial regression by least squares and statical methods: Pearson's correlation and coefficient of determination R^2. The patterns of the movement curves resulting from the data sets of the evaluated joints exhibit similarities between children of different ages. To corroborate the information obtained, a comparison was made with an already established database for the same measurement parameters, achieving similarity percentages of 78% for pelvic rotation as well as 53 and 79% for pelvic obliquity. Results suggest that is essential to extend the age range of the children, as well as to increase the numbers of participants evaluated to establish a follow-up in the complete maturation of the infantile gait.

Keywords: Infant gait · Infant locomotion · Kinematics · IMMU · Sensors

1 Introduction

Within infant locomotion, upright and independent walking is a great achievement and the opportunity to learn about the environment and adapt, as a result of improved locomotor efficiency [1], however, this efficiency requires coordinated movements of the limbs in a dynamic balance, even in stationary posture is necessary body stability that allows the movement of the arms and head [2], because of this children learn various possible combinations of body joints [3] that allow them to walk and develop physically, cognitively, perceptually and socially [4].

Infants, after going through several motor milestones as shown in Fig. 1, take their first steps, on average, at 12 months, or in a range from 8 to 18 months, these steps are oscillating, uneven, with long periods when both feet are on the ground and very short periods when they move one foot [5], the walking experience of 3 to 6 months causes infants to have better control of balance and upright position and complete development and control of gait, between 5 to 7 years of age [5].

V. Robles-Bykbaev et al. (Eds.): CITIS 2022, LNNS 607, pp. 3–12, 2023.
https://doi.org/10.1007/978-3-031-24327-1_1

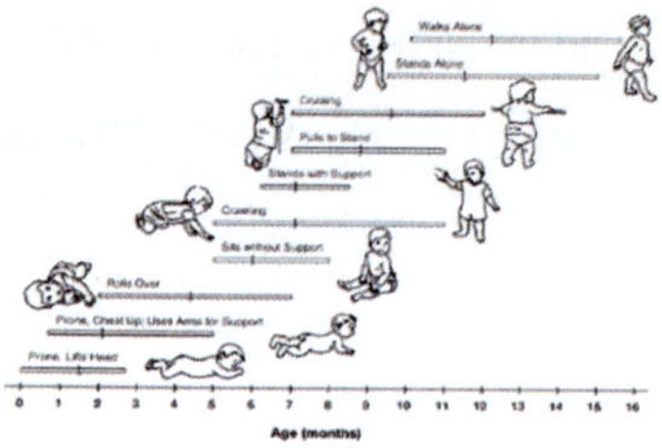

Fig. 1. Infant motor milestones chart showing the progression of postural and locomotor skills and age norms. Horizontal bars represent the normative range of skill onset; vertical lines show the average age of first onset [2].

Research or evaluations performed on gait in infants in general have increased considerably to analyze mainly the feasibility of early medical interventions to reduce or eliminate disabilities caused by motor deficits that some children present, because motor activity has been shown to directly influence the full typical development of infants. Neuromuscular coordination controlled by the central neural system (CNS) is complex [6].

Inertial sensors, introduced to the biomechanical evaluations of the human being, emerge as an alternative to those expensive methods that are even capable of capturing the body movement in 3D, or to studies that are limited because they are performed in laboratories.

The objective of this research was to analyze one of the most important milestones of infant locomotion by evaluating the biomechanics of independent gait in infants between 20 and 48 months of age to confirm their correct motor development or to infer the possibility of a neuromotor anomaly or deficiency, gait quantification was performed through the use of portable inertial sensors that analyze movement both kinetically and kinematically, generating a representative data set that can be studied and compared with infant locomotion parameters already established for children with typical physical and neural development.

The research paper below is made up of the characteristics of the participants, the methodology used to carry out the biomechanical analysis of independent walking in children, the final results obtained, and the corresponding conclusions.

2 Materials and Methods

2.1 Participants

Five children with TD, both physical and neural, born without complications, according to their pediatricians, and with differentiated characteristics described in Table 1, were recruited for the study, in the Metropolitan District of Quito, Pichincha, Ecuador.

Table 1. Infant characteristics.

P	Age (Months)	Sex	Pregnancy (Weeks)	Mass (kg)	Height (cm)	First steps (Months)
1	21	M	40	13.5	82	15
2	42	F	40	12.05	96.5	14
3	42	F	38	16.8	99	14
4	47	F	40	17.73	97	14
5	52	F	38	18.64	100	12

2.2 Method

The possibility of monitoring unrestricted human motion has popularized magnetic and inertial measurement units (MIMU or IMMU) that have the advantage of being small, portable, and wireless [7]; they are known as inertial sensors because they measure their own motion and consequently measure the motion of the rigid body to which they are attached [8].

The investigation was carried out using portable inertial sensors Xsens MTw Awinda (Xsens Technology B.V., Enschede, The Netherlands) [9], composed of 3D accelerometers, 3D gyroscopes, 3D magnetometers, a thermometer, and a barometer; they are small (47 mm × 30 mm × 13 mm), light (16 g) and with a wireless communication interface [10].

Data Collection. Data logging, consisting of triaxial acceleration (±16 g), triaxial angular velocity (±2000 deg/s) and triaxial magnetic fields (± 1.9 Gauss) was performed at a sampling rate of 100 Hz, data transmission from each of the five sensors was done wirelessly and in real time to the Awinda station, connected via USB to a laptop computer that stored all the collected data in the Xsens MT Manager software; data synchronization, with high accuracy, between the MTw sensors was less than 5 μs [10, 12–14].

For gait quantification, as shown in Fig. 2, the sensors, by means of Velcro straps, were distributed on the infant, on both ankles, exactly on the tibiae (distal tibiae or lateral malleoli); in the quadriceps (between the vastus lateralis and the iliotibial tract) of each leg near the knees and at the L4 or L5 level of the lumbar vertebrae, slightly above the pelvis or in the sacrum, specifically in the first or second sacral vertebra, S1 or S2 (in children under 24 months of life) [10, 11].

The standardized Belluscio et al. [11] process of three repetitions of a 10 m straight walking test, with a start line marked on the floor as well as a finish line, was used. This process was modified to be able to evaluate the 21-month-old infant P1 since he was the only one who presented problems following instructions.

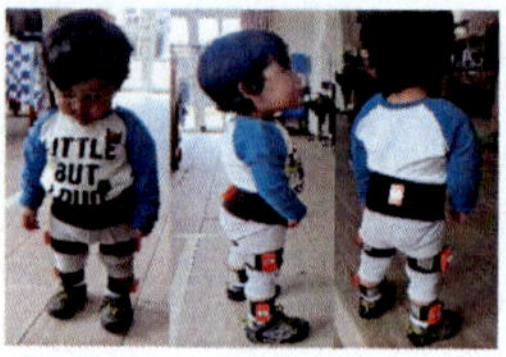

Fig. 2. Location of Xsens MTw Awinda sensors for gait assessment in a test infant: (a) front view, (b) side view, and (c) back view.

To compensate for the decrease of the 10 m of the walking test, the repetitions were increased and the children walked five times on a rigid surface of cement or tile of 7 m long at a walking speed selected by them, a start and finish line was also marked on the floor to establish the limits of the measurement.

2.3 Data Analysis

The sensors were programmed in incremental orientation to measure and store Euler angles, specifically the rotation in X axis known as Roll, ϕ, in Y axis called Pitch, θ, and in Z axis set as Yaw, Ψ [15]; of each of the extremities where the Xsens MTw Awinda were placed. With the objective of eliminating possible biases in the gait cycle measurements of each of the children, Eq. 1 implemented in the study of Garza-Ulloa [16] was resorted to.

$$\theta_{knee} = \theta_{shank} + (180 - \theta_{thigh}) \tag{1}$$

where θknee corresponds to the knee angle, θshank to the tibialis anterior angle and θthigh to the quadriceps angle, these three angles determined in degrees, ° and visualized in the schematic in Fig. 3 below.

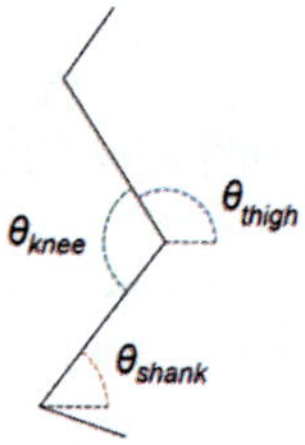

Fig. 3. Diagram of the knee, quadriceps, and tibialis anterior angle of movement in the sagittal plane.

To conclude the study, a comparative analysis was carried out between the graphs generated by the data obtained from the portable sensors and the graphs obtained from the investigations carried out by Calderón-Ulloa [17], to corroborate the validity of the data obtained by the Xsens MTw Awinda sensors.

3 Results and Discussion

The data generated by the IMMUs in the measurement of infant gait, after their location in the two quadriceps, the two distal tibiae or tibialis anterior and the pelvis either between the lumbar vertebrae L4/L5 or in the sacrum, between the sacral vertebrae S1 or S2.

Between 8 to 18 samples were determined for each child, for the two quadriceps, the two tibialis, pelvic obliquity, pelvic rotation, and the knees, the latter presenting the fewest samples, especially the right knee from which between 8 to 12 patterns were obtained.

The schematization of the data of the tests of each infant was performed in MATLAB software, each child was assigned a color according to his age and average curves were generated to perform the polynomial regression.

The number of samples obtained for the quadriceps and for the tibialis determined the number of samples that could be obtained for the knees, by using Eq. 1 already mentioned above.

From Fig. 4, it can be noted that there is a great similarity between the patterns of each of the participants for each of the limbs evaluated, the most notable example is that of Fig. 4(e), which demonstrates a similarity above 80% between the curves of each child throughout the gait cycle. On the other hand, it should be noted that the greatest variation between the patterns is found in Fig. 4 (d) corresponding to the pelvic rotation. This differentiation is generated by the increased dynamic instability in infants with less experience in walking [2], and which is reflected in the unequal patterns of P1 and P4, which correspond to infants with lower and greater physical and motor development respectively.

As shown in Fig. 4, all the graphs are result of the best adjustment of polynomial regression by least squares to which the average values of the data generated by the sensors were subjected, to carry out the regression we resorted to MATLAB software and the commands polyfit(x, y, n), which is in charge of finding the coefficients of the regression polynomial and polyval(p, x) which returns the values of the regression polynomial p(x) with degree n determined by polyfit and evaluated at x, that is, polyval evaluates the polynomial p at each point in x.

Finally, we proceeded to make a comparison of each of the measured limbs or joints with the data set resulting from the research in adults carried out by Calderón-Ulloa [17] to establish how similar they are.

Figure 5 shows the sagittal plane curves obtained in the gait cycle for pelvic obliquity (a), pelvic rotation (b), flexion and extension of the right quadriceps (c), flexion and extension of the left quadriceps (d), flexion and extension of the right tibialis anterior (e), flexion and extension of the left anterior tibialis (f), flexion and extension of the right knee (g) and flexion and extension of the left knee (h), of the five children evaluated and the final curves of the database belonging to the research carried out by Calderón-Ulloa [17].

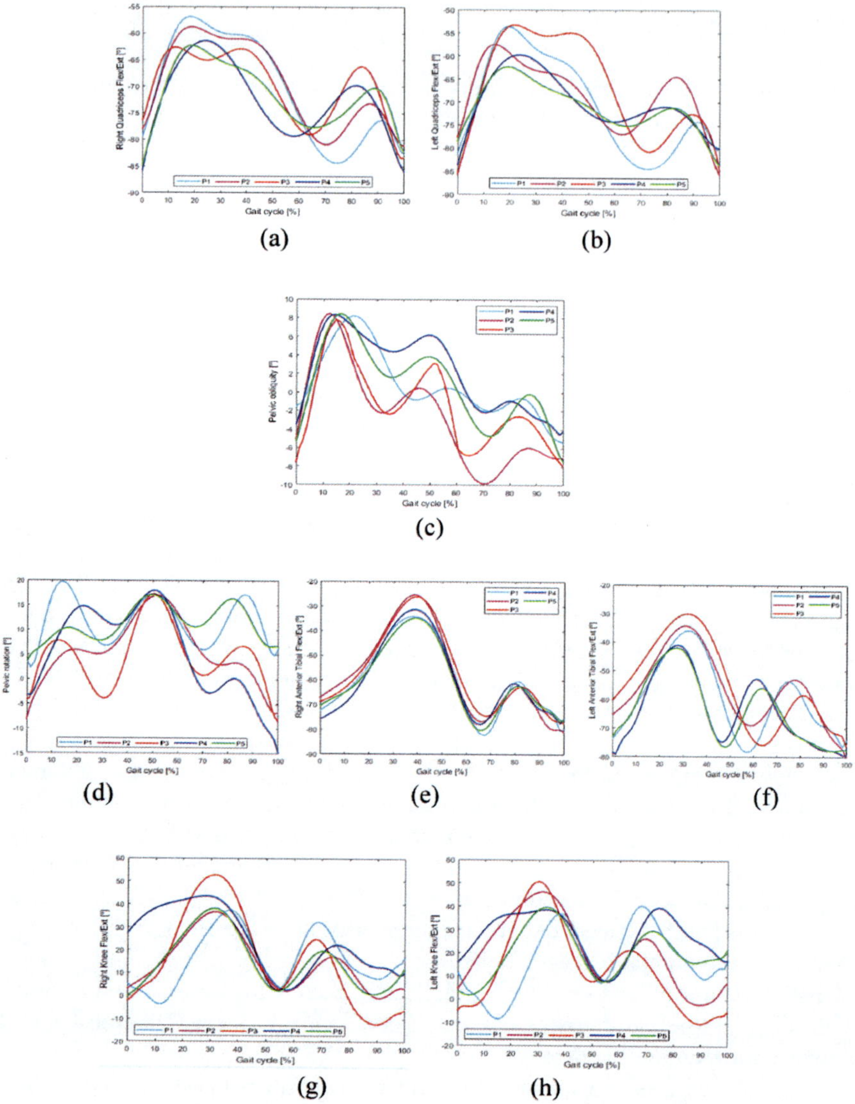

Fig. 4. (a) and (b). Angular displacement of the right and left quadriceps in the sagittal plane during gait of each of the infants. (c) and (d) Pelvic obliquity and pelvic rotation in the sagittal plane during gait of each of the infants. (e) and (f) Angular displacement of the right and left tibialis in the sagittal plane during gait of each of the infants. (g) and (h) Angular displacement of the right and left knee in the sagittal plane during gait for each of the infants.

The difference found when comparing the database provided by the sensors and adjusted by least squares and the data of Calderón-Ulloa [17] was expected, since even though it is the same gait cycle in sagittal plane, the age of the people evaluated is completely different, consequently, their development and maturation of independent

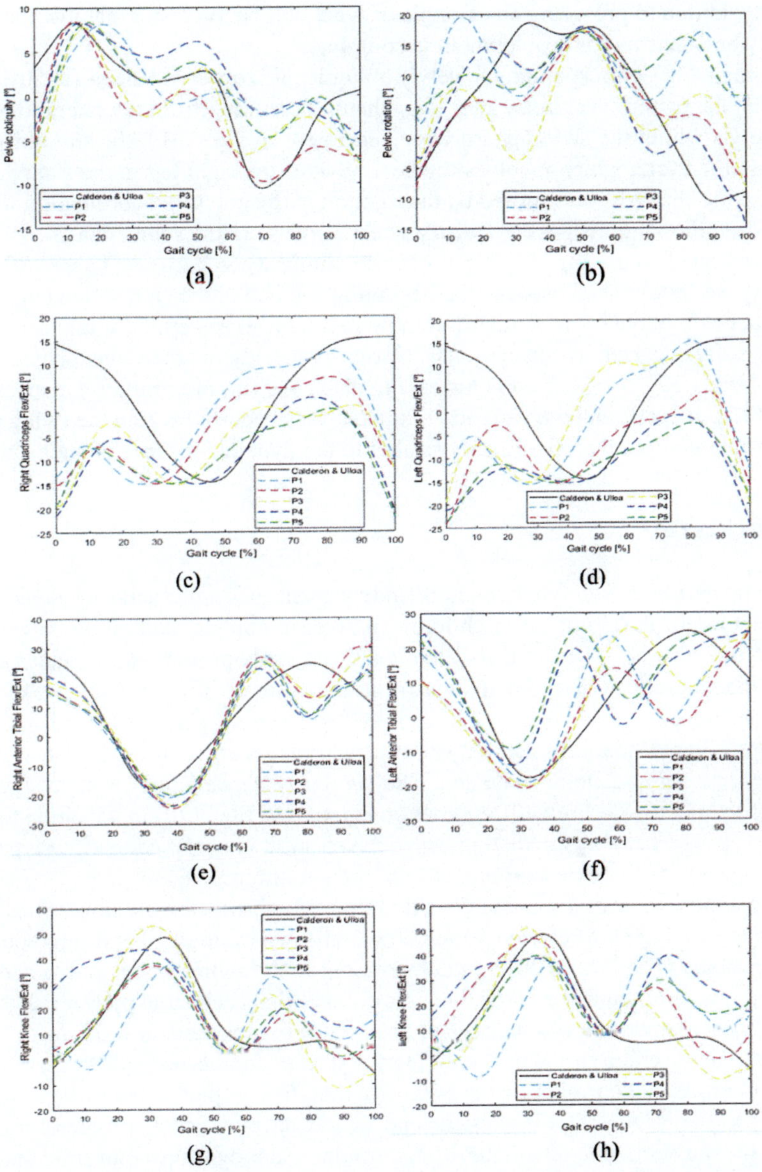

Fig. 5. Gait cycle's patterns comparison: (a) pelvic obliquity, (b) pelvic rotation, (c) right quadriceps flexion and extension, (d) left quadriceps flexion and extension, (e) right anterior tibialis flexion and extension, (f) left anterior tibialis flexion and extension, (g) right knee flexion and extension, and (h) left knee flexion and extension.

gait is also different. However, what is intended with this comparison is to explain what this differentiation between infants and Calderón-Ulloa represents [17], within the gait cycle. In addition to providing an example of what will be the result that infants should achieve when their motor development is complete.

From Fig. 5(a), corresponding to pelvic obliquity, it is observed that up to 80% of the gait cycle, the pattern is similar, after this, changes resulting from the lateral deviation of the pelvis within the swing phase were generated. In Fig. 5 (b), the similarity is up to 15% of the cycle, which involves the heel blow of the right leg in the stance phase. After this, the changes were caused by the rotation of the pelvis in coordination with the upper extremities and the passage from the stance phase to the swing phase.

For the quadriceps, Fig. 5 (c) and (d), the similarity is between 18 and 80%, the inequality was established between the beginning and end of the gait, which may be due to upright posture, bone structural differences in the lower extremity or the fact that the first inequality occurred within the stance phase and the other in the swing phase. For the anterior tibials, Fig. 5 (e), 5(f), and knees, Fig. 5 (g) and (h), the similarity reaches up to 60% of the gait cycle, and the variation of the patterns occurred within the swing phase, so the differentiation is proportionally related to the dynamic balance of each child.

4 Conclusions

The kinetic and kinematic parameters of independent gait are crucial for determining the correct motor development of children and their complete maturation between the ages of 5 to 7 years [5]. Due to this fact, the present study performed to infant gait was through the quantification of the angular position parameter for the pelvis, quadriceps, tibialis, and knee.

Figure 4 shows the similarities between the patterns generated from the data provided by the sensors for each joint evaluated. The similarity between the children's patterns is greater than 70%. The only differentiation is found in Fig. 4 (d) in which up to three different types of patterns are distinguished. Even so, it is observed that these patterns represent the progressive development of children at different ages.

The comparison with the Calderón-Ulloa database shows that the similarities of the patterns for each joint are between 50 and 60% of the gait cycle and that it corresponds to the stance phase. The largest differentiations occur in the swing phase, and as shown in Fig. 5 (b), it is a consequence of the dynamic balance that each child presents according to their physical, motor development and experience in independent walking.

The age of the children and the use of inertial sensors for this research has established important results that corroborate the statement that the complete maturation of the gait cycle depends on the infant's physical and neural developmental progression, since between the range of 20 to 52 months of age similar and progressive movement patterns have been generated that will continue to modify until reaching the patterns established by adults.

Acknowledgements. The authors would like to thank the volunteers who participated as study subjects, and the parents who gave their consent for their children to participate in this research.

References

1. Mangalindan, D.M.J., Schmuckler, M.A., Li, S.-A.: The impact of object carriage on independent locomotion. Infant Behav Dev. **37**(1), 76–85 (2014). https://doi.org/10.1016/j.infbeh.2013.12.008
2. Adolph, K.E., Rachwani, J., Hoch, J.E.: Motor and physical development: locomotion. In: Encyclopedia of Infant and Early Childhood Development, pp. 347–363. Elsevier (2020). https://doi.org/10.1016/B978-0-12-809324-5.05848-X
3. Hsu, W.-H., Miranda, D.L., Chistolini, T.L., Goldfield, E.C.: Toddlers actively reorganize their whole-body coordination to maintain walking stability while carrying an object. Gait Posture **50**, 75–81 (2016). https://doi.org/10.1016/j.gaitpost.2016.08.023
4. Looper, J., Chandler, L.S.: How do toddlers increase their gait velocity? Gait Posture **37**(4), 631–633 (2013). https://doi.org/10.1016/j.gaitpost.2012.09.009
5. Adolph, K.E., Franchak, J.M.: The development of motor behavior. Wiley Interdiscip. Rev. Cogn. Sci. **8**(1–2), e1430 (2017). https://doi.org/10.1002/wcs.1430
6. Fang, X., Liu, C., Jiang, Z.: Reference values of gait using APDM movement monitoring inertial sensor system. R. Soc. Open Sci. **5**(1), 170818 (2018). https://doi.org/10.1098/rsos.170818
7. Bergamini, E., Ligorio, G., Summa, A., Vannozzi, G., Cappozzo, A., Sabatini, A.M.: Estimating orientation using magnetic and inertial sensors and different sensor fusion approaches: accuracy assessment in manual and locomotion tasks. Sensors **14**(10), 18625–18649 (2014). https://doi.org/10.3390/s141018625
8. Iosa, M., Picerno, P., Paolucci, S., Morone, G.: Wearable inertial sensors for human movement analysis. Expert Rev. Med. Devices **13**(7), 641–659 (2016). https://doi.org/10.1080/17434440.2016.1198694
9. Al-Amri, M., Nicholas, K., Button, K., Sparkes, V., Sheeran, L., Davies, J.: Inertial measurement units for clinical movement analysis: Reliability and concurrent validity. Sensors **18**(3), 719 (2018). https://doi.org/10.3390/s18030719
10. Hardware, M., Manager, M.T., Protocol, A:. MTw Awinda User Manual. Xsens.com. [cited 2021 Apr 1]. Available from: https://documentation.xsens.com/mtw_user_manual
11. Belluscio, V., et al.: Dynamic balance assessment during gait in children with down and Prader-Willi syndromes using inertial sensors. Hum. Mov. Sci. **63**, 53–61 (2019). https://doi.org/10.1016/j.humov.2018.11.010
12. Zhou, Lin, et al.: How we found our IMU: guidelines to IMU selection and a comparison of seven IMUs for pervasive healthcare applications. Sensors **20**(15), 4090 (2020). https://doi.org/10.3390/s20154090
13. Auepanwiriyakul, C., Waibel, S., Songa, J., Bentley, A.P., Faisal, A.: Accuracy and acceptability of wearable motion tracking for inpatient monitoring using smartwatches. Sensors **20**(24), 7313 (2020). https://doi.org/10.3390/s20247313
14. Paulich, M., Schepers, M., Rudigkeit, N., Bellusci ,G.: Xsens MTw awinda: miniature wireless inertial-magnetic motion tracker for highly accurate 3D kinematic applications. Xsens Technologies B.V, Enschede, The Netherlands (2018). https://doi.org/10.13140/RG.2.2.23576.49929
15. Jouybari, A., Amiri, H., Ardalan, A.A., Zahraee, N.K.: Methods comparison for attitude determination of a lightweight buoy by raw data of IMU. Measurement **135**, 348–354 (2019). https://doi.org/10.1016/j.measurement.2018.11.061

16. Garza-Ulloa, J.: Experiment design, data acquisition and signal processing. In: Applied Biomechatronics using Mathematical Models, pp. 179–237. Elsevier (2018). https://doi.org/10.1016/B978-0-12-812594-6.00004-4
17. Calderón, D.M.J., Ulloa, J.R.A.: Cambios asociados al envejecimiento normal en los parámetros angulares de la marcha a una velocidad controlada. Rev. Med. Chil. **144**(1), 74–82 (2016). https://doi.org/10.4067/S0034-98872016000100010

Comparison of Biomechanical Conditions in Persons with Overweight Between 18 and 35 Years Old to Obtain Characteristic Curves with the Use of Inertial Sensors

Gerson Moreta, Diego Ortega, and Víctor Cabrera(✉)

Universidad Politécnica Salesiana, Quito, Ecuador
vcabrera@ups.edu.ec

Abstract. This article analyze gait in healthy subjects with a body mass index greater than 25, classifying them according to the World Health Organization as overweight people, and according to their range at different levels. The Xsens Awinda sensors are placed in 5 different parts of the lower body, these being: the center of gravity, quadriceps, and anterior tibialis, a 10-m walk test is performed. In this way it is possible to determine the characteristic curves for each part of the body and through the analysis, the knee graphs. The gait patterns are obtained for each subject, which are analyzed in conjunction with a database for people who do not present this condition, resulting in comparison graphs, and the error in their approximation, thus determining the existence of a gait pattern.

Keywords: Inertial · Posture · Overweight

1 Introduction

The overweight and obesity have become in recent years a global trend that worries doctors and researchers, this condition is directly related to the appearance and prevalence of some diseases. Sohrab [1] In his research considers that it is a great risk to the public health of the countries due to the fact of having a greater incidence in the appearance of cardiovascular diseases, as well as diabetes. On the other hand, Villarasa et al. consider that [2] the musculoskeletal system and the incidence of overweight at an early age should also be analyzed, since this is related to posture and related diseases.

The objective set for this research is the determination of the gait patterns obtained from people who have a body mass index greater than 25 with the use of commercial inertial sensors, based on balance and gait measured in the lower body. To check the impact that walking can have on this group that grows more and more every day. The expected results are the characteristic curves represented in graphs, as well as numerical results of connections that can determine prevalence and similarities with subjects who do not present the same condition, present in studies under similar conditions.

Gait as well as balance are important factors for clinical evaluations when diagnosing abnormalities in movement [3]. The traditional method to evaluate any anomaly in

V. Robles-Bykbaev et al. (Eds.): CITIS 2022, LNNS 607, pp. 13–23, 2023.
https://doi.org/10.1007/978-3-031-24327-1_2

posture is based on the preparation and experience that the medical provider may have [4], this is important for the early detection of postural disorders or possible signs of disease related to the motor skeleton system, so, having a specialist with little experience, the detection of the same can become erroneous or little anticipated.

Inertial sensors can be a practical way to measure movement, balance and posture, so they are used regularly in gait research, medicine, sports and rehabilitation [5], most of them are supported by large companies that use them regularly in physiotherapeutic diagnostic centers around the world.

Table 1 shows a compilation of studies in which inertial sensors are used as measuring instruments to determine gait and posture of the human being, among them are mentioned sensors for commercial use and their counterpart for free use.

The gait cycle or also known as locomotion consists of two phases, a rolling and support phase. Each of the phases is associated with a limb dividing into sub-phases. All the sub phases that can occur through the entire gait cycle. [6], the gait cycle ends the completion of all the sub phases detailed above.

Table 1. Related studies

Author	Application	Sensor type	Analysis method
Pollind and Soangra [7]	Posture measurement for fall prevention	ATMega-328 ICM-20498	Experimental
Talysson et al. [8]	Analysis of gait and posture curves	MPU9250	Analytic
Dinu et al. [9]	Postural tracking miniature sensors	MVN biomech Vicon	Experimental
Dugan et al. [10]	Gait and control after a concussion	Xsens	Experimental

In the present research we try to determine the characteristic curves of people who are overweight and compare them with an experimental database of people who present normality in weight, in order to observe variations that can help diagnose some future pathology.

2 Materials and Methods

2.1 Instruments

Mtw Awinda inertial sensors were used which are devices with measurements of (47 × 30 × 13 mm, 16 g), which serves to detect movement, inside it contains inertial sensor components such as: 3D speed gyroscope, 3D Accelerometer, Barometer, Thermometer. This equipment is used, since its software allows us to observe data collection in real time, correct errors and repeat the measurement, if necessary, in addition to being versatile, easy to apply and based on previous research [10].

The 3D orientation of the sensor on the unit quaternions can be represented by a normalized quaternion q = (WXYZ) with W the real component and X, Y, Z the imaginary parts. To visualize the 3D orientation and facilitate interpretation the quaternion is normally converted into Euler angles [11].

The body mass index was developed by Adolphe Quetelet at the end of the 19th century, this includes the weight and height of an individual in order to generate a risk indicator to relate it to the most common diseases and of higher degree of mortality such as cancer or diabetes [12]. This ratio is calculated using Eq. 1

$$BMI = \frac{Peso(kg)}{Altura^2(m^2)} \tag{1}$$

Table 2 shows the classification according to the WHO for adults over 20 years of age, showing how the BMI number and the degree of weight of a person are related.

Table 2. BMI classification [12].

BMI	State
<18.5	Low weight
18.5–24.9	Normal weight
25.0–29.9	Pre-obesity or overweight
30.0–34.9	Obesity class 1
35.0–39.9	Obesity class 2
>40	Obesity class 3

2.2 Method

The tests will be performed on 3 women and 2 men who are between 18 and 35 years old, for this age range they are called young adults. Body mass index (BMI) should be greater than 25, in men and women. The subjects who will be subjected to these tests must not have appreciable or previously detected physical diseases and can move under their own will without assistance.

Table 3 shows the individual characteristics of the participants in addition to their body mass. The variable is defined as the Individual to be analyzed and the BMI value is the product of the application of Eq. 1.

Table 3. Characteristics of the study population

I	Sex	Weight [Kg]	Height [m]	Age [Years]	BMI [Kg·m^{-2}]
1	F	77.9	1.56	32	32
2	M	134.6	1.78	26	42.5
3	M	99.5	1.70	25	34.4
4	F	76.5	1.70	22	26.4
5	F	78.9	1.64	24	29.3

To make the measurements, anticipatory postural movements (APAs) [14]. Are used, for which five inertial sensors were used as shown in Fig. 1 (a), (b), (c): the first of them is placed in a firm way on the back part with the help of elastic garters around the L3–L4 vertebrae, this specific part was chosen because that is where the center of mass of the body is located when standing.

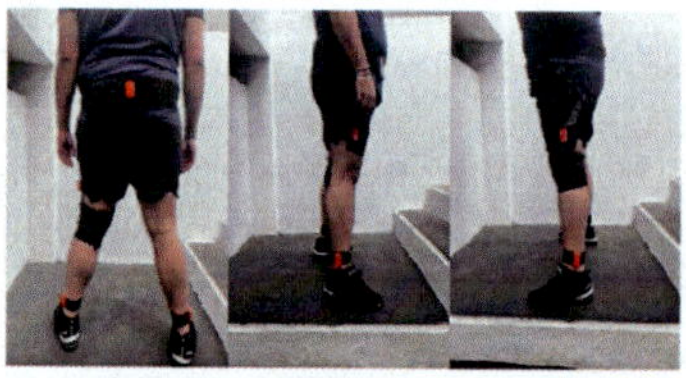

(a) (b) (c)

Fig. 1. (a) Posterior plane center of gravity, (b) Right lateral plane quadriceps and anterior tibial, (c) Left lateral plane quadriceps and anterior tibial

For the study the analysis of gait in 10 m is used, this method is used in conjunction with the locations of the sensors in the research of Donisi et al. [15]. In addition, Washabaugh et al. [16]. In their research used the 10 m on Foot (10 MWT) method as a method to measure gait parameters.

The Euler angles of the left and right quadriceps, the left anterior tibial and the right tibial obtained above can be used in Eq. 2 proposed by Garza-Ulloa [17] to obtain the angles relative to the rotation of the knee. The study is based on lower limbs in this case the left and right knee corresponding to rotational movement.

$$\theta\_Knee = \theta\_Shank + (180 - \theta\_Thingh) \quad (2)$$

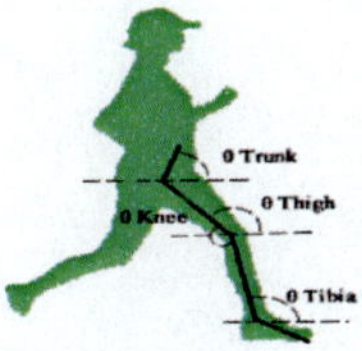

Fig. 2. Lower train joint angles [17].

Figure 2 graphically demonstrates the application of Eq. 2 that relates the joint angles to obtain the angle of rotation of the knee.

3 Results and Discussion

From the tests performed on the subjects, for the analysis the tests whose curves presented anomalies, as well as noises and sudden angular changes, were discarded, thus obtaining patterns as consistent as possible.

For pelvic obliquity, a total of 20 tests measured on a flat regular surface were obtained. Using the Matlab platform, 5 graphs of similar patterns and the polynomial adjustment curve are modeled, as shown in Fig. 3 (a).

In the same sensor from which the pelvic obliquity was obtained, the pelvic rotation during walking was also analyzed, modeling 5 graphs with similar patterns. Figure 3(b).

The sensor located in the right quadriceps yielded data with quite similar patterns from the march in the sagittal plane, called flexion and extension of the thigh. In Fig. 3(c) the gait cycle is observed. The left quadriceps was also analyzed, modeling 5 graphs with the most similar patterns, as shown in Fig. 3(d).

The sensors located in the ankle were analyzed in a particular way because when studying overweight people, the pressure in this area can affect the gait. The gait cycle begins because of the initial movement by an action of the anterior tibial muscles, as shown in Fig. 3(e). The left anterior tibial was analyzed in the same way because when performing the march, the left ankle receives the same pressure Fig. 3(f).

Using the results of the flexion/extension of the right quadriceps and the flexion/extension of the right anterior tibial, the angle that the knee performs through each walking cycle was calculated, thus presenting the angles that correspond to the flexion/extension of the right knee as shown in Fig. 3(g).

At the time of taking a stride the right and left knee perform an angle, previously the right knee was modeled, the angle of flexion / extension of the left knee was also analyzed because at the time of walking the elevation of the entire leg is not similar as shown in Fig. 3(h).

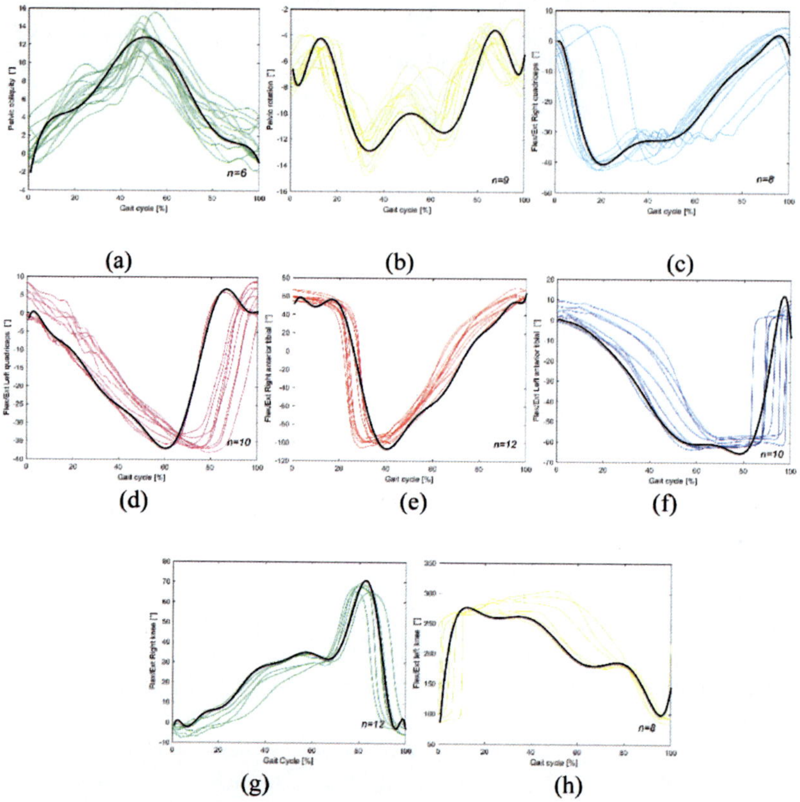

Fig. 3. Shows the results obtained for the different parts studied

The graphs in Fig. 3 show the behavior of the gait elements for one of the subjects, the graphs are a representation of all the tests taken from the subjects, all of them present a similar pattern for each part of the body studied. The polynomial graphs comparison are show in the Fig. 4.

The adjustment curves made by least squares were obtained polynomials of different degrees for the modeling of figures shown of each of the subjects. The selected grades were used for the approximation to the test data measured by the sensors.

From the polynomial adjustment, the result was the root of the mean square error (RSME) of the sensor data regarding the predictions for the parts of the body mentioned above. The RSME values of the 5 subjects are found in Table 5 corresponding to the gait on a flat regular surface (Table 4).

Table 4. Root value of the mean square error of the polynomial adjustment of curves

	Subject 1	Subject 2	Subject 3	Subject 4	Subject 5
Pelvic obliquity	7,08E−12	2,53E−08	3,54E−09	1,76E−10	5,50E−11
Pelvic rotation	1,06E−10	8,99E−11	3,12E−10	7,20E−11	6,75E−10
Flex/hex right quadriceps	4,02E−11	9,24E−11	1,81E−09	1,03E−09	4,15E−10
Flex/hex left quadriceps	1,36E−10	2,93E−11	1,78E−09	1,21E−09	1,46E−09
Right tibial flex/hex	3,41E−12	1,04E−07	5,05E−10	4,92E−11	2,25E−10
Left tibial flex/hex	5,87E−10	1,02E−10	6,59E−11	2,82E−09	5,27E−12
Flex/hex right knee	8,23E−10	2,27E−07	5,70E−09	1,80E−07	1,43E−10
Flex/hex left knee	1,68E−09	2,22E−11	9,81E−09	1,11E−11	8,62E−10

Once the gait cycles are adjusted, the predictions on the regular flat surface are compared with the database of the investigations carried out by Calderón and Ulloa [13]. To determine if there is any type of similarity with the experimental patterns.

Figure 4 modeled the resulting gait cycles of the 5 subjects. The legend sj1, sj2, sj3, sj4, sj5, shows the experimental data obtained for each subject analyzed in the study.

When analyzing the curves it must be determined if there is any type of relationship between the variables, for this the Pearson correlation coefficient is used which can take a range of +1 to −1 which represent the relationship that exists between the experimental data and the database.

Table 5 shows the values obtained from the Pearson correlation coefficient R and Table 6 shows the coefficients of determination R2 which correspond to the patterns and predictions made on a flat regular surface.

Table 5 shows the relationship between the database and the samples taken, having a relationship between medium and low according to the correlation criterion, but for the r2 test as seen in Table 6 the relationship is quite low, even reaching 0, and this is due to the variation of the maximum and minimum angles.

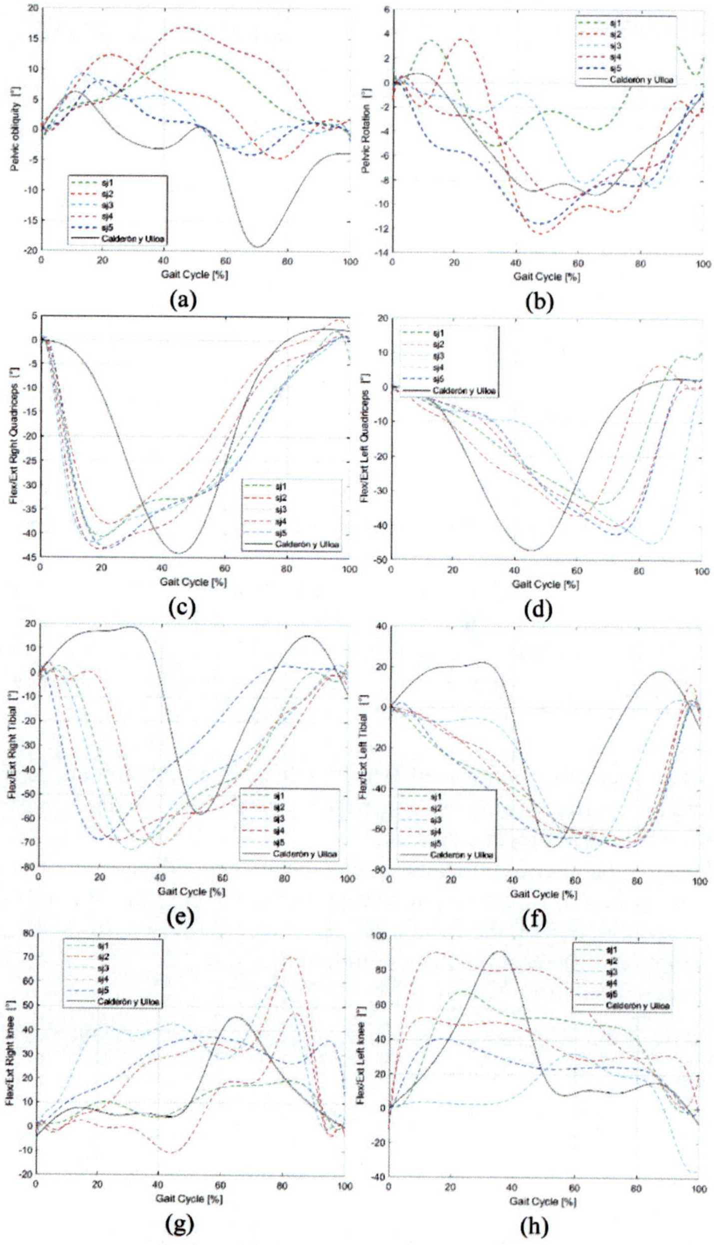

Fig. 4. Polynomial curves of least squares adjustment

Table 5. Pearson correlation coefficient

	Subject 1	Subject 2	Subject 3	Subject 4	Subject 5
Pelvic obliquity	−0,6404	−0,5956	−0,4564	−0,3277	−0,4656
Pelvic rotation	0,2796	−0,4888	−0,3141	−0,7533	0,0833
Flex/hex right quadriceps	−0,4352	−0,5116	−0,3708	−0,4503	−0,5882
Flex/hex left quadriceps	−0,7362	−0,7944	−0,7817	0,1570	0,1863
Right tibial flex/hex	−0,2787	−0,2587	−0,5005	−0,2618	0,1458
Left tibial flex/hex	0,1365	0,2263	0,1935	0,2868	−0,3756
Flex/hex right knee	0,4549	−0,1857	0,4360	0,0518	0,1981
Flex/hex left knee	−0,5778	−0,2254	−0,2694	−0,2513	−0,3462

Table 6. Pearson correlation coefficient R2

	Subject 1	Subject 2	Subject 3	Subject 4	Subject 5
Pelvic obliquity	41%	35%	21%	11%	22%
Pelvic rotation	8%	24%	10%	57%	1%
Flex/hex right quadriceps	19%	26%	14%	20%	35%
Flex/hex left quadriceps	54%	63%	61%	2%	3%
Right tibial flex/hex	8%	7%	25%	7%	2%
Left tibial flex/hex	2%	5%	4%	8%	14%
Flex/hex right knee	21%	3%	19%	0%	4%
Flex/hex left knee	33%	5%	7%	6%	12%

4 Conclusions

The characteristic gait curves for each person are obtained. It is shown that for each part of the body there is a gait tendency which can be adjusted polynomially, observing errors close to 0 using the mean square error method (RMSE).

In the study carried out, maximum and minimum values of the pelvic oblique degrees of 69°, −20°, pelvic rotation degrees 11°, −15°, right quadriceps 9°, −47°, left quadriceps 14°, −47°, right anterior tibial 10°, −73°, left anterior tibial 11°, −71°, Flex/ext. Right knee 68°, −20°, Flex/ext. Left knee 90°, −37° respectively. It can be concluded that during the 10 m of walking the patterns have similar angles in each of the lower limbs.

Considering the Pearson correlation coefficient (r) it is determined that: The pelvic oblique studied is in a range from −0.6404 to −0.3277, finding a mean negative relationship. The pelvic rotation on the other hand shows a range between −0.7533 to 0.2796 thus showing a considerable negative relationship. The right quadriceps presents data between −0.3708 to −0.5882 denoting a weak negative relationship. Opposite case of the left quadriceps showing a wider range between 0.1365 to −0.5, within the range for a weak ratio. In the same way in the right tibial there is a range from −0.7362 to 0.1863 being a wide range and within a considerable negative relationship. Otherwise the left tibial shows a short range from 0.1365 to −0.3756, being a low negative ratio. In the right knee a short range is also observed from −0.1857 to 0.4549 entering a medium ratio. In the case of the right knee, there is a range from −0.5778 to −0.2513, this being a mean ratio. It is thus concluded that the census data have average relationships with respect to the database.

For the analysis by areas of the body it is concluded that subject 5 has a stronger relationship since its maximum values are 35% for the right quadriceps, 14% for the left tibial and 12% for the left knee. These being the highest values for each dataset, this shows that subject 5 presents more graphs related to the database. Otherwise, both subjects 2,3 and 4 have the lowest relationship with the database, relating only to a graph being 63% for the left quadriceps of subject 2, right tibial with 25% for subject 3 and pelvic rotation with 57% for subject 4.

Based on the statistical analysis, it can be observed that the experimental curves as well as the standard curves differ discreetly, presenting considerable errors for each of the analyzed segments, these errors may be related to the study condition since they also represent a displacement in the angle of movement for each analyzed part.

It is observed that the curves maintain a defined pattern for each part of the body studied, being the variations for each subject that makes the curves differ with the base curves.

For future studies it is recommended to review the data shown in conjunction with a physiological and postural analysis to determine this condition.

The main goal of the current attempt demonstrate that people have a characteristic gait pattern and that it can be measured and analyzed, the equipment used was commercial inertial sensors which are placed in parts of the lower body to be used. These delivered a database which generates a measurable and comparable gait pattern with any related study to find similarities and differences between all population groups.

References

1. Sohrab, A.: Obesity and overweight prevalence in immigration: a meta-analysis. Obes. Med. **22**(100321) (2021
2. Villarrasa-Sapiña, I., Álvarez Pitti, J., Cabeza Ruiz, R., Redón, P., Lurbe, E., García Massó, X.: Relationship between body composition and postural control in prepubertal overweight/obese children: A cross-sectional study. Clin. Biomech. **52**, 1–6 (2018)
3. Ota, M., Tateuchi, H., Hashiguchi, T., Ichihashi, N.: Verification of validity of gait analysis systems during treadmill walking and running using human pose tracking algorithm. Gait Posture **85**, 290–297 (2021)
4. Yanmin, Z., Yang, Y.: Design and data analysis of wearable sports posture measurement system based on internet of things. Alex. Eng. J. **60**, 691–701 (2021)
5. Bötzel, K., Martinez Marti, F., Carvajal Rodríguez, M.Á., Plate, A., Olivares, A.: Gait recording with inertial sensors – how to determine initial and terminal contact. J. Biomech. **49**(3), 332–337 (2016)
6. Mohammed, S., Samé, A., Oukhellou, L., Kong, K., Huo, W., Amirat, Y.: Recognition of gait cycle phases using wearable sensors. Rob. Auton. Syst. **75**, 50–59 (2016)

7. Pollind, M., Soangra, R.: Development and validation of wearable inertial sensor system for postural sway analysis. Measurement **165** (2020)
8. Santos, T., et al.: A low-cost wireless system of inertial sensors to postural analysis during human movement. Measurement **148** (2019)
9. Dinu, D., Fayolas, M., Jacquet, M., Leguy, E., Slavinski, J., Houel, N.: Accuracy of postural human-motion tracking using miniature inertial sensors. Procedia Eng. **147**, 655–658 (2016)
10. Dugan, E., Shilt, J., Masterson, C., Ernest, K.: The use of inertial measurement units to assess gait and postural control following concussion. Gait Posture **83**, 262–267 (2021)
11. Paulich, M., Schepers, M., Rudigkeit, N., Bellusci, G.: Xsens MTw Awinda: rastreador de movimiento magnético inercial inalámbrico en miniatura para aplicaciones cinemáticas 3D precisas
12. Organización Mundial de la Salud: Índice de masa corporal – IMC. Temas de Salud Prevención de Enfermedades Nutrición (2020)
13. Mailyn, J., Calderon, D., Ricardo, A., Ulloa, J.: Changes associated with aging in angular kinematic parameters during a controlled speed walk. Revista medica de chile (2016)
14. Méndez, R.M., Huertas, M.R.: Uso de sensores inerciales en la medición y evaluación de movimiento humano para aplicaciones en la salud. México (2018)
15. Donisi, L., Pagano, G., Cesarelli, G., Coccia, A., Amitrano, F., D'Addio, G.: Benchmarking between two wearable inertial systems for gait analysis based on a different sensor placement using several statistical approaches. Measurement **173** (2021)
16. Washabaugh, E., Kalyanaraman, T., Adamczyk, P., Claflin, E., Krishnan, C.: Validity and repeatability of inertial measurement units for measuring gait parameters. Gait Posture 55, 87–93 (2017)
17. Garza Ulloa, J.: Experiment design, data acquisition and signal processing. de Applied Biomechatronics Using Mathematical Models, 179–237 (2018)

Inertial Sensors Based on Stabilometric Analysis for Postural Control in Elderly People: A Systematic Review

Byron Ricardo Zapata Chancusig(✉), José Luis Bucheli Naranjo, and Fabián Rodrigo Narváez Espinoza

Universidad Politécnica Salesiana, Quito, Ecuador
{bzapata,jbuchelin,fnarvaeze}@ups.edu.ec

Abstract. During last decades, many technological alternatives for posture assessment has been proposed for people with balance problems caused by several pathologies, these alternatives are commonly based on center of pressure (COP) analysis using strength platforms, being the most used quantitative technique in the current clinical environment. Nevertheless, COP is not the only parameter interfering in balance, recent related investigations on this subject has analyzed the center of mass (COM) in static and dynamic state of balance problems people.

Nowadays, inertial sensors based tools has been proposed for postural control of people on static and dynamic state, which have provided a suitable clinical use in non-specialized environments, this technology has been used to provide space-time parameters related to human movement. In order to improve COM's analysis and its effect in postural control many bio-mechanical models has been proposed, considering sensor arrays on upper only, lower only, or entire body. However, a suitable stabilometric analysis is still on research. The aim of this systematic review is to summarize the main Inertial sensors based stabilometric analysis for postural control in elderly people studies.

Keywords: Stabilometry · Inertial sensors · IMU · Elderly people · Postural control · Gait cycle · Balance · Body

1 Introduction

Aging is an innate characteristic in human nature. People decrease their walking capacity and their posture stability through the year's negative impacting their balance. Other aging related effects in the body affecting the balance are:

- Body strength decrease.
- Muscle mass and bone density decrease.
- Peripheral sensory functions decrease.
- Selective atrophy on central nervous system components in charge of balance and gait.
- Balance and gait problems due to the negative impact of elderly people medication.

V. Robles-Bykbaev et al. (Eds.): CITIS 2022, LNNS 607, pp. 24–35, 2023.
https://doi.org/10.1007/978-3-031-24327-1_3

On people over 60 years old, serious injuries and in some cases even death is mainly caused due to their stability decrease. A decreased postural balance causes 90% of hip fractures in adults around 80 years old. Seventy-five per-cent of older adults who have hip fractures become functionally dependent on a mechanism that allows them to walk after an injury [1].

As a result of aging, skeletal changes are made and hunched posture appears producing a center of mass shift, this displacement added to the muscle strength decrease, shortened stride and gait's fluidity loss affect negatively on balance [1].

Therefore, the study and research on the effects of aging on gait becomes of public interest to generate solutions to decrease falls in elderly people.

Nowadays, inertial measurement units (IMU) are used in stabilometric analysis due to their mobility, low cost and ease of implementation. IMU uses micro- electronics technology (MEMS) based on gyroscopes, accelerometers, magnetometers, to measure the body relative orientation change across time, allowing researchers to define acceleration, velocity and rotation angle of a study body. Thus, IMUs are highly useful for human body movement study in elderly people or post-operatory related disorders.

Several inertial sensors based bio-mechanic models have been developed to predict the gravity center of people in dynamic state by using an IMU on the pelvis obtaining a 90% of success with errors between 3.9 and 31 mm in healthy people [2]. Regarding people with gait issues, the gravity center does not match the one of a healthy person, therefore a bio-mechanic model based on only one IMU is not suitable to made a precise estimation of the gravity center and gait and balance parameters.

Some bio-mechanic models based on a 7 IMU network disposed on the pelvis and lower limbs studies the gait from static to dynamic state. This model has provided similar results comparing to camera-based systems with reflective markers [1].Nevertheless, this approach is focused on half of the body (pelvis and lower parts), considering that the gravity center varies according to the posture of upper and lower limbs, on the 7 sensor based model the upper limbs effect is not considered [2].

Thus, the following questions appears:

- ¿Could an inertial sensor based model provide precise time-space parameters of elderly people?
- ¿Could a bio-mechanic model be able to acquire the necessary parameters for a stabilometric analysis?

The purpose of this systematic review is to synthesize the currently developed studies, referring to the methods and tools used in the stabilometric analysis and postural control in older adults, with balance problems caused by some pathology, so this review starts with the methodology based on the PRISMA guidelines, for which the initial search is proposed without a specific method, then the results are presented in the search engines proposing a strategy based on keywords, later the filtering of documents is generated through the inclusion criteria and exclusion, finally the results of the different diseases that affect older men and women are presented, in addition the location of the inertial sensors in the different parts of the body is presented, in order to conclude the usefulness and most used areas for a stabilometric analysis based on inertial sensors.

2 Methods

This study provides a systematic review of the papers published in bio-medicine, mechatronics, electronics, mechanics, and bio-mechanics conferences based on the PRISMA (Preferred Reporting Items for Systematic Reviews and Meta-Analysis) directives. Next, the phases of this review are detailed.

2.1 Initial Search

The initial search was made in March 2021 by using the terms 'Balance' and 'Inertial sensor' obtaining a global reference on the subject with a considerable amount of information. This initial review allows us to start from a global search to deep dive into a systematic revision following PRISMA directives.

2.2 Systematic Search Strategy and Document Selection Process

The document selection strategy is based on the information acquired from ACM, Elsevier, IEEE, PubMed, Scopus and Web of Science search engines filtered from 2010 until present day and considering the strategy from Table 1. The consulted articles are filtered by title, abstract review and author keywords in order to obtain related documents to this systematic review, for which a synonym search strategy is used to provide a wider scope.

According to the formula depicted on Table 1, the following results were obtained: 13 in ACM, 113 in Elseivier, 58 in IEEE, 100 in Pubmed, 24 in Scopus and 171 in Web of Science, in addition to 10 articles from other sources. These results are filtered to select the most suitable articles according to inclusion and exclusion criteria.

Table 1. Search strategy

Search strategy used in this review	
Stabilometry	'Postural' OR 'Balance' OR 'Equilibrium' OR 'Posturography'
	'AND'
Inertial sensors	'Inertial Sensor' OR 'Wearable Sensor'

2.3 Inclusion Criteria

This review focused on the stabilometric analysis of older adults with pathological problems, through the use of inertial sensors for science and technology applications, for which the following inclusion criteria were proposed:

1. English written articles.
2. Peer-reviewed scientific articles.

3. Stabilometric studies based on one or more accelerometers, inertial sensors o portable sensors.

   (1) Elderly people gait disorder researches.
   (2) Studies where the location of the sensors on the body is mentioned.

2.4 Exclusion Criteria

Given to the large amount of literature about the use of inertial sensors for the study of postural control, the following exclusion criteria are proposed to reduce the articles and focus on the objective of this review.

(1) Articles published before 2010.
(2) Systematic reviews.
(3) Studies on people below 60 years old.
(4) Simulation or animal based studies.
(5) Invasive device or implant based studies.
(6) Sport technique focused studies.
(7) Non postural control related disease studies.
(8) Studies made on mechanisms or robots and on implant or prosthesis.

Using the search formula 489 articles were obtained and 23 of them met the inclusion and exclusion criteria set out for this systematic review.

In Fig. 1 the selection processes of the articles are described using the mentioned criteria and PRISMA guidelines.

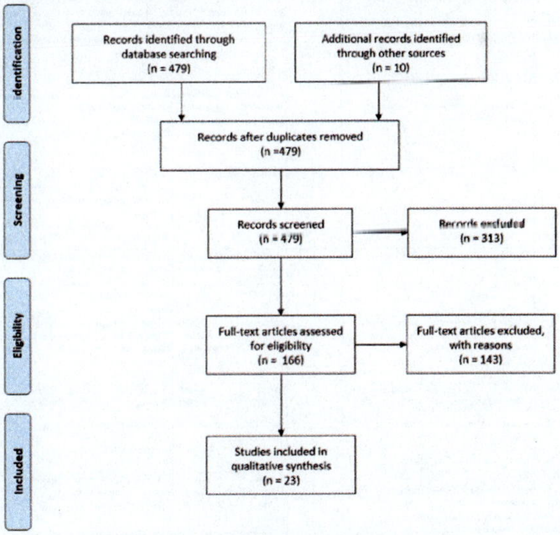

Fig. 1. PRISMA study flow diagram

2.5 Data Extraction

Table 2 and Table 3 shows the characteristics and data acquired from every study: author and year of publication, gender, disorder type of pathology, number of sensors, sensors type, specific sensors location, sensors variables, patient's mobility were the main data obtained from each study. These data allow us to know the postural control diseases on elderly people and becomes a starting point for the election of quantity and position of the inertial sensors in future works.

Table 2. Overview of the sensors positions, sensors setups and postural control variables utilized in each study

Author andyear of publication	Patient's age	Gender	Disorder typeof pathology	Number of sensors	Sensors type	Specific sensors location	Sensors variables	Patient's mobility
Kotas, R, 2019 [3]	50.92 ± 16.36	124 Female 81 Male	Peripheral dam-age of thevestib-ular sys-tem.	1	Designed inertial sensors: IMU sensor: LSM9DS1	Lumbar part (L5)	Center of mass	Static posturography
Veeravelli, S, 2016 [4]	57.2 ± 9.2	10 Persons. (Gender is not specified)	Older adults with HIV	5	Inertial sensors	(1 y 2) Legs (3 y 4) Thighs (5) Lower back	Center of Mass Oscilla-tion and Stride Speed	Semi-tandem balance posture
Zhou, He, 2018 [5]	45- (73.4 +- 6.8) Dia-betes (DM-) 78 -(57.2 +- 4.2) Dia-betes (DM+) 73 -(71.4 +- 5.4) Dia-betes (DM+)	(32 Female y 13 Male) Diabetes (DM-) (40 Female y 38 Male) Diabetes (DM+) (38 Female y 35 Male) Diabetes (DM+)	Diabetes pa-tients on hemodialysis	2	Wearable sensors (LegSys™, BioSensics, Watertown, MA, USA)	(1 y 2) Left and right lower shins. (1 y 2) Lower dominant shin y lumbar area	Stride speed Normalized stride speed Stride length Normalized stride length Gait cycle time Double bracket	Dynamic balance
Hasegawa, Naoya, 2019 [6]	(68.2 Average) Parkin-son's (68.4 Average) Healthy persons	(48 Male y 31 Female) Parkinson's (93 Male y 51 Female) Healthy persons	Parkinson's	3	Accelerometers Opal	(1) Lumbar region, (2y 3) Right and left feet	Balance, anticipatorypos-ture adjustments (APA), automatic postural re-sponses (APR), limits of stability (LOS).	Dynamic balance
Soangra, R, 2018 [7]	76,1 ± 3,6	5 Female 11 Male	Cardiovascular diseases	1	Inertia sensors based on smartphones. (IPhone 5)	(1) Lateral lumbar right region	Acceleration while walk-ing, and acceleration from sitting and standing	Dynamic balance, transi-tion from sitting to stand-ing
Schwenk, M, 2016 [8]	(68.73±8.72) Average Neuropathy (70.3 ± 8.7) Control group	(4 Male y 7 Female)pe-ripheral neuropathy (5 Male y 6 Female) Healthy persons	Peripheral neurop-athy	5	Inertial sensors, (LegSys TM ; BioSen-sics LLC, Cambridge, Mass., USA	(1 y 2) Legs (3 y 4) Thighs (5) Lower back	Center of mass, hip sway, gait speed,	Dynamic balance
Bonora, Gianluca, 2017 [9]	69.4 a 73.3	Gender is not specified	Parkinson's	3	IMUs (Opals, APDM Inc., Portland, OR, USA)	(1) Posterior of thetrunk at the level of L4-L5 (2 and 3) Left and right frontal tibiae	Trunk acceleration antic-ipatory spatial behav-ior of COM and balance Phase.	Dynamic balance and static
Agurto, Carla, 2021 [10]	(69 ± 8) Parkinson's (49±9) Control group	(16 Male y 17 Female) Parkinson's (20 Male y 13 Female) Control group	Parkinson's	6	Opal Version 1 Wear-able sensors (APDM Wearable Technolo-gies)	(1 y 2) Right and left feet (3 y 4) Right and left wrist, (5) Lumbar (6) Sternum	Speed and acceleration of each part of the body	Dynamic balance
Fino, P C, 2020 [11]	70 (7.3) Average	10 Female, 33 Male	Parkinson's	8	Inertial sensors (Opal, APDM Inc.)	Over the sternum, lumbar spine, and bi-laterally on the wrists, previous distal portion of the shank, and dorsum	Gait analysis: speed, stride length, step times	Dynamic balance
Rodríguez- Martín, Daniel,2015 [12]	21 a 56	22 Male 14 Female	Parkinson's	1	Inertial systems based on MEMS accelerometer (LIS3LV02DQ) and a dsPIC33F.	(1) Antero-lateral of the waist at 45 degrees and (1) Lateral 90 degrees (same sensor 2 tests)	Speed and acceleration	Sit-up transition
Schwenk, M, 2016 [13]	(77.8 ± 6.9) Mild cog-nitive impairment (79.0 ± 10.4) Controlgroup	(5 Male y 7 Female) Mild cognitive impairment (5 Male y 5 Female) Con-trol group	Mild cognitive impairment	5	Inertial sensors (LegSys, BioSen-sics LLC; Cambridge, Massachusetts)	(1 and 2) Upper and lower left leg. (3 and 4) Upper and lower right leg. (4) Low back. (4) Lower back	Center of gravity, gait speed	Dynamic balance
Grewal, Gurtej Singh, 2015 [14]	(62.58±7.98) Diabetic neuropathy (64.90±8.50) Control group	(8 Male y 11 Fe-male) Diabetic neuropa-thy (8 Male y 8 Female) Control group	Diabetic pe-ripheral neu-ropathy	5	Inertial sensors (LegSys, BioSen-sics LLC; Cambridge, Massachusetts)	(1 y 2) Left and right shins (3 y 4) Left and right thigh (5) Lower back	Balancing of the COG, roll and speed in the gait	Dynamic balance
Toosizadeh, N, 2015 [15]	(65 ± 8) Diabetic neu-ropathy (69 ± 3) Control group	(11 Male y 7 Fe-male) Diabetic neu-ropathy (7 Male y 11 Female) Control group	Diabetic pe-ripheral neu-ropathy	2	Triaxial Accelerometer and Triaxial Gyro-scope (BalanSens, BioSensics LLC, MA, EE. UU.)	(1) Anteroposterior and medial-lateral angles of the legs. (2) Hip	Balancing of the COG	Static balance

(continued)

Table 2. *(continued)*

Author andyear of publication	Patient's age	Gender	Disorder typeof pathology	Number of sensors	Sensors type	Specific sensors location	Sensors variables	Patient's mobility
Delrobaei, Mehdi, 2018 [16]	(65.28 Average) Parkinson's (64.9) Control group	(19 Male y 22 Female) Parkinson's (11 Male y 11 Female) Control group	Parkinson's	17	Portable motion capture system (IGS-180 Synertial Ltd., UK)	Whole -Body: Head, Right Clavicle, Left Clavicle, Chest, Pelvis, Right Wrist, Right Elbow, Right Shoulder, Left Wrist, Left Elbow, Left Shoulder, Right Hip, Right Knee, Right Ankle, Left Hip, Left Knee, Left Ankle	Tremor in limbs and effect on stability	Sitting Balance
Schwenk, M, 2015 [17]	(74.6±6.5) Control group (80.2±8.6) Pre Fragile (83.4±8.6) Fragile	(37 Female, 7 Male) Control group (45 Female, 15 Male) Pre Fragile (18 Female, 3 Male) Fragile	Fragility in the elderly	5	Inertial sensors (LegSys, BioSensics LLC; Cambridge, Massachusetts)	(1 y 2) Legs (3 y 4) Thighs (5) Lower back	Gait and balance parameters: walking speed and hip sway	Dynamic balance
Mosquera-Lopez, C, 2020 [18]	Year's rank (33 - 76)	(8 Male y 17 Female)	Multiple sclerosis	1	MotioWear (MotioSens, Portland, OR)	(1) Abdomen	Acceleration, orientation, and tilt Parameters	Dynamic balance
Moon, Sanghee, 2020 [19]	(66,73 ± 9,17) Parkinson's (66,98 ± 9,84) Essential tremor	Gender is not specified	Parkinson's	6	Sensor IMU (Opal, APDM, Inc., Portland, OR, USA)	(1) Sternum (2) Lumbar (back) (3 y 4) Wrists (5 y 6) Ankle	RMS balance Gait speed	Dynamic balance
Morris, Rosie, 2019 [20]	(67,6 ± 8,2) Average	(125 Male y 73 Female)	Parkinson's	6	Inertial sensors (Opal APDM Inc., Portland, OR.)	(1 y 2) Bilateral of the wrists. (3 y 4) Bilateral of the feet. (5) Sternum. (6) Fifth lumbar vertebra.	Gait speed Balance speed Balance frequency	Dynamic balance
Pérez-Cruzado, D, 2018 [21]	(72.33 (±3.97))Patients with stroke and healthy elderly people (73.04 (±3.58)) Healthy older adults	(3 Female, 2 Male)Patients with stroke and healthy elderly people (3 Female, 2 Male) Healthy older adults	Ictus, Stroke ictus	2	Inertial sensors InertialCube3 (Intersense Inc., Bedford, MA),	(1) Trunk (2) Lumbar region	Bending speed Extension speed Tilt speed Rotation speed	Static balance on one foot.
King, L A, 2012 [22]	(65.4 ± 5.5 (55-75)) Control Persons (64.0 ± 5.0 (57-75)) Middle Parkinson's (67.5 ± 8.5 (50-84)) Severe Parkinson's	(13 Male, 27 Female) Control Persons (10 Male, 14 Female) Middle Parkinson's (16 Male, 8 Female) Severe Parkinson's	Parkinson's	6	Inertial sensors XSens sensors	(1) Trunk, posterior to L5 level, near the center of the body of mass. (2 y 3) Anterior stem of each leg. (4 y 5) Side of the back of each arm, and (6) Sternum	Duration of the turn; Number of steps; Rotation speed; Total duration of Timed Up and Go.	Dynamic balance
Curtze, Carolin, 2015 [23]	(66.7 ± 5.5) Mild Parkinson's (66.3 ± 6.6) Severe Parkinson's	(34 Male, 18 Female) Mild Parkinson's (40 Male, 12 Female) Severe Parkinson's	Parkinson's	6	Inertial sensors MTX (Xsens)	(1 y 2) Right and left wrists. (3 y 4) Left and right ankle (5) Sternum (6) Fifth lumbar level	Gait rhythm Dynamic gait stability Postural balancing.	Dynamic balance
Parvaneh, Saman, 2017 [24]	(74.23±6.15) No Fragile. (80.70±8.68) Fragile/Pre Fragile	(2 Male) No Fragile. (18 Male) Fragile/Pre Fragile	Parkinson's	1	Inertial sensors BioSensics	(1) Sternum	Acceleration, orientation and position parameters.	Transition sit and stand up, Dynamic balance
Wuest, Seline, 2016 [25]	(64.7 ± 9.2) Stroke ictus patients 76.0 ± 5.7 (66.0–86.0) non-disabled elderly people	(2 Female, 2 Male)Patients withStroke ictus (17Female, 8 Male) non-disabled elderly peopie	Ischemic hem- or orrhagicstroke ictus	8	Inertial sensors (iTUG)	(1 y 2) Wrists (3 y 4) Legs y (5) Trunk (6 y 7) feet's (8) back	Gait speed Stride length.	Dynamic balance

3 Results

3.1 Elderly People Postural Control Diseases

People from 60 years old are considered elderly people in Ecuador [26], for this reason this systematic review studies the articles related to postural control diseases where the average age is 66.9 years.

In the bar chart depicted on Fig. 2, the pathologies considered on this systematic review (Table 2) are summarized.

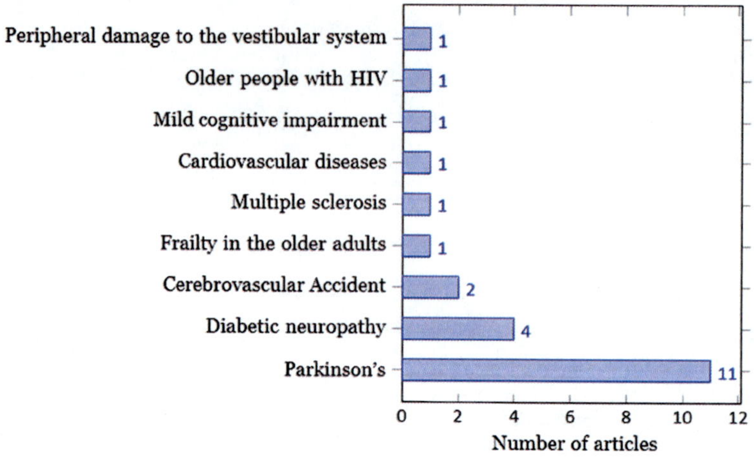

Fig. 2. Elderly people postural control diseases

In 11 of 23 articles, Parkinson disease effect is considered in postural control study as it affects women and men producing shakes, difficulty to walk and to maintain balance, thus decreasing postural control. Pathologies affecting women and men described on consulted articles are depicted in Fig. 3. As shown on Fig. 3, certain pathologies affect women and men differently. For example, men are more likely to have Parkinson disease as evidence suggest that the biological differences in the neuroprotective pathway of the nigrostriatal dopaminergic might be the trigger [27].

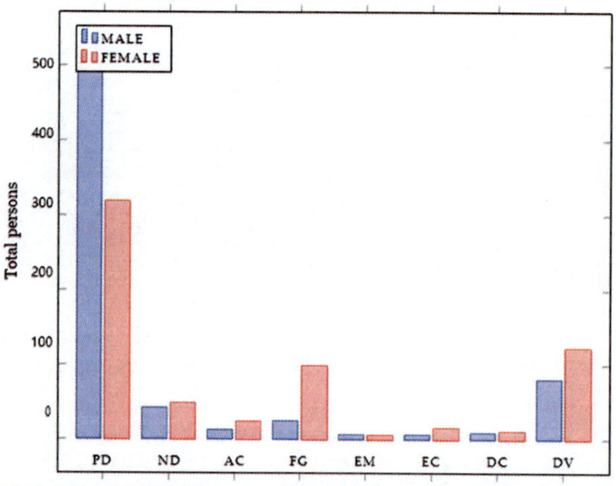

PD=Parkinson; ND=Diabetic neuropathy; AC=Cerebrovascular Accident; FG=Frailty in the older adults; EM=Multiple sclerosis; EC=Cardiovascular diseases; DC=Mild cognitive impairment; DV=Peripheral damage to the vestibular system;

Fig. 3. Elderly people postural control diseases for gender

3.2 Configuration and Sensor Variables

Quantity, type and positions of the inertial sensors for stabilometric study are shown in Table 2 and Table 3. The position of 105 sensors described in the 23 consulted articles were reviewed and Fig. 4 summarizes this information.

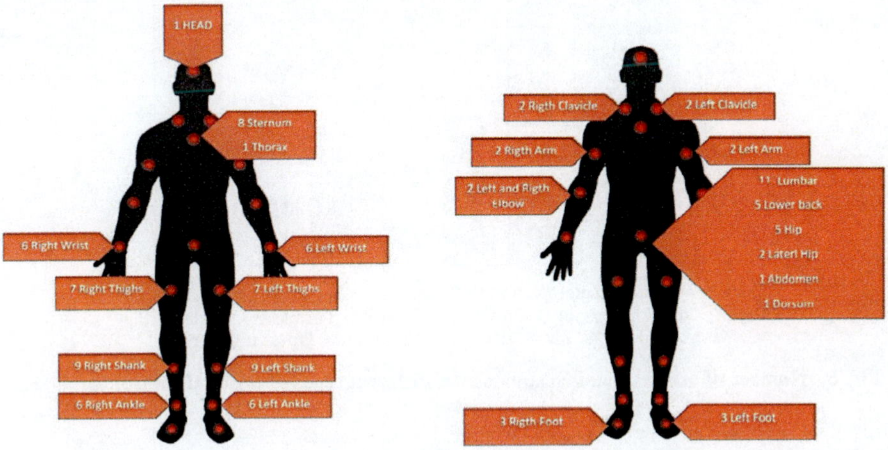

Fig. 4. Sensor-mounting region used in the studies included in this review.

In most articles, commercial inertial sensors are used as they incorporate accelerometers, gyroscopes and magnetometers and in certain cases a barometer, making possible the measurement speed, acceleration and roll in researches regarding human gait and static and dynamic balance. There are some cases in which sensors designed by the authors are used, as in: "Evaluation of agreement between static posturography methods employing tensometers and inertial sensors", which sensor is designed based on an ESP32 Wi-Fi module connected to an accelerometer LSM9DS1 and providing a correlation of 95%, and thus, an excellent method for gait analysis [3].

Another method included in this review is the one described in: "Inertial Sensor-Based Variables Are Indicators of Frailty and Adverse Post-Operative Outcomes in Cardiovascular Disease Patients" in which the sensors of a mobile device (iPhone 5) are used to obtain the velocity and acceleration of an elderly adult in static and dynamic balance. This method provides an objective and reliable measure but the compatibility with the applications and start and end times of the movement has to be considered as the velocity measures are based on these parameters [7]. Figure 5 shows the measurements acquired from the inertial sensors used in the articles to determine the stabilometry parameters in dynamic and static state. The most used measurements to determine postural control are stride length, gait speed, gait acceleration, swinging speed and swinging acceleration. Control groups (healthy people) are used to determine balance problems in people with different pathologies related with aging.

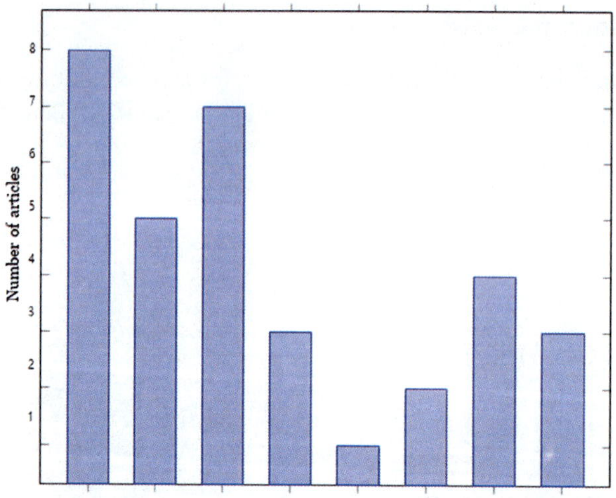

VAM = Speed and Acceleration in the gait, VAB = Speed and Acceleration of Balance; CoG = Center of Gravity; VG = Speed in Turns; A = Angles; CP = Number of Steps; LZ = Stride Length; Other = Other measures.

Fig. 5. Number of articles that investigated the different types of postural control variables.

4 Discussion

The aim of this systematic review was to analyze the scientific documents related to the implementation of inertial sensors in postural control studies in elderly people with different pathologies, Thus, relevant information related to this subject was summarized in order to provide a starting point for future investigations regarding bio-mechanic models and the quantity of inertial sensors used on the stabilometric analysis of elderly people. In addition, the quantity of papers implementing inertial sensors in the study of postural control due to different pathologies on elderly people were provided. This information can help future researches investigating different diseases using inertial sensors.

4.1 Sensor Location

The location of the 103 inertial sensors of the 23 consulted papers were analyzed, most of these sensors were located on the hip. In order to acquire a suitable information of the gravity center (CoG), 25 sensors were locating in the lumbar area to measure the angle, velocity and acceleration of the displacement of the CoG. In addition, 50 sensors were located in the lower limbs to measure the velocity and acceleration of the stride and support base. In the upper limbs, 18 sensors were located to measure the velocity and acceleration of the balance caused by arms movement.

A total of 9 sensors were placed in the sternum and rib cage to analyze the upright position of people to maintain balance.

Finally, 5 sensors were placed on the head and on the left and right clavicles, to analyze the position of the head in static and dynamic balance.

These data obtained will provide future researchers with an insight to rule out certain sensors that do not directly influence the analysis of balance and postural control.

5 Conclusion

The review focused on the analysis and feasibility of using inertial sensors for the study of pathologies that affect postural control in elderly people, through the study of articles published in the last 10 years. As a result, it can be seen that nowadays the use of inertial sensors draws the attention of several researchers due to its low cost compared to other methods, also because is a lightweight device that does not affect the equilibrium analysis process.

Despite the review of the articles included in this document, it is still difficult to define a single bio-mechanical model and the exact number of inertial sensors that should be used to study postural control in different pathologies. However, possible locations of the sensors are presented in order to determine the speeds, accelerations and angles associated with the gait and body balance, considering that the selected articles were based on older adults and with tests in static and dynamic state.

Compliance with Ethical Standards. Byron Zapata, José Bucheli and Fabian Narvaez were funded by the bio-mechatronics and bioengineer (GiByB) research group from Salesian Polytechnic University, Quito - Ecuador.

The funding group did not influence the collection, analysis and interpretation of data presented in this document, also the group does not influence in the approval or disapproval of this publication.

There was no interest conflict involved in this project.

References

1. Richards, J.: The Comprehensive Textbook of Clinical Biomechanics, 2nd edn. Elsevier, Preston, UK (2018)
2. Kirtley, C.: Clinical Gait Analysis. Theory and Practice, Churchill Living-Stone, United Kingdom (2006)
3. Kotas, R., Janc, M., Kamiski, M., Marciniak, P., Zamysowska-Szmytke, E., Tylman, W.: Evaluation of agreement between static posturography methods employing tensometers and inertial sensors. IEEE Access **7**, 164120–164126 (2019). https://doi.org/10.1109/ACCESS.2019.2952496
4. Veeravelli, S., Najafi, B., Marin, I., Blumenkron, F., Smith, S., Klotz, S.A.: Exergaming in older people living with HIV improves balance, mobility and ameliorates some aspects of frailty. J. Vis. Exp. **2016** (2016). https://doi.org/10.3791/54275
5. Zhou, H., et al.: Hemodialysis impact on motor function beyond aging and diabetes- objectively assessing gait and balance by wearable technology. Sensors (Basel) **18** (2018). https://doi.org/10.3390/s18113939
6. Hasegawa, N., Shah, V.V., Carlson-Kuhta, P., Nutt, J.G., Horak, F.B., Mancini, M.: How to select balance measures sensitive to Parkinson's disease from body-worn inertial sensors—separating the trees from the forest. Sensors **19**(15), 3320 (2019). https://doi.org/10.3390/s19153320
7. Soangra, R., Lockhart, T.: Inertial sensor-based variables are indicators of frailty and adverse post-operative outcomes in cardiovascular disease patients. Sensors **18**(6), 1792 (2018). https://doi.org/10.3390/s18061792

8. Schwenk, M., Grewal, G.S., Holloway, D., Muchna, A., Garland, L., Najafi, B.: Interactive sensor based balance training in older cancer patients with chemotherapy induced peripheral neuropathy: a randomized controlled trial. Gerontology **62**, 553–563 (2016). https://doi.org/10.1159/000442253
9. Bonora, G., et al.: Investigation of anticipatory postural adjustments during one leg stance using inertial sensors: evidence from subjects with parkinsonism. Front. Neurol. **8** (2017). https://doi.org/10.3389/fneur.2017.00361
10. Agurto, C., Heisig, S., Abrami, A., Ho, B.K., Caggiano, V.: Parkinson's disease medication state and severity assessment based on coordination during walking. PLoS ONE **16**(2), e0244842 (2021). https://doi.org/10.1371/journal.pone.0244842
11. Fino, P.C., Mancini, M.: Phase dependent effects of closed loop tactile feedback on gait stability in Parkinson's disease. IEEE Trans. Neural Syst. Rehabil. Eng. **28**, 1636–1641 (2020). https://doi.org/10.1109/TNSRE.2020.2997283
12. Rodríguez, D., Samà, A., Pérez-López, C., Cabestany, J., Català, A., Rodríguez Molinero, A.: Posture transition identification on PD patients through a SVM-based technique and a single waist-worn accelerometer. Neurocomputing **164**, 144–153 (2015). https://doi.org/10.1016/j.neucom.2014.09.084
13. Schwenk, M., et al.: Sensor based balance training with motion feedback in people with mild cognitive impairment. J. Rehabil. Res. Dev. **53**, 945–958 (2016). https://doi.org/10.1682/JRRD.2015.05.0089
14. Grewal, G.S., et al.: Sensor based interactive balance training with visual joint movement feedback for improving postural stability in diabetics with peripheral neuropathy: a randomized controlled trial. Gerontology **61**, 567–574 (2015). https://doi.org/10.1159/000371846
15. Toosizadeh, N., Mohler, J., Armstrong, D.G., Talal, T.K., Najafi, B.: The influence of diabetic peripheral neuropathy on local postural muscle and central sensory feedback balance control. PLoS ONE **10**(8), e0135255 (2015). https://doi.org/10.1371/journal.pone.0135255
16. Delrobaei, M., Memar, S., Pieterman, M., Stratton, T.W., McIsaac, K., Jog, M.: Towards remote monitoring of Parkinson's disease tremor using wearable motion capture systems. J. Neurol. Sci. **384**, 38–45 (2018). https://doi.org/10.1016/j.jns.2017.11.004
17. Schwenk, M., et al.: Wearable sensor-based in-home assessment of gait, balance, and physical activity for discrimination of frailty status: baseline results of the Arizona frailty cohort study. Gerontology **61**, 258–267 (2015). https://doi.org/10.1159/000369095
18. Mosquera-Lopez, C., et al.: Automated Detection of Real-World Falls: Modeled from People with Multiple Sclerosis. IEEE J. Biomed. Health Inform. **25**(6), 1975–1984 (2021). https://doi.org/10.1109/JBHI.2020.3041035
19. Moon, S., et al.: Classification of Parkinson's disease and essential tremor based on balance and gait characteristics from wearable motion sensors via machine learning techniques: a data-driven approach. J. Neuroeng. Rehabil. **17** (2020). https://doi.org/10.1186/s12984-020-00756-5
20. Morris, R., et al.: Cognitive associations with comprehensive gait and static balance measures in Parkinson's disease. Park. Relat. Disord. **69**, 104–110 (2019). https://doi.org/10.1016/j.parkreldis.2019.06.014
21. Perez, D., Gonzalez-Sanchez, M., Cuesta-Vargas, A.I.: differences in kinematic variables in single leg stance between patients with stroke and healthy elderly people measured with inertial sensors: a cross-sectional study. J. Stroke Cerebrovasc. Dis. **27**, 229–239 (2018). https://doi.org/10.1016/j.jstrokecerebrovasdis.2017.08.024
22. King, L.A., Mancini, M., Priest, K., Salarian, A., Rodrigues-de-Paula, F., Horak, F.: Do clinical scales of balance reflect turning abnormalities in people with Parkinson's disease? J. Neurol. Phys. Ther. **36**(1), 25–31 (2012). https://doi.org/10.1097/NPT.0b013e31824620d1

23. Curtze, C., Nutt, J.G., Carlson-Kuhta, P., Mancini, M., Horak, F.B.: Levodopa is a double-edged sword for balance and gait in people with Parkinson's disease. Mov. Disord. **30**, 1361–1370 (2015). https://doi.org/10.1002/mds.26269
24. Parvaneh, S., Mohler, J., Toosizadeh, N., Grewal, G.S., Najafi, B.: Postural transitions during activities of daily living could identify frailty status: application of wearable technology to identify frailty during unsupervised condition. Gerontology **63**, 479–487 (2017). https://doi.org/10.1159/000460292
25. Wuest, S., Masse, F., Aminian, K., Gonzenbach, R., de Bruin, E.D.: Reliability and validity of the inertial sensor-based Timed "Up and Go" test in individuals affected by stroke. J. Rehabil. Res. Dev. **53**, 599–610 (2016). https://doi.org/10.1682/JRRD.2015.04.0065
26. Gavilánez, E.L., Chedraui, P., Franco, K.G., Blum, D.M., Riofrío, J.P., Bajaña, A.S.: Osteoporotic hip fractures in older adults in Ecuador 2016. Rev. Osteoporos. Metab. Miner. **10**, 63–70 (2018). https://doi.org/10.4321/s1889-836x2018000200002
27. Gillies, G.E., Pienaar, I.S., Vohra, S., Qamhawi, Z.: Sex differences in Parkinson's disease. Front. Neuroendocrinol. **35**, 370–384 (2014). https://doi.org/10.1016/j.yfrne.2014.02.002

Graph Strength for Identification of Pre-training Desynchronization

Frank Yesid Zapata Castano[1], Óscar Wladimir Gómez Morales[2](✉), Andrés Marino Álvarez Meza[1], and César Germán Castellanos Domínguez[1]

[1] Signal Processing and Recognition Group, Universidad Nacional de Colombia, Manizales, Colombia
fyzapatac@unal.edu.co
[2] Instituto Superior Universitario Sucre, Quito, Ecuador
ogomezmo@unal.edu.co

Abstract. Motor processing can result in coordinated changes of ongoing/decreasing brain neural activity or event-related de/synchronization (ERD/S) that raises brain responses over either contralateral brain hemisphere. Because of the affordability and time resolution provided, Electroencephalographic (EEG) signals are commonly used to acquire motor imagery paradigms. However, the widely-known condition of low-noise signals makes detection and spatial localization of ERD/S challenging. Here, to deal with the high variability between subjects, we propose to perform group analysis of graph representations extracted from the weighted Phase lock Index. Statistical thresholding of the functional connectivity estimates is also accomplished to improve the assessments of phase synchronization between electrodes. The obtained results on a real-world database with 50 individuals show that the proposed methodology improves interpretation of ERD/S, allowing better prediction of motor imagery ability in subjects having low skills for practicing this paradigm.

Keywords: Event-related Synchronization/Desynchronization · Functional connectivity · Group analysis · wPLI

1 Introduction

Motor imaging (MI) is a dynamic mental state where an individual performs a mental repetition of the motor action with no manifest output. It is believed that actual movements and those performed mentally (imaginary movements) are functionally similar [1]. As a result, there is ample experimental evidence that MI contributes to substantial improvements in motor learning and performance [2], entertainment, sports conditioning, therapy to influence recovery and neuroplasticity in neurophysical regulation and rehabilitation, and enhancement of brain neural networks as the basis of motor learning [3]. These applications reinforce the importance of studying the evolution of brain organization for accurately modeling plastic changes.

V. Robles-Bykbaev et al. (Eds.): CITIS 2022, LNNS 607, pp. 36–44, 2023.
https://doi.org/10.1007/978-3-031-24327-1_4

Nevertheless, there is also a need to strengthen the dynamic modeling of temporal, spectral, and spatial features since most MI systems perform this task to identify distinctive neural activation patterns [4]. Systems dealing with MI use electroencephalography (EEG) to capture neural activation and interactions with low spatial resolution but encode brain dynamics with high temporal granularity [5]. Functional Connectivity (FC) networks are extracted to understand MI mechanisms fully, and their interaction with the co-activated brain regions is analyzed [6].

A related characterization of imaged hand movements is based on interpreting the frequency alterations in time-varying responses to a stimulus (event). Specifically, with the intention of moving one hand, the power of μ and β rhythms increases/decreases in the sensorimotor (C3 and C4) and premotor (Cz) areas located contralaterally to the hand engaged in the task [7]. However, feature extraction procedures are less effective in identifying distinct EEG spatio-spectral patterns because EEG data are not stationary. Among these factors are low EEG signal-to-noise ratio mainly due to inter-subject variability in EEG dynamics [8], and movement artifacts during recording [9]. In addition to the variability in the signal acquisition, low accuracy scores may also be the result of brain networks that are not well-developed for practicing MI tasks [10]. Due to this, MI systems perform in a highly variable manner across and within subjects, severely affecting their reliability.

To provide new insights into the evolution of collective mental imagery processes, we develop a dynamic model for estimating the mutual neural activity among subjects. According to the obtained validation results, the estimated collective dynamics reflect oriented activity in the sensorimotor cortex. As a result, the proposed model considers the inter-subject and inter-trial variability sources individually, depending on the engaged extraction method.

Here, we propose to perform the K-means algorithm for grouping the subject set according to their abilities in practicing motor imagery tasks. The results of the real-world database show that the functional connectivity better predicts the subject's ability from resting-state data during pretraining stages, outperforming the conventional power measures.

The paper is organized as follows: Sect. 2 describes the validated database and methods that are carried out and presents the experimental setup as well as the performed outcomes; Sect. 3 introduces a detailed discussion of the attained results; Sect. 4 providing the main conclusions and future work.

2 Methods and Experimental Setup

Weighted Phase Locking Index: This FC index is commonly used to estimate functional connectivity and quantifies the asymmetry of phase difference distribution between two specific channels due to its nonparametric nature and easy implementation:

$$\nu_2(c, c') = |\mathrm{E}\{\mathcal{F}\{\mathrm{S}(\mathrm{c}, \mathrm{c}'; \mathrm{n}, \mathrm{f})\} : \forall \mathrm{n} \in \mathrm{N}\}|\mathrm{E}\{\mathcal{F}\{\mathrm{S}(\mathrm{c}, \mathrm{c}'; \mathrm{n}, \mathrm{f}) : \forall \mathrm{n} \in \mathrm{N}\}|\}, \nu_2(,) \in [0, 1] \quad (1)$$

where $S(\mathrm{c}, \mathrm{c}'; \mathrm{n}, \mathrm{f}) \in \mathbb{C}$ is the cross-spectral density based on Morlet wavelets and $\mathscr{F}\{.\}$ stands for the imaginary part of a complex-valued function, $\mathrm{c} \in \mathrm{C}$.

Besides, a weighted network index is extracted from the phase synchronization measure. In particular, we compute the local-scale Node Strength that accounts for the number of links connected to each node, computed as follows:

Node Strength is a local-scale property that accounts for the number of links connected to each node, computed as follows:

$$\varphi_1(c) = CE\{\nu_2(\mathrm{c}, \mathrm{c}') : \forall \mathrm{c}' \in \mathrm{C}, \mathrm{c}' \neq \mathrm{c}\} \tag{2}$$

For evaluation purposes, we use the supervised piecewise network connectivity analysis pipeline, appraising two stages.

2.1 MI Databases Description and Pre-processing

Gigadb: We explore the collection publicly available at[1] that holds EEG data obtained from fifty-two subjects (although only $\mathrm{M} = 50$ are available) with $\mathrm{C} = 64$ channels, using a 10–10 placement electrode system. Each channel x(c) lasted $\mathrm{T} = 7\mathrm{s}$ and was sampled at $\mathrm{Fs} = 512$ Hz. At the beginning of the test, a fixation cross was displayed on a black screen during 2 s. Then, linked to either MI label $\lambda = [\mathrm{l} \text{ or } \mathrm{l}']$, a cue instruction appeared randomly on the screen within 3 s. The cue asked each subject to imagine moving his fingers, starting to form the index finger, and reaching the little finger, and touching each to his thumb. A blank screen was then displayed at the beginning of a break period, which ran randomly between 4.1 and 4.8 s. In addition, a single-trial resting-state recording, lasting 60 s, was collected from each subject.

Every raw EEG channel of either database was band-pass filtered in the frequency range $f \in [4\text{–}40]$ Hz, covering the sensorimotor rhythms considered (μ, β). Then, the band-passed EGG data are spatially filtered by a Laplacian filter centered on the selected electrode.

3 Results

3.1 Clustering of Intra/Inter-subject Variability

Parameter Tuning of Feature Extraction. Initially, the window length used to extract the functional connectivity measures considered is a critical parameter to fix. To this end, we search for a convenient trade-off between the reached accuracy and computational burden (the number of samples to be processed), adjusting the length of $\tau = 2$ s. At this value, we reach the separability for MI tasks with an accuracy value of 87.2 ± 11.6, averaged across the subject set and within $\Delta T3 \in [2.5\text{–}4.5]$ s, involving just 30 trials of resting state.

Next, we adjust the threshold q needed for preprocessing the functional connectivity measures to remove the false connections and noise as much as possible. Empirically, we found the best value for $q = 0.7$ selected as a comprise value to preserve an adequate confident level of FC measures (i.e., below $p < 0.05$) [11].

[1] http://gigadb.org/dataset/100295.

Clustering Strategies of Performance Variability. We also consider the influence of BCI illiteracy, meaning that the classifier framework may not be able to appropriately decode the brain responses captured from several subjects, resulting in poor accuracy. To deal with this issue, we cluster the individuals so that each partition holds points with related inter-subject variability of brain neural responses. For comparison purposes, we explore two strategies of clustering according to the accuracy: i) Baseline consideration a group with SMR [12, 13], ii) Group based on each extracted measure: graph strength and pre-training desynchronization.

Grouping Based on Accuracy Response and Neural Indicators. Baseline clustering. The rationale behind the proposed clustering of inter-subject variability is that the performance regularity may serve as an indicator of the individual ability to practice MI: the more accurate the subject is in distinguishing between MI tasks, the more efficient his brain network. As developed in [14], the performance irregularity is evaluated at different extraction windows. Also, the indicator values of neural desynchronization at rest over the sensorimotor area are considered. To this purpose, both sets feed the k-means algorithm with the accuracy sets, fixing the number of subject partitions to three according to cluster inertia estimated in (see Fig. 1).

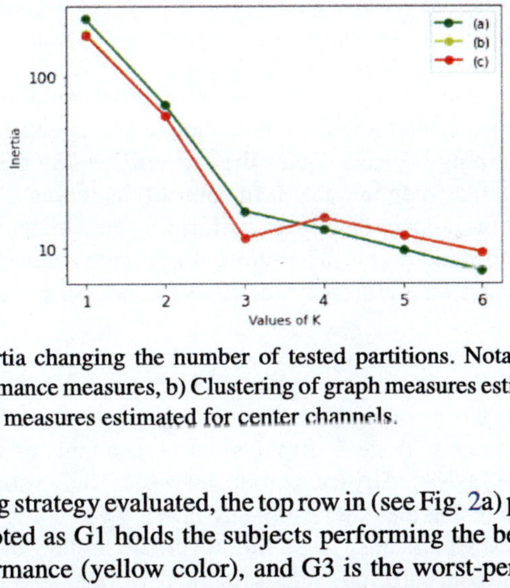

Fig. 1. Metric of inertia changing the number of tested partitions. Notations for a) stands for performance of performance measures, b) Clustering of graph measures estimated for all channels, c) Clustering of graph measures estimated for center channels.

For the clustering strategy evaluated, the top row in (see Fig. 2a) presents the resulting three groups: bar noted as G1 holds the subjects performing the best (green color), G2 is the middle-performance (yellow color), and G3 is the worst-performing individuals (red color). There is a relatively low overlapping ratio in the accomplished partitions distributed following the accuracy rank. To illustrate this fact, the bottom row of (see Fig. 2b) displays the subjects ranked in decreasing order by the achieved accuracy, showing that the partitions have not twisted subjects.

Subject Clustering by Graph Connectivity Measures. As regards the second strategy, the measures of computed graph strength are fed into the k-means algorithm together with the values of the accuracy classifier. As can be seen in (Fig. 2), the use of the graph

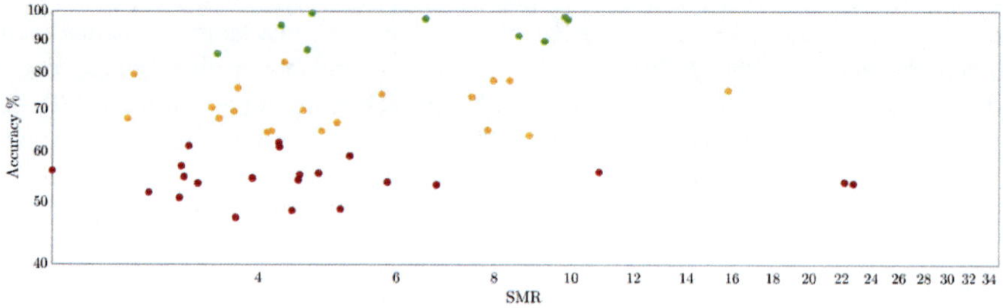

[a] Performance clustering

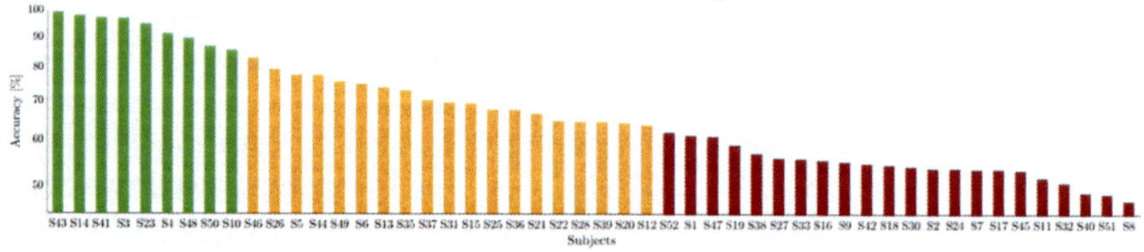

[b] Accuracy ranked in decreasing order of subjects

Fig. 2. Subject clustering using correlates between accuracy and power-based indicators extracted from sensorimotor rhythms.

measure allows partitions with a more clear distribution of subjects (see top row), also not having twisted points (bottom row).

One more aspect of consideration is the channel set configuration to extract the graph strength measure. Thus, if including the whole montage arrangement (left column), the number of bets performing subjects equals the one obtained by the baseline clustering strategy (nine subjects), as seen in Sect. 3. In contrast, including the sensorimotor area (C3 and C4) increases the number of the best-performing individuals while decreasing the number of worst-performing points. Therefore, this simple channel selection, together with the extracted graph strength measure, allows attenuating the effect of illiteracy subjects.

Relationship Between Resting-State and Neural Responses. We consider the relationship in brain neural activity between the resting-state and the neural responses elicited while performing MI tasks. Table 1 displays the assessments of correlation index p, showing that the values assessed by each group are r~0.45, being comparable with those published recently. Note that the best r estimates have a p-value less than 0.1, suggesting an acceptable statistical significance level. Moreover, the strategy of performing the correlation analysis within the partitions surpasses the performance achieved by the whole subject set.

Further, we visually inspect the spatial distribution of the brain neural activity estimated for two waveforms (low-frequency μ and high-frequency β) representative in MI analysis. In particular, (see Fig. 3) depicts the topograms that reflect the responses evoked right within the window of motor imagery and reveals that the best-performing group, G1, concentrates the brain responses within the sensory-motor are, as expected

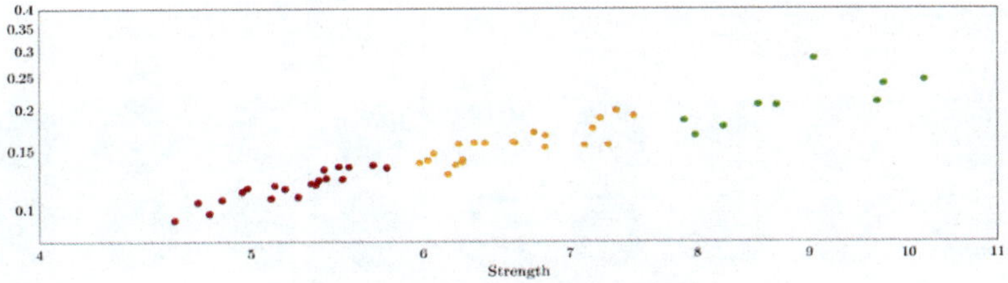

(a) Partitions clustering all channels

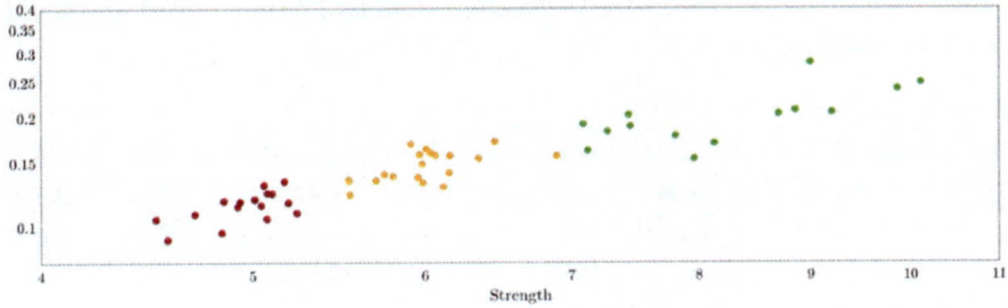

(b) Partitions clustering center channels

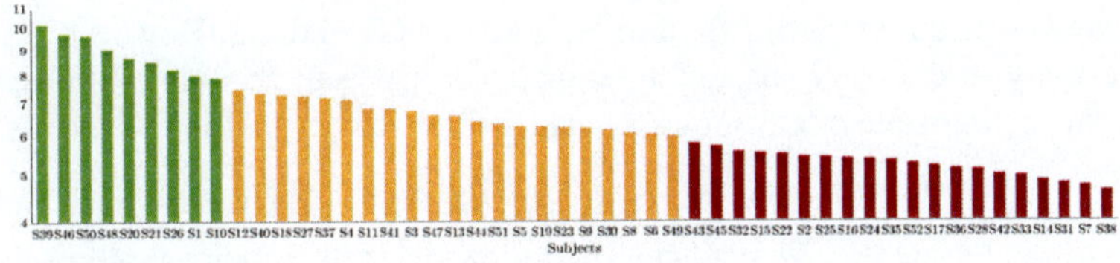

(c) All channels

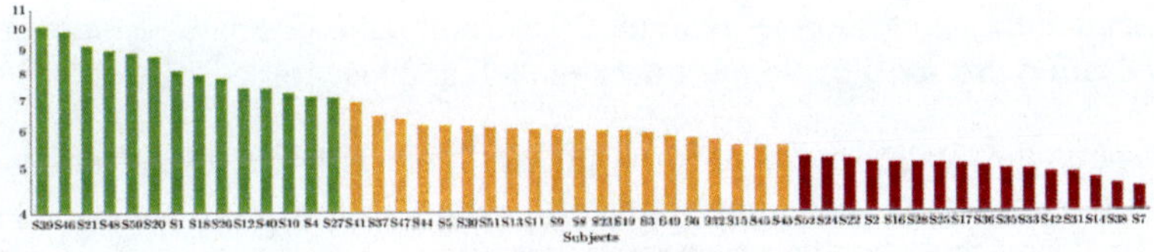

(d) Center channels

Fig. 3. Second clustering strategy through the extracted graph connectivity measure.

Table 1. Relationship between resting-state neural activity and neural responses in MI tasks. Notations *r* and *p* stand for the Pearson correlation and *p*-value, respectively. The best value of *r* each row is marked in black.

	Whole Subject set		G1		G2		G3	
	r	p	R	P	r	p	r	p
C_3-C_4	0.410	0.003	0.099	0.658	0.548	0.126	**0.450**	0.053
SMR	0.261	0.066	0.189	0.399	**0.597**	0.088	0.343	0.150
All channels	0.141	0.040	**0.440**	0.040	0.227	0.555	0.265	0.271

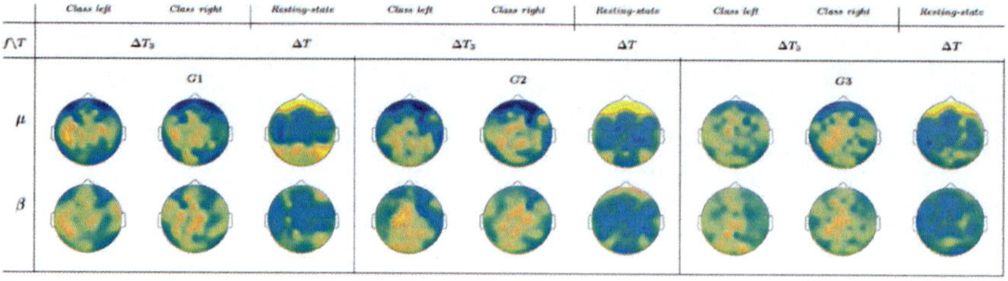

Fig. 4. Spatial distribution of the brain neural activity of MI (within the interval ΔT) and Resting-state estimated by wPLI.

in MI paradigms. Although G2 spreads the activity more in neighboring areas, similar areas are presented, such as the parietal, motor, and frontoparietal. At the same time, the worst-performing individuals hold neural activity spread all over the scalp, meaning that the sensorimotor cortex is weakly engaged. One more aspect to notice is the dominance of the right-hand present in all partitions.

On the other hand, the neural activity of the central region is small during resting-state regardless of the extracted rhythm, at least for G1 and G2. However, there is a notable activity in the posterior and anterior (frontoparietal) regions of the scalp. By contrast, the resting-state of G3 shows a small spatial activity distributed evenly across all areas.

Group Analysis Across the Subject Set. The following analysis point explores the quality of obtained partition in terms of the following aspects: i) Quality of clustering, ii) Discriminant ability present in all individual sets, aiming to estimate common areas of neural activation that most distinguish between left and hand tasks.

As regards the quality of obtained partitions, Table 2 shows several measures for validating the clustering, including Silhouette, mutual information, and score, homogeneity, similarity, and classifier accuracy that is estimated using an LDA algorithm, as performed in [18]. Using strength as a graph connectivity measure significantly improves the quality of clustering, resulting in more representative partitions.

The next aspect of consideration is only performed for the strength measure because of the improved clustering quality. Thus, we find the relevant connectivity measures using the group analysis method developed by [19], performed in two scenarios of comparison:

i. Estimation of the connectivity set present in every individual of the subject set (Using the rule AND), as shown in left column of each partition.
ii. Estimation of the connectivity set present in more than one individual (Using the rule OR) shown in the right columns (Fig. 5).

(See Fig. 4) depicts the results of group analysis obtained after performing the Wilcoxon test to get the most confident set of discriminant functional connectivity measures. In the case of the best performing subjects, the rule AND yields very focused activity located over the central and sensorimotor areas, at least for \$\mu\$ as widely

Table 2. Measurements of clustering quality: acc- classifier accuracy, MI - mutual information. Notation stands for graph strength measure.

	Acc			Metrics				
	G1	G2	G3	Silhouette	MI	Rand Score [15]	Homogeneity [16]	Similarity [17]
wPLI	71,90	74,78	69,01	0,22	0,47	0,47	0,47	0,44
wPLI + Acc	58,62	88,57	64,72	0,21				
$\varphi 1$	81,00	72,14	71,91	0,57	0,91	0,91	0,93	0,96
$\varphi 1$ + Acc	81,13	77,14	66,65	0,42				

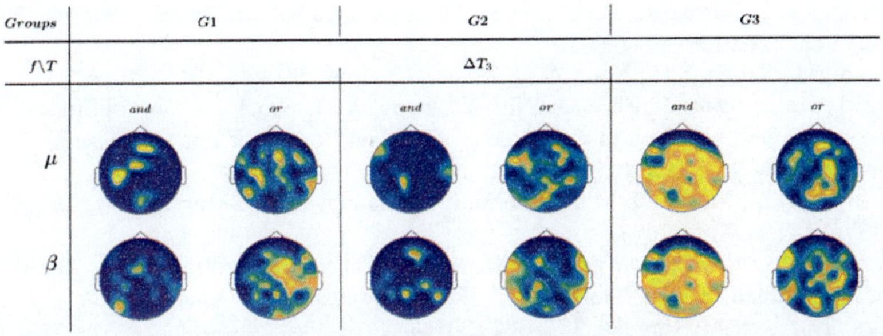

Fig. 5. Group analysis across the subject set performed to estimate the relevant functional connectivity that contributes to distinguishing between left and right hand tasks.

stated in the literature for MI paradigms. This result suggests high homogeneity in neural responses within the MI interval among the best-performing subjects. The rule OR results in less focused and more spread reactions of neural activity, meaning that the variability in performing the MI tasks rises among subjects. The OR rule for G2 also affirms this fact still. Both G1 and G2 groups localize the neural activity over the centro-parietal cortex, as mainly expected [20].

In partition G3, the variability of estimated common responses is so high that either rule extends the neural activity over those brain regions where no responses related to MI tasks should be expected [21]. For instance, μ waveform shows high activation computed by the rule AND in parietal and occipital, suggesting the presence of visual artifacts. The rule ON results in brain activity distributed all over the scalp, meaning no common responses among G3 subjects while practicing the MI paradigm [22].

4 Concluding Remarks

According to the above, classification tasks may benefit from the interpretation of electrical activity in the brain, including the evaluation of elicited MI responses, as well as other mental states in response to intrinsic neural activity, which are not elicited by a mental task. Here, K-means has been chosen as the grouping strategy to visualize the benefits of functional connectivity extracted from the neural response. Moreover, using a single graph measure computed for functional connectivity enhances the prediction ability of subjects in practicing motor imagery without losing explainability in the obtained spatial maps of brain activity.

As future work, the authors plan to include more robust graph measures with oriented properties of locality/globality between connectivity links. Also, more elaborate functional connectivity measures will be tested, like phase-amplitude connectivity and Gaussian Functional Connectivity, among others.

References

1. Stolbkov, Y.K., Moshonkina, T.R., Orlov, I.V., Tomilovskaya, E.S., Kozlovskaya, I.B., Gerasimenko, Y.P.: Human Physiology **45**, 104–114 (2019)
2. Aymeric, G., Ursula, D.: Frontiers in Physiology 396 (2019)
3. Machado, T.C., Carregosa, A.A., Santos, M.S., da Silva Ribeiro, N.M., Melo, A.: Topics in Stroke Rehabilitation **0**, 1–6 (2019)
4. Hamedi, M., Salleh, S.H., Noor, A.M.: Neural Comput. **28**, 999–1041 (2016)
5. Feng, J., Jin, J., Daly, I., Zhou, J., Niu, Y., Wang, X., Cichocki, A.: An optimized channel selection method based on multifrequency CSP-rank for motor imagery-based BCI system Comp. Int. and Neurosc (2019)
6. Stavrinou, M.L., Moraru, L., Cimponeriu, L., Della Penna, S., Bezerianos, A.: Brain Topogr. **19** 137–145 (2007)
7. Wierzgala, P., Zapala, D., Wojcik, G.M., Masiak, J.: Front. Neuroinform. **12**, 78 (2018)
8. Saha, S., Ahmed, K.I.U., Mostafa, R.: Hadjileontiadis L and Khandoker A. IEEE Trans. Neural Syst. Rehabil. Eng. **26**, 371–382 (2018)
9. Pattnaik, P.K., Sarraf, J.: Computer and Information Sciences **30**, 18–24 (2018)
10. Ahn, M., Jun, S.C.: J. Neurosci. Methods **243**, 103–110 (2015)
11. Adamovich, T., Zakharov, I., Tabueva, A., Malykh, S.: bioRxiv (2022)
12. Blankertz, B., Sannelli, C., Halder, S., Hammer, E.M., Kübler, A., Müller, K.R., Curio, G., Dickhaus, T.: Neuroimage **51**, 1303–1309 (2010)
13. Hammer, E.M., Halder, S., Blankertz, B., Sannelli, C., Dickhaus, T., Kleih, S., Müller, K.R., Kubler, A.: Biological psychology **89**, 80–86 (2012)
14. Velasquez-Martinez, L., Caicedo-Acosta, J., Castellanos-Dominguez, G.: Entropy **22**, 703 (2020)
15. Emmons, S., Kobourov, S., Gallant, M., Bőrner, K.: PloS one **11**, e0159161 (2016)
16. Rosenberg, A., Hirschberg, J.: V-measure: A conditional entropy-based external cluster evaluation measure Proceedings of the 2007 joint conference on empirical methods in natural language processing and computational natural language learning (EMNLP-CoNLL), pp. 410–420 (2007)
17. Franti, P., Rezaei, M., Zhao, Q.: Pattern Recognition 47, 3034–3045 (2014)
18. Velasquez-Martinez, L.F., Zapata-Castano, F., Castellanos-Dominguez, G.: Frontiers in Neuroscience **14**, 714 (2020)
19. Padilla-Buritica, J.I., Hurtado, J.V., Castellanos-Dominguez, G.: Biomedical Signal Processing and Control **52**, 341–346 (2019)
20. Kim, Y.K., Park, E., Lee, A., Im, C.H., Kim, Y.H.: PLoS ONE **13**, e0190715 (2018)
21. Sareen, E., Singh, L., Gupta, A., Verma, R., Achary, G.K., Varkey, B.: IEEE Trans. Neural Syst. Rehabil. Eng. **28**, 2420–2430 (2020)
22. Sun, S., et al.: IEEE Trans. Neural Syst. Rehabil. Eng. **27**, 429–439 (2019)

INFOTELECOM

Low-Cost Fault Prediction System for a Rolling System on an Augmented Reality Platform with Cloud Communication

Andrés Carvajal Andrade(✉), Kevin Pazmiño Tintín, Johanna Celi, and William Montalvo

Universidad Politécnica Salesiana, UPS, 170146 Quito, Ecuador
{cceli,wmontalvo}@ups.edu.ec

Abstract. The application of Augmented Reality (AR) technologies within the industry has allowed for improving preventive and predictive maintenance techniques, thanks to the fact that they allow information to be presented in real time and in the same analyzed environment, which ensures that maintenance operational processes improve. This work presents the development of a predictive system for thermal stress failures, by applying the analysis of the Weibull statistical model and the use of free AR tools and cloud technology, which allow determining the reliability and average lifetime of induction coils in real time of a rolling system. The temperature data were obtained through a wireless sensor network (WSN) that sends the data to an embedded system (Raspberry Pi 4), which behaves as the communication channel (Gateway) between the sensors and the cloud, through the MQTT server. The results are presented in graphs and are promised under the Weibull model.

Keywords: Augmented reality · WSN · Weibull · Internet of things (IoT)

1 Introduction

Most of the electrical and mechanical elements at the industrial level suffer wear from the first moment of use, therefore, the application of reliability techniques should be considered to determine the average lifetime before catastrophic failures occur, which consequently affect the resulting processes, thus avoiding the unnecessary expenditure of resources to perform corrective maintenance [1]. As a result, several reliability methods and techniques have been developed over time, however, these traditional methods require significant sample sizes, are time-consuming to evaluate, and often involve high costs [2].

In response to this, it is proposed to use the Weibull statistical model, which is widely used, as in the research of [3, 4] that performed similar reliability studies with Weibull's law to determine the useful life of motors. This allows modeling stochastic processes related to reliability vs. time, this exponential distribution is due to a more appropriate arrangement of the data, which reduces the dimensionality of the sample size to be studied [5]. For this reason, this study seeks to present an alternative for the

V. Robles-Bykbaev et al. (Eds.): CITIS 2022, LNNS 607, pp. 47–57, 2023.
https://doi.org/10.1007/978-3-031-24327-1_5

control of reliability in elements subject to thermal stress, taking a rolling system as a test plant. This equipment presents characteristics of failure in its windings which is a problem generating downtime, causing a significant decrease in the capacities of the equipment for which it was designed, given this problem, a reliability analysis was implemented using Weibull's law for which an algorithm was applied that allows to increase the reliability in addition to it is easy to execute in any microcontroller, thus giving a low-cost solution [6].

The results of the developed system were favorable, it is considered that if a correct Weibull analysis is performed, the reliability and average lifetime of the induction coils can be accurately determined, in addition, this system allows the data to be monitored in the AR application in real time.

The work done is divided as follows, Sect. 2 methodology design and implementation, Sect. 3 results of the failure prediction model, Sect. 4 discussion with other studies, and finally Sect. 5 conclusions.

2 Methodology

2.1 IoT Data Transmission

Data was transmitted via a wireless sensor network (WSN), with four MLX90614 sensors each connected to an ESP8266 WIFI NodeMCU for communication and data exchange via TCP/IP protocol [6], then each one is connected to the Raspberry Pi 4 to perform the data exchange through the MQTT protocol with the cloud server using Python scripts [7].

Topology. The developed system implements a star topology as shown in Fig. 1, where each sensor is connected to each node, which in turn is routed to a central node or Gateway, which serves as a border router (BR), through WIFI connection, the data transfer occurs because the client and the Gateway are linked to the same internet network. The Broker can receive all the information from the clients through the publication/subscription of messages, which are then sent to the cloud [6, 8].

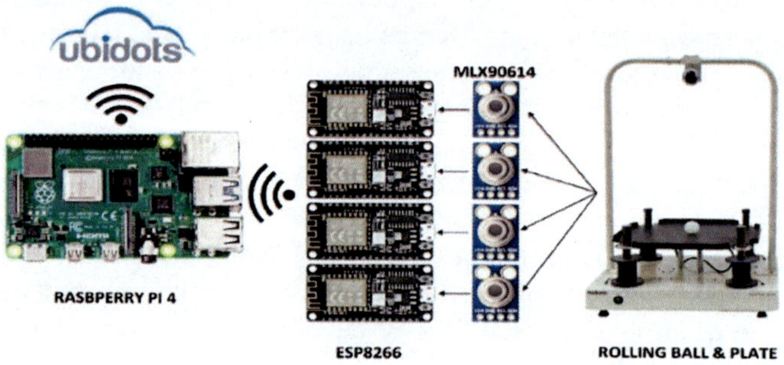

Fig. 1. Star topology

- ***Broker:*** In order to operate under the MQTT protocol, you need a server installed on the Raspberry Pi 4, in this case, Eclipse Mosquito, which works as an intermediary of the messages in the MQTT protocol [7].
- ***Client:*** The ESP8266 NodeMCU module has a library (PubSubClient) that allows it to be an MQTT client, publishing and subscribing messages to one or several topics via WIFI connection [7, 8].
- ***Cloud Server:*** The Ubidots platform allows the communication and transfer of data to and from the cloud, enabling its analysis and control in real time. The platform has APIs that allow data to be graphed and obtained in JSON format, facilitating communication for HTTP and MQTT protocols.

Prototype Design. This work proposed a practical approach for predictive maintenance, for which an AR application was developed, to have a perspective that combines the real world with a virtual interactive environment, in addition to the insertion of 3D virtual models' main features of augmented reality environments. This technology allows the user to physically interact with the equipment and at the same time visualize information in real time from the data obtained from the sensors [9, 10].

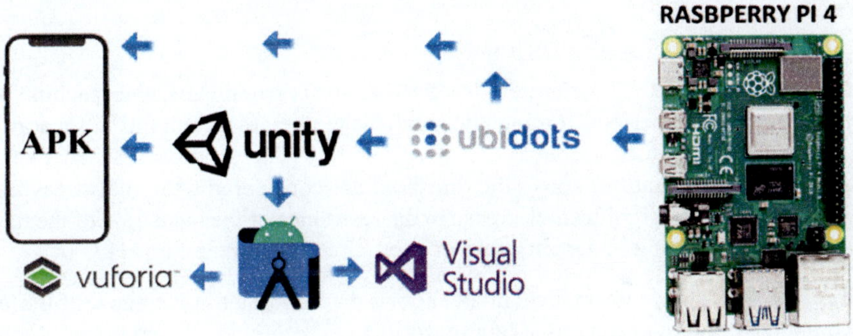

Fig. 2. Connection diagram

As shown in Fig. 2, several free AR tools were used, such as Vuforia Development, Unity, Visual Studio, and Android Studio. A target (QR) was used for object recognition and identification of areas of interest (coils), as shown in Figs. 3a, b. The application is developed in Unity using Visual Studio functions and scripts, the camera option recognizes the target, and once recognized the application shows the different options available as shown in Fig. 3c, the "Analysis Data" button opens an Excel macro where the Weibull reliability analysis is performed, another button where the system information and 3D simulation is found to provide guided maintenance and a button that allows viewing the data in the cloud.

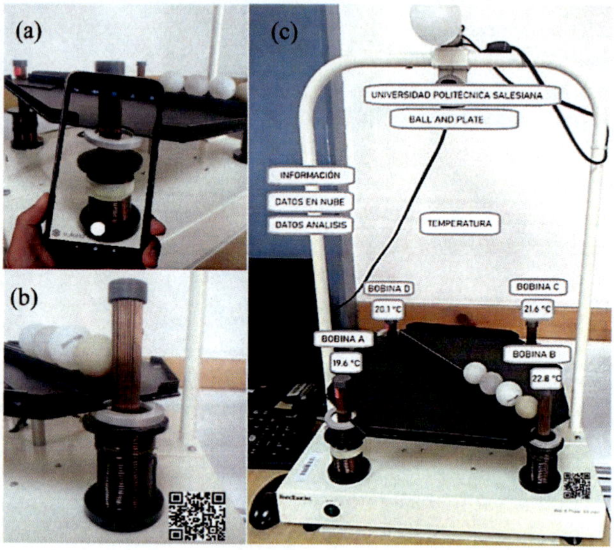

Fig. 3. (a) AR application. (b) QR code for target recognition on induction coil. (c) AR prototype on a rolling system.

System Identification. To determine the RBP analysis parameters, the machine was operated for constant periods, forcing it to reach the temperature limit (40 °C) according to its manufacturer, which produces a 30% reduction in the electrical resistance of the coils [11, 12]. Temperatures above the threshold are considered a thermal stress fault, and there is a risk of a significant decrease in the resistance of the insulation of the turns, leading to thermal aging, the main cause of short circuits between turns [13, 14].

Weibull's Law. This mathematical model allows determining the average useful life of the component before presenting a failure, for the case study the 2 parameters Weibull analysis (β and η) was used, taking the case of $t_0 = 0$ (position with the origin of distribution 0), this case was considered since the group of data follows the Weibull distribution and can be represented by a straight line. It was assumed that the origin is known and that the data agree with the experimental analysis [5, 15, 16].

The Weibull law is defined from the following equations [17, 18]:

$$F(t) = 1 - exp^{\left\{-\left(\frac{t-t_0}{n}\right)^{\beta}\right\}} \tag{1}$$

$$R(t) = exp^{-\left(\frac{t-t_0}{n}\right)^{\beta}} = 1 - F(t) \tag{2}$$

$$h(t) = \frac{\beta}{\eta}\left(\frac{t-t_0}{\eta}\right)^{\beta-1} \tag{3}$$

$$f(t) = \frac{\beta}{\eta}\left(\frac{t-t_0}{n}\right)^{\beta-1} exp^{\left\{-\left(\frac{t-t_0}{n}\right)^{\beta}\right\}} \tag{4}$$

where:

- $F(t)$: *Cumulative distribution function*
- $R(t)$: *Reliability function*
- $h(t)$: *Risk Function or Failure Ratio*
- β: *Shape parameter*
- η: *Scale parameter or characteristic life*
- $f(t)$: *Density function*
- t_0: *Initial location parameter*
- The analysis is based on the case of t0 = 0, where the component lifetime lacks intrinsic reliability [19]:
- β < 1 assumes that the component is in its juvenile stage, causing early failures, known as infant mortality.
- β = 1 Random or pseudo-random failures occur, considered as the component life stage.
- β > 1 Wear and tear is evident, producing failures due to aging.

Applying double Neperian logarithms:

$$\ln\left(\ln\left(\frac{1}{1-F(t)}\right)\right) = \beta\ln(t) - \beta\ln(\eta) \rightarrow Y = \ln\left(\ln\left(\frac{1}{1-F(t)}\right)\right) \quad (5)$$

Weibull's law can be used for the analysis of parameters related to cycles or work periods, start/stop operation time, time until an event occurs, kilometers traveled, etc., these data allow to determine the confidence level as a function of the equipment operating time. These events allow plotting the linear regression line and least squares.

Least Squares and Linear Regression. The shape and scale parameters are obtained by applying Eq. (5) using the least squares method, this equation can be obtained F(t) in a linear regression equation [5, 15, 20].

Analysis of the Data Obtained. TO obtain the Weibull parameters, a linear regression calculation is performed, corresponding to each coil, as shown in Table 1. The time data in the second column of Table 1 correspond to the times in which the equipment reached the maximum temperature of 40 °C, as explained in Sect. 2. To obtain the trace of the linear regression line, the Ber-nard median rank is applied as (Mean percentage line of the failure time) corresponding to the third column of Table 1, this non-parametric estimator is based on the order of the failures from lowest to highest (excessively organized).

Table 1. Linear regression coil A

Linear regression table			x	y	
# Failures	Time (ti)	Medium line %	Ln(ti)	Ln (Ln(1/1-F(ti)))	F(t)
1	20,4	2,30%	3,02	-3,76	0,977
2	21	5,59%	3,04	-2,86	0,944
...	...	...	...	...	...
29	44,6	94,41%	3,80	1,06	0,056
30	46,8	97,70%	3,85	1,33	0,023
$\sum$	964,3	15	103,23	-16,59	15

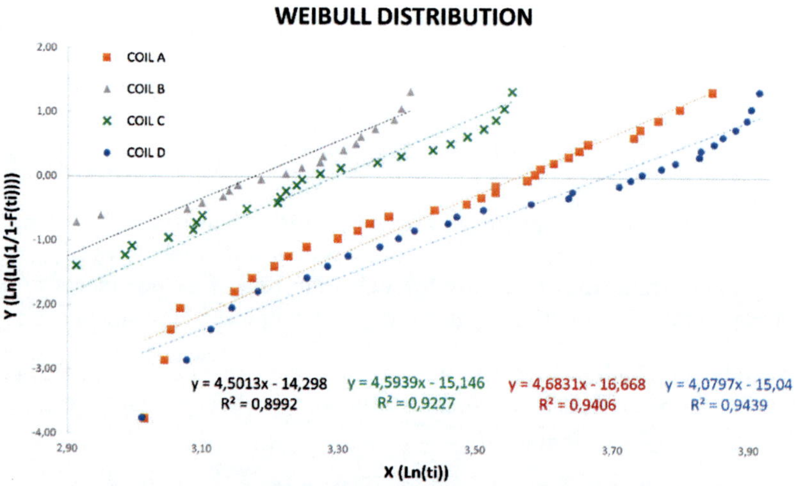

Fig. 4. Distribution of the data for the Weibull probability

For each fault, Eq. (1), (5), and (7) are applied to determine the summations that allow finding β, using the regression line and the slope Eq. (7).

$$\beta = \frac{\sum(x-\overline{x})(y-\overline{y})}{\sum(x-\overline{x})^2} = 4.6831 \tag{7}$$

the calculation of the scale parameter is given as:

$$\eta = \overline{y} - \beta\overline{x} = -16.6683$$

to obtain the correlation coefficient, we applied Eq. (8)

$$Correl(x, y) = R = \frac{\sum(x - \overline{x})(y - \overline{y})}{\sqrt{\sum(x - \overline{x})^2 \sum(y - \overline{y})^2}} = 0.96984 \quad (8)$$

where x- and y- are the mean values of the samples.

Table 2 shows the Weibull parameters and the coefficient of determination R^2 which shows an average greater than 92%, this determines that the data are linearly related, which can be seen graphically in Fig. 4, the data represent a linear trend.

Table 2. Weibull parameters (Coils A, B, C and D)

Parameters	Coil A	Coil B	Coil C	Coil D
β (*t*)	4,6831	4,5013	4,5939	4,0797
η (*t*)	35,13733	23,959007	27,02924	39,91124
R (*t*)	0,96984	0.94823	0.96056	0.97154
R^2 (*t*)	0,9406	0,8992	0,9227	0,9439

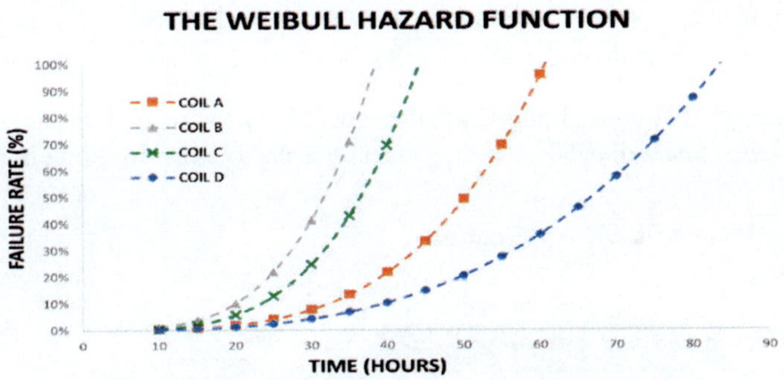

Fig. 5. Risk function for coils (A, B, C, and D)

The shape parameters in Table 2 determine that the coils are in the aging failure stage (β > 1). This indicates that the failure rate has increased with time. It is established that the sample values behave according to Eq. (4) and that the data are correlated allowing a reliability curve suitable for probability analysis.

In Fig. 5, it can be identified that the probability of failure is increasing with time, which allows for simulating the thermal stress suffered by the coils. To estimate the probability of a failure occurring over time, Eq. (3) was applied to obtain the risk function, as shown in Table 3.

Table 3. Risk function data for the coils (A, B, C and D)

	Coil A	Coil B	Coil C	Coil D
$h(t)$	9,6008512	13,5777456	12,2761727	7,56178265

3 Results

A hypothetical case was proposed to determine the useful lifetime of the system, for which it was proposed to use the RBP for one hour per day, for 25 days (1,500 min), the failure prediction system was applied and the following results were obtained:

$$R(1500) = exp^{-\left(\frac{1500-0}{-16.6683}\right)^{4.6831}} = 0.8162 = 81.62\%$$

Table 4. Failure probability hypothetical case

Probability of failure	Coil			
	A	B	C	D
Reliability	81.62%	29.79%	49.72%	86%
Probability of failure	18.38%	70.21%	50.28%	13.78%

Table 4 shows that coils B and C are the most likely to fail within this period, they have an efficiency lower than 50%, so they must be replaced to avoid short circuit failures.

3.1 IoT Platform and AR Application

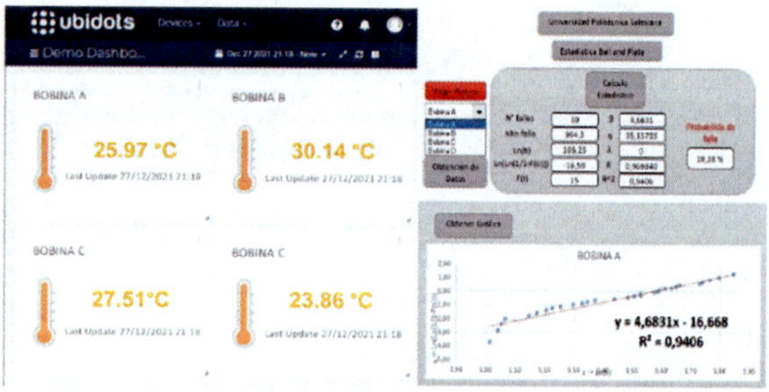

Fig. 6. (a) Ubidots display panel. (b) Excel Macro

Figure 6a shows the data sent to the IoT platform (Ubidots) for analysis, in Fig. 6b, Excel macro, where the data will be loaded and updated automatically, the data will be downloaded from the cloud and entered into the Weibull analysis, thus having the prediction of failures in real time.

4 Discussion

Contreras [1] establishes high-precision predictive maintenance under a Machine Learning (KNN) model; however, the analysis depends on certain specific characteristics of the equipment analyzed, which limits its use. Similarly, the study conducted by [22], establishes that the linear regression method for temperature failure prediction is reliable and effective. Nevertheless, its complexity increases when using linear and nonlinear models under customized equations, which implies that this model has a reduced range of applicability, since the method can only be used on equipment that complies with the same parameters with which the equations were designed, thus losing the scalability characteristic. In this work, the Weibull statistical model is used, which, in comparison to these studies, allows its global applicability, on any equipment regardless of its characteristics, since it is based on a comparative reliability analysis.

5 Conclusions

The Weibull statistical model allows determining the confidence level associated with the useful lifetime of a component, through parameters that evaluate the arbitrary failure rate when subjected to a given stress condition, establishing the maximum time intervals in which the element can work under continuous fatigue conditions before failure occurs.

The control of thermal aging is fundamental for the prevention of failures in induction coils, due to the wear in the insulation systems given the temperature variation dictated by the Arrhenius equation, for every 10 °C above the limit temperature, the insulation efficiency decreases by half.

The AR techniques make industrial maintenance more dynamic, allowing the sharing of the necessary information and the visualization of variables to carry out correct maintenance in real time, which reduces possible human errors.

The system developed in this study has scalability characteristics, the equipment used has several analog and digital inputs and outputs, which allows acquiring data of other variables, current, voltage, etc. In addition, the source code is elaborated in C and Python, simple languages that allow the implementation of more sensors.

The objective of this study was to develop a system on an IoT platform that can be visualized within an AR application, but under the slogan of "low cost", so affordable sensors and controllers were used, in addition to free software. These tools allowed the development of this research, which in turn will enable the system to be replicated in similar systems for fault prediction, data transmission, and cloud storage.

References

1. Contreras Alvarez, J.L.: DISEÑO DE UN MODELO PARA MANTENIMIENTO PREDICTIVO EN MOTORES DE INDUCCIÓN UTILIZANDO TÉCNICAS DE LA INDUSTRIA 4.0. Universidad Tecnológica del Perú, Lima (2020)
2. Johal, H., Mousavi, M.J.: Coil current analysis method for predictive maintenance of circuit breakers (2008). https://doi.org/10.1109/TDC.2008.4517278

3. Li, L., Guan, J., Yuan, P., Yin, Y., Li, Y.: A Weibull distribution-based method for the analysis of concrete fracture. Eng. Fract. Mech. **256**, 107964 (2021). https://doi.org/10.1016/J.ENGFRACMECH.2021.107964. Oct.
4. Méndez-González, L.C., Rodríguez-Picón, L.A., Pérez Olguin, I.J.C., Garcia, V., Quezada-Carreón, A.E.: Reliability analysis for DC motors under voltage step-stress scenario. Electr. Eng. **102**(3), 1433–1440 (2020). https://doi.org/10.1007/s00202-020-00966-z
5. Salazar, R., Fitz, E., López, I., Rojano, A.: Confiabilidad y Análisis de Fallas utilizando la Distribución Weibull. Congr. Int. Investig. EN CIENCIAS BÁSICAS, 5–10 (2017). [Online]. Available: https://dicea.chapingo.mx/wp-content/uploads/2018/05/MEMORIA_MESA_1A_CONGRESO2017.pdf
6. Krishna, G., et al.: Design and development of bi-directional IoT gateway using ZigBee and Wi-Fi technologies with MQTT protocol. Artic. Int. J. Eng. Technol. **7**(2), 125–129 (2018). https://doi.org/10.14419/ijet.v7i2.8.10344
7. Kashyap, M., Sharma, V., Gupta, N.: Taking MQTT and NodeMcu to IOT: Communication in Internet of Things. Procedia Comput. Sci. **132**, 1611–1618 (2018). https://doi.org/10.1016/J.PROCS.2018.05.126. Jan.
8. Upadhyay, Y., Borole, A., Dileepan, D.: MQTT based secured home automation system. In: 2016 Symp. Colossal Data Anal. Networking, CDAN 2016 (Sep. 2016). https://doi.org/10.1109/CDAN.2016.7570945
9. Fernández del Amo, I., Erkoyuncu, J.A., Roy, R., Palmarini, R., Onoufriou, D.: A systematic review of Augmented Reality content-related techniques for knowledge transfer in maintenance applications. Comput. Ind. **103**, 47–71 (2018). https://doi.org/10.1016/J.COMPIND.2018.08.007. Dec.
10. Kollatsch, C., Schumann, M., Klimant, P., Lorenz, M.: Industrial Augmented Reality: Transferring a Numerical Control Connected Augmented Realty System from Marketing to Maintenance. In: Adjunct Proceedings of the 2017 IEEE International Symposium on Mixed and Augmented Reality, ISMAR-Adjunct 2017, pp. 39–41 (Oct. 2017). https://doi.org/10.1109/ISMAR-Adjunct.2017.27
11. Feedback Instruments Ltd: Ball and Plate Control Experiments, vol. 33–240-Lab, no. 1160 (2008)
12. Núñez, H.: Development and implementation of a fuzzy controller using machine vision feedback applied to a ball-and-plate positioning system, p. 86 (2017)
13. Qiu, H., Zhao, X., Yang, C., Ran, Y., Wei, Y.: Influence of inter-turn short-circuit fault considering loop current on electromagnetic field of high-speed permanent magnet generator with gramme ring windings. Journal of Electrical Engineering & Technology **14**(2), 701–710 (2019). https://doi.org/10.1007/s42835-019-00122-z
14. Benchabane, F., Guettaf, A., Yahia, K., Sahraoui, M.: Experimental investigation on induction motors inter-turns short-circuit and broken rotor bars faults diagnosis through the discrete wavelet transform. e i Elektrotechnik und Informationstechnik 20181352 **135**(2), 187–194 (2018). https://doi.org/10.1007/s00502-018-0607-6. Mar.
15. Griggs, J.A., Zhang, Y.: Determining the confidence intervals of Weibull parameters estimated using a more precise probability estimator. J. Mater. Sci. Lett. 2003 2224 **22**(24), 1771–1773 (Dec. 2003). https://doi.org/10.1023/B:JMSL.0000005417.39465.35
16. Chaurasiya, P.K., Ahmed, S., Warudkar, V.: Study of different parameters estimation methods of Weibull distribution to determine wind power density using ground based Doppler SODAR instrument. Alexandria Eng. J. **57**(4), 2299–2311 (2018). https://doi.org/10.1016/J.AEJ.2017.08.008. Dec.
17. Mahmood, F.H., Resen, A.K., Khamees, A.B.: Wind characteristic analysis based on Weibull distribution of Al-Salman site, Iraq. Energy Rep. **6**, 79–87 (2020). https://doi.org/10.1016/J.EGYR.2019.10.021. Feb.

18. Strzelecki, P.: Determination of fatigue life for low probability of failure for different stress levels using 3-parameter Weibull distribution. Int. J. Fatigue **145**, 106080 (2021). https://doi.org/10.1016/J.IJFATIGUE.2020.106080. Apr.
19. Zhang, C.W.: Weibull parameter estimation and reliability analysis with zero-failure data from high-quality products. Reliab. Eng. Syst. Saf. **207**, 107321 (2021). https://doi.org/10.1016/J.RESS.2020.107321. Mar.
20. Annie Elisabeth Jebaseeli, E.: Monitoring the thermal behavior of induction motor using regression technique. In: 2017 Innovations in Power and Advanced Computing Technologies, i-PACT 2017, vol. 2017-January, pp. 1–4 (2017)

Nonlinear Fuzzy Control of Position and Velocity Using Crisp Outputs Implemented in a General-Purpose Microprocessor

Gabriel Moreano[1,2](✉), Julio Tafur Sotelo[1], and Jhonny Pancha[3]

[1] Pontificia Universidad Católica del Perú, Lima 32, San Miguel, Peru
gabriel.moreano@espoch.edu.ec
[2] Escuela Superior Politécnica de Chimborazo, Panamericana Sur Km 1 ½, Riobamba, Ecuador
[3] Universidad Politécnica Salesiana, Rumichaca y Morán Valverde, Quito, Ecuador

Abstract. A line follower robot is a classic prototype in automatic control learning and quite applicable in the industry for example in cargo transport, the common control strategy in this type of device is a PD linear controller since it produces a fairly stable response and also for its ease of tuning. The current work presents the design, analysis and implementation of a nonlinear fuzzy control strategy, which generates several operational benefits compared to PD control. For the analysis, the responses of the two types of controllers are compared through simulation algorithms and experimental tests, where parameters such as the response time or the effort in the system actuators are evaluated. In general, the fuzzy controller non-linearity generates some confidence when choosing a control tool for this type of device. Finally, a speed control strategy easily adaptable to the fuzzy position controller is presented in such a way that the travel time is further optimized.

Keywords: Nonlinear control · Fuzzy · Mamdani · Line follower · Singleton

1 Introduction

A line follower robot is defined as a mobile robotic platform that can follow a navigation path provided by a constant mark on the surface where it moves, this mark must have a color that contrasts with the surface for its correct identification, typically a mark black navigation pad on a white surface.

Line follower robots are commonly used as experimental platforms for tuning and testing automatic control algorithms and in the last 15 years they have become popular, gaining the prominence of several university-level robotics con-courses as can be seen in [1], in these contests the common objective is to travel a navigation route in the shortest time possible. Furthermore, due to its popularity, complications such as avoiding obstacles, solving intersections, among other skill challenges have been added.

Another factor that makes this type of robot popular is that it can be implemented with a wide range of budgets depending on the work to be done, in an educational environment prototypes can be built with extremely reduced budgets, while in an industrial

V. Robles-Bykbaev et al. (Eds.): CITIS 2022, LNNS 607, pp. 58–71, 2023.
https://doi.org/10.1007/978-3-031-24327-1_6

environment you can find platforms that exceed thousands of dollars. These robots typically operate with infrared sensors that detect the reflection of light from the surface on which the robot moves, opaque colors absorb light (less light reflection) and light colors reflect most of the light that influences on them, in [2–4] research works can be observed, where the authors explain in detail the components, as well as the implementation of a line follower robot with infrared sensors and an 8-bit microprocessor.

As it was already mentioned, this type of robot is used to understand how an automatic control algorithm operates. Likewise, the preferred control strategy for this type of system is a PID control as the authors show in [5] and [6], the PID controller generates a quite acceptable response for this type of system due to its linear characteristics. However, certain aids for optimal operation are needed such as a fast trajectory correction when the robot enters a curve that is too steep; Due to these operating difficulties, strategies have been used to improve the response of this type of robot, such as implementing arrangements with a greater number of infrared sensors. As it can be seen in [7], this strategy helped to have a greater range of robot position measurement, but due to the controller linear operation, the system still requires rapid trajectory correction because of motor blockage. In [8] and [9] another strategy for sensing the variable position using computer vision algorithms is shown. This strategy tries to anticipate any change in the trajectory, so that the controller can respond in advance by reducing actuator overload, energy consumption and the blocking of motors, in the pronounced corners on which the robot moves, the difficulty of this strategy is that its implementation cannot be carried out on an 8-bit processor, so a controller that include an operating system or an embedded computer vision library as explained in [10].

Although it is true that different measurement strategies have been tested to improve the response of this type of mobile robots, the testing of other control strategies that could be more powerful than the PID has not been documented. The only strategy on which information is found is the fuzzy logic control in a simulated way, as seen on [11–13], in these works favorable results are shown compared to PID in response time and improvement of system dynamics, the weakness is that no real lift is presented.

Fuzzy logic control systems use the "expert knowledge" of an operator to correct the operation of a system [14]. Most of the works cited present a control by Mamdani's inference method [15], which is a control method quite intuitive and that in theory would facilitate its tuning.

In this work the application of a fuzzy logic control system to a line follower robot is evaluated, the implemented controller will use the Mamdani's fuzzy inference method with the particularity that its output is formed only by fuzzy sets of the singleton type, a singleton set is one that gives a clear output value when it is activated by a fuzzy rule (it does not require defuzzification by areas) [16]; the use of this type of output sets considerably simplifies the computational cost of the controller and its response speed. The system will be implemented on an 8-bit microprocessor and its performance will be compared to a PID controller.

At the beginning of the work you will find a simplified model of a line follower robot, a platform that was used to evaluate the behavior of the aforementioned controllers, secondly, the form of construction and tuning of the FLC control with clear outputs is observed, in the third point we observe the comparative results obtained from simulated and real tests of the proposed system; At the end of Sect. 3, an innovative proposal for speed control of the line follower robot is included, finally, in the last section, the conclusions obtained from the investigation are shown.

2 Materials and Methods

2.1 System Model

There is a small number of works that present a dynamic model of the line-following robot, a model that draws attention is shown in [17] where the kinematic expressions of a bidirectional robot are used to obtain the equations of motion of the robot considering a coordinate axis coincident with the geometric center of the robot.

In this work a new dynamic model is proposed considering certain simplifications to have a linear model; Fig. 1a shows the physical characteristics of the system and its inputs, while Fig. 1b shows the output variable of the system. Note that the center of mass of the system does not coincide with its geometric center, this to bring the robot's center of mass closer to the traction axis and give it greater mechanical stability.

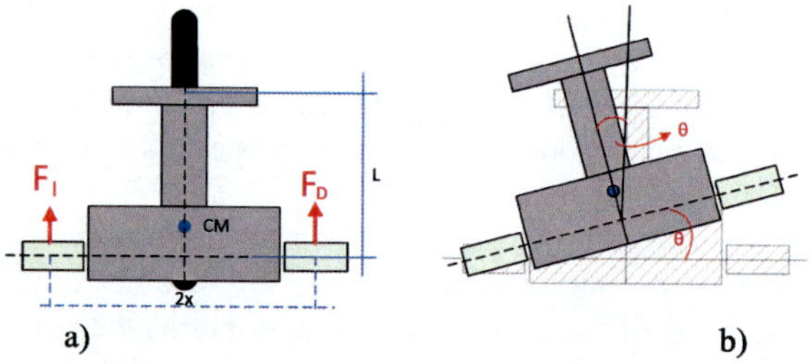

Fig. 1. Free-body diagrams of the line follower robot. a) System inputs. b) System output

In this work a new dynamic model is proposed, considering certain simplifications in a linear model; Fig. 1a shows the physical characteristics of the system and its inputs, while Fig. 1b shows the output variable of the system. Notice that the system's center of mass does not coincide with its geometric center, in order to bring the robot's center of mass closer to the traction axis and give it greater mechanical stability.

To simplify the model and due to its dimensions, it is assumed that the angle θ, that the traction axis rotates with respect to the horizontal, around a perpendicular axis passes through the center of the left tire. in Fig. 1b, it is exactly equal to the angle that the front axis of the robot rotates with respect to the vertical, about an axis that crosses through the robot center of mass.

To obtain the model, the sum of moments with respect to the center of mass of the robot is used, resulting in:

$$xF_D - xF_I - B\theta' = I\theta'' \tag{1}$$

where, x is the lever arm of each input force, B is the viscous friction constant of the system, I is the moment of inertia of the robot with respect to an axis that passes through its center of mass and θ is the exit angle of the system. Since the thrust forces are provided by the motors of each tire, the model can be approximated by the equations of a DC motor. If Ø is the angular displacement of each motor, we can write:

$$R_a i_a + L_a i_a' + kØ' = v(t) \tag{2}$$

where k is the numerical constant of the motor, R_a and L_a are the armature resistance and inductance and $Ø'$ is the angular velocity in rad/s. The mechanical ratio of the motor is:

$$ki_a - B_m Ø' - xF_D = I_m Ø'' \tag{3}$$

where I_m and B_m are the inertia and the rotational viscous friction constant of the motor. The angular displacement of the robot coincides with the length of the arc described by the tires, that is:

$$\theta L = Ør \tag{4}$$

where L is the length of the robot and r is the radius of the tires, with this expression the model system due to a single motor is:

$$\frac{\theta(s)}{V(s)} = \frac{k}{as^3 + bs^2 + cs} \tag{5}$$

where a, b and c are numerical constants resulting from the physical parameters of the system.

$$a = IL_a + I_m nL_a$$

$$b = BL_a + IR_a + B_m nL_a + I_m nR_a \tag{6}$$

$$c = BR_a + k^2 n + B_m nR_a$$

Finally, the system due to its two inputs, which is the voltage applied to each motor, can be expressed as:

$$\theta(s) = \left[\frac{k}{as^3+bs^2+cs} \;\; -\frac{k}{as^3+bs^2+cs} \right] \begin{bmatrix} V_D \\ V_I \end{bmatrix} \tag{7}$$

The single-engine system model, Eq. 6, will be used to tune and evaluate the performance of the control strategy as an independent linear system, prior to a full system simulation using the model in Eq. 7.

2.2 System Characteristics

The simulation of the model of Eq. 7 was carried out in order to evaluate the control algorithms. The physical construction of the real model is based on a black acrylic chassis, an 8-bit microprocessor, an array of 8 infrared sensors and a DC motor driver, the physical characteristics of the robot are shown in Table 1.

Table 1. Characteristics of the line follower robot.

Characteristic	Valor (Unidad)
Masa (M)	186 g
Inercia (I)	4.86×10^{-5} Kg m^2
CTE. Viscosa (B)	1.8×10^{-5} Kg/s
Longitud (L)	13 cm
Radio Neumáticos (R)	2.2 cm

The model is completed by the relation between the propulsion motors of each tire; the real model uses two high-speed micromotors with the physical characteristics presented in Table 2.

Table 2. Micromotors characteristics.

Characteristic	Valor (Unidad)
Voltage	6 V
Speed ($\emptyset'$)	3300 RPM
Moment	0.17 Kg $*$ cm
R_a	5.1 Ω
L_a	0.6 mH
Motor inertia	3.44×10^{-5} Kg m^2

2.3 Fuzzy Logic Control (FLC)

The control algorithm proposed in this work uses the position error of the robot and its derivative as inputs, each input variable is related to a universe of discourse through fuzzy membership subsets, as it can be seen in Fig. 2. The triangular set form is used to obtain a unit partition in the defuzzification process and reduce the computational cost. At the same time, the rule base of the system follows a common symmetric arrangement for fuzzy controllers by Mamdani inference, as it can be seen in [15]. For the activation of each rule, the T-norm of the membership product of each variable is used, as seen in Eq. 8.

$$\mu^{eUde} = \mu^{e} * \mu^{de} \tag{8}$$

The defuzzification method used in this control algorithm is the average center as seen in Eq. 9.

$$y = \frac{\sum_{i=1}^{R} q_i \mu_i(x)}{\sum_{i=1}^{R} \mu_i(x)} \tag{9}$$

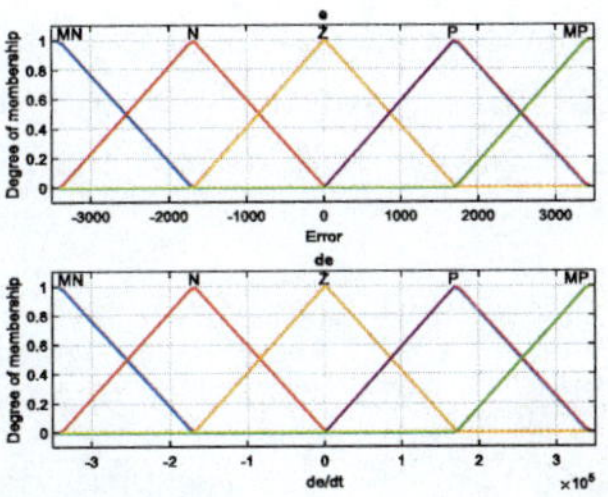

Fig. 2. Fuzzy sets of the input, error, and error derivative signals.

where q_i represents the average center of the activated output fuzzy set and $\mu_i(x)$ represents the degree of activation of each rule, since the average center method is used, singleton type fuzzy output sets can be used as shown in the Fig. 3, where it is not necessary to determine an average center, considerably reducing the number of operations and the computational cost.

Figure 3 shows the fuzzy sets of controller output distributed in such a way that there is less control action at the points closest to the null error and a more pronounced performance at the largest error points, the same happens with the derivative of the error, this generates a non-linear control action that favors the stability of the system.

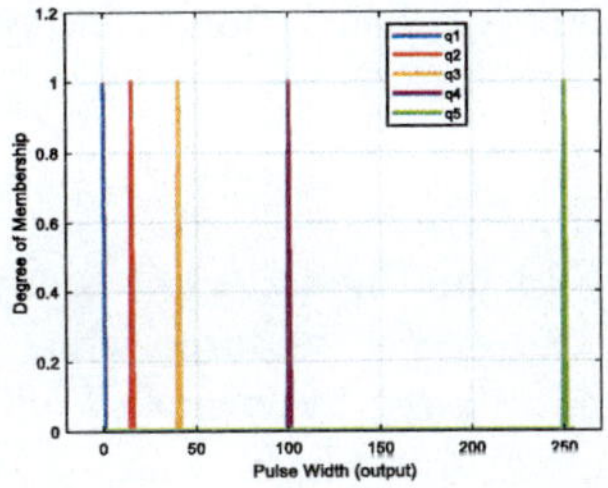

Fig. 3. Singleton fuzzy sets of the output signal (control action).

3 Results

3.1 Controller Non-linearity

Figure 4 shows a comparison between the action surfaces of the PD control and the FLC controller, where the control action of the PD is clearly linear; furthermore, due to the adjustment made for obtaining the best dynamic response of the system has a very reduced control action for the input signals, that is, it can be seen that with an error equal to or greater than a value of 1200 units (approximately 5.5°) the control action already reaches the maximum value of the control action (100% of the duty ratio of the PWM signal).

The PD control was tuned based on the model of Eq. 7 applying the analytical method of adjustment of static and dynamic characteristics, for this a maximum value of 2% of maximum overshoot and a response time of less than 0.3 s were established, these values have been chosen in advance.

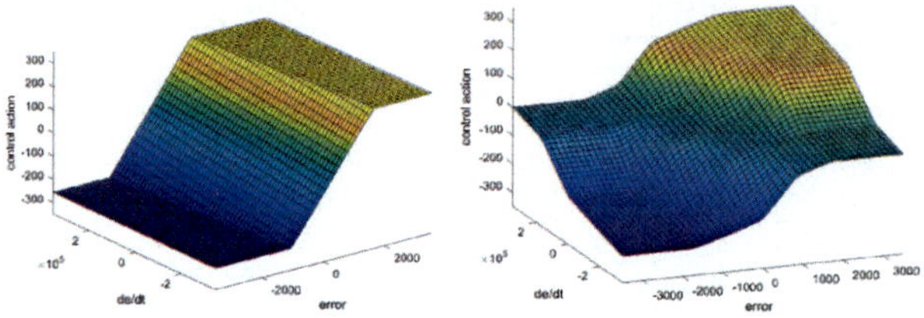

Fig. 4. Control surfaces of PD and FLC controllers.

On the other hand, the control surface of the FLC presents smoother curve in the areas close to the null error and a greater slope in the areas of larger errors, the same happens with respect to the rate of change of the error. The inflection points on the control surface clearly represent the non-linearities of the controller and the change that occurs in its operation for different operating conditions of the system.

Having small control actions around the null error will represent less effort on the part of the actuators and less energy consumption of the system in general.

3.2 SISO System

The first response comparison of the PD and FLC controllers is made based on the model of Eq. 6, where there is a single driving force. The FLC, in addition to its original fuzzy model, requires a fine adjustment by means of additional gains, therefore, the implemented model results as shown in Fig. 5, where *kc* is a gain proportional to the determined control action, this gain helps to eliminate the error in the stable state of the system and to reduce the stabilization time, while *kd* is the gain that acts on the derivative of the error and improves the dynamics of response.

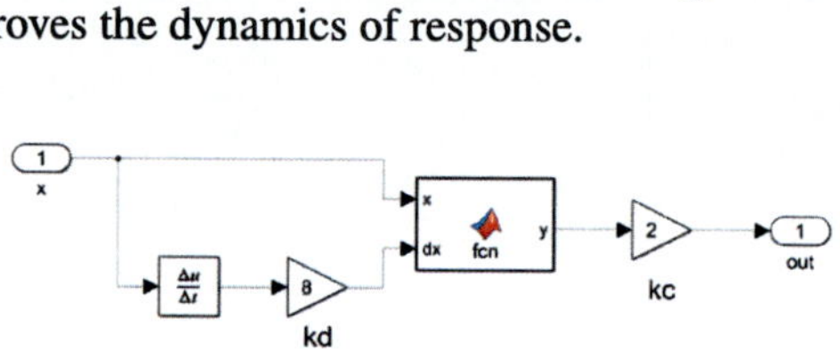

Fig. 5. FLC control model with fine adjustment by external gains.

FLC gain tuning is done by trial and error, the higher kc the system will be faster sacrificing stability and the non-linearity effect of the controller, on the other hand the higher kd the system will become slower improving system stability.

Figure 6 shows the responses obtained for the model established with the PD control and the FLC; It can be seen that the FLC controller generates a smoother response than that of the PD without sacrificing the dynamic characteristics of the system, that is, the system response has no overshoot and the stabilization time is very similar to that of the PD control.

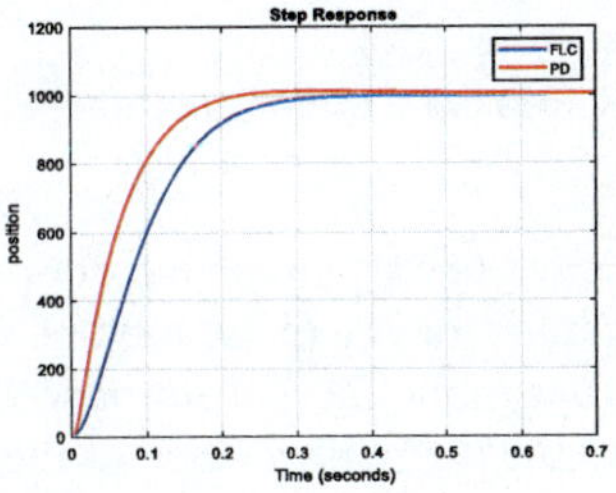

Fig. 6. Response of the SISO system (Eq. 6) to the step with the PD and FLC controllers.

The great difference between both controllers in this test stage, is presented in the control action generated by one and the other, in Fig. 7 the outputs of both controllers are shown.

The FLC presents an extremely lower control action than the PD, achieving the results of Fig. 6. The reduction of the control effort would be presented as a substantial improvement for the energy efficiency of the system and for the life time of the actuators.

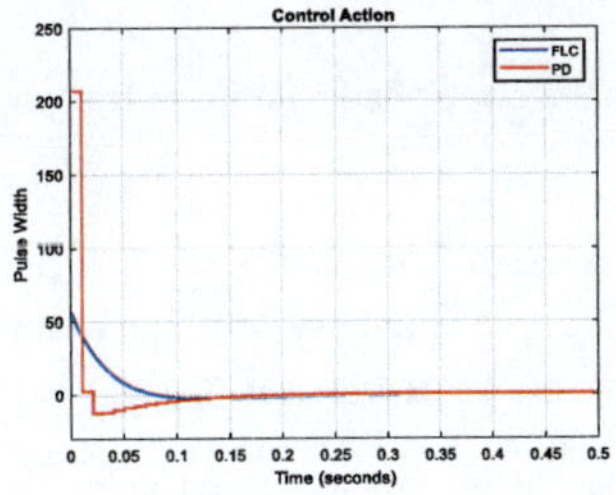

Fig. 7. Control output of the PD and FLC applied to the SISO system.

3.3 Complete System Model

The second comparison is made on the complete model obtained from Eq. 7, in addition, a known control strategy is added for this type of robot that is based on accelerating one wheel and braking another simultaneously to increase the response speed, the selection of which motor should be accelerated is made by means of the error sign as seen in Fig. 8, where input 1 is the control action that will be added or subtracted to a base speed, which the motors have to accelerate or brake, as the case may be, input 2 is the system error.

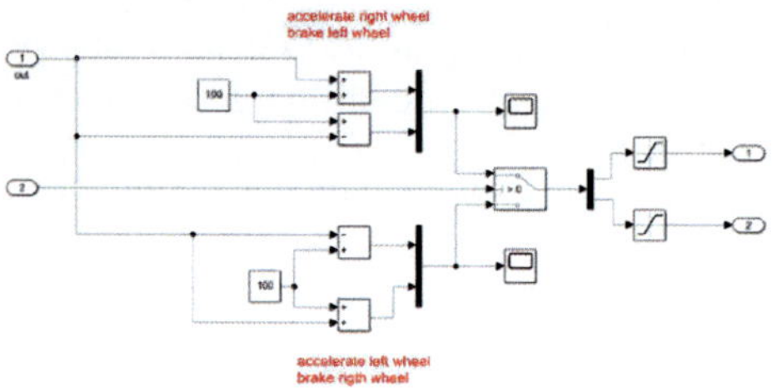

Fig. 8. Motors selection algorithm action.

Two disturbances were applied to the complete model of the system (Eq. 7) with the acceleration and braking selection strategy of the motors, which simulate curves within the trajectory of the line follower robot, these disturbances are reflected directly on the variable of output, and to evaluate the performance of the controllers, the error curves of each system, the outputs of each controller and the actions sent to each motor after the acceleration and braking stage were obtained.

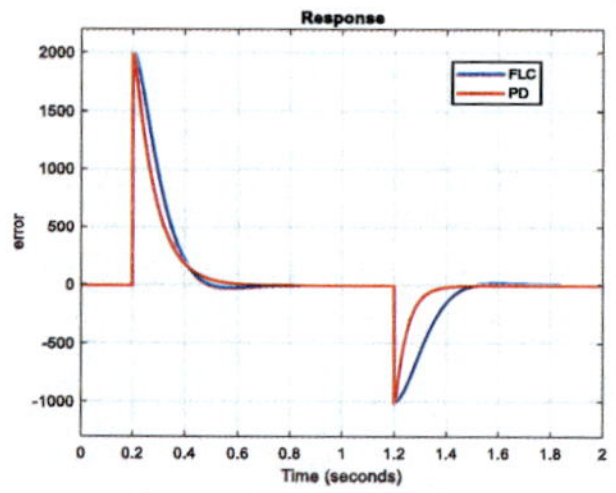

Fig. 9. Evolution of the error of the controlled system in the event of disturbances in its output.

Figure 9 shows the system error curves for each type of controller, it can be seen that both the PD and the FLC present a very similar response when the disturbance is large (approx. 9.14°) but when the error is small (approx. 5.5°) the response of the FLC is much slower than the response of the PD, the rapid correction of the PD in the event of a small error is due to the steep slope of its control surface unlike the FLC control, the PD is about 0.15s faster than FLC.

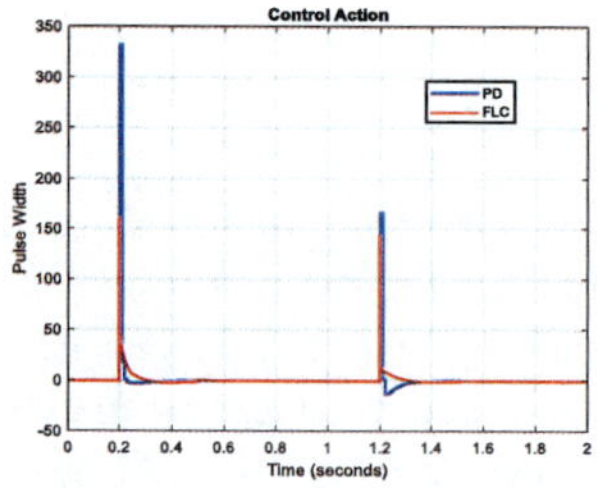

Fig. 10. Controller outputs.

For its part, in Fig. 10 the control actions generated by each of the controllers are observed, these curves present the pure output of the controller due to the error signals and their derivative, it is clearly observed that the PD output is much higher in both disturbances.

Finally, Fig. 11 shows the control signals sent to the motors after selection algorithms, the PD saturates the actuators in both disturbances, increasing the effort and its abraded, in addition to increasing battery consumption. For its part, the FLC saturates the actuators only in the first disturbance and for a shorter time than the PD. It is evident that the control surface acts as expected to reduce the control action in cases of small error and increases to correct the trajectory in cases of curves that are too stee.

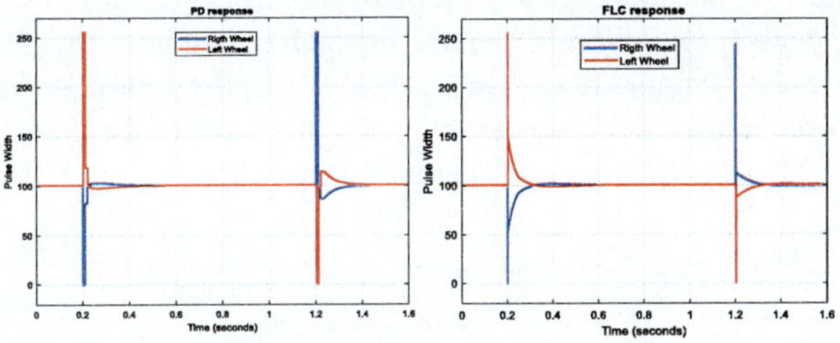

Fig. 11. Control signals sent to the actuators.

3.4 Experimental Results

In order to have a real evaluation of the performance of the FLC with its non-linear modifications, this algorithm was implemented in an experimental prototype and it was circulated on an oblong circular track (an elongated circular track where there are two straight lines and four curves in the same direction). In this test, data was taken from the error evolution and the prototype with the non-linear FLC and with the PD and the control actions.

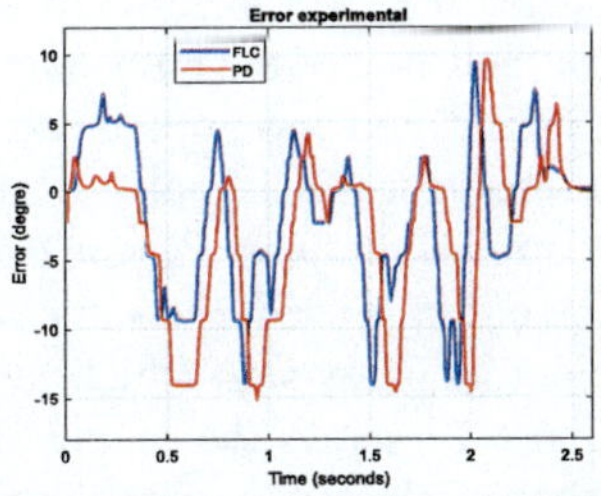

Fig. 12. Evolution of the robot error in a real experimental test.

The data obtained and presented are exclusively from one lap on the aforementioned track, Fig. 12 shows the error curves that the system presents with each controller, the first area of the graph (0–0.5s) represents the initial part of the path, a first straight area where the error is due to the initial position of the robot; For its part, each negative peak represents a curve in the left direction, that is, the robot moved counterclockwise, in zone 1 to 1.5s there is a tendency of the robot to maintain the error low, this is because the robot circled in the second straight zone.

FLC corrects the error faster than the PD, although the curves are very similar, the error remains less time in a peak when the FLC acts, in addition, in the first curve the error does not become as large as it is with the PD. Finally, the curves also indicate that the FLC makes the robot finish the lap approximately 0.2 s faster.

In Fig. 13 the control actions generated by the control algorithms are observed, it is observed that in general the actions generated by the FLC are lower than those produced by the PD, also in the straight area of the track (0–1.5 s) the action of the FLC is much less than the PD reducing energy consumption and the wear of the actuators, this effect can also be seen when leaving the fourth curve (time > 2 s).

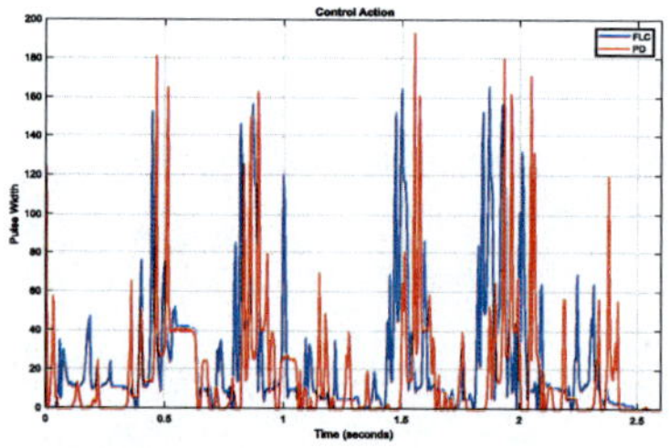

Fig. 13. Control action of the PD and FLC algorithms in an experimental test.

3.5 Base Speed Control

As mentioned before, the system works with a base speed, which marks the forward speed of the robot in a straight line, when the error is zero. To reduce the robot's circulation time, an algorithm that modifies the value of the base speed based on the same logic of the angular position controller but inverted was implemented, that is, if the error is large, the base speed is reduced and if the error is minimal, the base speed is increased, giving the system more thrust, mainly on straight lines. For the implementation of this system, the same position controller preceding is used, the same inference and defuzzification method, the only change is the output fuzzy sets, which for this algorithm represent the values of the base velocity. Figure 14 shows the output singleton sets that were used in the implementation of this control algorithm, with the difference that their output is inverse.

While Fig. 15 shows the response curves of the system error when it was tested with the FLC, and the FLC was applied plus the base speed control algorithm, the curves show a certain similarity but in general there is a greater error due to part of the system, with the speed algorithm that apparently is the price paid for reducing the time in which the robot completes a lap of the track by 0.2 s. While Fig. 15a shows the response curves of the system error when it was tested with the FLC, and when the FLC was applied in addition to the base speed control algorithm; the curves show a certain similarity but in general, there is a greater error on the part of the system with the speed algorithm, which apparently is the paid price for reducing the time in which the robot makes a lap of the track by 0.2 s.

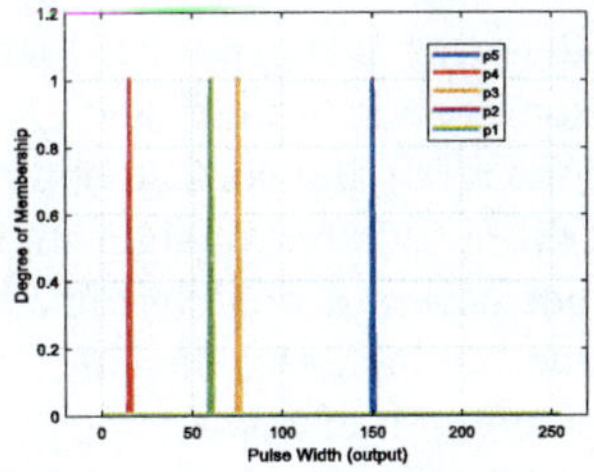

Fig. 14. Output fuzzy sets for the base speed control algorithm.

Finally, Fig. 15b shows that the control actions generated by the FLC when speed control is implemented increase considerably compared to when only the FLC was implemented.

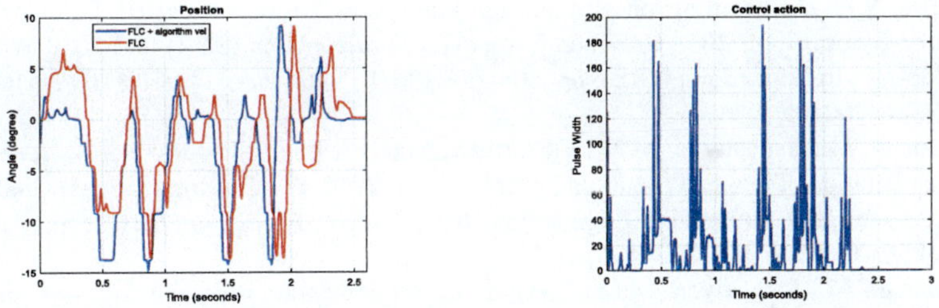

a) b)

Fig. 15. a) System position error in two experimental tests. b) FLC control action plus base speed control algorithm.

4 Conclusions

According to the simulated and experimental results, the non-linear control algorithm by fuzzy logic designed and implemented, stabilizes the system in an adequate way and corrects its error in a better way than a linear controller such as PD control.

The FLC presents a faster elimination of the error than the PD in the experimentation, thus improving the time in which a prototype can cover a pre-determined route; Its non-linear control surface gives the ability to push your control action up to the allowable limit on occasions when the error is too great, with a speed adequate enough not to require an additional braking or locking algorithm of motors, which causes a greater fluidity in the operation of the line follower robot.

FLC exhibits faster debugging than PD in experimentation, thus improving the time a prototype can cover a predetermined path; its non-linear control surface provides the ability to raise its control action to the allowed limit. Sometimes when the error is too large, with an adequate speed, an additional braking or motor blocking algorithm is not required, which causes greater fluidity in the operation of the line follower robot.

In addition, the FLC presents much lower control actions than the PD controller, due to its rapid elimination of the error, this generates less wear on the actuators of the system, giving them an extension in their lifetime and also generating energy savings. Providing greater autonomy to the robot and consequently economic savings.

Based on experimentation, the speed control algorithm manages to reduce the time in which the line follower robot covers a predetermined route, in its commission it increases the error reduction time and increases the control actions generated to correct the robot position; The greatest strength of this strategy is that the computational cost of the algorithm is not increased since it uses the same fuzzy antecedents and the same inference method of position control. It avoids uncertainties in selecting a curve that slows down as it is based on the empirical knowledge of the designer.

References

1. Reis Caitité, V.G., Maradona Gonçalves dos Santos, D., Gregório, I.C., da Silva, W.B., Mendes, V.F.: Diffusion of robotics through line follower robots. In: 2018 Latin American Robotic Symposium, 2018 Brazilian Symposium on Robotics (SBR) and 2018 Workshop on Robotics in Education (WRE), pp. 604–609 (2018). https://doi.org/10.1109/LARS/SBR/WRE.2018.00109
2. Pakdaman, M., Sanaatiyan, M.M., Ghahroudi, M.R.: A line follower robot from design to implementation: Technical issues and problems. In: 2010 The 2nd International Conference on Computer and Automation Engineering (ICCAE), vol. 1, pp. 5–9 (2010). https://doi.org/10.1109/ICCAE.2010.5451881
3. Pakdaman, M., Sanaatiyan, M.M.: Design and implementation of line follower robot. In: 2009 Second International Conference on Computer and Electrical Engineering, pp. 585–590 (2009). https://doi.org/10.1109/ICCEE.2009.43
4. Tayal, S., Rao, H.P.G., Bhardwaj, S., Aggarwal, H.: Line follower robot: design and hardware application. In: 2020 8th International Conference Reliability Infocom Technologies Optimization (Trends Future Directions, pp. 10–13 (2020). https://doi.org/10.1109/ICRITO48877.2020.9197968
5. Joni, K., Ulum, M., Abidin, Z.: Robot line follower berbasis kendali proportional-integral-derivative (PID) Untuk Lintasan Dengan Sudut Ekstrim. J. Infotel **8**(2), 138–142 (2016). https://doi.org/10.20895/INFOTEL.V8I2.129
6. Babu, D.V., Jenniffer, D.C., Karthikeyan, R.: Line follower robot & obstacle detection using PID controller. AIP Conf. Proc. **2271**(1), 030005 (2020). https://doi.org/10.1063/5.0024760
7. Bajestani, S.E.M., Vosoughinia, A.: Technical report of building a line follower robot. In: ICEIE 2010 – 2010 International Conference Electronics Information Engineering Proceedings, vol. 1 (2010). https://doi.org/10.1109/ICEIE.2010.5559826

8. Gomes, M.V., Bássora, L.A., Morandin, O., Vivaldini, K.C.T.: PID control applied on a line-follower AGV using a RGB camera. In: IEEE Conference Intelligent Transportation System Proceedings, ITSC, pp. 194–198 (2016). https://doi.org/10.1109/ITSC.2016.7795553
9. Dilaver, M.B., Uslu, E., Cakmak, F., Altuntas, N., Amasyali, M.F., Yavuz, S.: Robocup arama kurtarma ligi diş ortam kategorisi çizgi izleyen robot platformu tasarimi ve gerçklenmesi. In: 2017 25th Signal Processing Communication Applications Conference SIU 2017 (2017). https://doi.org/10.1109/SIU.2017.7960294
10. Sanchez, L.A.P., Moreno, C.A.A., Daza, H.A.B.: Design and construction of a line follower robot guided by pixels values of a camera connected to an FPGA. In: 2015 20th Symposium on Signal Processing Images Computer Vision, STSIVA 2015 – Conference Proceedings (2015). https://doi.org/10.1109/STSIVA.2015.7330453
11. Proaño, V., Gualsaquí, M.: Line follower simulator with fuzzy control. In: 2017 CHILEAN Conference Electrical Electronics Engineering Information Communication Technologies CHILECON 2017 – Proceedings, vol. 2017, pp. 1–6 (2017). https://doi.org/10.1109/CHILECON.2017.8229638
12. Sanjaya, A., Mawengkang, H., Efendi, S., Zarlis, M.: Stability of line follower robots with fuzzy logic and kalman filter methods. J. Phys. Conf. Ser. **1361**(1), 012016 (2019). https://doi.org/10.1088/1742-6596/1361/1/012016
13. Wajiansyah, A., Supriadi, S., Nur, S., Wicaksono, A.B.P.: Implementasi fuzzy logic pada robot line follower. J. Teknol. Inf. dan Ilmu Komput. **5**(4), 395 (2018). https://doi.org/10.25126/JTIIK.201854747
14. Zadeh, L.A.: Quantitative fuzzy semantics. Inf. Sci. (Ny) **3**(2), 159–176 (1971). https://doi.org/10.1016/S0020-0255(71)80004-X
15. Mamdani, E.H.: Application of fuzzy algorithms for control of simple dynamic plant. Proc. Inst. Electr. Eng. **121**(12), 1585–1588 (1974). https://doi.org/10.1049/piee.1974.0328
16. Lilly, J.: Fuzzy Control and Identification, 1st edn. John Wiley & Sons Inc, Hoboken, New Jersey (2010)
17. Saadatmand, S., Azizi, S., Kavousi, M., Wunsch, D.: Autonomous control of a line follower robot using a Q-learning controller. In: 2020 10th Annual Computing Communication Workshop Conference CCWC 2020, pp. 556–561 (2020). https://doi.org/10.1109/CCWC47524.2020.9031160

Analysis of a Network Interface for an On-Chip Network Architecture

Byron Ojeda[1], Mayerly Saenz[2], Verónica Alulema[1], and Darwin Alulema[1](✉)

[1] Universidad de las Fuerzas Armadas ESPE, Sangolquí, Ecuador
doalulema@espe.edu.ec
[2] Universidad Politécnica Salesiana, Cuenca, Ecuador

Abstract. This document details a study on the Network Interface (Network on Chip), focused on performance parameters, by injecting traffic of diverse types, CBR traffic and multimedia traffic, which are the most used today. For this, a NIRGAM Network simulator is used as a tool, which has the necessary configuration parameters to obtain the performance, latency, and power values to analyse the proposal. The results show that performance is a parameter that can be optimized by increasing the flit and/or operating frequency value, but without neglecting the power consumption that it would generate.

Keywords: Network on Chip (NOC) · Field Programmable Gate Arrays (FPGA) · System on Chip (SoC)

1 Introduction

The design of network interfaces (network interface) in conventional packet switching networks has generated great interest and has managed to propose very efficient designs with high throughput and low latency, however, these characteristics have been obtained with the use of processors. Dedicated to communication and large storage buffers. There are several design proposals for network interfaces with small implementation areas, but these still do not provide the throughput and latency desirable for complex SoC structures such as an MPSOC [1, 2].

Communication in old devices was simple, it was enough to have a hierarchy of buses. However, when presenting the option of integrating several devices in a single MPSOC chip (Multi-Processor-System-on-chip) [7, 9], the bus hierarchy presents several drawbacks that arise from the convergence of applications that it demands the combination of various types of traffic in the same SoC (System on Chip) [3] and the necessary QoS (Quality of Service) within the network. For this, a complete change must be made in the on-chip interconnection that must ensure a flexible, reconfigurable, and reprogrammable structure, in addition to allowing interconnection between the different modules within the same chip, for this the NOC (Network on-chip) [8]. A NOC architecture is based on packet switching networks and consists of three fundamental blocks: switches, links, and Network interfaces. Therefore, to improve the performance

V. Robles-Bykbaev et al. (Eds.): CITIS 2022, LNNS 607, pp. 72–80, 2023.
https://doi.org/10.1007/978-3-031-24327-1_7

of the NOC, the performance of both the switches and the Network interface must be improved, with emphasis on the optimization of the network interface [4–6].

To improve the NOC study, it is important to analyse the behaviour of the most important network performance parameters with different configurations, such as flit size, operating frequency and circulating traffic. This analysis is conducted throughout this article, which is made up of five parts. The first of them constitutes the introduction, the second presents the state of the art, in the third, the NOC proposal and its parameters are indicated, the results obtained when performing simulated experiments in NIRGAM [27] are detailed in the fourth part, to, at the end and as a fifth part, present the respective performance analysis of the architecture and in the final part, the conclusions are detailed.

2 Embedded Systems

This section describes the evolution, switching and routing elements and techniques of a NOC, all this, from a theoretical point of view, as a starting point for understanding its main characteristics.

2.1 NOC

Today's MPSOCs contain thousands of cores, so maintaining communication between them requires a high-performance interconnect. The first generation of on-chip interconnects was developed using a conventional bus interface, which currently has many limitations. This is due to latency problems when there is a high priority of accesses, which are stalled by the number of transactions that are in progress on the bus. Hence, designers face new challenges concerning on-chip interconnect design that goes beyond the increase in IP cores and performance requirements, including power management, multiple clock domains that may be present, requirements security, error control among others, whose physical implementation through layout, becomes increasingly difficult [10, 11].

Due to these interconnection limitations, a new interconnection paradigm arises between the different modules within the same chip, using the NOC architecture. The architecture of a NOC is based on packet switching networks and can be described by the strategy it has for routing, flow control, switching, latency and throughput [12]. The main components of the NOC architecture are [13, 14]: (a) Network Interface NI: It connects all the IP Cores of the network, that is, it maps the bus-type transitions coming from the IP Cores, in packets that can be propagated within the NOC as well as on the opposite side. Additionally, it builds the necessary transactions so that the packets can leave the NOC; (b) Switches or routers: They are used to send packets within the network, they generally have internal buffers, where information is stored, to reduce the risk of data loss due to congestion problems; and (c) Links: Responsible for connecting the Switches with the Network Interface, or in turn with other switches, they can be insensitive to network latency, depending on the application, they can have buffers.

2.2 Switching Techniques for a NOC

Routing in a NOC is quite like that of any network, the routing algorithm refers to how the data is routed from the sender to the receiver, that is, it is responsible for the correct and efficient way of routing packets. The source to the destination. The NOC is based on packet switching, for which there are switching techniques, the most used today is: (a) Wormhole: In this technique, the buffer requirements are reduced to a flit, which is the unit of minimum transmission within the NOC in a clock cycle instead of a complete packet. A flit of a packet is sent to the receiving router, if the space for it is sufficient, the entire packet is stored, otherwise, part of the packet is distributed in two or more routers, if the distribution is made between two or more routers, it can occur a packet lock. To mitigate this problem, this technique has adopted virtual channels to make efficient use of physical channels [15]; (b) Store and Forward Networks (SAF): It is the simplest packet switching technique, a packet is sent from one router to the next only if the receiving router has buffer space for the entire packet, so the packet transmission cannot stall. Routers send a packet only when it has been received in its entirety, which means that the latency per router and the size of the router are at least the size of the packet, due to this restriction few NOCs use this switching technique [14, 16]; and (c) Virtual Cut-off Switching (VCT): This technique reduces per-router latency by forwarding the first flit of a packet as soon as there is room on the next router for the entire packet, thus all other flits. Follow without delay, otherwise, the entire package is stored. In this technique the time in which the package is stored is reduced, the disadvantage lies in the cost of implementation [17].

Routing Techniques for a NOC. TO choose a routing algorithm, certain parameters must be considered, such as: reducing the power required for routing, which is achieved by minimizing the logic and routing tables; increasing performance by reducing delay and maximizing network traffic utilization; and finally improving the robustness to better adapt to the needs of traffic changes [18].

3 NOC Proposal and Its Parameters

The main parameters to be considered when designing a NOC are detailed below:

- *High performance and low latency:* The performance and latency depend clearly on the routing and switching that it has, as well as the topology that is being used. Within a NOC, it is necessary to have the optimal values of certain parameters that describe its performance, that is, the NOC must offer guarantees on latency, throughput, and power, considering that it supports diverse types of traffic.
- *Network topology:* Mesh topology is where all the links have the same length, the connection to the node and the routing algorithm is simple, which facilitates the physical implementation.
- *Switching protocol:* With the detailed information regarding the communication protocols and focusing on the requirements of the desired network interface, it was decided to use the Wormhole algorithm. This is because it conducts the transmission of the

packet through flits, which helps to reduce network latency. The main advantage of this algorithm is that it uses physical channels efficiently since it has virtual channels that prevent blocking situations from occurring, these blockages occur when the size of the packet sent is too large, which is why it must be distributed to two or more routers, that is, the packet remains in the network occupying resources, if another message wishes to occupy the same resource it will also be blocked [19].

- *Routing algorithm:* For the choice of the routing algorithm, it focused on obtaining optimal latency values, throughput, and channel utilization, the Odd-even routing algorithm was used in this project, since it has an improvement of 18 and 25 per cent compared to the other algorithms. This algorithm provides the highest throughput values per packet when compared to other routing algorithms such as: the origin algorithm and XY algorithm. Additionally, the routing technique performed by Odd-even helps to avoid deadlock, which is the stagnation of packets because they wait for each other to release resources [20].
- *Flit size:* The size of the flit influences the performance parameters of the network. If it is high, it offers greater network performance, but its cost also increases, so the size of the flit increases. Flit must be chosen according to the needs of the NOC, without neglecting the issue of cost. The cost of improving performance by increasing the size of the flit is not offset, so it is difficult to justify an increase in the size of the flit. To improve NOC throughput and latency, the flit value should be equal to or less than the smallest packet size [21].
- *Clock frequency:* The frequency at which the NOC works directly affects the performance of the network, but the effects of working at a high frequency must be considered. Undoubtedly, a high working frequency would provide greater performance. However, both the power consumption and the wiring complexity would increase, it must be considered that if the NOC studied is to be implemented, the maximum operating frequency of the FPGA must be used. The clock frequency at which the Virtex-6 ML605 card used is 200 MHz, it can also work with a maximum external frequency of 700 MHz, which will not be considered in this study, because a new module that generates it and what is required is to optimize the use of necessary hardware [22].
- *Network traffic:* Several types of traffic can flow over the NOC. This work focuses on the traffic most used today: (a) CBR traffic (Constant Bit rate): Intended for real-time applications, since it has delay restrictions and has a different pattern than burst traffic that varies in time. The CBR exclusively reserves the network resources, that is, a constant bandwidth is made available during the connection, in case of having traffic that presents the variation in its bandwidth, it is not recommended to use it, since it would be wasted the same; and (b) Multimedia Traffic: At present, the use of multimedia applications, specifically, video transmission, has increased, this has given rise to various hardware implementations of video and audio codec for faster processing, for which evaluating the NOC under multimedia traffic becomes an important aspect when studying the performance parameters of a NOC. There are limitations when transmitting this type of traffic. Due to the video quality, the transmission of information through the network is conducted through a greater number of bits, if there is not an optimal infrastructure, the video quality will be affected causing failures in the images, pixelation or freezing causing delays in transmission [23].

This research simulates MPEG-4 video traffic, which is a standard designed to support the transport, synchronization, encoding, and processing of multimedia content transmitted online. Some areas of application are: digital TV, multimedia content for mobile phones, network video, video conference, among others. MPEG-4 offers high data compression occupying less memory space without neglecting video quality [24]. To analyse this type of traffic in the NOC, it is necessary to analyse the Hurst parameter. The Hurst parameter measures the degree of self-similarity of the signals through stochastic processes, which allows representing video traffic to simulate it within a NOC. The value of the Hurst parameter varies between 0 and 1. Table 1 details the type of series and the equivalent range for it.

Table 1. Hurst parameter ranges

Hurst parameter value range	Series type
0.8–1	Movies, series, and cartoons
0.6–0.8	Sports transmissions
> 1	Video conference

It must be considered that when the scenes of a video have a lot of variation in movement, brightness, colour, among others, the Hurst parameter increases, a high-definition video has a higher parameter than a lower quality video. The calculated values of the Hurst parameter have been taken, which represent the traffic of well-known television series and movies that will cross the NOC [25, 26].

4 Results

The development of this section consists of two parts. The first describes the scenarios, parameters and results obtained when performing the simulation in the NIRGAM software; In the second part, the results obtained for network performance are analysed.

4.1 Simulation

Once the design parameters of the NOC were determined, a simulation was conducted, obtaining data on latency, performance, and power consumed for both CBR and multimedia traffic. To simulate the NOC with the mentioned configurations, the NIRGAM simulator version 2.1 [27] was used. This simulator was chosen because it meets the required topology specifications, switching algorithms and routing. Table 2 details the simulation configuration for CBR traffic and Multimedia traffic.

Table 2. Simulation configuration parameters with CBR traffic and multimedia traffic

Parameter	CBR traffic	Media traffic
Topology	mesh 5 × 5	mesh 5 × 5
Switching algorithm	Wormhole	Wormhole
Routing algorithm	odd-even	odd-even
Flit size	6/11/21 bytes	9/16/26 bytes
Package size	1 kbytes	1 kbytes
Bandwidth usage	75%	75%
Operating frequencies	500 MHz–700 MHz–1 GHhz	500 MHz–700 MHz–1 GHhz
Clock cycles that the simulation lasts	1000	20000

The 5 × 5 mesh topology that was simulated is displayed in Fig. 1. a, which consists of twenty-five tiles that exchange data with each other. Figure 1. a show the source and destination tiles of the network, as well as the trajectory to get from one to the other, respectively. Figure 1. b details the elements contained in each tile.

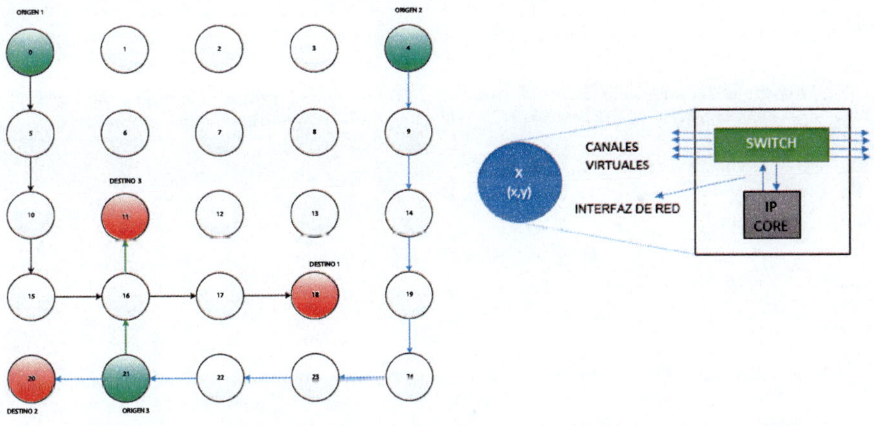

a. Simulated NOC with CBR traffic *b. NOC tile elements*

Fig. 1. Simulated NOC

Table 1 shows the parameters obtained from the simulation varying the size of the flit with a frequency of 500 MHz, 700 MHz and 1 GHz respectively [28]. The second simulation was performed with multimedia traffic representing several television series. Once the simulation with multimedia traffic was conducted, the data tabulated in Table 1 were obtained; for this, the value of the operating frequency is varied: 500 MHz, 700 MHz and 1 GHz, respectively.

5 Analysis of Results

Performance, latency, and power: The size of the flit influences these parameters, if the size of the flit increases, better performance is observed, as well as an increase in the power dissipated by the network. For this reason, the value of this parameter must be considered in such a way that it does not affect the optimal working values of the NOC. The latency, in turn, does not undergo considerable changes. The frequency at which the network works is 700 MHz (maximum frequency supported by the Virtex-6 ML605 FPGA card), it is verified that for every 100 MHz the power consumed by the network increases by 0.2142 watts, for this reason, if you want to have high performance which would imply high frequencies, power is a parameter to consider. Latency in this case also does not show behaviour that checks for changes.

Figure 2.a and Fig. 2.b show that the performance behaviour varying the size of the flit with both CBR traffic and multimedia traffic is proportional. Figure 2.c and Fig. 2.d show that having a higher performance power consumption increases proportionally, so it is important to analyse these two parameters at the same time. In Fig. 2.e and Fig. 2.f it

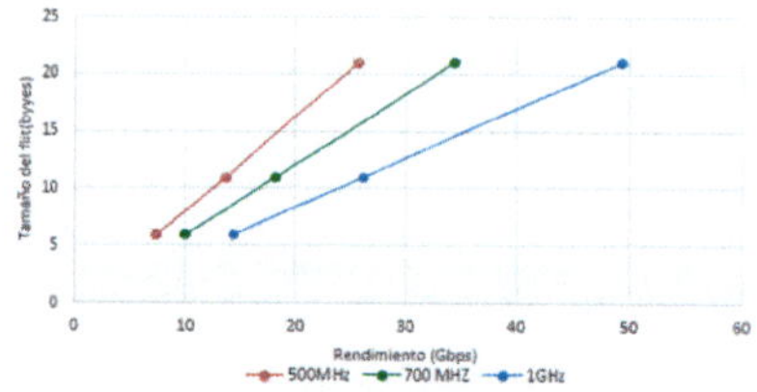

(a) Flit size vs Throughput in CBR traffic with different operating frequencies.

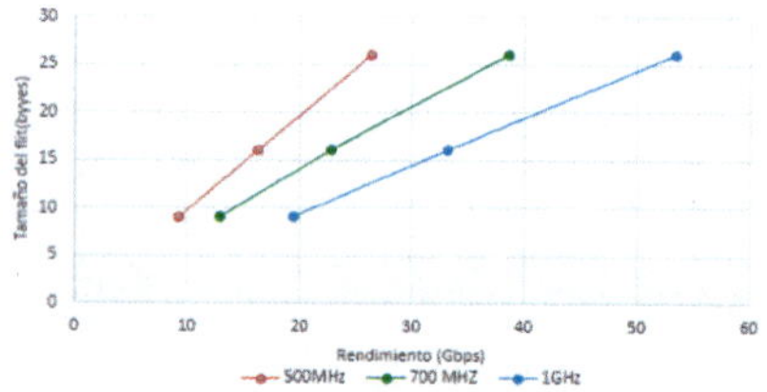

(b) Flit size vs Throughput in multimedia traffic with different operating frequencies.

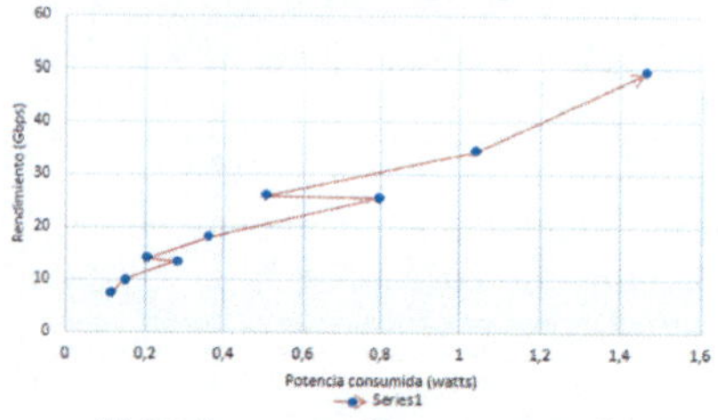

(c) Performance vs Power consumed in CBR traffic

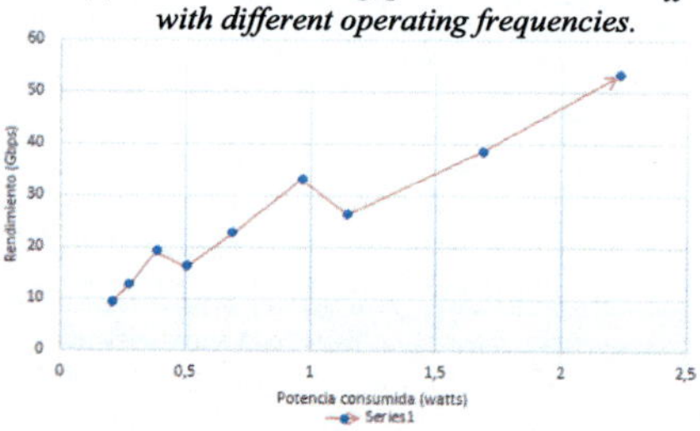

(d) Performance vs Power consumed in multimedia traffic

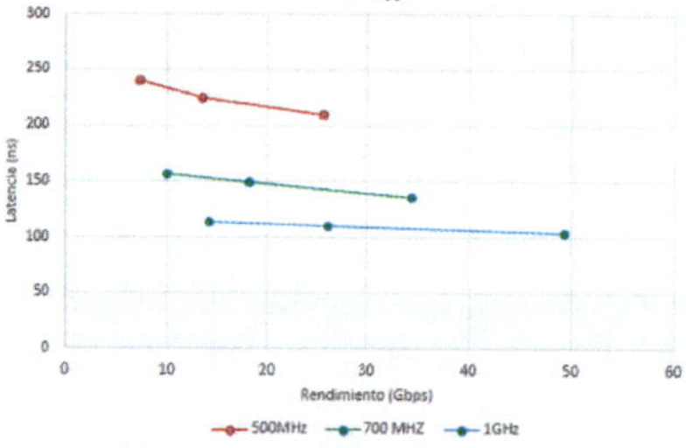

(e) Latency vs Throughput in CBR traffic

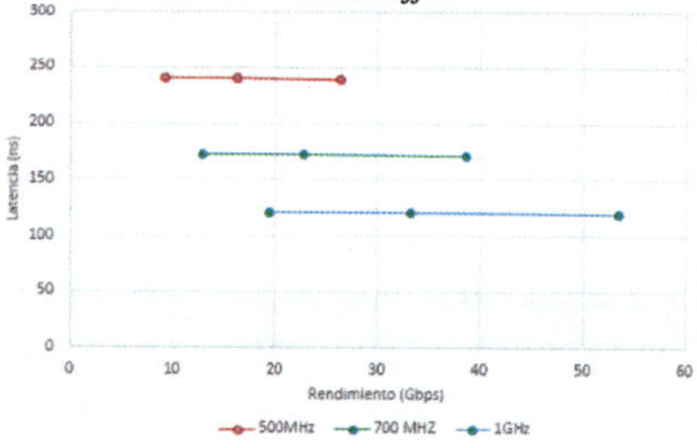

(f) Latency vs Throughput in multimedia traffic

Fig. 2. Latency vs Throughput

is verified that the latency varies based on the performance of the network in a minimum proportion, with a higher performance the latency is reduced, therefore, by having a higher frequency of operation, the latency of the network decreases.

6 Conclusions and Future Work

The WormHole switching algorithm is the most feasible to use since it reduces the size of the buffers because the packets are transmitted in units called flit (flow control units) which are the smallest units for flow control. Flow. The optimal routing algorithm is the odd-even, considering: the power required to perform the routing, which must be the least possible. The performance of the NOC can be increased by maximizing the use of traffic in the network, as well as reducing the delay, and finally, the robustness to adapt to changes in traffic is a parameter.

The simulation results verify that the size of the flit and the frequency of operation are directly proportional to the performance of the NOC. It must be considered that, if the performance increases, the implementation cost behaves proportionally, since the performance of the network is closely related to the size of the flit, the operating frequency, and the power consumption, which It must be considered if it is feasible to vary any of these parameters.

The maximum operating frequency at which the Virtex-6 ML605 FPGA card that was used works is 700 MHz, which was considered when simulating the NOC, an FPGA card option is Altera, which has several models that can be adapted to implementation needs.

Some lines of future work emerge from the work: (a) An in-depth study regarding the Hurst parameter should be carried out, to analyse the behaviour of other high-quality video formats currently used within a NOC using the NIRGAM simulator; and (b) The implementation on an FPGA card of the study carried out would contribute to corroborate the results obtained in the simulation of the main performance parameters of a network, it must be taken into account that the characteristics of the card support the design that is to be implemented., such as the maximum operating frequency it supports, the power it consumes, RAM blocks, among others.

References

1. Attia, B., Chouchene, W., Zitouni, A., Turki, R.: Network Interface Sharing for SoCs based NOC. IEEE, p. 6 (2011)
2. Balasubramanian, D., Nagda, T., Bhatia, D., Sighn, P., Bhoj, S., Balsara, P.: Network interface for NOC based architectures. University of Texas and Dallas, p. 18 (2007)
3. Al-Hashimi, B., Zwolinski, M., Singh, M., Rosinger, P., Laxmi, V., Jain, L.: NIRGAM A Simulator for NOC Interconnect Routing and Application Modeling (01 09 2007)
4. Metra, C., Chung Lo, J., Lombardi, F.: Special Section on Design and Test of Systems-on-Chip (SoC). IEEE, p. 2 (2006)
5. Saleh, R., Wilton, S., Mirabbasi, S., Hu, A., Greenstreet, M., Ivanov, A.: System-on-Chip: Reuse and Integration. IEEE, p. 20 (2006)
6. Mollocana, J., Cadena, J.: DISEÑO DE HARDWARE Y SOFTWARE DE SYSTEM ON CHIP EMPLEANDO TECNOLOGIA XILINX EDK. ESPE, p. 13 (2012)

7. Susanto, K.W.: A Verification Platform for System on Chip. University of Glasgow, p. 166 (2003)
8. Hübner, M., Becker, J.: Multiprocessor System-on-Chip: Hardware Design and Tool Integration. Springer, New York (2011)
9. Reza, M.: Reliable and Variation-tolerant Reliable and Variation-tolerant Power MPSoCs. Universita' di Bologna (2012)
10. Wolf, W., Jerraya, A., Martin, G.: Multiprocessor System-on-Chip (MPSoC) Technology. IEEE, p. 13 (October 2008)
11. ASTERIS: A comparison of Network-on-Chip and Busses. ASTERIS S.A, p. 11 (2005)
12. ASTERIS: From "Bus" and Crossbar" to "Network-On-Chip". ASTERIS S.A, p. 10 (2009)
13. Kumar, S., Jantsch, A., Soininen, J.P., Forsell, M.: A Network on Chip Architecture and Design Methodology. IEEE, p. 8 (2002)
14. Atienzaa, D., Angiolinic, F., Muralia, S., Pullinid, A., Beninic, L., Michelia, G.D.: Network-on-Chip design and synthesis outlook. Science Direct (2007)
15. Pasricha, S., Dutt, N.: On-Chip Communication Architectures System on Chip Interconnect. Elsevier, Burlington (2008)
16. Lu, Z.: Using Wormhole Switching for Networks on Chip: Feasibility Analysis and Microarchitecture Adaptation. Royal Institute of Technology, p. 70 (2005)
17. Benini, L., De Micheli, G.: Networks on Chips Technology and Tools. Elsevier, San Francisco (2006)
18. Mohapatra, P.: Wormhole routing techniques for directly connected multicomputer systems. Lowa State University, p. 45
19. Mubeen, S.: Evaluation of source routing for mesh topology network on chip platforms. Jönköping Institute of Technology, p. 109 (2009)
20. Sánchez García, J.L.: Análisis y evaluación de técnicas de reconfiguración dinámica de la red de interconexión en sistemas masivamente paralelos. Universidad de Castilla-La Mancha, p. 245 (1998)
21. Shrimawale, A.K., Gaikwad, M.A.: Review of Odd-Even Routing Algorithm for 2D Mesh Topology of Network-on-Chip Architecture for Bursty Traffic. International Journal of Computer Applications, p. 12 (2013)
22. Lee, J., Park, S.J., Swaminathan, M., Kim, J.: Do We Need Wide Flits in Networks-On-Chip?. University of Cyprus, p. 6
23. XILINX: Virtex-6 CXT Family Data Sheet. XILINX, p. 52 (2011)
24. Rivas Ramírez, J.E.: Análisis de autosimilaridad para fuentes sintéticas de video. UNIVERSIDAD CATÓLICA DE MANIZALES, p. 74 (2013)
25. Castro Gil, M.A., Comenar Santos, A.: Tecnología Multimedia-Video Digital. Universidad Nacional de Educación a Distancia, p. 116
26. Casilari Pérez, E.: Caracterización y modelado de tráfico de vídeo VBR. UNIVERSIDAD DE MÁLAGA, p. 378 (1998)
27. Fitzek, F., Reisslein, M.: Telecommunication Networks Group (TKN) (02 2003)
28. Jain, L.: NIRGAM: A Simulator for NOC Interconnect Routing and Application Modeling. NIRGAM Authors, p. 28 (2007)

Credit Default Risk Analysis Using Machine Learning Algorithms with Hyperparameter Optimization

Juan Inga(✉) and Erwin Sacoto-Cabrera

Universidad Politécnica Salesiana, Cuenca, Ecuador
jingap@est.ups.edu.ec, esacoto@ups.edu.ec

Abstract. Machine learning models are an important tool that provide a scientific method to identify potential debtors early and predict which clients are more likely to default on their debts, improving the accuracy of assessment in credit risk analysis in financial companies. The purpose of this study was to analyze the performance of gradient boosting machine learning algorithms (CatBoost, LightGBM, and XGBoost) in predicting customer default risk, and the ability of the *RandomUnderSampler* sampling technique to address unbalanced categories of credit risk. The exploratory analysis of the data set was carried out, then the data preprocessing, finally the training with hyperparameter adjustments with the GridSearchCV method to identify the largest number of clients with credit risk. The model is evaluated based on metrics of sensitivity, specificity and precision, on a set of consumer credit data. Among the proposed algorithms, XGBoost outperformed the LightGBM and catBoost models. Experimental results confirmed that the XGBoost model performs better for credit risk prediction with historical data.

Keywords: Credit risk · Machine learning · Binary classification · Gradient boosting

1 Introduction

The application of artificial intelligence tools in credit risk analysis processes in financial institutions is a pillar of the Digital Transformation in the financial sector [1, 2]. The use of these tools brings numerous benefits from customer interaction experiences or market analysis to processing functions, including activities related to risk control and management [3].

In recent months, the coronavirus pandemic has caused an unprecedented economic upheaval, the impact of which has led to an unprecedented drop in the number of customers able to obtain and repay credit and a significant increase in non-performing loans. This has led to dramatic changes in loan portfolios in relatively short periods of time.

In [4, 5] the authors mention that minimizing risk at the time of granting loans is an important concern for financial institutions, in order to avoid witnessing defaults by a borrower. In the same sense, in [6, 7], it is indicated that the digitization of systems

V. Robles-Bykbaev et al. (Eds.): CITIS 2022, LNNS 607, pp. 81–95, 2023.
https://doi.org/10.1007/978-3-031-24327-1_8

to measure risks for the provision of mortgage loans, credit cards among others, using machine learning techniques are of great help.

Therefore, the use of tools based on machine learning to reduce credit risk has become very relevant in recent times. In this regard, there are several studies related to the use of machine learning tools in the financial sector. In this regard, in [8] describes the use of binary classification models based on automatic deep learning models on real data to predict the probability of default on loans, which indicates that tree-based models are more stable than models based on multilayer artificial neural networks. In the same line of research in [9], the authors used Deep and Machine Learning models on several tabular datasets, related to data from different sources of banking customers, in the mentioned study, it is concluded that tree set models (XGBoost) outperform deep learning models, in addition to requiring fewer adjustments in their implementation. In the same sense, in [10] the authors use state-of-the-art machine learning techniques such as Extreme Gradient Boosting (XGBoost) and neural deep learning networks (MXNET) on a large dataset of corporate credit loans. In the aforementioned study, it is concluded that these algorithms provide better classification accuracy compared to traditional classification methods, such as logistic regression and linear discriminant analysis. On the other hand, in [11] models are described for the analysis of defaulted loans, in the cited document, it is concluded that for these cases the decision tree algorithm presents better results than the logistic regression and gradient boosting algorithms, because it allows to identify the variables that can be modified in the different nodes, allowing to analyze different assumptions and to manage the credit risk in a better way.

In previous studies it was possible to verify how machine learning algorithms are increasingly used in the area of risk management and its corresponding function as vital as the rating and granting of credit, based on the context of predicting the risk of default of customers, in this paper we propose the training of gradient boosting machine learning models (CatBoost, LightGBM and XGBoost), on a historical database of consumer loans, which was analyzed, treated and trained in order to obtain a balanced rate between specificity and sensitivity metrics minimizing false positives and false negatives respectively.

RandomUnderSampler was used to deal with imbalanced classes and grid search for adjustment of the main hyperparameters, in order to avoid that customer are erroneously identified as delinquent and to identify the maximum number of customers who are really likely to default on their debts [12]. The proposed model will provide lending institutions with a tool for credit analysis, especially for clients with unsatisfactory payment history, which is an indispensable requirement for this type of institutions, as indicated by the authors in [13, 14].

The main contributions of this paper are the following.

1. Analyze the different boosting algorithms with hyperparameter optimization used for credit default risk prediction.
2. The output of the proposed credit risk model is analyzed in terms of sensitivity, specificity and accuracy metrics seeking to minimize false negatives on a set of consumer credit data.
3. The conditions are established under which the model proposed are feasible.

This article is organized as follows: in Sect. 2 we cover the description of the model used. In Sect. 3 we present the training results and their analysis. Finally, in Sect. 4 we provide conclusions and future work.

1.1 Related Work

The objective of the Machine Learning model is to help in the decision making in credit risk analysis, as well as to predict the default in the payment of installments.

In this regard, in [15], a comparison is made between several Machine Learning models applied to credit risk, which were selected, trained and evaluated, using various libraries, for the level of assertiveness using metrics such as Accuracy, Precision, Recall, F1 Score and AUC ROC, finding the best model with the best performance in the evaluation of credit granting to Artificial Neural Network (ANN), with an assertiveness level of 93.72%, over the Logistic Regression (86.07%), Random Forest (66.35%), Support Vector Machine (84.44%), Decision Tree (88.80%) and k-Nearest Neighbor (65.98%) models, which allowed it to reduce the level of risk in the granting of microcredits. Another particular case, in [16], models based on machine learning algorithms such as: AdaBoost, Random Forest, SVM, and Naive Bayes, which are favorable alternatives in the estimation of a credit scoring, obtaining the best metrics in Accuracy, Sensitivity, Specificity and AUC, are analyzed. The aforementioned paper concludes that the most regular model of all was AdaBoost, since it obtained the best results in three of the metrics (Sensitivity, Specificity and AUC) and has a fairly high sensitivity and an acceptable specificity given the large imbalance in the data.

Likewise, in [17, 18], the authors describe the XGBoost model, as a scalable supervised learning algorithm for sequential trees and works with multiple decision trees, the trees are added sequentially (Boosting) in such a way that for each new iteration it learns the error of the previous trees and thus correct the error, until no more corrections can be made, this method is known as “downward gradient”. In the same sense, the authors in [19, 20] mention that the XGBoost algorithm has a fast computational speed by using the maximum depth of the tree as a parameter to prune the tree and a better performance by optimizing the classification through parallel processing. Meanwhile, the complexity of the tree model is added to the optimization objective as a regularization term to avoid overfitting.

The process starts with a tree A_0 to predict the target variable y, and the result obtained is associated with a residual (y–A_0), thus obtaining a new tree H_1 which adjusts the error of the previous process and is subsequently combined with A_0 to obtain a new tree A_1. This process iterates until the mean square error is minimized to the maximum, as indicated in [18, 21].

Moreover, the authors in [18] indicate that the Light Gradient Boosting Machine (LightGBM) algorithm adapts the Gradient Boosting Decision Tree (GBDT) framework and improves the learning efficiency of a decision tree while maintaining the accuracy of the learned decision trees. Likewise, the authors in [22, 23] introduce two novel techniques, Gradient-based One-Sided Sampling (GOSS) and Exclusive Feature Clustering (EFB), in the process of finding the splitting points in decision tree learning. On the other hand, in [24, 25] mention that the main feature of the LightGBM algorithm lies in its decision trees that grow by leaves vertically, by growing the same leaf, the loss associated with it is much lower compared to other boosting algorithms that use a horizontal approach (by levels). It is noted for having higher accuracy, faster training speed, with an ability to handle large scale data (generally more than 10,000 data values) and is compatible with GPU learning. In [26] specifies that it has been shown to significantly reduce overfitting.

Likewise, in [22] it is stated that LightGBM has similar parameters to XGBoost, except for some special parameters in GOSS and EFB, such as the maximum number of containers in which the feature values will be collected, the maximum conflict rate for the packets.

Regarding the CatBoost algorithm, the authors in [27] indicate that CatBoost is an automatic learning algorithm based on gradient boosting developed by Yandex researchers that is suitable in multiple applications. Gradient boosting is a technique that is applied in multiple types of problems such as fraud detection, recommendation engines and predictions. In addition to obtaining good results with relatively less data than Deep Learning models. In the same sense, in [28, 29], the authors indicate that the strengths of this algorithm are: Produce excellent results with relatively smaller data sets than other machine learning methods and Work automatically with categorical features. It is possible to use both numerical and categorical data with CatBoost without performing any previous preprocessing.

In [30] mentions that it improves its handling of categorical features and is also faster compared to other boosting algorithms, since it implements symmetric trees. It works well with large datasets and has low latency requirements.

2 Model Description

The conceptual framework of the proposed model is presented in the Fig. 1 and is based on similar research [13, 31], where they analyze machine learning algorithms.

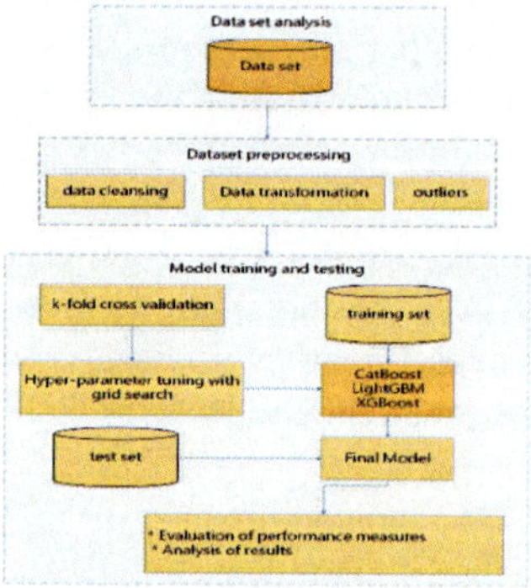

Fig. 1. Conceptual framework

The presented workflow consists of three stages Data Set Analysis, Preprocessing of the Data Set and Training and testing of the model. It will allow us to thoroughly investigate the performance of the model in order to draw reliable conclusions.

2.1 Data Set Analysis

In this phase, the exploratory analysis of the data is carried out in order to know how the data set is made up in relation to quantitative and qualitative variables and the target variable. It is also necessary to know the most important variables that will contribute to the model.

Data Set: The data set was obtained from clients with credit history with consumer loans, current in the last three years, provided by the financial institution, the information was consulted in its database with a total of 7984 records composed of 18 variables. The summary of the consumer credit data set is shown in Table 1.

Table 1. Summary of the consumer credit data set

Description	Value
Number of records	7984
Clients at risk	6823
Clients without risk	1161
Number of features	18
Defaulter	14%

In order to determine the most relevant characteristics involved in the credit risk analysis model, similar studies [18, 22] and together with the risk personnel, data variables and client credit information were selected as shown in Table 2.

Table 2. Description of the data set under study

Features	Description
Age	Client's age
Marital status	Client's marital status
Type address	Type of client's home
Economic activity	Activity performed by the client
Education level	Level of customer education
Profession	Customer's profession
Type credit	Type of credit
Interest rate	Interest rate of the requested amount
Monthly income	Customer's monthly income
Assets	Amount in customer's assets

(continued)

Table 2. (*continued*)

Features	Description
Outgoings	Customer's monthly outgoings
Working time	Length of time working at current job
Quota value	Value of the installment to be paid
Amount	Amount of credit
Fees	Number of installments of the credit requested
Purpose	Destination of funds
Genre	Customer gender
Risk	Variable to predict

In the analysis performed, the target variable "Risk" was considered (see Table 2), which contains two categories: value = 1, in which customers with risk are considered, and value = 0 for customers without risk. In this regard, for the analysis we performed a grouping by class and concluded that the data set presents a large imbalance between customers with credit risk and customers without credit risk, see Fig. 2. The ratio between samples with risk and without risk in the data set is approximately 5 to 1, in this sense, the business indicator indicates that we have to look for 5 cases to find a risky customer.

2.2 Preprocessing of the Data Set

The preprocessing was performed with the objective of obtaining data that are representative and consistent, helping to improve the performance of the model in terms of accuracy and time complexity [31, 32]. For the preparation of the data for a classification algorithm, the following was contemplated:

Dataset Cleaning: The consumer loan data set contains a large proportion of atypical values (an unusually large or small observation), for which the procedure of dimensionality reduction was applied by 5%. This is to avoid having perverse consequences on the estimation stage of the parameters, estimates that will be biased, the bias being the relationship between the outlier value and the explanatory variables included in the model, as described in [33].

Each characteristic of the data set presented in Table 2, was verified in order to determine if there are null values, finding the variable working time with 196 missing values. The imputation technique of randomly selecting a value from a confidence interval (the mean minus the standard deviation and the mean plus the standard deviation) was used to replace the missing values. The characteristics (gender and purpose) were eliminated because they did not contribute to the training of the model since their categories presented percentages similar to class with value = 1 of the target variable risk with percentage 14%.

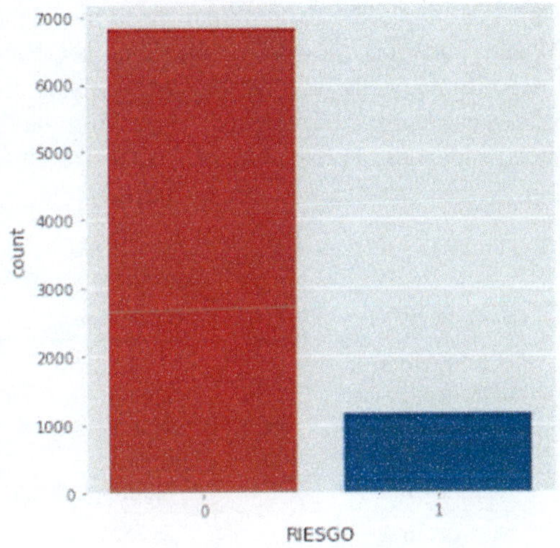

Fig. 2. Imbalance between classes with risk (1) and without risk (0)

Data Transformation: To establish the categorical characteristics, the OneHotEncoding strategy was used, in which a new binary characteristic created for each category [34].

To establish the numeric characteristics, we normalized from a narrow, fixed range [31].

2.3 Model Training and Testing

At this stage, the dataset was partitioned into training and test datasets with 80% and 20% respectively.

Sampling: Given the nature of the data set under study, there is a clearly unbalanced distribution, which means that we have considerably more cases that are not at risk than cases at risk. This leads to low accuracy for the at-risk samples, but high accuracy for the non-risk samples, resulting in an assessment biased towards the majority class and poor performance in the minority class [35]. To avoid influencing the outcome of our models and to solve this problem, the RandomUnderSampler sampling technique was applied on the training dataset, which randomly eliminates instances of the majority class, until it equals the amount of the minority class.

Hyperparameter Optimization: For hyperparameter optimization, k-fold cross-validation was used as a data partitioning strategy with an approach for hyperparameter fitting in combination with the method GridSearchCV, which will search over specified parameter values for each estimator. The hyperparameters selected for fitting in gradient boosting models are presented in Table 3.

Classification Models: WE applied the gradient boosting decision tree models: XGBoost, LightGBM and CatBoost, on the training data set. Finally, we performed the evaluation of the algorithms on the test data set for analysis. In [36] specifies that the goal of a learning algorithm is to learn from the training data to predict class (0) risk-free customer labels, and class (1) risk customer labels for unseen data (test data).

Table 3. Parameters selected to fit in gradient boosting models

Method	Parameters	Grid search
XGBoost	N estimators	0, 1000, 50
	Max depth	1, 3, 5
	Min child weight	1, 3, 6
	Gamma	0, 1, 5
	Learning rate	0.0001, 0.001, 0.01, 0.1
LightGBM	Max depth	5, 75, 10
	Learning rate	0.001, 0.01, 0.1
	Num leaves	20, 220, 50
	Min data in leaf	100, 1000, 100
CatBoost	Depth	6, 8, 10
	Learning rate	0.03, 0.1
	12 leaf reg	1, 5, 10

Performance Metrics: Based on the confusion matrix 1, different performance metrics such as accuracy, sensitivity, specificity and F1 score [36] are used to evaluate classification algorithms. The confusion matrix consists of four basic measures that are used to define the classifier performance metrics. For the analysis, the following measures were considered, as described in [31]:

- True positive (TP) and true negative (TN) represent the number of good (non-defaulting) and bad (defaulting) borrowers, respectively, classified correctly.
- False negative (FN) and false positive (FP) represent the number of good and bad borrowers respectively, classified incorrectly.

Due to the unbalanced distribution for the data set under study (as indicated in subsection II-A) of risky and non-credit-risky samples. The performance metrics adopted for the analysis are: Precision in Eq. 1, Sensitivity in Eq. 2, Specificity in Eq. 3, F1 score in Eq. 4, which are calculated on the basis of the TP, TN, FP and FN mentioned above.

Precision. of an algorithm is represented as the proportion of clients correctly classified with credit risk (TP) with respect to the total number of clients with predicted credit risk (TP + FP).

$$Precision = \frac{TP}{TP + FP} \tag{1}$$

Sensitivity: Defined as the proportion of clients with correctly classified credit risk (TP) divided by the total number of clients actually at risk.

$$Sensitivity = \frac{TP}{TP + FN} \tag{2}$$

Specificity: Is defined as the proportion of clients without credit risk (TN) that are correctly classified, divided by the number of clients that are not at risk.

$$Specificity = \frac{TN}{TN + FP} \tag{3}$$

F1 Score: Is the weighted average of accuracy and sensitivity [38].

$$F1Score = \frac{2 * precision * specificity}{precision + specificity} \tag{4}$$

In [37] the authors define a confusion matrix as a table that visualizes and summarizes the performance of a classification algorithm.

ROC Curve: The ROC curve allows us to graphically visualize the results of the classification algorithms, the closer it is to the left, the better the performance of the model will be. The y-axis of the ROC graphs is the true positive rate (TPR), while the x-axis is the false positive rate (FPR) [39]. The TPR and FPR are defined as follows:

$$TPR = \frac{TP}{TP + FN} \tag{5}$$

$$FPR = \frac{FP}{TN + FP} \tag{6}$$

3 Analysis and Results

In this section, we present the training results of the gradient boosting machine learning algorithms CatBoost, LightGBM and XGBoost on a tabular consumer credit data set.

To improve the training performance of the models, the training dataset was balanced with the RandomUnderSampler technique and optimized with the values of Table 4, which were found by grid search in order to obtain a minimum of false positives and false negatives.

Table 4. Optimal parameters of gradient boosting models

Method	Parameters	Optimal parameters by grid search
XGBoost	N estimators	200
	Max depth	3
	Min child weight	3
	Gamma	1
	Learning rate	0.1
LightGBM	Max depth	15
	Learning rate	0.01
	Num leaves	20
	Min data in leaf	500
CatBoost	Depth	6
	Learning rate	0.03
	l2 leaf reg	10

The model is validated on the test data set on different metrics, each reflecting different aspects of model performance:

1. Specificity measures the proportion of risk-free clients that are correctly identified.
2. Sensitivity measures the proportion of risky customers that are correctly classified.

3.1 Confusion Matrix

Figure 3 shows the confusion matrix of different gradient boosting machine learning methods, among the three algorithms tested, XGBoost and CatBoost gave the best results, with a sensitivity rate of 67% and 66% respectively, and with a similar rate 36% of false positives. On the other hand, LightGBM provided a better false positive count at 24%, but its false negatives were substantially higher than those of XGBoost and CatBoost, resulting in a weaker sensitivity rate. The coverage of real cases with risk in XGBoost (67%) is higher than the other models, which indicates that XGBoost outperforms CatBoost and LightGBM.

Through the prediction model, what we want is to detect if a client is going to fall into arrears, that is, if he is going to default on the payment of installments, or if he is a solvent client, for this we can verify how accurate the model is through the metrics of Fig. 4 where it presents the sensitivity, specificity and precision score of different gradient boosting machine learning methods. A score closer to 1 indicates that the model performs better. Figure 4 shows that XGBoost achieved a higher score in sensitivity than CatBoost, although not in specificity and precision (0.67, 0.67 and 0.26, respectively), and CatBoost (0. 66, 0.67, and 0.26, respectively) and LightGBM performed the worst in terms of sensitivity, specificity, and precision scores (0.54, 0.76, and 0.29, respectively).

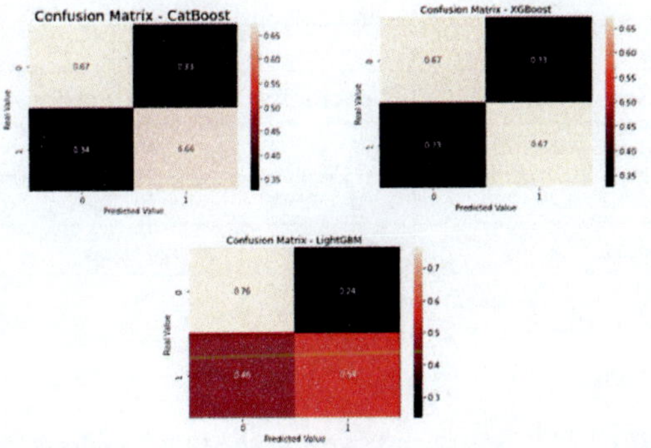

Fig. 3. Confusion matrix of different machine learning methods: CatBoost; XGBoost; LightGBM.

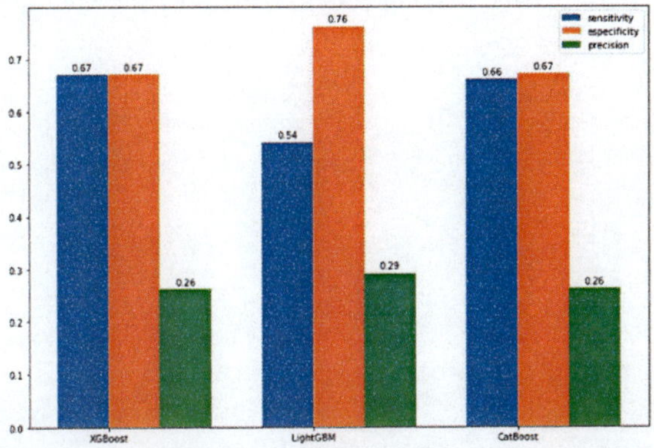

Fig. 4. Sensitivity, especificity, precision of different machine learning method.

3.2 Comparison of Metrics

Table 5, shows a comparison between the CatBoost, LightGBM and XGBoost models.

The comparison between the metrics of the different algorithms indicates that the performance of the CatBoost and XGBoost models, presents similar results for the metrics of sensitivity and specificity, the metrics of the XGBoost model is superior to the CatBoost model, in sensitivity, which determines the XGBoost model as the best model in the estimation of credit risk with the data set under study. This represents an ideal model, as it covers the largest number of cases with risk, reaching minimum values of false positives and false negatives.

Table 5. Comparison of metrics between CatBoost, lightgbm and XGBoost algorithms

Algorithm	Sensitivity	Especificity	Precision	NPV
XGBoost	0.67	0.67	0.26	0.92
LightGBM	0.54	0.76	0.29	0.90
Catboost	0.66	0.67	0.26	0.92

3.3 ROC AUC Metric

Figure 5 shows the ROC AUC score and, we can see that XGBoost works better, since it has the highest AUC with respect to LightGBM and with a minimum value of difference with respect to CatBoost and, therefore, it is the best classifier to distinguish between at-risk and non-risk classes and making better predictions.

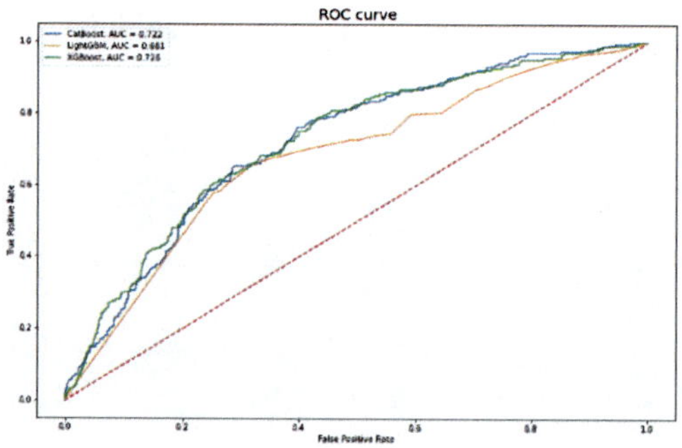

Fig. 5. Area under ROC curve

3.4 Reliability

To evaluate the performance of each model in predicting the probability that the sample belongs to the positive class, that is, the probability of default. A reliability plot and a Brier score were used, where the former creates a plot of actual versus predicted probabilities in a test set and the latter calculates the mean squared error between the predicted probabilities and their respective positive class values. Since the Brier score is a function of cost, a lower Brier score indicates a more accurate prediction. We can see in Fig. 6 the Brier Score of the XGBoost algorithm, where it performs better once again, compared to other gradient boosting models. From this score and the plot, we conclude that our model is well calibrated for probability prediction, meaning that the predicted probabilities closely match the expected distribution of probabilities for each class, and thus does not require additional calibration.

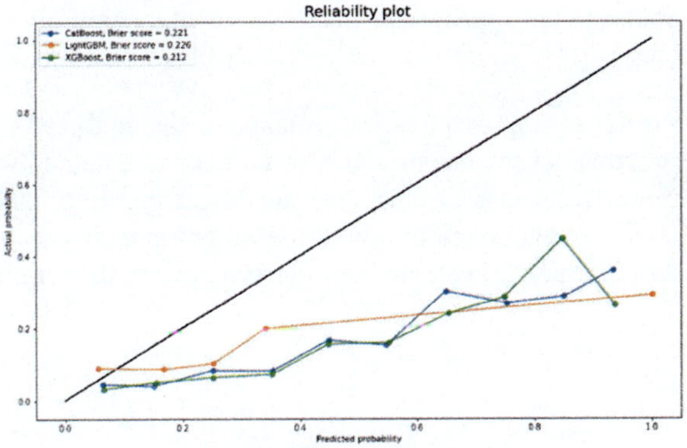

Fig. 6. Reliability

4 Conclusions

In this paper, gradient boosting machine learning models XGBoost, LightGBM and CatBoost were applied to verify which of them presents a better estimation in the prediction of default risk, these models were trained and evaluated on a set of consumer credit data from customers with valid credit history from three years ago.

The data set by its nature presented a large imbalance between the classes with risk and without risk, for which the RandomUnderSampler sampling technique was used to equalize the values towards the minority class.

Additionally, the models were optimized with the main hyperparameters, the same that were searched with the grid search technique, in order to improve the coverage metrics for both risky and riskless clients, the parameter to be taken into account was the sensitivity, in order to minimize false negatives, However, it was also necessary for the specificity to have a high value in order to obtain a minimum value in false positives and thus stop going to a client who is solvent, which would generate a loss for the financial institution. Therefore, the best model would be one that could minimize false negatives, identifying all defaulters among the customer base, and at the same time minimize false positives, preventing customers from being misclassified as defaulters. Considering the importance of predicting the probability of customer default, we decided to obtain balanced values between specificity and sensitivity, looking for the best hyperparameters that could increase the percentage of these metrics.

Among the three gradient boosting algorithms trained, XGBoost and CatBoost presented the best results in sensitivity and specificity metrics, XGBoost being superior with a minimal difference of 1% in sensitivity metric.

In the evaluation of the LightGBM model, it obtained a better false positive count but the false negative value was very high, so it was discarded to analyze the risk of credit default.

In the application of the gradient boosting algorithms, it was possible to verify that the good performance of a classification algorithm depends on factors such as: exploratory analysis, data quality, selection of the characteristics that will move the target variable, data balance and applying the appropriate classification algorithm for the problem to be solved.

Future Work. In order to improve the performance of the analyzed models, in future work we propose to consider the use of a greater number of data and with higher quality, also to implement credit history characteristics of external companies for a better classification of risky customers and finally to use other balancing techniques different from the one studied, in order to evaluate their performance in the training of the model.

References

1. Tames, L.A.L., Campos, B.C., Navarro, F.A.C.: Inteligencia artificial para la transformaci on digital en toma de decisiones. Tecnologıa Vital **4**(7) (2020)
2. Redero Juanes, B., et al.: La transformaci on digital de las empresas. estudio del caso del sector financiero. el banco santander (2020)
3. Borrero-Tigreros, D., Bedoya-Leiva, O.F.: Predicci on de riesgo crediticio en colombia usando t ecnicas de inteligencia artificial. Revista UIS Ingenier ıas **19**(4), 37–52 (2020)
4. Shwartz-Ziv, R., Armon, A.: Deep Learning in not all you need (2021)
5. Pinedo Borobio, P.: Modelos recientes de la estad ıstica y el aprendizaje m aquina para la valoración del riesgo de incumplimiento crediticio.
6. Saqib Aziz, M.D.: Ai and machine learning for risk management. SSRN Electronic Journal (2018)
7. Guiterrez-Portela, F., Moreno-Hernandez, J.J., Echeverry, B.D., Jaramillo, A.S.: Uso de los sistemas inteligentes para la deteccion de fraudes financieros. Revista Sinergia **1**(6), 6–30 (2019)
8. Peter Martey Addo, D.G., Hassani, B.: Credit risk analysis using machine and deep. Computational Methods for Risk Management in Economics (2018)
9. Ravid Shwartz-Ziv, A.A.: Deep learning is not all you need (2021)
10. Anastasios Petropoulos, E.S., Siakoulis, V., Klamargias, A.: A robust machine learning approach for credit risk. Ninth IFC Conference (2018)
11. Alvarez, J.G.: Machine learning y riesgo de crédito. Facultad de Ciencias Económicas y Empresariales (2020)
12. Bastos, R.: Credit risk analysis with machine learning. Towards Data Science (2020)
13. Ala'raj, M., Abbod, M.F., Majdalawieh, M.: Modelling customers credit card behaviour using bidirectional lstm neural networks. Journal of Big Data 8 (2021)
14. Sacoto Cabrera, E.: Análisis basado en teoría de juegos de modelos de negocio de operadores m oviles virtuales en redes 4g y 5g, Ph.D. dissertation, Universitat Polit'ecnica de Valéncia (2021)
15. Rojo, M.R.A.: Modelo predictivo de análisis de riesgo crediticio usando machine learning en una entidad del sector microfinanciero (2019)
16. Fernandez, D.T.: Aplicacion de metodologias machine learning en la gestion de riesgo de crédito. Universidad Politecnica de Madrid (2017)
17. Tianqi Chen, A.G.: XGBoost: A scalable tree boosting system (2016)
18. Malbrn, A.E.: Modelo de predicción de operaciones de crédito con posible default financiero (2020)
19. Wu, S.: How to choose between different boosting algorithms. Towards Data Science (2021)
20. Li, Y.: Credit risk prediction based on machine learning (2019)

21. Sacoto Cabrera, A.J., Palaguachi, S., Leon-Paredes, G.A., Gallegos-Segovia, P.L., Bravo-Quezada, O.G.: Industrial communication based on mqtt and modbus communication applied in a meteorological network. In: The International Conference on Advances in Emerging Trends and Technologies, pp. 29–41. Springer (2020)
22. Ligang Zhou, H.D.R.M., Fujita, H.: Credit risk modeling on data with two timestamps in peer-to-peer lending by gradient boosting (2021)
23. Sacoto-Cabrera, A.J., Leon-Paredes, G., Verdugo-Romero, W.: Lorawan: Application of non-linear optimization to base stations location. In: Communication, Smart Technologies and Innovation for Society, pp. 515–524. Springer (2022)
24. Daoud, A.A.: Comparison between xgboost, lightgbm and catboost using a home credit dataset. International Journal of Computer and Information Engineering (2019)
25. Naik, K.: Predicting credit risk for unsecured lending: A machine learning approach (2021)
26. Coser, A.: Predictive models for loan default risk (2019)
27. Hancock, J.T., Khoshgoftaar, T.M.: Catboost for big data: an interdisciplinary review. Journal of big data **7**(1), 1–45 (2020)
28. Liudmila Prokhorenkova, A.V.A.V.D.A.G., Gusev, G.: Catboost: unbiased boosting with categorical features (2017)
29. Biarnes, A.: How catboost encodes categorical variables? Towards Data Science (2021)
30. Sujoy Barua, P.S.L.S.J.R., Gavandi, D.: Predicting the probability of loan defaults using CatBoost algorithm. In: 5th International Conference on Computing Methodologies and Communication (ICCMC) (2021)
31. Wirot Yotsawat, A., Wattuya, P.: Improved credit scoring model using xgboost with bayesian hyper-parameter optimization. International Journal of Electrical and Computer Engineering (IJECE) (2021)
32. Fisnik Doko, S.K., Mishkovski, I.: Credit risk model based on central bank credit registry data (2021)
33. de Juan Fernandez, A.: Importancia de los valores atípicos en el modelo de regresión causas, consecuencias, detección y tratamiento (1995)
34. Arias, L.A.P.: Evaluación de modelos de machine learning para sistemas de detección de intrusos en redes iot (2021)
35. Tarawneh, A.S., Hassanat, A.B., Altarawneh, G.A., Almuhaimeed, A.: Stop oversampling for class imbalance learning: A review. IEEE Access **10**, 47 643–47 660 (2022)
36. Tharwat, A.: Classification assessment methods. Applied Computing and Informatics **17**, 168–192 (2021)
37. Pushpa Singh, K.K.S.A.S., Singh, N.: Diagnosing of disease using machine learning (2021)
38. Nitesh, L.O.H.W.P.K., Chawla, V., Bowyer, K.W.: Smote: Synthetic minority over-sampling technique. Journal of Artificial Intelligence Research (2002)
39. Ma, T., Wu, L., Zhu, S., Zhu, H.: Multiclassification prediction of clay sensitivity using extreme gradient boosting based on imbalanced dataset. Applied Sciences **12**(3) (2022)

Social, Labor and Cultural Integration of Immigrants: A Quali-Quantitative Approach to the Venezuelan-Ecuadorian Case

Candice Medina[1], Nicole Tapia[2], Marco Velásquez[3], and Fabricio Zanzzi[3](✉)

[1] New York University, New York, USA
[2] Saint Peter's University, Jersey City, USA
[3] ESPOL Polytechnic University, Guayaquil, Ecuador
pzanzzi@espol.edu.ec

Abstract. Between the two extremes of the social exclusion-integration spectrum, here we addressed the second since the social integration of immigrants must be assisted by the host country. In this paper, we analyzed this process as a combination of cultural integration and labor integration, through in-depth interviews complemented with non-parametric methods, Bayesian statistics, and Boolean algebra applied to small data. In the empirical approach, we used the Venezuelan-Ecuadorian case. We found that this relationship is fully satisfied, therefore social integration can be evaluated in a more tangible manner. The mechanism that accelerates it is the network of friends. The absence of this network affects women more intensely. We recommend that governments encourage its conformation, evidence, and reduce the naturalization of precarious jobs of immigrants. Our results opened a line of research on differential factors with Venezuelan immigration in other countries and perceptions of immigrant inequality.

Keywords: Multiculturalism · Migration · Venezuela · Ecuador · Small data

1 Introduction

Immigrants must be assisted in the evolution of their own processes of social integration in the host country. The purpose of this article is to show the relationship between cultural integration (this is intangible) and labor integration (this is tangible), through an analysis that combines quantitative and qualitative approaches to understand the emergence of spontaneous mechanisms to optimize an adequate adaptation.

In this section, due to lack of space, we present the two most important antecedents, in section two we present the concepts that frame our research. The methods are in section three, the results of each instrument are in section four, and the conclusions, recommendations and limitations are in section five.

We have three motivations: in a theoretical view, this allows us to answer our research question: Can the social integration of immigrants be explained by observing their processes of cultural integration and labor integration? In a methodological view, we want

V. Robles-Bykbaev et al. (Eds.): CITIS 2022, LNNS 607, pp. 96–112, 2023.
https://doi.org/10.1007/978-3-031-24327-1_9

to use quantitative methods for small data. In a practical view, this would entail a reduction in state spending on social assistance, which is crucial in countries with budget financing issues, such as Ecuador, which has welcomed more than 400,000 Venezuelans and experienced a fiscal deficit of 11% in 2020.

Venezuela is going through an economic, political, social, and humanitarian crisis, which has caused a mass exodus to other countries, such as Ecuador. Currently, migrants represent 3% of the world population [1], almost the same percentage of Venezuelan immigrants with respect to the population of Ecuador. They are a minority group vulnerable to all kinds of abuses since only 125 thousand of them have some type of visa; the rest are not in the public records [2].

On the economic health of Ecuador: before the great Venezuelan immigration wave, Ecuador had received a Colombian immigration wave that during 2011 involved 24,615 people [3] that, to a large extent, were assisted by the UN Refugee Agency, UNHCR [4]. Additionally, a slowdown caused a 4% of GDP fiscal deficit in 2019 [5] that has been deepened and accelerated by the COVID-19 pandemic.

2 Theoretical Framework

There are two extremes in the social spectrum: exclusion, through genocide, expulsion, and marginalization; and integration, to maintain unity and continuity [6].

The exclusion of immigrants happens because of fiscal tightness in the provision of social assistance and because of the economic recession that reduces jobs, cultural heritage, and sovereignty. This undermines the social system.

Social integration is the process of inclusion of very vulnerable groups (drug addicts, ex-convicts, people in extreme poverty, etc.) in society, to avoid the transmission of their inherited inequalities. In the case of immigrants, the difficulty of obtaining it contributes to the spatial social problem and leads to segregation, ghettoization, gentrification, among others, which force society to run at three speeds: a) that of immigrant neighborhoods; b) that of "urbanized towns"; and, c) that of the metropolis [7]. All waiting for opportunities in the latter, which can even launch them into invalidation and social non-existence [8].

We consider that social integration is reflected in cultural integration and labor integration, which are easier to define. The first, due to the everyday perception of the individual (this is intangible); and the second is because its role within the economic process builds the most important supports of the individuals in the social structure (this is tangible). It grants them stable links, networks, and a protection system -including social security- against risks. It generates them nodes of social cohesion (integration) and forecasting the future with greater serenity [8, 9]. However, social integration can also become precarious [9].

2.1 Cultural Integration

The concept of "integration of immigrants" assumes that before their arrival there was already a harmonious, linked, united, and integrated society whose network runs the risk of being destroyed or overwhelmed by the in-migratory process [10]. Thus, when individuals migrate from a poor nation to a successful one and integration is not enough, "parallel societies" would emerge and deteriorate the "social model" [11]. They "import" the failed social models of their failed democracies and destroy the reasons of their migration [12]. Therefore, integration seeks to strengthen society through social cohesion [13] and this refers to conditions of social trust and participation that generate a sense of common identity and solidarity across differences [6]. We will take two forms of cultural integration: assimilation and multiculturalism [14].

The Assimilation. Here, the actions are to incorporate immigrants into society through unilateral processes of adaptation [15]. It is individualized, very close to republican civism that emphasizes the unity of the state and the acculturation of citizens in a melting pot [16]. Then, equality will be achieved by sacrificing cultural diversity and the formation of ethnic communities. Such is the case of the majority of third-generation Tibetan-Canadians [6]. Assimilation model peaked in the 60 s.

The Multiculturalism. It is on the opposite side of assimilation. It is more oriented toward social cohesion policies than cultural differentiation policies [17]. There are no impositions. It is about legitimization of the social identity of a people minority group [18]. Here immigrants are guaranteed all rights in all sectors without relegating their cultural diversity, framing themselves in minimal state records.

Since access to public education, the ethnic diversity is a very important tool to reduce the intergenerational transmission of inequalities [19]. In the second-generation social mobility is already notable [6].

This model emphasizes the cohabitation of ethnic minority groups to maintain their cultural and linguistic differences for generations, through recognition and accommodation [16], as in the case of American Indians [6] and as in the case of Tibetan refugees in South Asia [20].

2.2 Labor Integration

Before talking about labor integration, it is necessary to differentiate two concepts: job and employment. "Job" refers to the effort made, the activity to produce welfare for other people; while "employment" refers to a contractual situation, the salary [21]. Both could be precarious.

A precarious job is characterized by dissatisfaction with conditions and the environment, generally through situations in which the worker does not have tangible and/or symbolic recognition for his labor activity [22]. These workers work harder and in worse conditions than the permanent employees, which is why they are more prone to turnover, illnesses, and more serious accidents [9].

Precarious employment is characterized by instability and insecurity in hiring, which neglects social and citizen rights and does not allow the employee to foresee his or her professional future. It is done through "learning organizations", "zero waste production organizations", "Taylorist/Fordist workforce organizations" (an intense division of labor), and organizations with a paternalistic tradition (authoritarian, moral and benevolent). The employment contracts are of limited duration, which includes Sundays and holidays without extra pay; seasonal, part-time jobs [23], internships, and in companies of temporary workers, repositioning the activity in the commercial sphere with its respective non-labor regulation [9]. This leads to the emergence of private pension funds, prepaid medicine, private insurance, outsourcing, etc.

Precariousness is based on forms of flexibilization imposed, managerially, to those hired [24]; therefore, it devalues, blames, and gives the employee a feeling of useless and disposable [9].

We considered the four types of labor integration from the individual's perspective towards the firm/organization according to Paugam [25], based on the concepts of job satisfaction (not precarious job) and stable/secure employment (not precarious employment). The order and some names shown in Table 1 are ours:

Table 1. Types of labor integration.

Categories	Job satisfaction (a respectable job, and value-adding tasks. No physiological or psychological suffering)	Stable employment (Permanent contract, social protection, indemnities)	Frequent business status
Full integration	Yes	Yes	Well-established organization
Uncertain integration	Yes	No	Business uncertainty; or employee evaluation processes
Exhausting integration	No	Yes	Internal conflicts and union protests
Disqualifying integration	No	No	Prolonged and unavoidable exposure to chemicals or radiation [26]; firm in closure, restructuring, merger

3 Methodological Strategy

To understand the social integration of immigrants through the definitions of cultural integration and labor integration, we combine them in Table 2, where the cell made up of the "full" labor category and the "multiculturalism" cultural category is called "total social integration"; the others made up the "partial social integration" zone. Each immigrant, according to the information collected, was included in this classification.

Due to circulation restrictions resulting from the COVID-19 pandemic, the sample consists of 15 immigrants from central and western Venezuela, mainly from Caracas, Barquisimeto, Maracay, La Victoria, Maracaibo, Guacara, and Isla de Margarita. They have been based in Ecuador for one to five years and were interviewed between May and October 2020. Of them, half made the trip alone, 10 are women, and 5 are alone (divorced, widowed, or single). Almost all of them have a college degree, and more than

Table 2. Social integration as a construct.

		Cultural integration	
		Assimilation	Multiculturalism
Labor integration	Full		Total social integration
	Uncertain	Partial	
	Exhausting	Social	Integration
	Disqualifying		

half are business-related. In addition, half have more than 10 years of work experience. Regarding their residence preferences, only 2 stayed in Ecuador for unintended reasons, 9 settled in the big cities: Guayaquil, Quito, and Cuenca. At present, almost all of them have lived in Ecuador for more than 2 years, 11 admit to having a network of friends who support each other, 12 have incomes of less than two legal minimum wages, and 5 live in poor neighborhoods. Regarding the family, 6 have already brought their children, 9 send remittances to their extended family that is still in Venezuela, and 12 think not to return to settle in their country of origin.

To operationalize the variables necessary to establish the categories of labor integration and cultural integration, answers were required on the following aspects.

About the job (effort made): durability of the job, enjoyment of social benefits and social security, promotion in the job role, compliance with the remuneration about the market, length of the working day, weekly rest days, accident assistance, and occupational diseases, harassment situations of all kinds, availability of residence by the employer, financing of the trip by the employer, change of economic sector when migrating, time off work [27].

Employment (contractual status) should have been analyzed from two very close dimensions, stability and security [9]. The first refers to the time an employee stays in an organization/firm. Variables such as type of contract, the average number of hours/days worked, number of months in Ecuador, the average number of months per organization/firm were defined. The second refers to the possibility of remaining employed without lasting interruptions based on his abilities/skills, not necessarily in the same organization/firm. Variables such as length of time of unemployment, time to first job, etc.

About assimilation and multiculturalism: Venezuelan networks; perceptions about the equal treatment of Ecuadorians, acceptance/tolerance of native customs, imposition of local customs, marginalization/segregation situations, episodes of discrimination/racism, protection of the laws; family reunification; etc.

Then, in the following demographic and socioeconomic variables, we seek the explanation of each category already mentioned: sex, marital status, university degree, business knowledge, labor experience (10+ years), destination country, settlement in main cities, not poor neighborhood, not low-income person, a network of friends, living in Ecuador (2+ years), children in Ecuador, sending remittances, willing not to return to Venezuela.

Almost all the determining aspects of the categories were qualified in binaries: presence "1", absence "0"; the few exceptions were the result of mathematical operations. A "satisfactory" job was considered if the score was greater than 50. It was considered "stable" employment if the score was greater than 50. It was considered "multi-cultural" if the score was greater than 80. The results were then weighted and standardized between zero and one hundred. In our online supplementary material can be seen the qualifications of each category, in addition to the demographic and socioeconomic variables.

We formalized the analysis with non-parametric methods that are valid for small data. This allowed us to characterize robustly each category of labor and cultural integration, and to reveal the mechanism that triggers social integration. We used the following statistical tools:

1. The Kruskal-Wallis test to differentiate the four categories of labor integration (full, uncertain, exhausting, and disqualifying). It is equivalent to a one-way analysis of variance and tests the hypothesis (null) that the medians of three or more samples are identical, through a Chi-square distribution "X^2". The test statistic is calculated as $H = \frac{12}{N(N+1)} \sum_{i=1}^{k} \frac{R_i^2}{n_i} - 3(N+1)$, where "N" is the total data of the sample, "R_i" is the sum of ranks of each sample, "n_i" is the number of data in each sample, and "k" is the number of samples. If ($H > X^2$) the null hypothesis is rejected [28].
2. The Wilcoxon-Mann-Whitney test to differentiate the two categories of cultural integration (assimilation and multiculturalism). Tests the hypothesis (null) that the medians of two samples are identical, through a normal distribution. The test statistic is calculated as $U = n_1(n_1 + n_2 + 1) - U_T$, where "$n_1$" and "$n_2$" are the sizes of both samples, and "U_T" is the sum of the ranks of the largest group. If ($-Z > U > Z$) the null hypothesis is rejected [29].
3. Bayesian statistics to verify if the distributions in the categories of cultural integration are different. This approach allowed us to adhere to our "small data", through the averages and standard deviations of our variables, "prior" information from a probabilistic distribution. The result will be a subsequent "posterior" distribution that will reflect a balance between both that, in our case, will not be influenced by the prior, but only by our observed small data, since we use a normal distribution [30]. This can be done because an ANOVA is, functionally, equivalent to a Simple Linear Regression of categorical predictors, even the F statistic of an ANOVA will be the same as the F statistic of a Linear Regression for a model that only uses categories as predictors [31].
4. The k-nearest neighbors "KNN" algorithm to find the relationship between labor integration and cultural integration. Here we settled the category "total social integration". The KNN algorithm is of supervised classification (it implies that we know the dependent variable) and estimates the value of the posterior probability distribution function of an element belonging to a certain category, based on the information provided. This implies that the element will be assigned to a certain category if it is the most frequent among the k closest examples using the Euclidean distance [32]. Normalization of the data is advisable to improve the accuracy of the classificatory forecasting.
5. The correlation matrices to establish the characteristics most associated with the individuals who enjoy total social integration. This allowed us to summarize data,

as an input into a more advanced analysis, since it is a square and symmetrically reflected table that shows in each of its cells the correlation coefficient between two variables. This coefficient can be Pearson's Product-Moment Correlation "R", Spearmen's, or Kendall's Tau-b. The last two are non-parametric and less susceptible to outliers than "R" [33].

6. The boolean algebra for qualitative comparative analysis to find the configurations before which total social integration arises and, from there, deduce the mechanisms that catapult it. This method does not contain formulations but requires the following steps [34]: 1) binary coding of the presence of the causal factors and the presence of the desired situations; 2) organization of data; 3) construction of a truth table with the frequency of the results; 4) elimination/aggregation of causal factors in it; 5) simplification of factors in positive combinations; 6) elimination of superfluous or redundant combinations; and 7) specification of the final model, in which "and" is represented in a multiplicative way, "or" in an additive way, and the symbol " ~" represents the absence of the characteristic.

4 Data Analysis, Triangulation, and Results

With the data from the interviews, qualified and organized, (see our online supplementary material), the cells of Table 2 were filled in, synthesizing in Table 3.

Table 3. Social integration in our sample.

		Cultural integration		Horizontal total
		Assimilation	Multiculturalism	
Labor integration	Full	4	6 (Total social integration)	10
	Uncertain	1	0	1
	Exhausting	1	0	1
	Disqualifying	3	0	3
	Vertical total	9	6	15

Below, we show the verification of this proposed association.

4.1 The Categories of Labor Integration

The first part of our analysis consisted of testing whether there are statistically significant differences between the scores of the constitutive variables (job satisfaction and stable employment) of each category of labor integration (full, uncertain, exhausting, and disqualifying). If there are, the demographic and socioeconomic characteristics of each one can then be analyzed.

We applied the Kruskal-Wallis Test in the variables: job satisfaction and stable employment. We used the free Statistics Kingdom online platform [35]. For job satisfaction, the p-value was equal to 0.02523, $(p(x \leq 9.3281) = 0.9748)$; and the test statistic equals 9.3281, which is not in the 95% region of acceptance: [0: 7.8147]. The null hypothesis was rejected; job satisfaction was statistically different in each category of labor integration, as seen in literal a of Fig. 3.

For stable employment, the p-value was equal to 0.02997, $(p(x \leq 8.9498) = 0.97)$; and the test statistic equals 8.9498, which is not in the 95% region of acceptance: [0: 7.8147]. The null hypothesis was rejected; stable employment was different in each category of labor integration, as seen in literal b of Fig. 1.

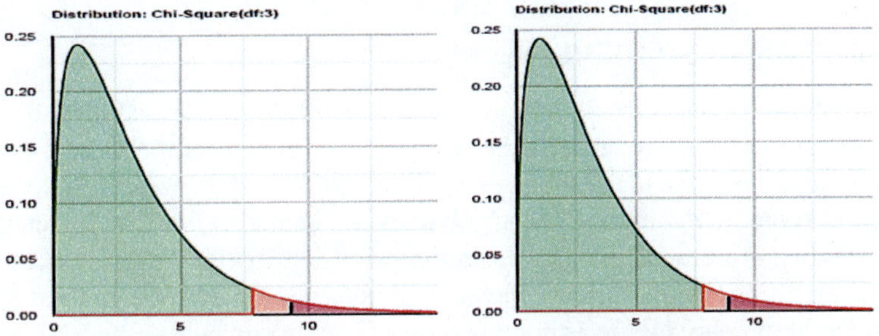

a) Job satisfaction b) Stable employment

Fig. 1. Significance in differences by labor integration.

4.2 The Categories of Cultural Integration

The second part of our analysis consisted of testing the statistical significant differences between the scores of the constitutive variables of labor integration (job satisfaction and stable employment) in each category of cultural integration (assimilation and multiculturalism). If there were any, they would be an indication of grouping that will later make it possible to analyze the demographic and socioeconomic characteristics of each one.

We applied the Wilcoxon-Mann Whitney Test in the variables: job satisfaction and stable employment. We used the free Statistics Kingdom online platform [36]. For job satisfaction, the p-value was equal to 0.0255692, $(p(x \leq Z) = 0.0127846)$; and the test statistic equals −2.232692, which is not in the 95% region of acceptance: [−1.96: 1.96]. The null hypothesis was rejected; job satisfaction was different in each category of cultural integration, as seen in literal a of Fig. 2.

For stable employment, the p-value was equal to 0.0514799, $(p(x \leq Z) = 0.0257399)$; and the test statistic equals −1.95, which is just in the 95% region of acceptance: [−1.96:1.96]. The null hypothesis could not be rejected, the difference in stable employment was not sufficiently robust in each category of cultural integration, as seen in literal b of Fig. 4. However, it was in a 94% region of acceptance: [−1.88:1.88]. This small lack of definition, perhaps, is due to small data. In the next section, we show how we robust our analysis with Bayesian statistics.

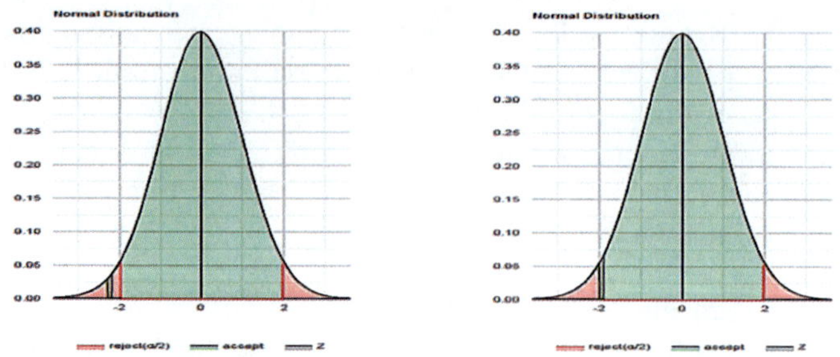

a) Job satisfaction b) Stable employment

Fig. 2. Significance in differences by cultural integration.

4.3 Bayesian Statistics to the Rescue

In this third part, we analyze the distribution of our small data on cultural integration using Bayesian statistics. We ran a Bayesian analysis with two parallel Markov chains of three thousand samples each, in Python using the PyMC3 module [37]. We adhered to the averages and standard deviations of our job satisfaction and stable employment scores in each category of cultural integration to the prior distributions and we obtained the probability distributions for each coefficient and determine if the 90% highest posterior densities (HPD) overlap. Figure 3 shows that the HPDs do not overlap, which is evidence that the average scores of job satisfaction and stable employment were not equal for each category of cultural integration (see our full code for frequentist and Bayesian statistics in the online supplementary material).

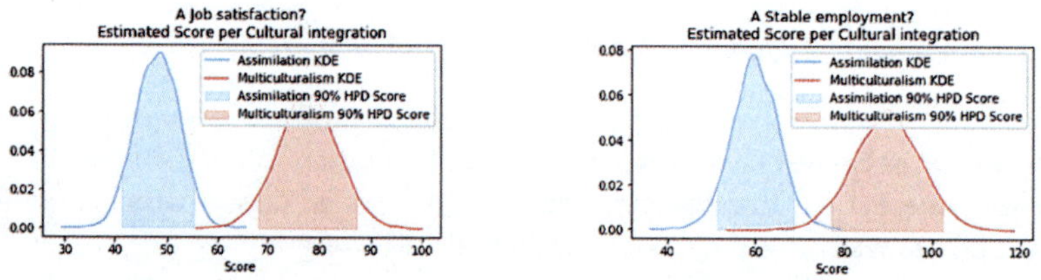

a) Job satisfaction b) Stable employment

Fig. 3. Posterior densities of cultural integration.

Then, we calculated that there is a 99% and 98% probability that job satisfaction and stable employment scores in the assimilation category are lower than in the multiculturalism category, respectively (see our R full code in the online supplementary material).

4.4 Clustering Social Integration

Once we have defined that each category of labor integration and each category of cultural integration are clearly differentiated, we use the K-nearest neighbors algorithm implemented in a supervised Machine Learning environment, in Python, to define whether it is possible to generate a cluster (the total social integration) from them. (See our full code in the online supplementary material).

The result of our algorithmic application was correct (accuracy), at least, in 80% (in 100% of the test data set and in 73% of the training data set), as can be seen in Fig. 4, where the red area modeled represents the possible combinations between job satisfaction and stable employment for each individual who considers himself to be in an "assimilation" situation; while the green area modeled represents such combinations when an individual considers to be in a situation of "multiculturalism". The red and green dots correspond to the combinations of the observed data. Thus, a red dot in the red area means that the prediction of an individual in "assimilation" was correct; a similar interpretation occurs with the green dots of "multiculturalism".

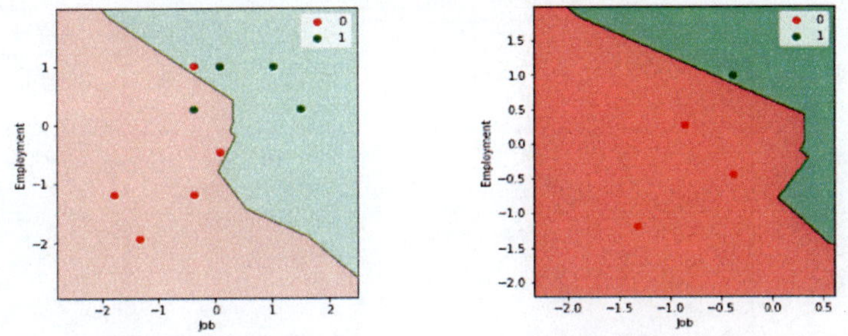

a) KNN training data set.   b) KNN test data set.

Fig. 4. Contour plot applied to data.

In this way, we show that the categories of labor integration and the categories of cultural integration are related. An important specification: in our data, all the observations of "multiculturalism" coincide with the observations of "full", that is to say, that total social integration of the immigrants require that they enjoy high levels of multiculturalism and job security, which includes a high labor market mobility against discrimination and achieve optimal integration of these new population groups [38].

4.5 The Most Frequent Characteristics

Certain that cultural integration and labor integration are closely related in shaping social integration, we proceeded to differentiate the demographic/socioeconomic characteristics that are common to each category (see our online supplementary material). For this, we use contingency tables and correlation matrices.

In this associative analysis we included the categories of "total social integration" which contains those individuals in "full" and "multiculturalism"; "partial social integration" which contains the others; "NOfullNOmulti" which contains those who are not in "full" but in "assimilation"; and "DisqNOmulti" which contains those who are in the worst situation: "disqualifying" and "assimilation". Our full code can be seen in the online supplementary material.

High associations "H" and medium "M", whether positive or negative, can be seen in Table 4 and each one can be interpreted as follows: in general terms, an immigrant located in "DisqNOmulti" is a married woman who lives in a poor neighborhood; while an immigrant located in the category "total social integration" is alone, without children, lives in a non-poor neighborhood and has established a network of friends.

Table 4. Common characteristics of the individuals in each category.

Variable	DisqNOmulti	NOfullNOmulti	Partial social integration	Total social integration
Poor neighborhood not	(−) M	(−) M	(−) H	(+) H
Alone	(−) M	(−) H	(−) H	(+) H
Network of Friends		(−) H	(−) M	(+) H
Children in Ecuador			(+) M	(−) M
Woman	(+) M	(+) H		
Low-income person not		(−) M		

As the category gets better, the communalities increase even if the direction of the variable is opposite (this differentiates them). Likewise, in the less favorable categories, most of the effects are negative; while in "Total" they have become positive (See Table 3).

Two aspects to highlight: 1) the following characteristics were not determining factors to differentiate the integration categories: university degree (bachelor), business knowledge, labor experience (10+ years), destination country, settlement in main cities, living in Ecuador (2+ years), sending remittances, willing not to return to Venezuela, legal documents; 2) the variable network of friends is important for all categories, except for "DisqNOmulti"; however, only under "Total" is it consolidated.

4.6 The Mechanisms of Social Integration

Finally, the qualitative comparative analysis allowed us to combine the common characteristics detected in the previous section and configure the rise of total social integration, driven by a social mechanism.

A mechanism is a causal, frequent, and recognizable pattern that arises under certain conditions and acts, simultaneously or sequentially, as a system of levers and pulleys at certain moments and social spaces. Only its overall effect is visible. It opposes natural laws and helps to understand social phenomena. It explains possibilities but does not accurately predict results. The more mechanisms that have to be explained, the greater the indeterminacy in their predictions [39].

Our estimation was carried out by entering our data from our online supplementary material in the free software fs/QCA 3.1b, [40, 41]. Since the common characteristics of the Total Social Integration category were four, according to Table 4, with binary responses, the possible combinations were $2^4 = 16$; however, that category was only present in 3 of our samples. Table 5 shows the configurations that explain it in a "complex solution":

Equation (1) specifies the total social integration combinatorial model.

$$\text{Total social integration} = \text{f (Network\_of\_Friends, Poor\_neighborhood\_not, Alone, Children\_in\_Ecuador)} \quad (1)$$

Table 5. Configurations for total social integration.

Assumptions	Raw coverage	Unique coverage	Consistency
Alone*~ChildrenInEcuador*PoorNeighborhoodNot*NetworkOfFriends	0.6667	0.6667	0.8
~Alone*ChildrenInEcuador*PoorNeighborhoodNot*NetworkOfFriends	0.1667	0.1667	1.0
Solution coverage: 0.8333			
Solution consistency: 0.8333			

As can be seen, the model covered 83.33% of the sample with only two configurations associated with total social integration:

1. A man or woman alone, without children, who lives in a non-poor neighborhood and who has forged a network of friends; or,
2. A not-alone man or woman, with children, who lives in a non-poor neighborhood and who has forged a network of friends.

Living in a non-poor neighborhood does not imply that the immigrant is rich.

The common characteristics of the Partial Social Integration category were four, according to Table 4, with binary responses. Therefore, the possible combinations were $2^4 = 16$; however, that category was present in 7 of our samples. Table 6 shows the configurations that explain it in a "complex solution":

Equation (2) specifies the partial social integration combinatorial model.

$$\text{Partial social integration} = \text{f(Network\_of\_Friends, Poor\_neighborhood\_not, Alone, Children\_in\_Ecuador)} \quad (2)$$

As can be seen, the model covered 78% of the sample with only two configurations associated with partial social integration:

1. A not-alone man or woman, with children, who has not forged a network of friends; or,

Table 6. Configurations for partial social integration.

Assumptions	Raw coverage	Unique coverage	Consistency
~Alone*~PoorNeighborhoodNot	0.5556	0.4444	1.0
~Alone*ChildrenInEcuador*~NetworkOfFriends	0.3333	0.2222	1.0
Solution coverage: 0.7778			
Solution consistency: 1			

2. A not-alone man or woman, who lives in a poor neighborhood.

The Raw Coverage, Table 5 y Table 6, shows the percentage of cases contained in each configuration, but that may be contained in other resulting configurations in the same category. The Unique Coverage is lower, as it shows the percentage of cases explained only by the respective configuration. In the next column, Consistency (of the configuration) shows the percentage of cases that meeting a certain configuration result in the analyzed category. The Solution Coverage shows us that the component variables (characteristics) of our configurations are not the only ones that lead us to a certain category, but only to a proportion of the total. Thus, for example, the analysis of the first configuration of Table 6 tells us that 56% of the cases that lead to this category are contained in it or other configurations, but that 44% only contains them, and that 100% of the observations of this configuration led to partial social integration.

From the configurations before which total social integration arises, it can be deduced that the mechanism that generates it faster is dominated by the network of friends, since -according to the data- as long as the immigrant has not forged it, it remains in the category of partial social integration, which coincides with job categories where wages are low. The absence of this network affects women more intensely.

5 Conclusions and Recommendations

Our most important conclusion is theoretical since we show that cultural integration is related to labor integration, and that by combining the two we can have a more concrete idea about the degree of social integration of the immigrant (see Table 2). Both components are also simple in their operationalization and contain an approximation between the immaterial and material aspects of the individual, so we propose them as an indicator. We consider that this was reflected in the Venezuelan-Ecuadorian case since six of the fifteen Venezuelan immigrants are in a situation of "full" labor integration (see Table 3). This 40% was very close to the 36% of full employment of the Ecuadorian population [42] at the time of our information gathering.

The mechanism that drives the arrival to the category of total social integration is the network of friends. In other words, immigrants must enlarge their social networks in the territory [43] in all possible ways, to reduce their high risk of exclusion. This can be alleviated, in solidarity, with the necessary commitment of the host government [44], in: a) permanent flexicurity in their hiring; b) the permitting of family reunification and remittances; and, c) the issuance of public policies for participation in the affairs of the communities where they live.

This network allowed them not only to get a job but to carry out activities of greater value addition and with greater work roots since almost all of the Venezuelans interviewed have a university degree. However, it is not automatic; it requires time to reinforce the lines of trust and a permanent re-calculation of personal goals in the short and long term.

The use of the combination of cultural-labor integration can be used by the government to establish controls that prevent the naturalization of precarious jobs that immigrants accept to have financial support or marginal strategy to get revenues [45].

As a consequence of the rapid total social integration (it does not imply high salaries) of immigrants, we can intuitively expect a reduction in the national budget for social assistance costs, which would reduce the pressure of nationalist groups and appease outbreaks of xenophobia. This is most urgent in times of national emergency, such as the COVID-19 pandemic in countries with high levels of poverty, such as Ecuador.

We consider that our approach to the problem opens up an interesting line of research that links and complements perspectives on the differentiation factors of Venezuelan migration to other countries of the Andean area; the differentiation factors with other immigrant groups in Ecuadorian territory, such as the Colombian inmigrants; the time of successful work adaptation of Venezuelan immigrants; the perception of inequality and the social mobility of Venezuelans in foreign territory.

We think the most important limitation in our research is the sample size; but it is also the reason for the use of Bayesian statistics and non-parametric methods. It is the reason why we combine quantitative and qualitative methods.

Acknowledgments. We are grateful to Mariana Busso and Byrone Tomalá for their helpful amendments to our theoretical framework of labor and culture, respectively. Any remaining errors are ours. We thank Aleyda Lastre and Jéssica Toledo for their assistance in collecting and organizing the data.

Funding. This work was supported by the ESPOL Research Dean's Office (FCSH-15-2021).

Conflict of Interest Statement. None declared.

Supplementary Data. Supplementary data is available with this article.

References

1. IPU Inter-Parlamentary Union: Migración, derechos humanos y gobernanza (Migration, human rights and governance). In: Manual Para Parlamentarios (Handbook for parliamentarians), vol. 24 (2015)
2. El Comercio: Ciudadanos venezolanos tienen hasta fin de marzo de 2020 para pedir la visa humanitaria en Ecuador (Venezuelan immigrants may apply for humanitarian visa in Ecuador until the end of March 2020). Quito (2020)

3. Carreño, A.: Situación de refugiados colombianos en Ecuador (Colombian refugees in Ecuador). Justicia (Justice) **17**(22), 20–46 (2012). https://doi.org/10.17081/just.17.22.706
4. UN United Nations: Análisis estadístico de las fichas de atención a personas con necesidades de protección internacional en Guayaquil de 2011 a 2014 (Statistical analysis of persons in need of international protection assisted in Guayaquil from 2011 to 2014). Guayaquil (2014)
5. Expreso: El déficit fiscal cerró el año en $ 4043 millones (At the end of the year, the fiscal deficit came to $ 4,043 million dollars). Guayaquil (2020)
6. MacPherson, S.: Ethno-cultural diversity education in Canada, the USA and India: the experience of the Tibetan diaspora. Compare: A J. Comp. Int. Educ. **48**(6), 844–860 (2017). https://doi.org/10.1080/03057925.2017.1362547
7. Menéndez, C., García, S.: De la invención de lo social a la ciudad asediada, entrevista a Jacques Donzelot (From social behavior to the city under siege: an interview with Jacques Donzelot). Cuadernos de Trabajo Social (Soc. Work Lett.) **30**(2), 273–284 (2017). https://doi.org/10.5209/CUTS.55353
8. Castel, R.L.: Metamorfosis de la cuestión social, una crónica del salariado (The metamorphosis of the social question, a chronicle of salaried work). Paidós, Barcelona (1995)
9. Neffa, J., Oliveri, M., Persia, J., Trucco, P.: Empleo, desempleo y políticas de empleo (Employment, unemployment and employment policies). CONICET, Buenos Aires (2010)
10. Joppke, C., Morowska, E.: Integrating immigrants in liberal nation-states: policies and practices. In: Joppke, C., Morowska, E. (eds.) Toward Assimilation and Citizenship, Immigrants in Liberal Nation-States, pp. 1–36. Palgrave Maximilian, New York (2003)
11. Collier, P.: Exodus, immigration and multiculturalism in the 21st Century. Allen Lane, London (2013)
12. Osterloh, M., Frey, B.: Cooperatives instead of migration partnerships. Anal. Krit. **40**(2), 201–225 (2018). https://doi.org/10.1515/auk-2018-0010
13. Joshee, R.: Citizenship and multicultural education in Canada: from assimilation to social cohesion. In: Bank, J. (ed.) Diversity and Citizenship Education: Global Perspectives, pp. 127–156. Jossey-Bass, San Francisco (2004)
14. Bouchard, G., Taylor, C.: Building the Future: A Time for Reconciliation. Government of Quebec (2008)
15. Castles, S.: Ethnicity and globalization. In: Torres L., Integration of migrants in Chile. Multicultural assimilation and rhetoric, Migraciones Internacionales **10**(36), (2000). https://doi.org/10.33679/rmi.v1i1.2068
16. Tremblay, L.: The Bouchard-Taylor report on cultural and religious accommodation: multiculturalism by any other name? EUI Working Papers Law (18) (2009)
17. Magsino, R.: The Canadian multicultural policy: a pluralist ideal revisited. Canadian Int. Educ. **28**(2), 33–51 (1999)
18. Verkuyten, M.: Multicultural recognition and ethnic minority rights: a social identity perspective. Eur. Rev. Soc. Psychol. **17**(1), 148–184 (2006). https://doi.org/10.1080/10463280600937418
19. Corak, M.: Income inequality, equality of opportunity, and intergenerational mobility. J. Econ. Perspect. **27**(3), 79–102 (2013). https://doi.org/10.1257/jep.27.3.79
20. MacPherson, S., Buthi, D.: Multilingual practices in emerging diasporas: a Tibetan case study. Diaspora, Indigenous Minor. Educ. **2**(3), 188–216 (2008). https://doi.org/10.1080/15595690802145414
21. Neffa, J., Panigo, D., Pérez, P.: Actividad, empleo y desempleo, conceptos y definiciones (Job, employment and unemployment, concepts and definitions). CONICET, Buenos Aires (2000)
22. Longo, J.: Las fronteras de la precariedad: percepciones y sentidos del trabajo de los jóvenes trabajadores precarios de hipermercados (The borders of precariousness: perceptions and

meanings of work for young workers precarious of hypermarkets). Trabajo y Sociedad (Labor Soc.) **19**, 375–392 (2012)
23. Battistini, O.: La precariedad como referencial identitario. Un estudio sobre la realidad del trabajo en la Argentina actual (Precariousness as an identity referential. A study on work in Argentina today). Psicoperspectivas (Psychoperspectives) **8**(2), 120–142 (2009). https://doi.org/10.5027/psicoperspectivas-Vol8-Issue2-fulltext-65
24. Diana, N.: La múltiple dimensión de la precariedad laboral: el caso de la administración pública en Argentina (The multiple dimension of labor precariousness: the case of the public administration in Argentina). Revista de Ciencias Sociales (J. Soc. Sci.) **II–III**(128–129), 119–136 (2010)
25. Paugam, S.: Formes d'intégration professionnelle et attitudes syndicales et politiques (Different types of professional integration and the attitude from unions and politicians). Revue Française de Sociologie (French J. Sociol.) **40**(4), 715–751 (1999). https://doi.org/10.2307/3322856
26. Zanzzi, F.: Riesgo laboral ¿abuso o descuido? evidencias desde la ausencia de control estatal en el diagnóstico médico complementado con Rayos X (Occupational risk: abuse or negligence? Evidence from the absence of government control in medical diagnosis complemented with X-rays). CienciaUnemi **5**(8), 9–20 (2012). https://doi.org/10.29076/issn.2528-7737vol5iss8.2012pp9-20p
27. Bardomás, S., Blanco, M.: Condiciones laborales, riesgo y salud de los trabajadores forestales de Misiones, Corrientes y Entre Ríos (Argentina), 2010–2014 (Labor conditions, risk and health of forestry workers in Misiones, Corrientes and Entre Ríos (Argentina), 2010–2014). Salud Colectiva (Collective Health) **14**(4), 695–711 (2018). https://doi.org/10.18294/sc.2018.1564
28. Amat, J.: Test Kruskal-Wallis? Ciencia de datos (Data science). https://www.cienciadedatos.net/documentos/20_Kruskal-Wallis_test.html
29. León, W.: Métodos no paramétricos, parte II (Nonparametric methods, part II). Universidad Nacional Mayor de San Marcos, Lima (2014)
30. Van de Schoot, R., Kaplan, D., Denissen, J., Asendorpf, J., Neyer, F., Van Aken, M.: A gentle introduction to Bayesian analysis: applications to developmental research. Child Dev. **85**(3), 842–860 (2014). https://doi.org/10.1111/cdev.12169
31. Lora, P.: What about small data? A Bayesian ANOVA solution. Towards Data Science. https://towardsdatascience.com/what-about-small-data-e8993e06eaca (2019)
32. Aler Mur, R.: KNN vecinos más cercanos (K nearest neighbors). Universidad Carlos III, Madrid (2014)
33. Bock, T.: What is a correlation matrix? Display R blog. https://www.displayr.com/what-is-a-correlation-matrix/
34. Ragin, C.L.: construcción de la investigación social, introducción a los métodos y su diversidad (The construction of social research: methods and diversity). SAGE Publications, Bogotá (2007)
35. Statistics Kingdom: Kruskal-Wallis test. https://www.statskingdom.com/kruskal-wallis-calculator.html (2017)
36. Statistics Kingdom: Mann Whitney U test calculator (Wilcoxon Rank-Sum). https://www.statskingdom.com/170median_mann_whitney.html (2017)
37. Salvatier, J., Wiecki, T., Fonnesbeck, C.: Probabilistic programming in Python using PyMC3. PeerJ. Comput. Sci. **2**, e55 (2016). https://doi.org/10.7717/peerj-cs.55
38. De Freitas, D., et al.: Psychological correlates of perceived ethnic discrimination in Europe: a meta-analysis. Psychol. Violence **8**(6), 712–725 (2018). https://doi.org/10.1037/vio0000215
39. Elster, J.: Egonomics. Gedisa S.A., Barcelona (1997)
40. Ragin, C., Davey, S. Fuzzy-set/qualitative comparative analysis 3.0. University of California, Irvine (2016)

41. Ragin, C.: User's Guide to Fuzzy-set/Qualitative Comparative Analysis 3.0. University of California, Irvine (2018)
42. INEC Instituto Nacional de Estadísticas y Censos del Ecuador (National Institute of Statistics and Census of Ecuador): Empleo septiembre 2020 (Employment September 2020). Quito (2020)
43. Kim, A.: Social exclusion of multicultural families in Korea. Soc. Sci. **7**(63), 1–16 (2018). https://doi.org/10.3390/socsci7040063
44. Room, G.: Social exclusion, solidarity and the challenge of globalization. Int. J. Soc. Welf. **8**(3), 166–174 (1999). https://doi.org/10.1111/1468-2397.00080
45. Busso, M., Bouffartigue, P.: ¿Más allá de la «precariedad» y la «informalidad»? Aportes para el debate desde una perspectiva comparada (Beyond "precariousness" and "informality"? Contributions to the debate from a comparative perspective). In: Del Bono, A., Quaranta, G. (eds.) Convivir con la incertidumbre: Aproximaciones a la flexibilización y precarización del trabajo en la Argentina (Living with uncertainty: approaches to labor flexibilization and precariousness in Argentina). CEIL-PIETTE, CONICET, Buenos Aires (2010)

Study Case: LPWAN Propagation Power Estimation in Outdoor Scenarios, Based on the Okumura-Hata Model

Alejandra Pinto-Erazo(✉), Luis Suárez-Zambrano, Fabián Cuzme-Rodríguez, and Edgar Jaramillo-Vinueza

Universidad Técnica del Norte, Ibarra, Ecuador
{ampinto,lesuarez,fgcuzme,edjaramillo}@utn.edu.ec

Abstract. Three case studies are proposed to analyze the power of Low Power Wide Area Networks (LPWAN) technologies (LoRa, NB-IoT, Sigfox) in outdoor environments in lbarra city, specifically in "El Olivo" Campus at "Universidad Técnica del Norte". To develop the project, the Okumura-Hata loss model is used in urban, suburban, and rural scenarios through the development and simulation of a graphical interface made in MATLAB software. The developed interface has working sections, such as: input parameters, data analysis and graphical results of power versus distance, and finally, the best LPWAN technology option is evaluated according to the proposed scenario.

Keywords: LPWAN · LoRa · Sigfox · Narrowband · NB-IoT · loT

1 Introduction

Currently, IoT radio technologies and their devices, known as LPWAN[1], are booming because there is a demand for applications to be developed without the need for large deployment of its infrastructure, being scalable according to the need; in this sense LPWAN technologies are presented as an economical and effective solution in the implementation and maintenance of long distance wireless links with sensors that work mainly with batteries which must be of long duration[1].

There are many LPWAN networks, however this study focused mainly on three technologies, such as LoRa, NB-IoT[2] and Sigfox, considered by the market as the most deployed networks for application development [1].

To conduct the power analysis, the initial metric is given by the MCL[3] which is the maximum coupling loss value [2]. In the case of LoRa, the allowed MCL is 154 dB [3], which represents the theoretical loss limit of this technology. For Sigfox the MCL value is around 158 and for NB-IoT 164 [4]. In the first stage of this project, which evaluated indoor and outdoor propagation models; It was determined that, for outdoor,

[1] LPWAN Low Power Wide Area Network.

[2] NB-IoT NarrowBand-Internet of things.

[3] MCL Maximum Coupling Loss.

V. Robles-Bykbaev et al. (Eds.): CITIS 2022, LNNS 607, pp. 113–123, 2023.
https://doi.org/10.1007/978-3-031-24327-1_10

where there are interferences rather than obstacles, the application of the Okumura Hata model is the best option. Also, this model contemplates the range of distances of the proposed scenarios, it is a model that allows evaluating the coverage and propagation loss in different types of areas, being these: urban, suburban, and rural areas through a set of equations proposed by Masaharu Hata from the measurements of Yoshihisa Okumura [5].

When working with this type of technology, the frequency of operation varies depending on the region in which you work; for Ecuador, where the cases to be analyzed are located, the manufacturer's table shows the value of 902 MHz to 920 MHz for LoRa and Sigfox, which are within the frequency values of the ISM band, which is oriented to research scientific and medical. The manufacturer suggests using the 920 MHz frequency for Equipment configuration and the use of the spectrum does not represent any additional cost; on the contrary, in the case of NB-IoT, it works mainly in the 2G bands, and the operating frequency is around 800 MHz [6], representing an additional cost or based on the network infrastructure.

The document is structured as follows: first, the main considerations of the Okumura-Hata model are described, which serve as the basis for the implementation of the graphical interface developed in MATLAB software, in addition, the accepted metric ranges are configured, and the user can enter the initial data of the proposed scenarios. The following section presents the results according to the equations developed by Okumura-Hata and the corresponding graphs. Subsequently, the results obtained are analyzed according to the proposed scenarios and the best alternative and technology were evaluated. Finally, the conclusions and recommendations of the work developed are generated.

2 Methods

2.1 Modelo Okumura-Hata

The mathematical model for power calculation in urban scenarios is given by the following equation described by [7] and [5]:

$$\mathrm{Lur} = 69.55 + 26.16\log(\mathrm{f}) - 13.82\log \mathrm{h\_1} - \mathrm{a(h\_2)} + (44.9 - 6.55\log \mathrm{h\_1})\log(\mathrm{d}) \tag{1}$$

where:

f = frequency (MHz)
h_1 = transmission antenna height (m)
h_2 = receiving antenna height (m)
d = distance (Km)

h_2 is a correction factor that depends on the size of the city. For this case medium-small and is defined by:

$$\mathrm{a(h\_2)} = (1.1\log \mathrm{f} - 0.7)\mathrm{h\_2} - 1.56\log \mathrm{f} + 0.8 \tag{2}$$

Also, for suburban scenarios this is given by:

$$\mathrm{Lsu} = \mathrm{Lur} - 2[\log\{f\}/\{28\}]^2 \tag{3}$$

and for rural scenarios:

$$\mathrm{Lru} = \mathrm{Lur} - 4.78(\log f)^2 + 18.33 \log f - 40.94 \tag{4}$$

The Okumura-Hata model establishes the necessary conditions for its use, among them are the heights of the antennas, both transmitting and receiving, as well as the range of frequencies under which the model works, for an already established distance range.

f = Between 150 MHz and 1500 MHz.
h1 = Transmission antenna height 30 m–200 m.
h2 = Receiving antenna height 1 m–10 m.
d = distance of 1 km–20 km.

3 Proposed Scenarios

Once the characteristics of the model to be used were defined, the scenarios for the corresponding analysis were described:

3.1 Urban Scenario

The "El Olivo" Campus of the Universidad Técnica del Norte has buildings of 5 floors for each school. The urban scenario analyzed is identified between two of the buildings, where the offices are located, as follows: point 1, considered the fifth floor of the Engineering in Applied Sciences School (FICA) and point 2, considered the second floor of the Central Building, located approximately 0.13 km between both points without line of sight, as shown in Fig. 1.

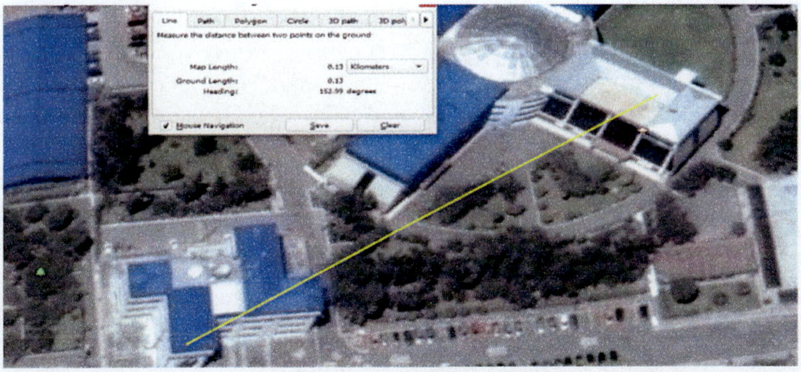

Fig. 1. Location of FICA Building and Central Building, in "El Olivo" Campus of "Universidad Técnica del Norte"

In the urban scenario, the height of the transmitting antenna is 30 m above ground level, while the height of the receiving antenna is 5 m above ground level; both antennas are located outside the corresponding building. It should be noted that the values are approximations and are subject to variations.

3.2 Suburban Scenario

For the suburban scenario, a link is proposed between two campuses of the University, point 1 is the "Antiguo Hospital San Vicente de Paúl" Campus and point 2 is the "Colegio Universitario" Campus, located on Ulpiano de la Torre street, with approximately 1.86 km between these points. At point 1, the height of the transmitting antenna with respect to the ground is 30 m and at point 2, the height of the receiving antenna is 4 m. The link distance is shown in Fig. 2.

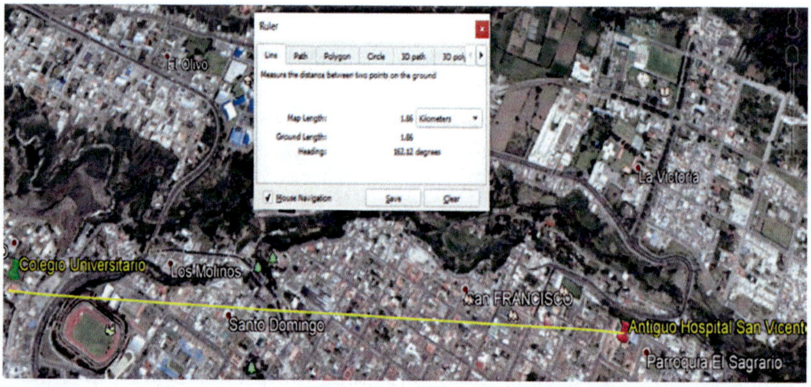

Fig. 2. Distance between "Old Hospital San Vicente de Paúl" Campus and "Colegio Universitario" Campus, in the City of Ibarra.

3.3 Rural Scenario

The rural scenario is given by point 1 is the "La Pradera" Farm located on the road to the Ranch around "San José de Chaltura", and point 2 is a house located in the same area of San José de Chaltura, where the distance between the points is about 1.02 km. The mentioned places are far away and are adapted as a rural area. It is proposed that the location of the transmitter be at the farm since from that point information can be obtained through sensors such as humidity, temperature, pressure, etc., of the crops grown in that area, while the data is sent through the proposed link and generate results without having to be on site. For this purpose, the height of the transmission antenna at point 1 is 30 m and at point 2, the antenna height is 2 m, both distances are with respect to the ground. It is shown in Fig. 3.

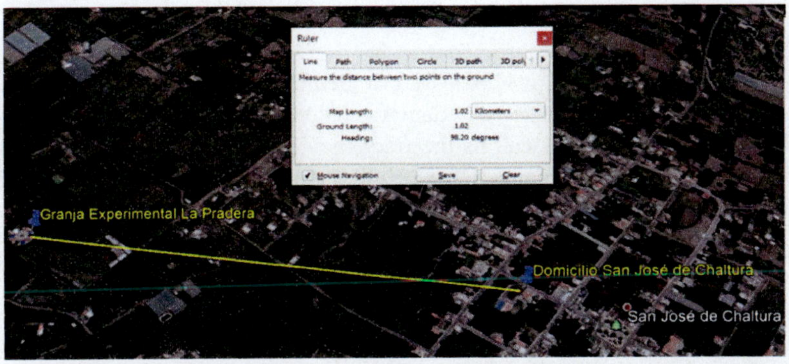

Fig. 3. Distance between La Pradera farm and the suggested address.

A summary of the initial parameters according to the urban, suburban, and rural scenarios is presented in Table 1 and Table 2 below.

Table 1. Reference of outdoor scenarios

Link	
Urban	FICA Building – Central Building
Suburban	San Vicente Campus – Colegio Universitario Campus
Rural	La Pradera farm – suggested address

Table 2. Outdoor scenario data

Link	Distance	Tx height	Rx height
Urban	0.13 km	30 m	5 m
Suburban	1.86 km	30 m	4 m
Rural	1.02 km	30 m	2 m

3.4 Software Used

To represent the links, a graphic interface was designed based on the parameters requested by the Okumura-Hata model as MCL of each study technology and those described in the urban, suburban, and rural scenarios.

The MATLAB software with its App Designer tool, allowed to generate in an uncomplicated way the data entry, development of formulas and obtaining graphs of results. It is a simple interface, but it addresses all the parameters of the evaluated model and was coupled to the scenario requirements.

4 Results and Discussion

Prior to obtaining the results, it is essential to identify the content of the interface, as shown in Fig. 4, which is finalized and data was entered for the three technologies such as LoRa, NB-IoT, Sigfox; as well as the frequency range and the scenario for which the coverage graph is to be obtained. The result obtained provided the propagation loss according to the height and distance in the link.

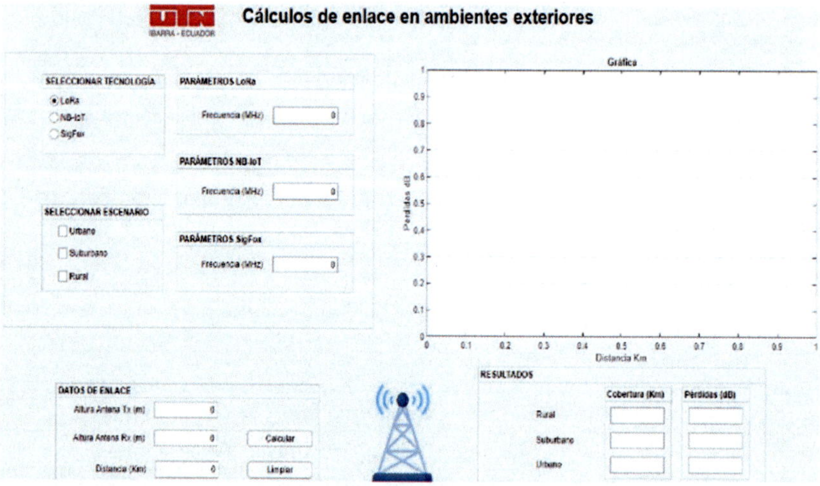

Fig. 4. Interface – Outdoor propagation losses.

In the graph, the reference is given by blue color for urban scenario, orange color for suburban scenario and yellow color for rural scenario. In addition, the red dashed line represents the MCL allowed for a given technology.

4.1 LoRa Technology

The first choice was LoRa. A check was performed in the corresponding scenario, whether urban, suburban, or rural, with the proposed data. The frequency considered was 920 MHz. The results presented in Fig. 5 and Table 3 were obtained.

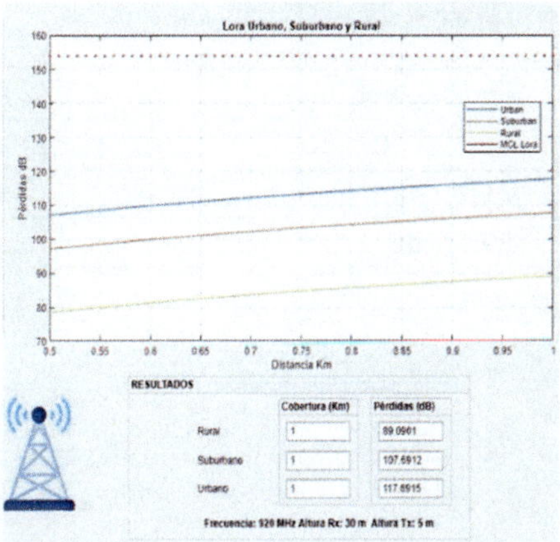

Fig. 5. Coverage vs. propagation loss with LoRa.

Figure 5 presents the resulting graph, which shows that the scenario that best fits this technology is the rural scenario with low propagation loss; although there is coverage for all three scenarios, there is a range of attenuation to be considered.

Table 3. Results LoRa technology

Scenario	Propagation loss	Coverage
Urban	117.69 dB	0.13 km
Suburban	107.67 dB	0.13 km
Rural	89.09 dB	0.13 km

As shown in the table, an urban scenario evaluated with LoRa technology yields a higher value of propagation losses, mainly due to the presence of buildings which attenuate the signal and hinder the connection as in [3] where in the most remote areas the connection was made after several attempts.

4.2 NB-IoT Technology

The next choice was NB-IoT. A check was performed in the corresponding scenario, whether urban, suburban, or rural, with the proposed data. The frequency entered, within the allowed range, was 800 MHz. The results presented in Table 4 were obtained.

Table 4. NB-IoT technology results

Scenario	Propagation loss	Coverage
Urban	125.83 dB	1.86 km
Suburban	116.19 dB	1.86 km
Rural	97.82 dB	1.86 km

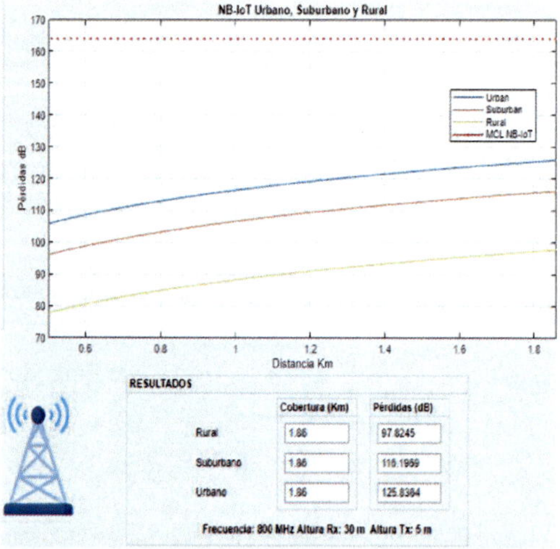

Fig. 6. Coverage vs. propagation loss with NB-IoT

Again, the result is shown; with NB-IoT technology, the urban scenario presented the highest propagation loss with 125.83 dB, while the rural scenario was the best option for deploying NB-IoT technology, since there were no obstacles. Figure 6 shows the results obtained with NB-IoT, where the technology provides coverage in the three scenarios, and is within the MCL margin.

4.3 Sigfox Technology

The last selection was Sigfox. A check was made in the corresponding scenario, whether urban, suburban, or rural, with the proposed data. The frequency entered, within the allowed range, was 920 MHz. The results presented in Table 5 were obtained.

Table 5. Sigfox technology results

Scenario	Propagation loss	Coverage
Urban	131.70 dB	2.5 km
Suburban	121.70 dB	2.5 km
Rural	103.10 dB	2.5 km

Figure 7 shows with Sigfox technology, the propagation loss in the urban scenario was obtained as the highest, with a value of 131.70 dB; however, the technology covers all three outdoors.

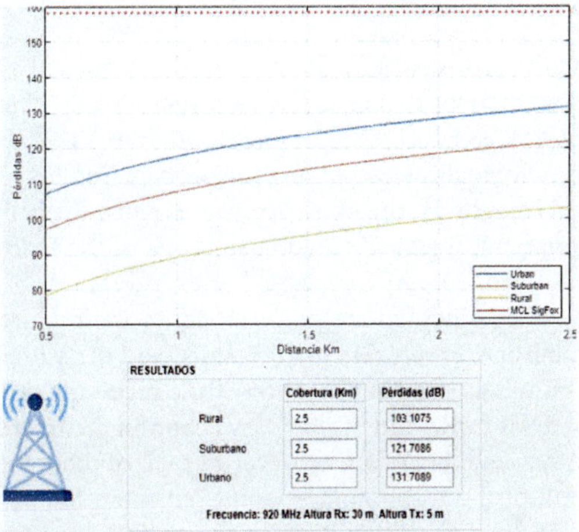

Fig. 7. Coverage vs. propagation loss with Sigfox

Once the simulations were developed with the parameters according to the university campuses, urban, suburban, and rural, the summary of losses per propagation was presented in Table 6, to draw conclusions and study the feasibility of implementing these technologies in the scenarios proposed, as well as the possibility of generating extreme environments that can be covered by certain technology.

Table 6. Summary of technologies vs scenarios

Scenario	LoRa	NB-IoT	Sigfox
Urban	117.69 dB	125.83 dB	131.70 dB
Suburban	107.67 dB	116.19 dB	121.70 dB
Rural	89.09 dB	97.82 dB	103.10 dB

The propagation loss values for each technology are shown; in the urban scenario, both LoRa, NB-IoT and Sigfox present a higher power loss value than other scenarios; and that its growth is directly proportional to the operating frequency. On the contrary, the proposed rural scenario, developed with the three technologies, provides a much lower attenuation than expected, considering that there are no obstacles.

In general, LoRa technology shows lower attenuation than the other two technologies analyzed, regardless of the scenario applied. In addition, the applied model fully covers according to the input data.

The greater the distance between the transmitting and receiving antennas, the higher the attenuation level of the link, at one point exceeding the MCL and not being an adequate design.

5 Conclusions

It is essential to mention that this paper is the result of a second phase of research. In the first phase, the characteristics such as frequency, coverage, obstacles of low power wide area network (LPWAN) technologies were studied; as well as the propagation model configurations, which allow estimating the propagation loss in several scenarios. Through the presented case study, the objective was to show the propagation estimation in an academic environment, for the different campuses of the Universidad Técnica del Norte.

The propagation losses obtained are proportional to the required coverage, as was the case for urban, suburban, and rural environments; and the values obtained from the graphical interface assume the outdoor location of the receiving devices.

The operating bands of LoRa and Sigfox vary according to the region and continent, whether European, American, or Asian, however in each of them there is an established range for the operation of the technologies, and even more the configuration value in the equipment for its optimal operation. In the case of LoRa and Sigfox the configured value was 920 MHz, while for NB-IoT it was configured at 800 MHz.

MCL values are used as a reference to find the maximum possible coverage within an environment given the parameters of antenna height, operating frequency, and link distance.

Narrowband (NB-IoT) working through a licensed frequency band is subject to less interference, specific to the environment, to which technologies such as LoRa and Sigfox are exposed.

The results obtained show that the highest number of losses occur in urban environments due to the presence of buildings that attenuate the signals as in the case of study 1, where line of sight is minimal, while the environments with the lowest losses are rural scenarios due to the low presence of obstructions.

6 Recommendations

It is advisable to work with distances greater than 1 km to take advantage of the capacity of the technologies which can cover much greater distances.

Review the regulations on the use of the radioelectric spectrum as there may be certain clauses regarding the work cycles of any technology.

In the case of NB-IoT, it is possible to work with frequencies up to 2600 MHz; however, due to the limitation of the Okumura-Hata model, it is advisable to work with frequencies within the established range, such as 800 MHz, which is also close to the range available for Lora and Sigfox.

When calculating the link budget based on the developed interface, it is necessary to delimit a small margin of error because the results are only estimates of reality.

References

1. Bou, M.B.: Feasibility study and actual deployment of an IoT network (2020)
2. Techplayon, Maximum CouplingLoss (MCL) and Maximum Path Loss (MPL)-Techplayon, http://www.techplayon.com/maximum-coupling-loss-mcl-and-maximum-path-loss-mpl/ (2018)
3. Andrés, M., Quimbita, M., Carlos, M., Salvador, E.P.: LoRa/LoRaWAN gateway evaluation in urban environments. Politécnica de Valencia University, Valencia, Tech. Rep. https://riunet.upv.es/bitstream/handle/10251/109791/Moya-Evaluaci%C3%B3n depasare-laLoRa/LoRaWANenentornosurbanos.pdf/?sequence=1&isAllowed=y (2018)
4. Vejlgaard, B., Nguyen, H., Kovacs, I.Z., Mogensen, P., Sorensen, M., Lauridsen, M.: Coverage and Capacity Analysis of Sigfox, LoRa, GPRS, and NB-IoT. In: IEEE Vehicular Technology Conference, vol. 2017, pp. 1–5. Institute of Electrical and Electronics Engineers Inc.. http://ieeexplore.ieee.org/document/8108666 (2017)
5. Pinto Mangones, A., Torres, J., García Bello, A., Pérez, N., Uzcátegui, J.: Model for propagation loss estimation in digital television broadcasting system. Ingeniería Electrónica, Automática y Comunicaciones **37**(2), 67–81 (2016)
6. Ballart, D.A.J.: Study of lora modulation coverage in the 915 [mhz] band in an urban environment. Director **1**(2), 2–3 (2018). https://www.researchgate.net/profile/JuanAparicio7/publication/253571379Losestudiossobreelcambioconceptual
7. P.529: Prediction methods for the terrestrial land mobile service in the VHF and UHF bands (2022). https://www.itu.int/rec/R-REC-P.529/en

Industry

Nonlinear Identification and Position Control of a Pneumatic System

Víctor Huilcapi[1(✉)], Ricardo Cajo[2], Jorge Orellana[1], and Alex Cascante[1]

[1] Universidad Politécnica Salesiana, Guayaquil, Ecuador
vhuilcapi@ups.edu.ec

[2] Facultad de Ingeniería en Electricidad y Computación, Escuela Superior Politécnica del Litoral, ESPOL, Campus Gustavo Galindo Km 30.5 Vía Perimetral, P.O.Box 09-01-5863, Guayaquil, Ecuador
rcajo@espol.edu.ec

Abstract. In this work a pneumatic positioning system is studied. The system presents high nonlinearities since it is subjected to static friction, dead bands, and dead times. Firstly, the system identification is done using nonlinear identification based on the Hammerstein-Wiener model. Then, some control strategies such as Proportional-Integral-Derivative (PID) controller, Fuzzy-PID controller and Model Predictive Control (MPC) are designed to handle these nonlinearities. The main goal is to control the displacement of the pneumatic cylinder to reach any position or trajectory tracking in the shortest possible time and with the least steady state error. The results show that the Hammerstein-Wiener model identified for the system satisfactorily characterizes its nonlinear dynamics. The MPC is more efficient to control the system compared to the other controllers as it has less steady state error and stabilizes the system faster.

Keywords: Pneumatic system · PID · Fuzzy-PID · Nonlinear Identification · Model Predictive Control (MPC)

1 Introduction

Nowadays, pneumatic actuators are widely used in the industry in many processes [5–7]. The main reasons are because pneumatic positioning systems feature some advantages, such as low cost of purchase, environmentally clean, mechanical simplicity, durability, and robustness [3, 20]. These features can make them favorable for applications of servomechanisms [1, 13]. Precise control of the force and position of actuators is complex since air compression is a highly nonlinear flow that circulates through pneumatic system components and causes significant friction [15, 17]. Although servo pneumatic systems have been extensively studied in the control literature [23]. The performance achieved by controlling them can be further improved, for example in the case of continuous motion or trajectory control. As a result, there are many research based on mathematical modeling of pneumatic systems such as [12, 15, 22] and parametric identification based on models [2, 11]. Many control strategies have been studied to improve the efficiency of pneumatic

V. Robles-Bykbaev et al. (Eds.): CITIS 2022, LNNS 607, pp. 127–138, 2023.
https://doi.org/10.1007/978-3-031-24327-1_11

positioning systems like the one analyzed in this paper. For example, the Proportional-Integral-Derivative (PID) control structures are the most used to control these systems, due to their ease of tuning and maintenance [4, 16]. Although, the use of this control strategy has several disadvantages, since it will not always work well and will have poor performance when is applied to highly complex, nonlinear, constrained systems [10]. On the other hand, the use of advance control strategies instead of PID such as sliding mode control [19, 21, 24], adaptive control [17] among others [9, 14, 18] have shown a better performance dealing with these nonlinearities present in the system. The dynamics of a pneumatic positioning system presents high nonlinearities, therefore its control is complex. In this research, to try to solve the problem, three scenarios with different types of controllers are proposed (PID, PID-Fuzzy, and MPC). Likewise, to obtain a model of the system that captures its nonlinearities, the identification is performed based on a Hammerstein-Wiener model. This is summarized in the conceptual graph shown in Fig. 1. It is important to indicate that all controllers have been simulated and implemented in real time using a PCI-1711 data acquisition card with MATLAB/Simulink. The remainder of this research is structured in five sections. Where, Sect. 2 describes the pneumatic system. Section 3 describes the methodology and methods applied to identify a nonlinear model (Hammerstein-Wiener model) and tune three different types of controllers (PID, Fuzzy-PID, and MPC) for the system. Section 4 corresponds to the analysis of results, where the validation of the nonlinear model of the system and of each proposed controller are shown. Finally, the conclusions are shown in Sect. 5.

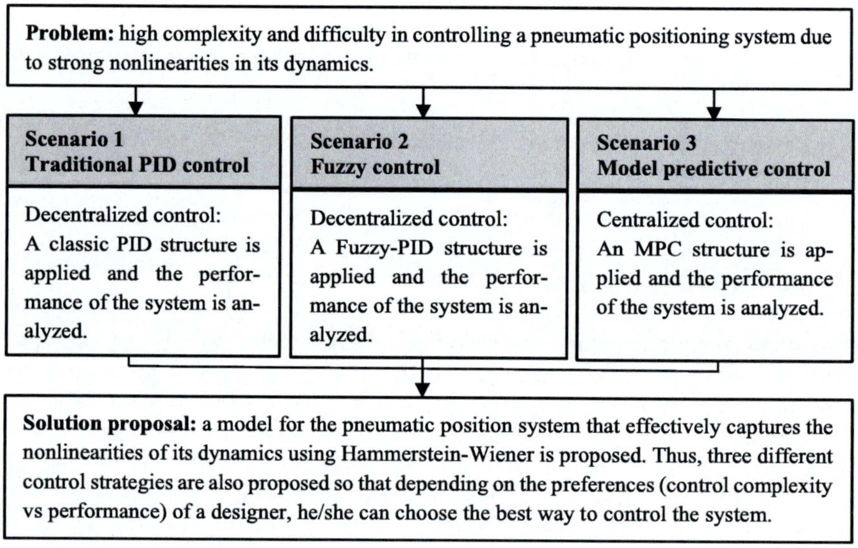

Fig. 1. Conceptual graphic of the research problem and the proposed solution.

2 System Description

The Pneumatic System is a rodless linear actuator with guide manufactured by FESTO Company as shown in Fig. 2. The main function of this system is to control the air pressure to stabilize the movement of a pneumatic cylinder. Both the transmission and the dissipation of forces are involved in this process. The servomechanism system is composed of a 5/3-way proportional valve, a linear potentiometer, and a pneumatic cylinder (double acting) with two air intakes to produce the output displacements in both directions. For the experimentation, the supply pressure was regulated to 0.6 MPa and the system has been incorporated a computer with a PCI 1711 acquisition card, MATLAB was used to monitor and control the system.

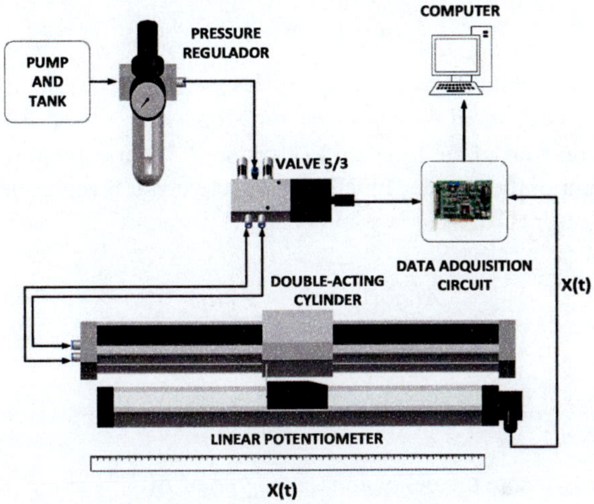

Fig. 2. Position control for the pneumatic system.

3 Methodology and Methods

This section shows the methodology and methods applied to identify a nonlinear model and tune the controllers for the pneumatic positioning system analyzed in this paper.

3.1 Hammerstein-Wiener Modelling

A Hammerstein-Wiener model represents the dynamics of a process using a diagram with two nonlinear blocks cascaded with one linear block (see Fig. 3), [8]. In this work, a Hammerstein-Wiener model is used, which is preceded by a nonlinear block and followed by a linear dynamical system and a nonlinear output block as shown in Fig. 3.

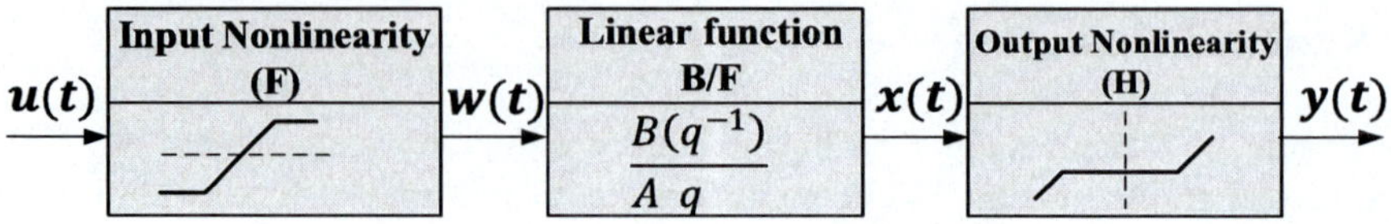

Fig. 3. The Hammerstein-Wiener model [8].

The Hammerstein-Wiener model is represented as shown in (1)–(3).

$$y(t) = H(x(t)) \tag{1}$$

$$A\left(q^{-1}\right) \cdot x(t) = B\left(q^{-1}\right) \cdot w(t) \tag{2}$$

$$w(t) = F(u(t)) \tag{3}$$

where, $u(t)$, $y(t)$ are input and output respectively, $w(t)$ and $x(t)$ are internal variables. $F(u(t))$ and $H(x(t))$ represent the nonlinear functions, $A\left(q^{-1}\right)$ and $B\left(q^{-1}\right)$ are polynomial functions expressed in the time-shift operator q^{-1}, corresponding to the denominator and numerator of the transfer function, respectively. It is represented by the model structure Output-Error (OE) as shown in (4) and (5).

$$A\left(q^{-1}\right) = 1 + \sum_{i=1}^{n_A} a_i q^{-i} \tag{4}$$

$$B\left(q^{-1}\right) = q^{-d} \sum_{j=1}^{n_B} b_{j-d} q^{-j+d} \tag{5}$$

where d, n_A and n_B denote the delay and the degree of $A\left(q^{-1}\right)$ and $B\left(q^{-1}\right)$.

The functions $F(u(t))$ and $H(x(t))$ are a dead zone nonlinearity function and piecewise linear function parameterized by breakpoint locations, respectively. To estimate the Hammerstein-Wiener model, search algorithms are used to minimize the loss function. In this case, Gauss-Newton's method for minimization of prediction or simulation errors is applied.

3.2 Controllers Design

In this section, the design of the proposed controllers is presented such as PID, PID-Fuzzy and Predictive Controller Model (MPC).

3.3 PID Controller

The PID control structure is shown in Fig. 4. To design the PID controller it is necessary to linearize the Hammerstein-Wiener model around the position $x(t) = 0$. The transfer function is shown in (6).

$$G(s) = \frac{0.02292s^2 + 112.2s - 0.001673}{s^3 + 9.822s^2 + 4.132s} \tag{6}$$

The PID controller parameters are tuned using the MATLAB Control System Toolbox and are shown in (7).

$$C_1(s) = \frac{k_{d1}s^2 + k_{p1}s + k_{i1}}{s} \quad (7)$$
$$k_{d1} = 0.012;\ k_{p1} = 0.836;\ k_{i1} = 0.735$$

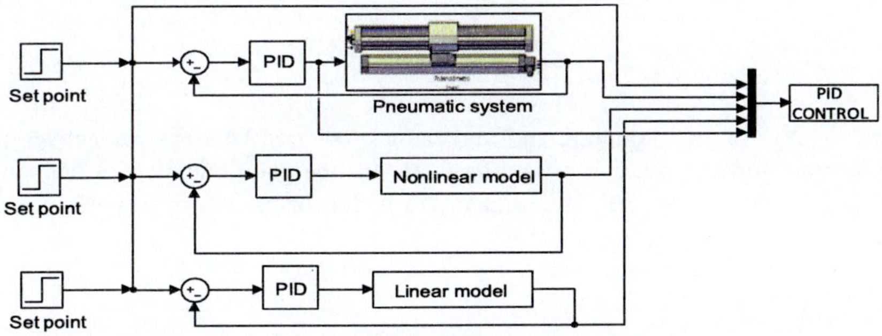

Fig. 4. PID control structure.

3.4 PID-Fuzzy Controller

The design of the proposed PID-Fuzzy controller is shown in Fig. 5. The input and output signals of the system can vary between ±5 Vdc. With this information, we proceed to create the membership functions for the input error signal, derivative of error and output of the controller.

Finally, the rules of inference for the output of the PID-Fuzzy controller are created. The parameters of the PID-Fuzzy are tuned experimentally for the best response, obtaining the values shown below.

$$K_p = 0.8;\ K_I = 0.1;\ K_D = 0.002$$

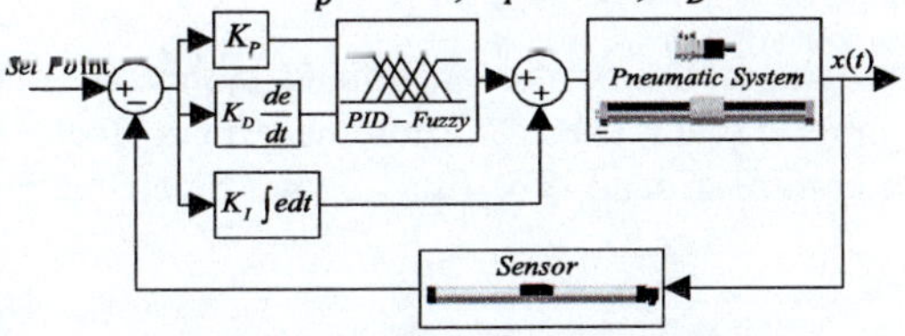

Fig. 5. PID-Fuzzy control structure.

3.5 MPC Controller

The structure of the MPC used to control the cylinder position of the pneumatic system is shown in Fig. 6. Thus, the manipulated variable (MV) used for this purpose corresponds to the analog signal applied to the 5/3-way proportional valve, while the position of the cylinder is measured through the linear potentiometer and corresponds to the controlled variable (CV). The purpose of the MPC is to calculate the future value for the manipulated variable. The control objective function of the MPC is shown in (8).

$$J(N_1, N_2, N_u) = \sum_{j=N_1}^{N_2} \delta(j)\left[\hat{y}(t+j|t) - w(t+j)\right]^2 + \sum_{j=1}^{N_u} \lambda(j)\left[\Delta u(t+j-1)\right]^2 \quad (8)$$

where, N_1, N_2 are the minimum, and maximum prediction horizons respectively; N_u is the control horizon; δ is the weighting factor for the plant output; λ is the control weighting factor; w is the reference signal; Δu is the control effort; $\hat{y}$ is the predicted output.

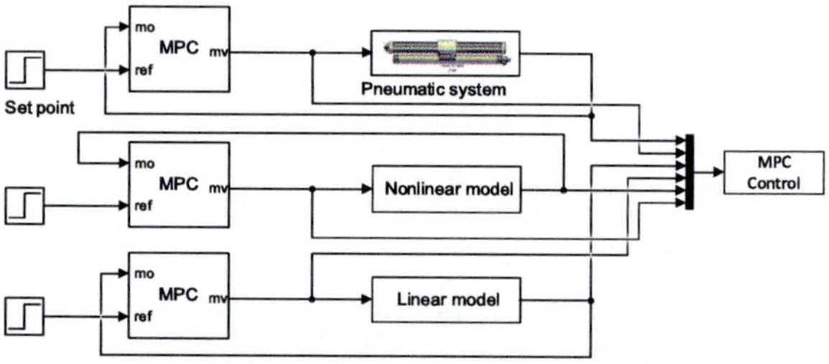

Fig. 6. MPC control structure.

4 Results and Discussion

4.1 Model Validation

To validate the nonlinear model of the pneumatic system a pseudorandom binary sequence (PRBS) as input signal is used. The nonlinear input block of the Hammerstein-Wiener model uses a nonlinear dead zone function whose characteristics are shown in (9)–(11).

$$a < x < b;\ F(x) = 0 \quad (9)$$

$$x < a;\ F(x) = x - a \quad (10)$$

$$x > b;\ F(x) = x - b \quad (11)$$

where, $a = -0.1561$, and $b = 0.1979$.

x is an internal variable used by the nonlinear identifier and corresponds to the output of the linear block of the Hammerstein-Wiener model (see Fig. 3).

The linear block of the Hammerstein-Wiener model is shown in (12).

$$\frac{B}{F} = \frac{-0.9269s + 1}{s^3 - 1.1249s^2 - 0.7127s + 0.8377} \tag{12}$$

The nonlinear output block of the Hammerstein-Wiener model uses a piecewise function to identify the output of the system. This nonlinear estimator is formed with 10 break points as shown in Table 1.

Table 1. Break points of the nonlinear output block estimator.

i	1	2	3	4	5	6	7	8	9	10
x_i	−287.53	−258.07	−93.29	−50.80	79.1	134.5	241.3	259.27	377.1	377.4
y_i	−3.043	−2.7828	−1.142	−0.801	0.46	0.95	2.04	2.22	3.32	3.32

Figure 7 shows the validation on the experimental system of the nonlinear model identified for the position of the pneumatic system. The average absolute position error obtained with this model is 4.1101 mm.

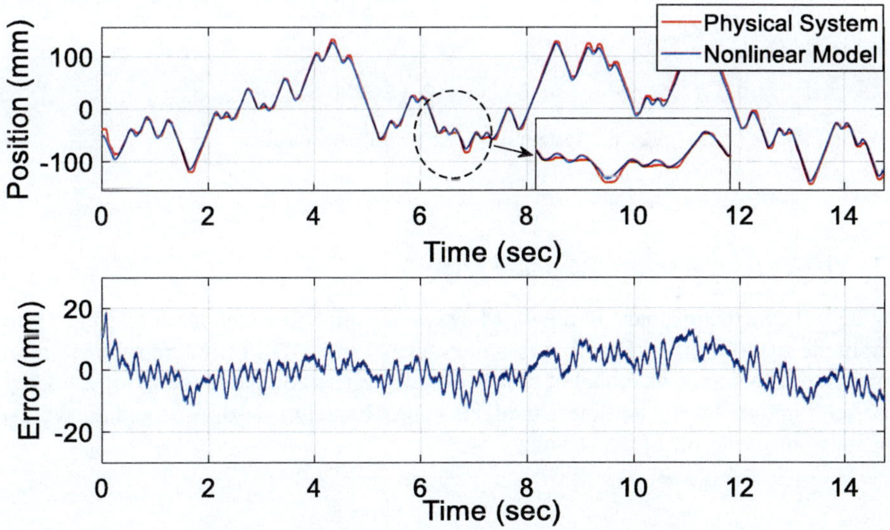

Fig. 7. Validation of the pneumatic system model.

4.2 PID Controller Response Analysis

Figure 8 shows the step response of the PID controller for the position of the pneumatic system (reference with amplitude of 46 mm). In Fig. 8 it is observed that the PID controller is more efficient on the nonlinear model of the pneumatic system compared to the linear model. The settling time for the nonlinear model is approximately $t_s = 0.41$ s and an average absolute position error was 0.8994 mm.

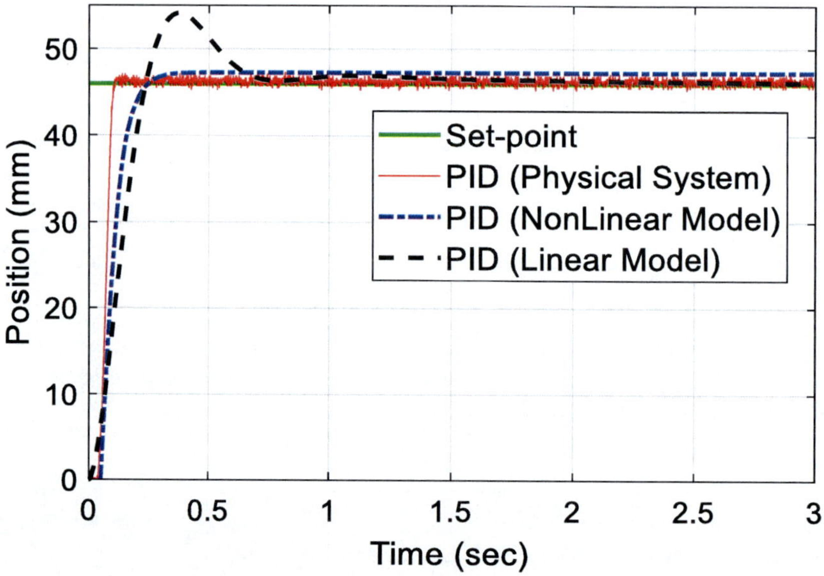

Fig. 8. System response with PID controller.

4.3 PID-Fuzzy Controller Response Analysis

Figure 9 shows the response of the PID-Fuzzy controller for a step input reference with amplitude of 46 mm. In Fig. 9 it is observed that the PID-Fuzzy controller is more efficient on the nonlinear model of the pneumatic system compared to the linear model. The settling time for the nonlinear model is approximately $t_s = 0.397$ s and an average absolute position error of 0.5420 mm.

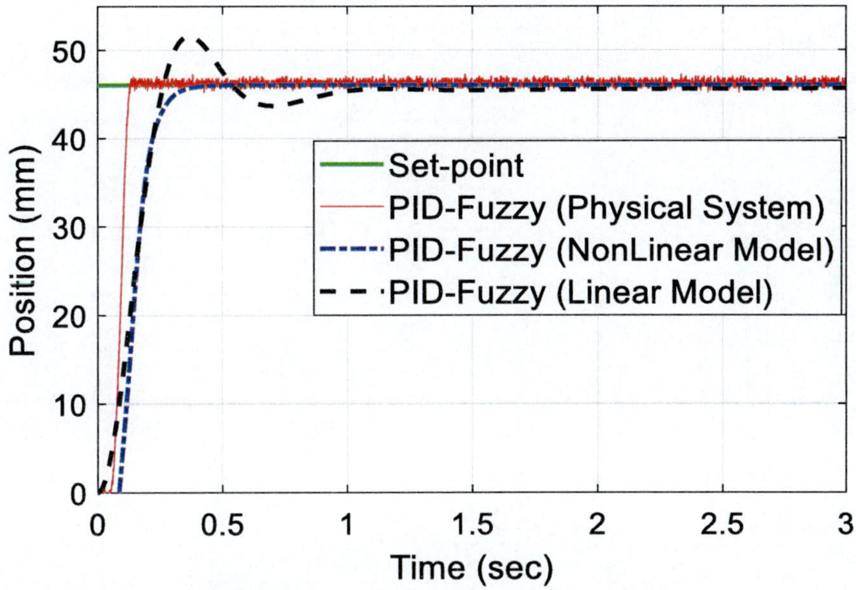

Fig. 9. System response with PID-Fuzzy controller.

4.4 MPC Controller Response Analysis

The response of the MPC controller to control the position of the pneumatic system for a step input of 46 mm amplitude is shown in Fig. 10. The settling time for both controllers is approximately $t_s = 0.338$ s and the average absolute position error was 0.5306 mm. Likewise, the MPC controller is more efficient in the nonlinear model of the pneumatic system. An analysis of the performance of each controller proposed for the pneumatic system is shown in detail in next section.

4.5 Performance Evaluation of Controllers in the Pneumatic System

To evaluate the performance of the controller different performance indexes are used such as *IAE* (integral absolute error) and *ISU* (integral of the control action value). The IAE allows evaluating the efficiency of each controller to reference input tracking. The *ISU* makes it possible to analyze the performance of each controller in terms of its control effort (energy used by the controller). Performance indexes are shown in (13), (14).

$$IAE = J_e = \sum_{k=1}^{N} |r_k - y_k| \tag{13}$$

$$ISU = J_u = \sum_{k=0}^{N} (u(k) - u_{ss})^2 \tag{14}$$

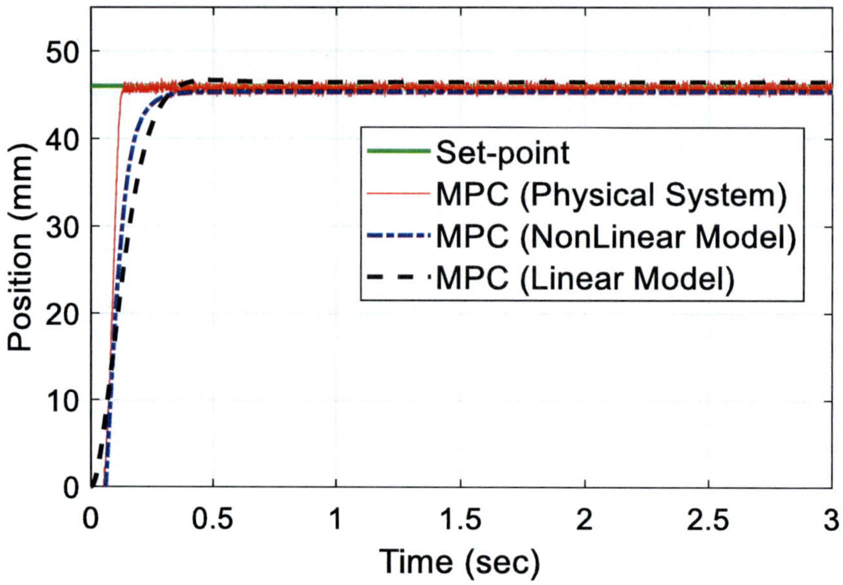

Fig. 10. System response with MPC controller.

Table 2. Evaluation of performance indexes on system controllers.

Controllers	IAE (position error)	ISU (control effort)
PID	1.7989×10^4	120.1082
PID-Fuzzy	1.0841×10^4	**50.1944**
MPC	**1.0613×10^4**	57.6572

The evaluation of the performance indexes on each controller of the pneumatic system is shown in Table 2.

In Table 2 it is shown in bold that the PID-Fuzzy controller has better performance in control effort (energetic uses), while the MPC controller has better reference tracking (the smallest steady state error). However, both controllers demonstrate a better performance over the PID controller with both criteria analyzed.

5 Conclusions

In this work, nonlinear identification was performed to characterize a pneumatic position system using a Hammerstein-Wiener model. The results show that the proposed nonlinear model (Hammerstein-Wiener model) for the system is satisfactory, since the approximation error between the model and the real data does not exceed 4 mm on average in a range from -100 mm to 100 mm. Moreover, three types of controllers are proposed for the system (PID, Fuzzy-PID, and MPC). The results show that all controllers achieve to stabilize the system in a successful way. However, through a detailed analysis using performance indexes (IAE, ISU), we found that the MPC is the most efficient controller for reference tracking when compared to the Fuzzy-PID controllers and the traditional PID controller. The Fuzzy-PID controller is the most efficient controller in terms of its control effort, but it loses reference tracking when compared to the MPC. The traditional PID controller costs approximately twice (control effort) to stabilize the system when compared to the other controllers. If a designer wants to prioritize reference tracking, he/she should choose the MPC controller. If a designer wishes to choose a less aggressive controller (softer control effort) he/she should choose the Fuzzy-PID controller (although this controller loses in reference tracking performance when compared to the MPC controller).

In this work, a versatility of controllers for the pneumatic positioning system was proposed. Two are decentralized controllers (PID, Fuzzy-PID) and one is a centralized controller (MPC) in such a way as to offer the possibility to a control engineer to choose the best available controller based on his/her preferences.

References

1. Azahar, M.I.P., Irawan, A., Raja Ismail, R.M.T.: Self-tuning hybrid fuzzy sliding surface control for pneumatic servo system positioning. Control Eng. Pract. **113**, 104838 (2021)
2. Azira, A.R., Osman, K., Hammami, Sulaiman, S.F., Samsudin, S.I.: Predictive functional controller with reduced-order observer design for pneumatic positioning system. In: 2019 IEEE 10th Control and System Graduate Research Colloquium (ICSGRC), pp. 66–70 (2019)
3. Bogatu, L., Rizescu, C., Grămescu, B., Rizescu, D.: Innovative solutions for pneumatic positioning systems. In: Machado, J., Soares, F., Veiga, G. (eds.) HELIX 2018. LNEE, vol. 505, pp. 530–536. Springer, Cham (2019). https://doi.org/10.1007/978-3-319-91334-6_72
4. Chacón, J., Vargas, H., Dormido, S., Sánchez, J.: Experimental study of nonlinear pid controllers in an air levitation system. IFAC-PapersOnLine **51**(4), 304–309 (2018)
5. Dong, J., Shi, J., Liu, C., Tianbiao, Y.: Research of pneumatic polishing force control system based on high speed on/off with pwm controlling. Robot. Comput.-Integr. Manuf. **70**, 102133 (2021)
6. Du, H., Liu, W., Bian, X., Xiong, W.: Energy-saving for industrial pneumatic actuation systems by exhausted air reuse based on a constant pressure elastic accumulator. Sustainability **14**(6), 3535 (2022)
7. Fan, C., Soon Hong, G., Zhao, J., Zhang, L., Zhao, J., Sun, L.: The integral sliding mode control of a pneumatic force servo for the polishing process. Precis. Eng. **55**, 154–170 (2019)
8. Fouad, G., Er-Wei, B.: Block-Oriented Nonlinear System Identification, vol. 1. Springer (2010)
9. Huilcapi, V., Blasco, X., Herrero, J.M., Reynoso-Meza, G.: A loop pairing method for multivariable control systems under a multi-objective optimization approach. IEEE Access **7**, 81994–82014 (2019)
10. Huilcapi, V., Blasco, X., Herrero, J.M., Reynoso-Meza, G.: A loop pairing method for nonlinear multivariable control systems under a multi-objective optimization approach. IEEE Access **8**, 41262–41281 (2020)

11. Lafmejani, A.S., Masouleh, M.T., Kalhor, A.: Dynamic modeling, identification, and a comparative experimental study on position control of a pneumatic actuator based on soft switching and backstepping–sliding mode controllers. In: Backstepping Control of Nonlinear Dynamical Systems, pp. 261–289. Elsevier (2021). https://doi.org/10.1016/B978-0-12-817582-8.00019-2
12. Lin, W., Dong, W., Deng, Y., Qian, C., Qiu, J.: Contact force modelling and adaptive control of pneumatic system. In: 2016 31st Youth Academic Annual Conference of Chinese Association of Automation (YAC), pp. 357–362. IEEE (2016)
13. Lin, W., Guan, R., Yuan, L., Li, Z., Tong, M.: Positionfeedback dynamic surface control for pneumatic actuator position servo system. Syst. Sci. Control Eng. **6**(1), 388–397 (2018)
14. Liu, Z., Yin, X., Peng, K., Wang, X., Chen, Q.: Soft pneumatic actuators adapted in multiple environments: a novel fuzzy cascade strategy for the dynamics control with hysteresis compensation. Mechatronics **84**, 102797 (2022)
15. Nikitin, A.A., Sorokin, E.A., Nikitina, T.N., Sokolov, D.A., Andreychikov, I.V.: Mathematical model of a single-acting pneumatic cylinder actuator. J. Phys.: Conf. Ser. **1889**(4), 042079 (2021)
16. Osman, K., Mohd Faudzi, A.A., Rahmat, M.F., Mustafa, N.D., Abidin, A.F.Z., Suzumori, K.: Proportional-integrative controller design of Pneumatic system using particle swarm optimization. In: 2013 IEEE Student Conference on Research and Developement (2013), pp. 421–426
17. Ren, H.-P., Jiao, S.-S., Wang, X., Li, J.: Adaptive rbf neural network control method for pneumatic position servo system. IFAC-PapersOnLine **53**(2), 8826–8831 (2020)
18. Ren, H.-P., Wang, X., Fan, J.-T., Kaynak, O.: Adaptive backstepping control of a pneumatic system with unknown model parameters and control direction. IEEE Access **7**, 64471–64482 (2019)
19. Ren, H.-P., Wang, X., Fan, J.-T., Kaynak, O.: Fractional order sliding mode control of a pneumatic position servo system. J. Franklin Inst. **356**(12), 6160–6174 (2019)
20. Saravanakumar, D., Mohan, B., Muthuramalingam, T.: A review on recent research trends in servo pneumatic positioning systems. Precis. Eng. **49**, 481–492 (2017)
21. Sulaiman, S.F., Rahmat, M.F., Faudzi, A.M., Osman, K.: Disturbance rejection using model predictive control for pneumatic actuator system. In: 2016 IEEE 12th International Colloquium on Signal Processing & Its Applications (CSPA), pp. 285–290 (2016)
22. Szakacs, T.: A matlab/simulink® dynamic model of a pneumatic piston and system for industrial application. In: 2020 4th International Symposium on Multidisciplinary Studies and Innovative Technologies (ISMSIT), pp. 1–8 (2020)
23. Walker, J., et al.: Soft robotics: a review of recent developments of pneumatic soft actuators. In Actuators **9**, 3 (2020)
24. Zhao, L., Xia, Y., Yang, H., Zhang, J.: Sliding mode control. In: Pneumatic Servo Systems Analysis, pp. 71–84. Springer (2022)

After-Sales and Customer Loyalty Strategies for Fixed Internet Through the Implementation of Virtual Assistance in the Ecuadorian Context

Bruno Alcívar-Cruz[1,2,3] and Joe Llerena-Izquierdo[1,2,3](✉)

[1] GIEACI Research Group, Universidad Politécnica Salesiana, Guayaquil, Ecuador
balcivarc2@est.ups.edu.ec, jllerena@ups.edu.ec
[2] GieTICEA Educational Innovation Group, Universidad Politécnica Salesiana, Guayaquil, Ecuador
[3] Universidad Politécnica Salesiana, Guayaquil, Ecuador

Abstract. The ability of companies to provide quality products and services motivates organizations to develop business strategies with investment in technology. This research presents the analysis of the incidence of virtual assistance with the use of the chat bot tool to contribute to customer loyalty in companies providing fixed Internet service. The objective is framed in establishing post-sales strategies in the fixed Internet service, by means of customer retention and satisfaction tools. The type of research is descriptive, and the work contributes with the situational explanation of the problem and possible ways to solve them. The structured survey technique was used to 406 users of the Internet service in the home modality in the provinces of Ecuador. The study variables determined are customer loyalty, after-sales strategies, and virtual assistance, according to the national context. The results show that 45% of users remain with the same provider for more than 12 months. It is concluded that the virtual assistant strategy reaches an acceptance rate of 71% and generates the beginning of customer loyalty for after-sales services in fixed Internet companies.

Keywords: Virtual assistance · After-sales strategies · Customer loyalty · Fixed internet

1 Introduction

The ability of companies to provide quality products and services and corporate competitiveness oriented towards higher profits with maximum efficiency motivates organizations to develop business strategies that lead them to increase their profitability. In previous work, [1] emphasize the innovation of modern project management practices to drive business development.

At present, when users have a problem with the service, they must go to the different customer service centers, causing discomfort to users due to the different transfers, the wait depending on the shift and the total capacity of the public spaces, due to the restrictions during the year 2020 because of the pandemic.

V. Robles-Bykbaev et al. (Eds.): CITIS 2022, LNNS 607, pp. 139–149, 2023.
https://doi.org/10.1007/978-3-031-24327-1_12

The motivation for this analysis focuses on the need to create immediate solutions between fixed Internet service providers and their corresponding users. All these inconveniences can be avoided by implementing a system that allows solving customer requirements from the comfort of their homes or mobile devices, significantly transforming the communication between providers and customers.

Likewise [2] in its analysis, evidences the importance of virtual assistance to astronauts. It aims to solve conflicts with the operating systems or the type of assistance that solves the conflicts and attacks the problem at the source, avoiding repetition of such anomalies.

According to ARCOTEL[1], for the year 2020, the number of lines providing fixed internet service in the country reached a total of 2,241,836 in contrast to the 472,429 fixed internet accounts corresponding to the year 2010, which evidences the sustainable growth of this segment in home internet access [3].

Another important factor for the sustained increase of fixed Internet service in the country was the arrival of the COVID-19 virus, in March 2020, forcing government authorities and the Emergency Operations Committee (COE) to take measures to halt the advance of the pandemic, isolation for travelers, class suspensions, confinement, teleworking, etc., massified Internet service in all modalities (see Fig. 1).

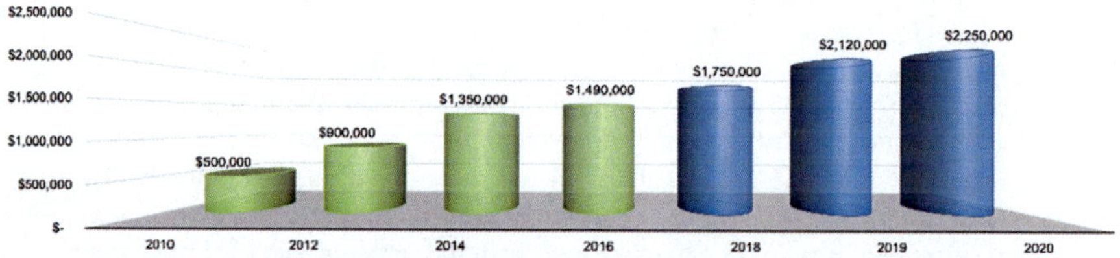

Fig. 1. Nationwide fixed internet accounts

Machine learning is understood as the action where machines are learning situations already produced generally in past experiences, without the need to be programmed for that purpose. Automatic learning is related to artificial intelligence where machines or computer programs have been developed to learn and optimize actions from their experiences (supervised or unsupervised training), predicting possible future problems for a better decision [4].

Chat Bots are programs or applications that use artificial intelligence to develop conversations. They perform various tasks and queries in the areas of sales, marketing, and customer service. These queries can be basic, as well as complex information, including organizing meetings [5].

1.1 Customer Loyalty and Factors to Be Considered

Marketing analysts make various efforts and strategies to achieve customer loyalty [6], this is due to the fact that having a large portfolio of profitable customers in the long term

[1] Available at https://www.arcotel.gob.ec/.

is significant for the continuity of companies. That is to say, selling at the moment is very important for the company, however, it is not everything, it is necessary to make sure that the customer returns and will only do so if he/she obtained an excellent product and good service [7, 8]. To obtain sympathy from customers, it is obtained through two routes: emotional and economic. Organizations that work with their customers on emotional as well as other aspects are difficult to change, even if there are new and better conditions in the competition [1].

Establishing a bond of loyalty between suppliers and their customers is not a simple task, it is necessary to work with tools that leverage and streamline this task. Chat Bots or virtual assistants are software applications created for electronic devices [5]. They maintain a dynamic communication between users and service providers, who answer in a similar way to a human. This interactive chat is successfully developed by the improvement of computer programs or software (see Fig. 2).

These resources are created by managing patterns that customers will use depending on the service they need. The purpose of chatbots is to execute important tasks that are somewhat complex for humans to accomplish or at least take more time to execute [9]. Another objective is to save resources in this type of tasks, usually more than one person is needed. This translates into efficiency in the optimization of resources for fixed Internet providers in the country. It should develop acts of customer care [10], i.e., develop specific rules, which points out the cases in which the virtual assistant should solve the whole problem or those that need human intervention.

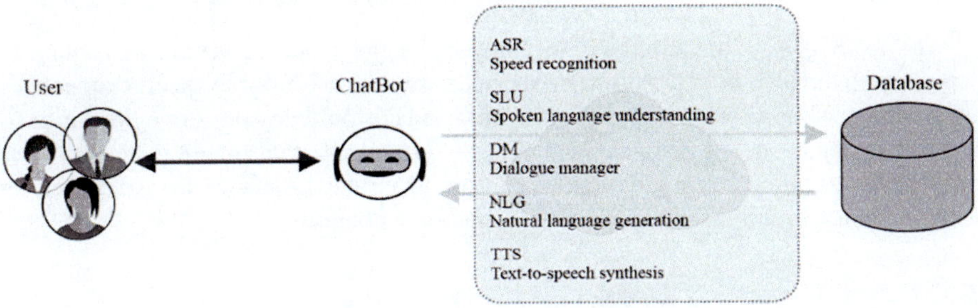

Fig. 2. Virtual assistance structure

1.2 The Structure of Virtual Assistants

Virtual assistants have a great projection in all markets. The research network CHM Research states that by the year 2030, almost all services will be through human-machine interaction. There are various models of dialog systems, depending on the degree of complexity that Internet service providers wish to program. In several studies [11–14], the after-sales service is a strategy to build customer loyalty, and that organizations develop to increase the good image in front of them, according to the author cited, customers who purchase products or services are the most important conquest, due to its recurrence, therefore, care and understand the world of after-sales.

Among them, strategies such as monitoring customer perception, scheduling scheduled surveys, and training the after-sales team, lead the strategies according to the service provided by the respective company. The literature review presents a summary of after-sales and customer loyalty strategies according to previous research work, see Table 1.

Table 1. Strategies of fixed Internet providers.

Strategies	After sales	Loyalty	Reference
Replenishment of goods	X		[11]
Customer perception monitoring		X	[15]
Scheduled surveys	X	X	[16]
After-sales team training	X		[17]
Discount Coupon Box		X	[18]
Digital media interaction	X		[19]
Customer service through TIC		X	[20]
Cloud-based banking assistant		X	[21]
Interactive chat		X	[22]
Addressing and resolving complaints	X	X	[23]

The main goal of this study is to analyze the incidence of virtual assistance to contribute to the loyalty of fixed Internet service providers. While the specific objectives are: To identify the different types of after-sales and customer loyalty strategies for fixed Internet; to determine the main characteristics of virtual assistants in after-sales services and to diagnose the current situation in relation to virtual assistance for fixed Internet service and its level of acceptance in the Ecuadorian context.

2 Materials and Methods

In this work, a descriptive and explanatory research study is carried out, which provides a situational description of both the problem and the possible ways to solve it. The present study details the strategies that allow to create affinity and loyalty among the providers that offer the internet service, in the same way it explains the virtual assistance or chatbot, which are computer software that offers the opportunity to interact with the users in need of customer service, this virtual assistance has a purpose, the satisfaction of the service [24, 25].

For the study of after-sales strategies whose purpose is to translate into customer loyalty of fixed Internet service providers, a qualitative-quantitative approach is required to make numerical measurements and collect information from customers regarding the perception of the quality of service.

The study variables are considered to determine the degree of influence, the same that are established from the bibliographic background, in this case, the variables post-sales strategies and customer loyalty are related to the variable of implementation of virtual assistance, whose purpose is to determine the impact they have between one variable and the other.

The approach of the study allows measuring the issue of virtual assistance in companies aligned to the telecommunications service in the country and users of the same service whose responses serve to validate the idea of implementing virtual assistants according to their experiences with the customer loyalty strategy. The modality applied by the study that seeks to build customer loyalty of fixed Internet customers is non-probabilistic sampling. For [26] in this type of sampling, there is no probability of identifying the sample without conditioning.

To identify the results, the survey instrument is used, which is made up of 12 questions that are obtained using the indicators of the dimensions of the assigned variables, of which 6 questions are configured under the Likert scale with a score from 1 to 5, where 1 is assigned to the assessment of less validity and 5 to the assessment of greater validity. The unit of analysis is the customers of all fixed Internet providers in the country, i.e., 1,538,545 inhabitants with Internet at home.

The population is made up of users in the provinces with the highest representation of fixed internet service, which are Pichincha, Guayas, Azuay, Manabí, and El Oro: Pichincha, Guayas, Azuay, Manabí, and El Oro, among the 5 districts, reaching 72% of the total fixed internet in the country. The finite statistical formula is used, since the size of the population is known with precision, considering the data of the public company ARCOTEL whose scope is for the year 2019.

The number of 406 users of fixed internet service is established as the size of the representative sample of the population under study. To determine the reliability of the research instrument, Cronbach's Alpha formula is used according to [27]. The result of the reliability coefficient is 0.84, which, according to the theory, is within the range of excellent reliability, which translates into internal consistency in the elaborated questionnaire.

3 Results

According to the questions applied in the survey, for the question "Do you have fixed Internet service in your residence or domicile?", it is evident that, of the 406 participants, 99% indicate that they do have fixed Internet service in their residence or domicile, while 1% indicate that they do not. This result confirms the data provided by ARCOTEL in the year 2020 regarding the sustained growth of the country's inhabitants to have this service, mainly due to the measures adopted to combat COVID-19, where many tasks such as work, study, meetings, etc., are carried out in virtual platforms.

The participants are asked in what age range they are according to the standards of an economically active population. The results show that 33% of the respondents are between 25 and 34 years old, 46% are between 35 and 44 years old, 16% are between 45 and 54 years old and 5% are over 55 years old. The results show that the respondents who use the internet service the most are between 25 and 44 years old, representing 79% of the total sample, which is like the data available from the Institute of Digital Economy ICEMD[2].

[2] Available in https://icemd.esic.edu/.

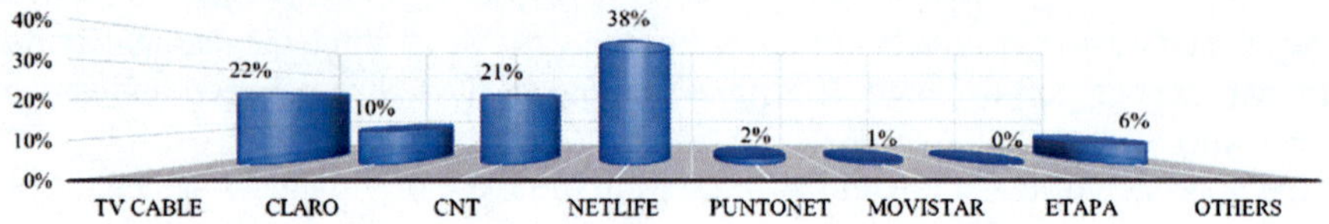

Fig. 3. Percentage of participants that have a contractual relationship with an Internet provider

Participants were asked to mention the name of the provider that offers fixed Internet service; the companies that lead the market share of this segment at the national level are: Netlife, CNT and Tv Cable with 38%, 22% and 21% respectively. Claro, Movistar, Etapa and Others with 10%, 1%, 0% and 6% (see Fig. 3).

The private company Megadatos (Netlife) leads the market share of fixed Internet service nationwide, followed by the public company CNT, according to data from the Telecommunications Regulation and Control Agency (ARCOTEL) as of December 2021[3]. This validates this article, which refers to the development of after-sales strategies to build customer loyalty through virtual assistants. The CNT Public Company did not carry out this strategy, which led it to lose the first place in the fixed Internet market share.

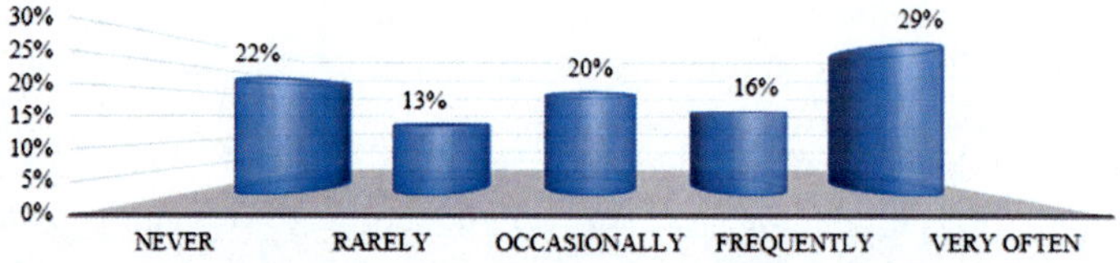

Fig. 4. Percentage of participants who perceive after-sales strategies of their Internet providers.

The results show that 22% never use after-sales strategies, 13% rarely, 20% occasionally, 16% frequently and 29% very frequently (see Fig. 4).

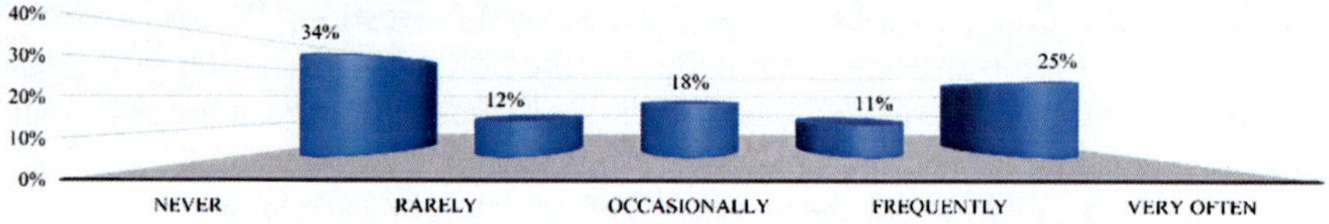

Fig. 5. Percentage of participants using a virtual assistant

In response to the question, "Have you used virtual assistants to request assistance services from your provider?", 34% indicated that they have never used virtual assistants, 12% said rarely, 18% occasionally, 11% frequently and 25% very frequently (see Fig. 5). This shows that one third of the participants do not use virtual assistance and this is a determining point in the study, knowing the current changes in a digitalized society, i.e., 36% are familiar with virtual assistance, a relatively low percentage.

[3] Available in http://www.arcotel.gob.ec/servicio-de-acceso-a-internet-sai2/.

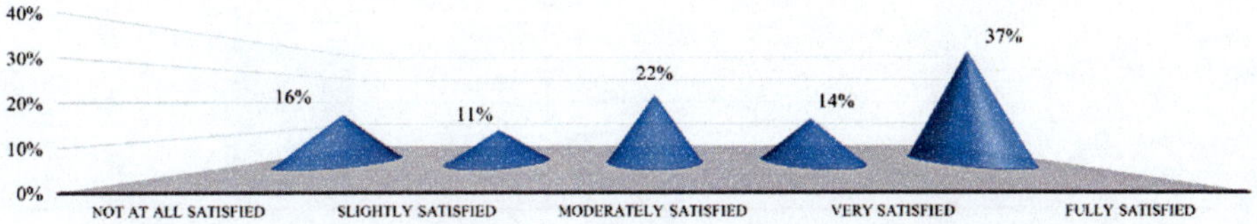

Fig. 6. Percentage of participants' satisfaction with the use of a virtual assistant

Participants' experiences of working with virtual assistants showed that 16% were not at all satisfied, 11% not very satisfied, 22% moderately satisfied, 14% very satisfied and 37% totally satisfied (see Fig. 6). This contrasts with the previous question and shows the level of satisfaction with the use of virtual assistants.

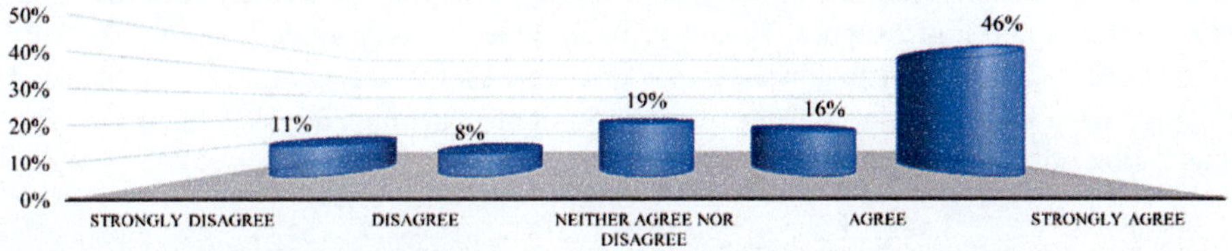

Fig. 7. Participants who consider that after-sales strategies, through the implementation of virtual assistance, lead to customer loyalty.

The answer to the question, "Do you consider that after-sales strategies through the implementation of virtual assistance lead to customer loyalty?", shows that 11% totally disagree, 8% disagree, 19% neither agree nor disagree, 16% agree and 46% totally agree (see Fig. 7). With this, it is evident that 62% agree that the after-sales strategy through the implementation of virtual assistance leads to customer loyalty.

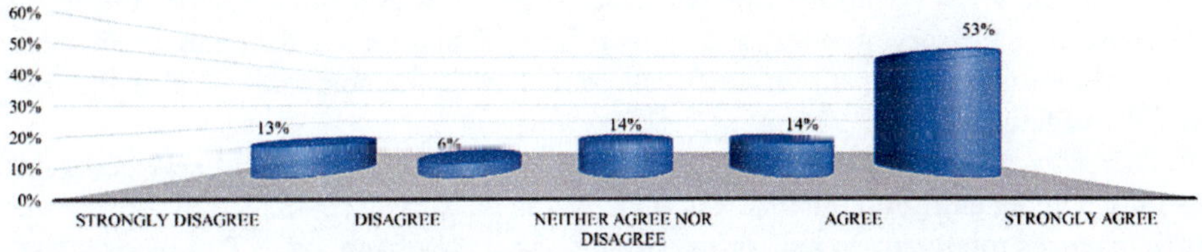

Fig. 8. Percentage of participants who agree that their Internet provider implements virtual assistants for solving technical problems

Participants were asked, "Do you agree that your Internet provider implements virtual assistants to solve technical problems?", 13% indicated that they totally disagreed that their Internet provider implements virtual assistants to solve technical problems, 6% disagreed, 14% neither agreed nor disagreed, 14% agreed, and 53% totally agreed (see Fig. 8). It is evident that 67% agree with the use of virtual assistants to solve technical problems related to the fixed Internet service.

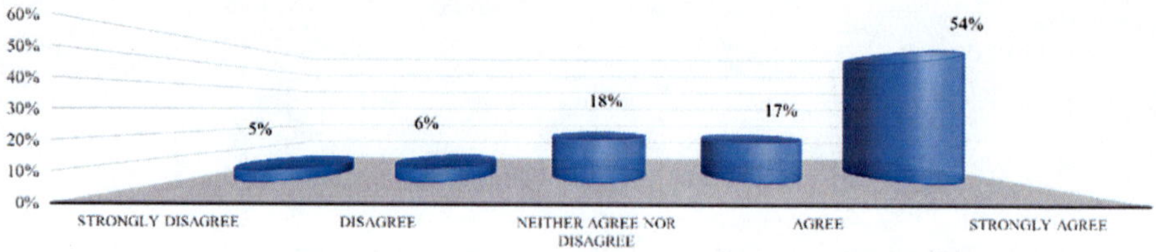

Fig. 9. Percentage of participants who agree that their Internet provider implements virtual assistants to promote products or services that may be of interest to them

To the question, "Do you agree that your Internet provider implements virtual assistants to promote products or services that may be of interest to you?", the result shows that 5% of respondents strongly disagree that their Internet provider implements virtual assistants to promote products or services that may be of interest to them, 6% disagree, 18% neither agree nor disagree, 17% agree and 54% strongly agree (see Fig. 9). This confirms that the use of virtual assistants, 71%, are a useful strategy that should be used by companies that provide internet service to initiate customer loyalty, having a high penetration post-sales medium.

4 Discussion

The purchasing decision of the "Y" generation (41 to 27 years old) [28] and its relationship with marketing, whose results consider this generation segment as a group of interest for organizations, which translates to these generational segments as the most avid in using the Internet service, which is similar to the result of this study in relation to the question of the age range of respondents.

Similarly, the Public Company ARCOTEL, makes a monthly detail of internet service providers with cut to the year 2021, which highlights the company Megadatos whose trade name is Netlife, as the first provider in the country of fixed internet service, followed by the company CNT who is in second place, this result published by ARCOTEL, is corroborated in the survey conducted in this article. What differs in the publication made by the same Public Company in 2019, where it showed the company CNT as the first provider of fixed internet.

Therefore, we can infer that, by implementing the virtual assistant service to the institution under study, option that was answered in the survey conducted by 71% of the participants as totally agree and agree, could serve as leverage to CNT to recover part of the customers who changed provider.

The referenced antecedents showed that including after-sales service is an efficient strategy for customer loyalty developed by the organizations to increase the good corporate image, these assertions are confirmed when 62% of the respondents consider that they agree and totally agree that after-sales strategies through the implementation of virtual assistance entail to customer loyalty [2].

5 Conclusions

This study concludes by evaluating the fulfillment of the specific objectives previously determined, which have been developed in the context of the analysis related to after-sales strategies and customer loyalty of fixed Internet customers through the implementation of virtual assistance in the country.

The main characteristics of virtual assistants in after-sales services are established, which are essential in different organizations that already execute this service, with favorable results for these companies. This is evidence that the use of virtual assistants is a useful strategy that should be extended by companies that provide Internet service for a high penetration post-sales medium.

A diagnosis is made of the current situation, in relation to virtual assistance for the fixed inter-net service and its level of acceptance in the Ecuadorian context, using surveys with Likert scales, and subjected to validity by means of the Cronbach's Alpha process.

The results of the surveys allow to determine the acceptance of the implementation of virtual assistance for the fixed internet service, which will serve to receive after-sales strategies and whose purpose is the customer loyalty of the studied segment. This demonstrates the need to implement virtual channels for customer loyalty, through innovative technological means such as social networks or telephone calls with mass-use software.

References

1. Veselova, N.A., Tarasova, V.A., Kossukhina, M.A.: The application of customer loyalty management methods to the management of innovative project. In: 2017 IEEE Conference of Russian Young Researchers in Electrical and Electronic Engineering (EIConRus), pp. 1371–1373 (2017)
2. Josan, P.K., et al.: Experimental design & pilot testing for ECLSS anomaly resolution using daphne-AT virtual assistant. In: 2021 IEEE Aerospace Conference (50100), pp. 1–13 (2021)
3. ARCOTEL: Boletín Estadístico del Sector de las Telecomunicaciones – Agencia de Regulación y Control de las Telecomunicaciones (2020). https://www.arcotel.gob.ec/boletines-estadisticos/
4. Reshma, Sharma, K.K.: Artificial intelligence techniques-boon for industry 4.0. In: 2021 2nd International Conference on Smart Electronics and Communication (ICOSEC), pp. 457–461 IEEE (2021). https://doi.org/10.1109/ICOSEC51865.2021.9591690
5. Sharath, J.S., Banafsheh, R.: Conversational question answering over knowledge base using chat-bot framework. In: 2021 IEEE 15th International Conference on Semantic Computing (ICSC), pp. 84–85. IEEE (2021). https://doi.org/10.1109/ICSC50631.2021.00020
6. Alcázar, P.: Fidelizar clientes en la nueva economía. Emprend. las claves la Econ. y el éxito Prof. 69–80 (2018)
7. Alcázar, P.: El cliente. Relación Empres. Unidad. **9**, 206–209 (2015)
8. Melendrez-Caicedo, G., Llerena-Izquierdo, J.: Secure data model for the healthcare industry in ecuador using blockchain technology. In: Rocha, Á., López-López, P.C., Salgado-Guerrero, J.P. (eds.) Communication, Smart Technologies and Innovation for Society. SIST, vol. 252, pp. 479–489. Springer, Singapore (2022). https://doi.org/10.1007/978-981-16-4126-8_43
9. Stoicu, A., Ionica, D.-G.: Social media avatar: my dear virtual assistant. In: 2018 IEEE 16th International Conference on Embedded and Ubiquitous Computing (EUC), pp. 117–122. IEEE (2018). https://doi.org/10.1109/EUC.2018.00024

10. Michaud, L.N.: Observations of a new Chatbot: drawing conclusions from early interactions with users. IT Prof. **20**, 40–47 (2018). https://doi.org/10.1109/MITP.2018.053891336
11. Januantoro, A., Sarno, R.: Strategy for Optimizing Cost and Time of After Sales Using Integer Programming. In: 2018 International Seminar on Application for Technology of Information and Communication, pp. 49–54. IEEE (2018). https://doi.org/10.1109/ISEMANTIC.2018.8549733
12. Bao, D., Hu, Y., Shen, D.: Service employees use emoticons of different emotion intensity in online after-sales service. In: Proc. – 2021 13th Int. Conf. Intell. Human-Machine Syst. Cybern. IHMSC 2021, pp. 35–38 (2021). https://doi.org/10.1109/IHMSC52134.2021.00016
13. Akbar, R., Silvana, M., Hersyah, M.H., Jannah, M.: Implementation of business intelligence for sales data management using interactive dashboard visualization in XYZ stores. In: 2020 International Conference Information Technology System Innovation ICITSI 2020 – Proceedings, pp. 242–249 (2020). https://doi.org/10.1109/ICITSI50517.2020.9264984
14. Mydyti, H., Kadriu, A.: The data mining approach: a case study – clustering algorithms for after sales service. In: 2021 10th Mediterranean Conference Embedded Computing MECO 2021, pp. 1–6 (2021). https://doi.org/10.1109/MECO52532.2021.9460149
15. Rodas, C.A.V., Bravo, A.G.Y.: Calidad del servicio y fidelización del cliente: Plan estratégico de CRM para la Universidad Politécnica Salesiana Sede Guayaquil, http://dspace.ups.edu.ec/handle/123456789/19495 (2020)
16. Thakkar, M., Pise, N.: Survey of available datasets for designing task oriented dialogue agents. In: 2019 International Conference on Mechatronics, Remote Sensing, Information Systems and Industrial Information Technologies (ICMRSISIIT), pp. 1–10. IEEE (2019). https://doi.org/10.1109/ICMRSISIIT46373.2020.9405898
17. Singh, S.S., Sen, R., Borle, S.: Online training of salespeople: impact, heterogeneity, and spillover effects. J. Mark. Res. **59**, 230–249 (2022)
18. Romero, A.F.M., Pulido, K.D.S., Others: Planear el montaje de un spa para vehículos en la ciudad de Fusagasugá (2021)
19. Alam Kazmi, S.H., Wahab, A., Zaman, S.I., Kou, Y.: Rise of digital media to triumph brand loyalty. In: 2018 10th International Conference on Intelligent Human-Machine Systems and Cybernetics (IHMSC). pp. 297–300. IEEE (2018). https://doi.org/10.1109/IHMSC.2018.00075
20. Sharma, E., Ashwarya, A.: ICT Enabled customer services: way towards smart retailing. In: 2020 8th International Conference on Reliability, Infocom Technologies and Optimization (Trends and Future Directions)(ICRITO), pp. 738–743 (2020)
21. Mandlekar, A., et al.: What matters in learning from offline human demonstrations for robot manipulation. arXiv Prepr. arXiv2108.03298 (2021). https://doi.org/10.48550/arXiv.2108.03298
22. Shabana, A.: Dynamics of Multibody Systems. Cambridge University Press (2020)
23. Dias, M.P.M., Udunuwara, U.K.A., Weragama, R.L., Senadeera, P.M.: Customer Experience Management Platform (CEMP). In: 2020 11th IEEE Annual Information Technology, Electronics and Mobile Communication Conference (IEMCON), pp. 677–680 (2020)
24. Cabezas, E., Andrade, D., Torres, J.: Introducción a la metodología de la investigación científica (2018)
25. Ayala Carabajo, R.: Formación de investigadores de las ciencias sociales y humanas en el enfoque fenomenológico hermenéutico (De van manen) en el contexto hispanoamericano. Educ. XX1 **19**, 359–381 (2016). https://doi.org/10.5944/educXX1.13945

26. López-Roldán, P., Fachelli, S.: Metodología de la investigación social cuantitativa. Bellaterra (Cerdanyola del Vallès) Dipòsit Digit. Doc. Univ. Autònoma Barcelona (2015)
27. Tuapanta Dacto, J.V., Duque Vaca, M.A., Mena Reinoso, A.P.: Alfa de Cronbach para validar un cuestionario de uso de TIC en Docentes Universitarios (2017)
28. Rodríguez, J.M.P., Estrada, J.M.C., Nacipucha, N.A.S., Castillo, A.N.L.: Decisión de compra de la generación Z y su relación con el Marketing Ecológico. J. Acad. **5**, 62–77 (2021)

Synthesis of Nanometric and Submicron Particles of Titanium Dioxide for the Formation of Nanostructured Films

Daysi Baño Morales[1](✉), Nelly Rosas-Laverde[2], and Cristian Santacruz[3]

[1] GIAUTO, Universidad Politécnica Salesiana, Quito 170702, Ecuador
dbano@ups.edu.ec
[2] Facultad de Ingeniería Mecánica, Escuela Politécnica Nacional, Quito 170525, Ecuador
[3] Facultad de Ciencias, Escuela Politécnica Nacional, Quito 170525, Ecuador

Abstract. Dye-sensitized solar cells (DSSCs) are solar devices that have attracted the attention of renewable energy researchers in the two last decades. Every component of DSSC has been an improvement to obtain a satisfactory efficiency. DSSC consists of three principal layers, two of them act as load carriers and the other keeps the electrolyte responsible for producing a charge flow to the electrodes of the solar cell. In this work, the intermediate layer was made from nanometric and submicron particles of TiO_2 synthesized by the sol-gel method and assembled by the screen-printing technique. The screen-printing pastes of TiO_2 consisted of a mixture of polyethylene glycol, acetic acid, α-terpineol, and water, where the porosity depends on the percentage of polyethylene glycol/TiO_2 in the film. The characterization of the particles of TiO_2 based on Raman spectroscopy and DLS indicated that the film was anatase phase and the particle size was nanometric and submicron respectively, while the characterization of the porous film consisted of the determination of the crystalline structure, morphological characteristics, and absorbance capacity. The results proved that the TiO_2 films built with MD paste showed 35.94% of porosity, the highest porosity of all pastes, and absorbance of 0.04835, a good absorbance of low energy UV compared to the absorbance of commercial CO film, absence of cracks, and cracking.

Keywords: Nanostructures films · Titanium dioxide · Screen printing

1 Introduction

Nanoscience and nanotechnology are broad, interdisciplinary areas of research and development that have evolved worldwide in recent years due to their impact on human endeavors [1]. Nanoscience studies the phenomena of materials at the nanoscale, which means that at least one dimension (width, length, or thickness) of the material is less than 100 nm. These materials known as nanomaterials exhibit unique and better properties than macro-scale materials [2, 3].

Nanomaterials have been studied for photovoltaic applications mainly because they possess unique physical properties such as significantly enhanced photon capture and

V. Robles-Bykbaev et al. (Eds.): CITIS 2022, LNNS 607, pp. 150–160, 2023.
https://doi.org/10.1007/978-3-031-24327-1_13

photo compressor collection capability, size-dependent bandgap, and multiple exciton generation (MEG) which allows the photovoltaic mechanism to break the current thermodynamic limit [2].

Titanium dioxide TiO_2 nanomaterials exhibit unusual physical, chemical, electronic and optical properties that depend on the shape, size, organization, and surface properties [3]. The morphology of nanostructured TiO_2 materials has attracted considerable attention mainly in the fabrication of solar cells because the interaction between devices and media is easier when the specific surface area and surface-to-volume ratio increase as well as the size of the material decreases [4, 5]. This large surface area obtained improves the adsorption of dye molecules and the motion of electrolyte, facilitating electron transport in the device [6]. Moreover, titanium dioxide TiO_2 is highly used for dye-sensitized solar cells (DSSCs) due to its energy band gap of 3.2 eV, non-toxicity, and stability over a wide range of pH [7].

The methods of obtaining nanoparticulate material are developed under bottom-up as well as top-down approaches, i.e., atoms or molecules are assembled to obtain particles of nanometer size or large solid materials are disassembled until they are constituted of only a few atoms [8]. Some methods of TiO_2 synthesis for dye-sensitized solar cells (DSSCs) have been studied such as spray pyrolysis, solvothermal method, precipitation method, and sol-gel deposition; however, the last one, sol-gel synthesis, is the most widely used method to prepare TiO_2 nanomaterials due to its simple operation, moderate reaction conditions, and scalable production [3], it is developed under bottom-up approach.

Titanium dioxide TiO_2 has four common polymorphs: anatase, rutile, brookite, and TiO_2 (B). In a dye-sensitized solar cell (DSSC), anatase TiO_2 has been used widely because it has a high surface area and helps to easy electron transfer [6]. Park [9] showed that the short-circuit photocurrent in a cell based on the rutile phase is only about 30% lower than in a cell based on the anatase phase because a retile-based cell has a small surface area compared with an anatase-based cell, in consequence, a retile-based cell has a low amount of adsorbed dye.

There are several methods for the building of thin films such as spin coating, doctor blade, and screen printing. In the screen-printing technique, the process is performed on a mesh of woven material (i.e., steel mesh or synthetic fiber) that has been glued onto a frame [10]. The film thickness depends on the volume of the paste, the volume, and thickness of the mesh, and the force with the squeegee pressed on the mesh [11]. Screen printing is a very versatile printing technique that produces very thick wet films and there is no loss of coating solution during printing.

In this work, we describe a method for preparing TiO_2 nanometric and submicron particles and assembly TiO_2 thin films by the screen-printing technique.

TiO_2 particles were made by sol-gel and used to prepare pastes. The particle size distribution profile of the TiO_2 solution will be determined by DLS (Dynamic Light Scattering) analysis. The thin films were obtained by screen printing using the TiO_2 pastes, and their properties were studied by X-ray diffraction (XRD), Scanning Electron Microscopy (SEM), and UV–visible Spectroscopy.

2 Materials and Methods

2.1 Materials

The materials used to obtain TiO_2 nanoparticulate material were Titanium (IV) oxysulfate ≥ 29% Ti (as TiO_2) base from Sigma-Aldrich brand as precursor, nitric acid by Panreac, sulfuric acid 69%, hydrogen peroxide 30% by Emsure, and purified water type 2 as chemical reagents. On the other hand, the porous films were prepared with TiO_2 nanoparticle material synthesized by the sol-gel method, commercial diffuse titanium dioxide (Ti – Nanoxide D/SP) Solaronix, polyethylene glycol (PEG) UltraSigma Av. Mol. Wt.: 3350, 600 and 400 for different film, α-terpineol, acetic acid GR. The TiO_2 films were deposited on common glass of 2 mm of thickness.

2.2 Synthesis of TiO_2 Particles

The process for obtaining TiO_2 nanoparticles consisted of hydrolyzing the titanium oxysulfate by adding water. For the preparation, 2.5 g of titanium oxysulfate was dissolved in 300 ml of distilled water, and the solution remained under constant stirring at room temperature until reaching a pH of approximately 8.5, by the addition of 10% w/w ammonium hydroxide. Once hydrolysis has occurred, the formation of a sol (precipitate) was visualized, which was washed and filtered three times to remove excess reagent and impurities [12].

After, two different methods were carried out to reach the nanometric and submicrometric particles. The first method consisted of adding the precipitate in 40 ml of absolute ethyl alcohol and 100 ml of distilled water, stirring constantly at 80 °C until obtaining a homogeneous mixture for approximately 1 h. The second method consisted of mixing the precipitate with 7.4 ml of hydrogen peroxide at 30% w/w and 70 ml of distilled water and subjecting the mixture to a reflux process for 2 h at 100 °C. This solution was left for an aging time of 48 h to obtain the submicron particle size. Finally, each solution (sol-gel) was annealed at 500 °C for 1 h [13].

2.3 Preparation of TiO_2 Films

The preparation of the screen-printing pastes performed by a stirring process, consisted of TiO_2 nanometric and submicron particles in a proportion of 50% w/w, polyethylene glycol with different molecular weights (Wt.: 3350, 600, and 400), acetic acid, and α-terpineol, according to Table 1. The porosity of the film was modified from the variation of the polyethylene glycol/titanium dioxide ratio [14, 15]. Meanwhile, glasses of 2 × 1 cm^2 were washed with water and Micro 90 detergent, then placed in sonication for 15 min with distilled water and isopropanol.

The TiO_2 films were deposited on glass by the screen-printing technique using a SEFAR PET 1500 mesh, N°77/195–55 PW. Each sample was made by applying the TiO_2 paste on 5.5 × 5.5 mm^2 squares of the mesh and sliding a rubber squeegee for the paste to pass through the mesh and the glasses are impregnated. Then, the glasses were placed on a heating plate for 5 min at 120 °C for every deposited layer, but the last layer was sintered at 450 °C for a time of 30 min, with a heating ramp of 50 °C/min; and then slowly cooled in the oven. This process was carried out similarly with a commercial TiO_2 paste (Ti-Nanoxide D/SP Diffusing active layer, Solaronix).

Table 1. Chemical composition of five types of TiO_2 screen printing paste obtained.

Components	Screen-printing pastes				
	Samples				
	MA	MB	MC	MD	ME
TiO_2 [g]	1	1	1	1	1
PEG [g]	1 Wt.:3350	1 Wt.:3350	0.5 Wt.:3350,	0.5 Wt.:600,	0.5 Wt.:400,
H2O [μl]	1000	680	170	170	50
Ethylene glycol [μl]	1000	-	-	-	-
Acetic acid [μl]	-	170	170	170	120
α-terpineol [μl]	-	170	170	100	50

The particle size and the crystalline structure of TiO2 were analyzed by Dynamic Light Scattering (DLS Brookhaven 90 Plus) and Raman microscopy (LabSpec6 (Horiba Scientific) spectroscope using a 633 nm laser with a 600 gr/mm of grating, and 10% ND filter), respectively. The characterization of the porous TiO2 film consisted of the determination of the morphological characteristics by a Scanning Electron Microscope (SEM), and the crystalline structure by a diffractometer Philips Norelco X-ray (XRD), and the absorbance capacity by a Varian Cary 50 Bio UV-Visible Spectrophotometer.

3 Results and Discussion

3.1 DLS and Raman Analysis

The effective diameter of the TiO2 particles obtained by the synthesis method 1 is 22.5 nm. Whereas the effective diameter obtained by the synthesis method 2 is 355 nm. The diameter size obtained for each of the methods coincides with the particle size reported by Mosquera [13].

The particle size of TiO_2 depends on the pH value reached by the sol-gel, so Su [16] in their experimentation indicates that the zone of maximum TiO_2 particle size is located within the pH range between 5 and 8, which practically coincides with the pH of the isoelectric point of TiO_2 [17]. Therefore, the addition of nitric acid in synthesis 1 and hydrogen peroxide in synthesis 2, marked the difference in pH between both substances and therefore in obtaining the nanometric and sub micrometric particle size, respectively.

The crystal structure of TiO_2 was determined by Raman analysis, it confirms the existence of an anatase phase in the particles obtained with both synthesis methods, as shown in Fig. 1. Anatase TiO_2 has six fundamental Raman active peaks in the vibrational spectrum: three Eg modes centered around 144, 197, and 639 cm^{-1}, two B1g modes at

399 and 519 cm^{-1} called B1g (1) and B1g (2d), and one A1g mode at 513 cm^{-1} [4]. The displacement of the peaks may be due to the presence of impurities that prevent crystal growth and therefore decrease the stability of the anatase phase, or to the calcination temperature since at a temperature of 600 °C, there is the anatase-rutile phase transition [18]. The A1g mode cannot be observed in the two spectra due to the calcination temperature, since three factors depend on it: relative intensity, frequency, and line width of the Raman mode.

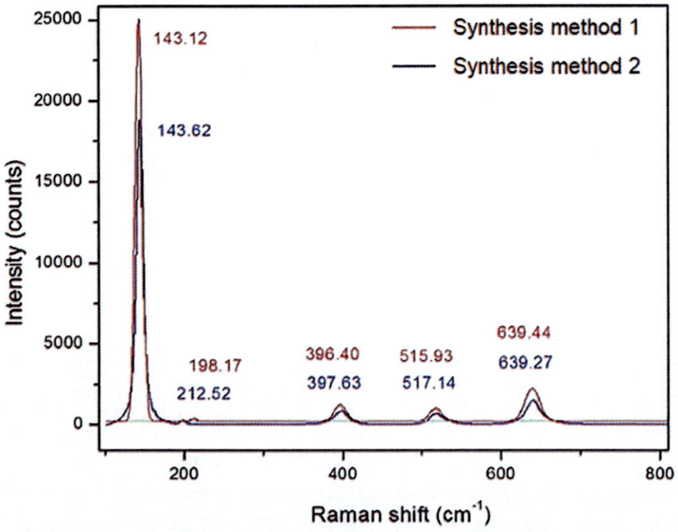

Fig. 1. Raman spectra of TiO_2 particulate material prepared by synthesis methods 1 and 2.

3.2 SEM Analysis

An important parameter considered in the formation of porous films is the evaluation of their morphological structure. Initially, the films obtained with each of the TiO_2 pastes (MA, MB, MC, MD, and ME,) obtained by screen printing are stable and without the formation of cracks or flaking, due to the coexistence of water and acid. Acetic acid [19]. Furthermore, α-terpineol and PEG were added to adjust the viscosity of the pastes and the porosity of the fabricated film.

Figure 2 shows the morphology for each of the films obtained by screen printing using the scanning electron microscope. In the micrograph of paste A in which particles of irregular shapes are visualized, the shapes adopted by the particles in this film are mostly triangular and quadrangular; in this case, the formation of fibers is not observed, which may be due to the lack of acetic acid, causing instability in the film.

In Figs. 2 (b), (c), (d), and (e) the formation of fibers that are randomly oriented, and non-uniform sizes, is observed. You have many empty spaces, which are interconnected with the fibers. Figure 2 (b) shows the presence of fibers with the formation of small agglomerations of TiO_2 particles in a triangular shape (indicated within the circles); unlike Fig. 2 (c) in which fibers of uniform size are observed, and without the presence of particle agglomerations. Figure 2 (d) and (e) were made with PEG 600 and 400, respectively, and show the formation of fibers of heterogeneous size. The film obtained with the commercial CO paste is homogeneous, the TiO_2 particles are not visible, however, the formation of cracks on its surface is visible.

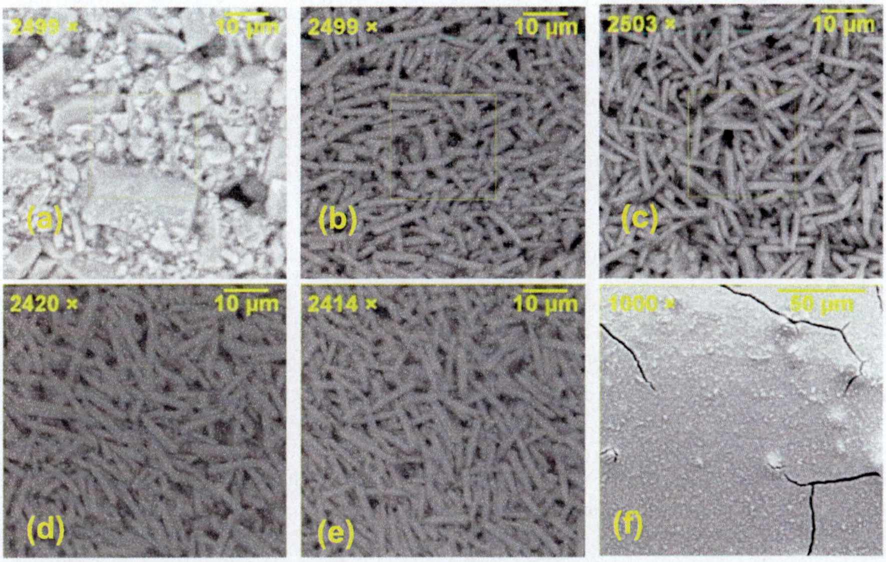

Fig. 2. SEM micrographs of the TiO_2 film were obtained by screen-printing three-layer deposition, (a) with MA paste, (b) with MB paste, (c) with MC paste, (d) with MD paste, (e) with ME paste, (f) with commercial CO paste.

3.3 Porosity Analysis

Porosity is an important characteristic of TiO_2 films because they allow greater dye absorption in the solar cell assembly. The porosity was determined based on the percentage of porosity at the surface level using the Threshold graphic tool of the ImageJ software in the micrographs obtained by SEM microscopy as shown in Fig. 3.

In addition, Table 2 shows the variation of porosity in the function of the paste used, observing that the film with the highest percentage of surface porosity is MB with 47.79%, this is due to the amount of PEG used to the amount of TiO_2 particles (1/1 w/w), In this way, after calcination, the PEG volatilizes and the area it occupied gives rise to the formation of porosities. While in the films with a PEG/TiO_2 ratio of 0.5 w/w, lower surface porosity is observed [20].

In this way, it can be determined that the number and density of the pores present in the films increase as more PEG is added, that is, PEG plays a template role in the

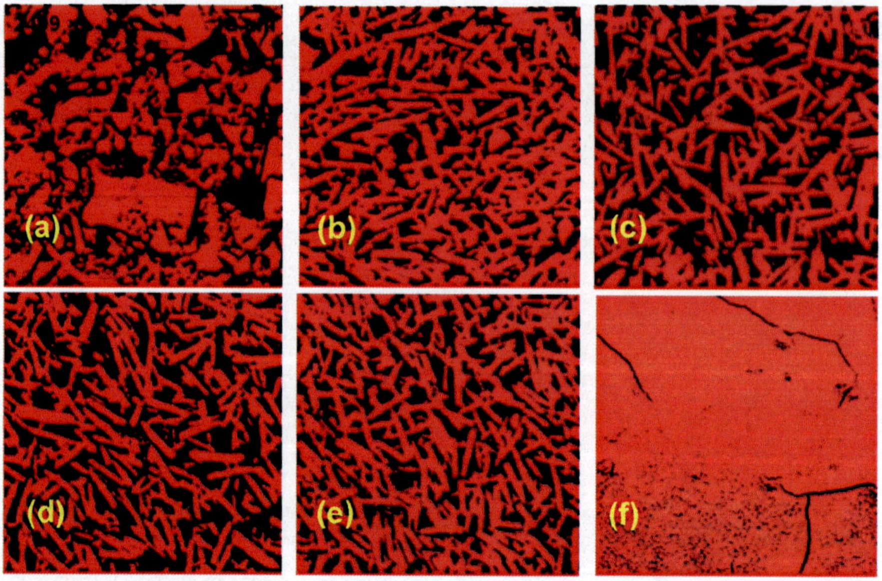

Fig. 3. Porosity analysis of TiO_2 films, (a) with MA paste, (b) with MB paste, (c) with MC paste, (d) with MD paste, (e) with ME paste, and (f) with commercial CO paste.

formation of the inorganic porous structure [21]. Also, the molecular weight of PEG influences the resulting morphology of the TiO_2 films, for example, the surface porosity of the MC (PEG Wt:3500) is greater than that of the MD sample (PEG Wt:600), and this, in turn, is greater than that of the ME sample (PEG Wt:400), which agrees with what was reported in the work of Ramírez [22] and Bu [21].

Table 2. Surface porosity of TiO_2 films.

	Screen-printing pastes					
	MA	MB	MC	MD	ME	CO
Surface porosity [%]	39.48	47.79	45.60	35.94	33.95	6.85

3.4 XRD Analysis

The crystal structure of TiO_2 was determined from X-ray diffraction (XRD) analysis for each of the films. Figure 4 presents a comparison of the graphs obtained by XRD. The peaks at diffraction angles of 25.36°, 37.88°, 48.1°, 53.95°, 55.16°, and 62.8°, correspond to reflections from the anatase crystal planes, (101), (004), (200), (105), (211) and (204), respectively [23]. These peaks have been compared with those obtained for each of the TiO_2 films as shown in Table 3.

The MB film is the one that presents the greatest noise, which may be due to the high porosity that this film presents and therefore there is the presence of air inside it. In addition, the diffraction lines widen with the increase in the molecular weight of PEG, which indicates that the size of the nanocrystals decreases, therefore the intensity of the peaks in the diffraction pattern decreases [24].

Table 3. Comparison of the diffraction angles of the anatase phase for the TiO_2 films.

Anatase phase 2θ (°) *	Diffraction angles					
	25.36	**37.88**	**48.10**	**53.95**	**55.16**	**62.80**
MA	25.16	37.79	47.99	54.16	54.98	62.89
MB	25.40	38.03	48.09	54.17	55.33	62.46
MC	25.20	37.78	47.87	53.98	55.10	62.95
MD	25.17	37.51	47.76	54.00	54.85	62.54
ME	25.29	37.84	48.12	53.90	54.96	62.63
CO	25.28	37.88	47.90	53.84	55.01	62.82

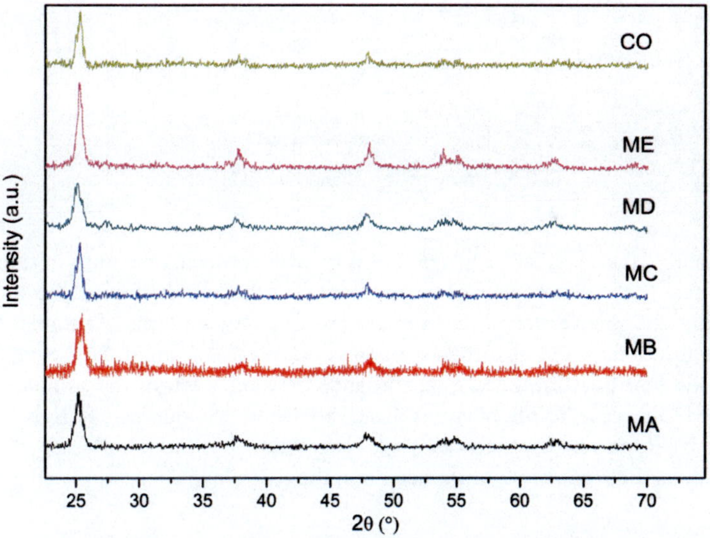

Fig. 4. XRD patterns for TiO_2 films getting by screen printing.

3.5 Absorbance Capacity

In DSSC, the dye sensitizer is responsible for absorbing photons. In this work, we use a natural dye obtained from a flower due to easy synthesis, biodegradability, and availability [25]. The roselle (Hibiscus Sabdariffa) has been used as a sensitizer in DSSC and efficiencies of around 0.107% have been reported [26]. The pigment is extracted from the roselle, where the largest number of anthocyanins present in this plant are found, which give it its characteristic intense red color and must be anchored with the DSSC semiconductor through their -OH bonds [27].

The absorbances at different known concentrations of Jamaica's flower dye in the TiO_2 films were obtained, compared to the absorbances of the samples of unknown concentration, relating them through Beer's Law and obtaining a linear behavior in the calibration curve. Next, the parameters of the line were obtained, and the absorbances of the known samples were adjusted, obtaining the results shown in Fig. 5.

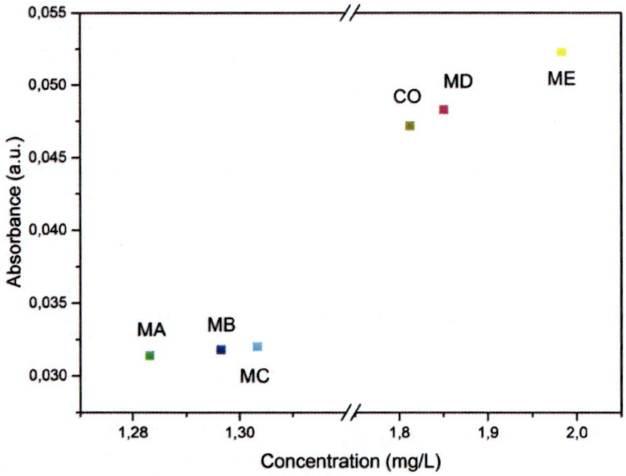

Fig. 5. Pigment absorbance in TiO_2 films.

It is observed that the absorbance of the pigment absorbed by the films of samples MD and ME is near to the absorbance of the pigment absorbed by the film of the commercial sample, while the absorbance of the pigment absorbed by the films of samples MA, MB and MC is lower than the absorbance of the commercial sample. However, it should be mentioned that the difference in absorbance between each of the samples does not exceed 1%. In this case, this behavior is observed due to the addition of PEG in the films, as reported by Ramírez [22] the PEG/TiO2 ratio influences the absorbance of the films, for example: as the PEG/TiO2 ratio increases, the films lose absorption capacity [28].

4 Conclusions

The screen-printing technique was a good method for depositing TiO_2 films, given its good adaptability to the pastes and the substrate used. In addition, it allowed obtaining TiO_2 thin films with uniform morphology, as well as controlled particle size and surface area, which is important when assembling solar cells since its adequate control determines its good performance.

The use of polyethylene glycol PEG provided viscosity to the screen-printing paste, while the TiO_2 films presented a higher porosity as the PEG- TiO_2 ratio increased, this is because the PEG is eliminated when it arrives at a calcination temperature of 450 °C, leaving empty spaces and forming a porous membrane.

Films made with MD and ME pastes have a higher absorbance capacity than the absorbance of the commercial sample, despite having a lower porosity compared to samples MB and MC because the amount of pigment absorbed is independent of the thickness and porosity of the film.

References

1. Vaidyanathan, R., Gopalram, S., Kalishwaralal, K., Deepak, V., Pandian, S.R.K., Gurunathan, S.: Enhanced silver nanoparticle synthesis by optimization of nitrate reductase activity. Colloids Surf. B Biointerf. **75**(1), 335–341 (2010). https://doi.org/10.1016/j.colsurfb.2009.09.006
2. Yu, R., Lin, Q., Leung, S.-F., Fan, Z.: Nanomaterials and nanostructures for efficient light absorption and photovoltaics. Nano Energy **1**(1), 57–72 (2012). https://doi.org/10.1016/j.nanoen.2011.10.002
3. Yan, X., Chen, X.: Titanium dioxide nanomaterials. In: Encyclopedia of Inorganic and Bioinorganic Chemistry, pp. 1–38. Wiley, Chichester (2015)
4. Chen, X., Mao, S.S.: Titanium dioxide nanomaterials: synthesis, properties, modifications and applications. Chem. Rev. **107**(7), 2891–2959 (2007). https://doi.org/10.1021/cr0500535
5. Bai, Y., Angelis, F.D., Bisquert, J., Wang, P.: Titanium dioxide nanomaterials for photocatalytic applications. Chem. Rev. 10131–10176 (2014). https://doi.org/10.1021/cr400606n
6. Shah, A.A., Umar, A.A., Salleh, M.M.: Porous (001)-faceted anatase TiO2 nanorice thin film for efficient dye-sensitized solar cell. EPJ Photovoltaics **7**, 70501 (2016). https://doi.org/10.1051/EPJPV/2015009
7. Jahromi, H.S., Taghdisian, H., Afshar, S., Tasharrofi, S.: Effects of pH and polyethylene glycol on surface morphology of TiO2 thin film. Surf. Coatings Technol. **203**(14), 1991–1996 (2009). https://doi.org/10.1016/j.surfcoat.2009.01.034
8. Guerrero, V.H., et al.: Nuevos Materiales: Aplicaciones Estructurales E Industriales 2015 (2011)
9. Park, N.G., Van De Lagemaat, J., Frank, A.J.: Comparison of dye-sensitized rutile- and anatase-based TiO2 solar cells. J. Phys. Chem. B **104**(38), 8989–8994 (2000). https://doi.org/10.1021/JP994365L
10. Chopra, K.L., Paulson, P.D., Dutta, V.: Thin-film solar cells: an overview. Prog. Photovoltaics Res. Appl. **12**(23), 69–92 (2004). https://doi.org/10.1002/pip.541
11. Krebs, F.C.: Fabrication and processing of polymer solar cells: a review of printing and coating techniques. Sol. Energy Mater. Sol. Cells **93**(4), 394–412 (2009). https://doi.org/10.1016/j.solmat.2008.10.004
12. Giraldo, M., Robledo, L.: Estudio por Microscopía de Fuerza Atómica (AFM) del Soporte de TiO2 sobre Vidrio para la Degradación de Contaminantes por Fotocatálisis Heterogénea, Universidad Tecnológica de Pereira (2015)
13. Mosquera, E., Rosas, N., Debut, A., Guerrero, V.H.: Síntesis y Caracterización de Nanopartículas de Dióxido de Titanio Obtenidas por el Método de Sol-Gel. Rev. Politécnica **36**(3) (2015)
14. Benkstein, K., Kopidakis, N., Van de Lagemaat, J., Frank, A.: Influence of the percolation network geometry on electron transport in dye-sensitized titanium dioxide solar cells. J. Phys. Chem. B **107**(31), 7759–7767 (2003). https://doi.org/10.1021/Jp0226811

15. Wang, Z.-S., Yanagida, M., Sayama, K., Sugihara, H.: Electronic-insulating coating of CaCO3 on TiO2 electrode in dye-sensitized solar cells: improvement of electron lifetime and efficiency. Chem. Mater. **18**(12), 2912–2916 (2006). https://doi.org/10.1021/cm0603102
16. Su, C., Hong, B.-Y., Tseng, C.-M.: Sol–gel preparation and photocatalysis of titanium dioxide. Catal. Today **96**(3), 119–126 (2004). https://doi.org/10.1016/j.cattod.2004.06.132
17. Bahnemann, D., Henglein, A., Spanhel, L.: Detection of the intermediates of colloidal TiO2-catalysed photoreactions. Faraday Discuss. Chem. Soc. **78**, 151 (1984). https://doi.org/10.1039/dc9847800151
18. Agudelo, L., Escobar, D., De La Roche, J., Restrepo, E., Arango, P.: Producción de películas delgadas de TIO 2 obtenidas por la técnica de arco catódico. Sci. Tech. **19**(1), 84–88 (2014)
19. Ito, S., et al.: Fabrication of screen-printing pastes from TiO2 powders for dye-sensitised solar cells. Prog. Photovoltaics Res. Appl. **15**(7), 603–612 (2007). https://doi.org/10.1002/pip.768
20. Fan, K., Liu, M., Peng, T., Ma, L., Dai, K.: Effects of paste components on the properties of screen-printed porous TiO2 film for dye-sensitized solar cells. Renew. Energy **35**(2), 555–561 (2010). https://doi.org/10.1016/j.renene.2009.07.010
21. Bu, S., Jin, Z., Liu, X., Yang, L., Cheng, Z.: Synthesis of TiO2 porous thin films by polyethylene glycol templating and chemistry of the process. J. Eur. Ceram. Soc. **25**(5), 673–679 (2005). https://doi.org/10.1016/j.jeurceramsoc.2003.12.025
22. Ramírez, A.: Desarrollo De Películas Porosas De Tio2 Sobre Vidrio Conductor Ito Por El Método Sol-Gel Para Su Uso Como Fotoánodo En Celdas Fotoelectroquímicas. Universidad Industrial de Santander (2009)
23. Lei, J., Li, X., Li, W., Sun, F., Lu, D., Yi, J.: Arrayed porous iron-doped TiO2 as photoelectrocatalyst with controllable pore size. Int. J. Hydrogen Energy **36**(14), 8167–8172 (2011). https://doi.org/10.1016/j.ijhydene.2011.03.121
24. Calderon-Moreno, J.M., et al.: Effect of polyethylene glycol on porous transparent TiO2 films prepared by sol–gel method. Ceram. Int. **40**(1), 2209–2220 (2014). https://doi.org/10.1016/j.ceramint.2013.07.139
25. Bhogaita, M., Yadav, S., Bhanushali, A.U., Parsola, A.A., Pratibha Nalini, R.: Synthesis and characterization of TiO2 thin films for DSSC prototype. Mater. Today Proc. **3**(6), 2052–2061 (2016). https://doi.org/10.1016/J.MATPR.2016.04.108
26. Calero, M., Santacruz, P.: Construcción y caracterización de celdas solares sensibilizadas por pigmentos naturales extraídos de plantas existentes en el Ecuador y determinación de la eficiencia de conversión energética. Escuela Politécnica Nacional, Quito (2016)
27. Cid-Ortega, S., Guerrero-Beltran, J.A.: Roselle Calyces particle size effect on the physicochemical and phytochemicals characteristics. J. Food Res. **3**(5), p83 (2014). https://doi.org/10.5539/JFR.V3N5P83
28. Xie, D., et al.: Preparation of porous nanocrystalline TiO2 electrode by screen-printing technique. Chin. Sci. Bull. **52**(18), 2481–2485 (2007). https://doi.org/10.1007/s11434-007-0372-0

Classification of Mechanical Failures in Provoked Ignition Engine by Means of ANN and SVM

Rafael Wilmer Contreras Urgilés(✉), José Maldonado Ortega, Esteban Rocano, and Jorge Chiluisa

Universidad Politécnica Salesiana, Cuenca, Ecuador
rcontreras@ups.edu.ec

Abstract. This paper presents the methodology applied to determine the mechanical failures in an internal combustion engine caused by the application of artificial intelligence in the classification of mechanical failures associated with the cancellation of cylinder work, that is to say this methodology is applied on the data obtained from the signal of the KS sensor (Knock Sensor) and the CMP sensor (Camshaft Position Sensor) during engine operation. To evaluate the data obtained, the acquisition of samples applied to different operating conditions is carried out, after which an attribute matrix is created that allows a selection and reduction of variables with the application of methods based on the Random Forest architecture. Subsequently, an ANN (artificial neural network) and an SVM (support vector machine) was created and trained, from which a classification error value of 0.1267% and 0.0067%, respectively, was obtained.

Keywords: Diagnostic · Mechanical failures · K-sensor · CMP-sensor · Artificial neural networks · Support-vector machines

1 Introduction

The internal combustion engine in its operation generates different wear of elements and subsequent breakdowns as a consequence of these wear failures are generated which end in energy wear. This wear and tear begin to be evaluated from the point of view of artificial intelligence, in turn, through Artificial Neural Network (ANN) techniques, it allows the construction of unsupervised mathematical and statistical systems capable of managing, diagnosing and acting on maintenance and fault detection systems witch rate of 90% in the precision [1, 2].

Due to previous research, B. Samanta [2–4] compares the performance of fault detection in bearings and gears, using two classifiers, obtaining test success in a range of (85.06%–100%) for SVM (support vector machines) and (98.61%–100%) for ANN (artificial neural networks), research carried out in 2003 and 2004. Each time this type of methodology, fast and precise, allowed researchers to obtain results with a high precision in the classification for the monitoring and diagnosis of the machines. This is the conclusion of David Antory [5] in 2005 and Achmad Widodo, Bo-Suk Yang, Tian Han

V. Robles-Bykbaev et al. (Eds.): CITIS 2022, LNNS 607, pp. 161–172, 2023.
https://doi.org/10.1007/978-3-031-24327-1_14

[6, 7] in 2007, during their investigations where they used a variant of auto-associated neural networks (AANN) and the use of SVM.

In 2010, with the use of artificial intelligence for fault diagnosis of a bevel gearbox, N. Saravanan [8] concluded that the global average efficiency of the entire classification using ANN was 97.5% and the efficiency of classification using PSVM was 97%. During 2015, the diagnosis of faults not only transcended in the part of engines or machines, but it was also applied to faults in the hydraulic brake system of automobiles with R. Jegadeeshwaran and V. Sugumaran [9] determining that the 98.36% of failures are correctly classified by the Nu-SVM method. For 2017, Prasad V. Kane and Atul B. Andhare [10], show that the accuracy of fault classification is 95.65% for the ANN and 93.44% for SVM, methodology used for the inspection of gearbox failures at the end of the assembly line. For the same year Sánchez R., Vásquez R., among others [11] acquired a classification accuracy of 98.7% with SVM for the diagnosis of multiple failures of rotating machinery. After this investigation in 2019 others were added in the 2019 applied to bearings carried out by Agrawal P. and Jayaswal P., reaching a precise classification, with ANN and SVM results of 96% and 100%, respectively [12]. In the same year, Contreras W., achieved a precise identification of the different types of mechanical conditions of the MEP by applying feed-forward backpropagation neural network for fault diagnosis, with which a classification error of 1.89e–11 is obtained. [13]. Shin et al. [14], in 2020, carried out a study with neural networks and deep learning to predict the phenomena of a motor and achieved great accuracy, similarly Hashim et, al. [15], developed a technique for detecting faults in an internal combustion engine by extracting the characteristics of the vibration signal and concluded that neural networks have great potential to analyze and identify faults that can occur in the engine.

The use of artificial intelligence spread over SVMs, developing studies with classification results with minimal errors [16, 17]. All the studies and investigations carried out previously provide greater knowledge about the application of artificial intelligence for the diagnosis of faults in systems within the automotive branch, demonstrating the precision of its results, however, they do not go beyond the application of a classification technique, therefore the present work is developed with the purpose of using two widely applied techniques for the classification of failures such as ANN and SVM, in addition to comparing and finding the error between the two methods, in order to verify which method has less error in the diagnosis of failures for a provoked ignition engine.

2 Materials and Methods

This section explains the instruments and the methodology used for the application of mechanical failures in the experimental element, the conditions for taking samples in the KS and CMP sensors, and the methodology for obtaining the matrix of attributes for its analysis, selection and reduction and its application in ANN and SVM. The KS and CMP sensors are applied since the first one captures the vibration produced by the engine in the different operating conditions and the second identifies the phases of the engine to obtain the order of failure. The other sensors are not addressed because they do not correspond to the end of this research.

2.1 Experimental Unit

The engine used to simulate the faults is a Hyundai SONATA 2.0 Engine, on which the data acquisition instruments are installed. Table 1 describes the characteristics of the motor used for the experimentation.

Table 1. Technical characteristics of the engine.

Description	Characteristics
Number of cylinders	4
Displacement (cm^3)	2000
Valve train	DOCH
Compression	10.5:1
Power [CV/rpm]	175/6000
Pair [Nm/rpm]	168/4000
Combustiblc	Gasoline
Transmission	Automatic

2.2 Sampling

To obtain the samples of the electrical signals from the sensors, a DAQ card is used. The samples are obtained at a sampling frequency of 200 kHz established according to the Nyquist criterion, for the KS sensor and the CMP sensor, with a time of 3 s. The samples are acquired with the motor at normal operating temperature, turning on the fan at least twice, avoiding erroneous samples.

2.3 Conditions for Data Acquisition

As the objective of the applied method is to reduce the diagnostic time and a minimally invasive intervention in the engine, we proceed to take the signals provided by the KS

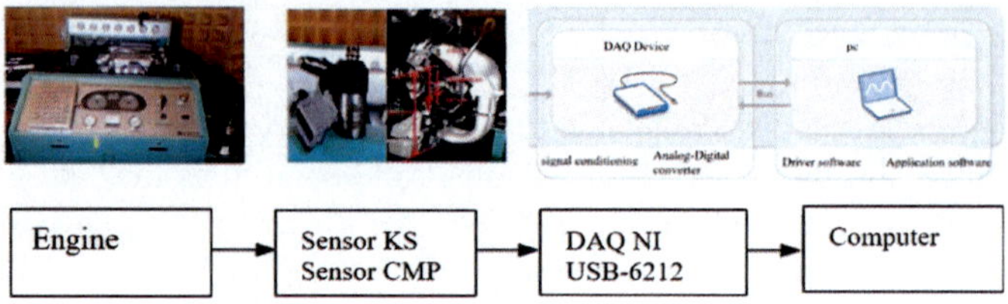

Fig. 1. Process for acquisition samples

sensor and the CMP sensor installed on a Hyundai Sonata 2.0 DOHC engine. The process for obtaining data is described below Fig. 1.

The process begins with the experimental unit, which must have optimal operation [19] as well as the sensor signal (see Fig. 2). With the respective verification, the signal from the sensors is acquired, recorded and saved in a general matrix for fault checking. To determine the second state of the experimental unit, controlled failures are caused, as described in Table 2.

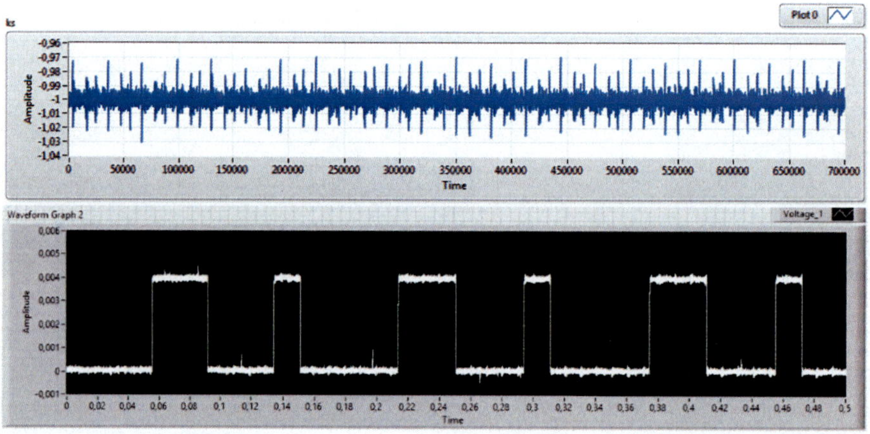

Fig. 2. Sampling signal, upper KS, lower CMP.

2.4 Methodology for Data Acquisition

To obtain the data the process of Fig. 3 is followed. The experimental test begins with the motor in correct operation and later with the supervised faults. With the correct connection of the sensors, the electrical signal is recorded with the software Labview and stored in a spreadsheet [13]. To record the signals, the procedure described above is applied both for the case of the motor in good operation and for the motor with a supervised fault. In order to obtain a sufficiently large database, the data acquisition process is performed 30 times for each of the engine conditions.

Table 2. Conditions for the experimental unit.

Nº	Mechanical failure	Code
0	Engine in correct operation (OK)	0
1	Injector 1 fault (finy1)	100
2	Injector 2 fault (finy2)	200
3	Injector 3 fault (finy3)	300
4	Injector 4 fault (finy4)	400
5	Spark plug failure 1 (bujcer1)	500
6	Spark plug failure 2 (bujcer2)	600
7	Spark plug failure 3 (bujcer3)	700
8	Spark plug failure 4 (bujcer4)	800
9	Coil failure 1–4 (bob1–4)	900
10	Coil failure 2–3 (bob2–3)	1000

2.5 Obtaining the Attribute Matrix

The attributes obtained correspond to 72 in total, and are the results of the descriptive statistics of the signal as observed in [16]: energy, power, asymmetry, kurtosis factor, maximum, minimum, median, average, mode, trimmed median, root mean square, crest factor, standard deviation, variance, range, coefficient of variance, Fourier energy (signal Fourier transform energy), Fourier power (signal Fourier transform power), Max index (maximum position of the Fourier transform; for KS and CMP sensor.

2.6 Attribute Reduction

For the selection and reduction of attributes, the general matrix is analyzed through 3 methods: Curvature Test (Fig. 4), Standard CART (Fig. 5) and Interaction-curvature (Fig. 6), which are based on the Random Forest architecture.

2.7 Selection of Attributes for the Training of the Artificial Neural Network

The attributes are selected based on the highest importance of the predictors, seen in Figs. 4, 5 and 6. The Pareto tool is applied so as to order the attributes and those with the highest prediction effectiveness are obtained. The attributes used in the input of the neural network are described in Table 3, for case of this experimental test there are 11 attributes, it should be noted that these attributes represent approximately 31% and that based on the Pareto principle this percentage is convenient to analyze mechanical failures.

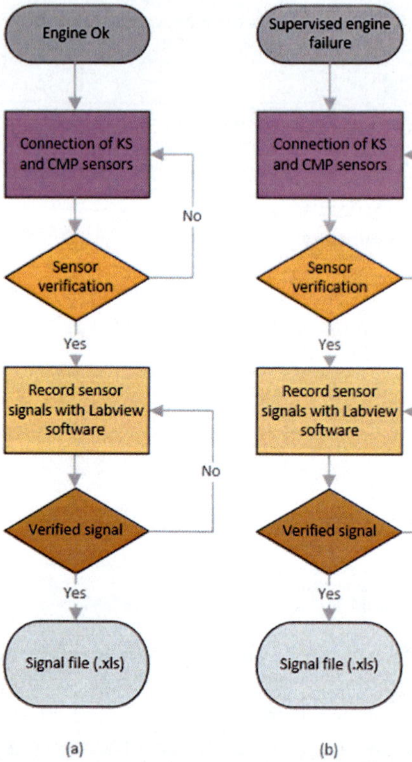

Fig. 3. Process flowchart for data acquisition (a) Motor OK, (b) Motor with fault supervised.

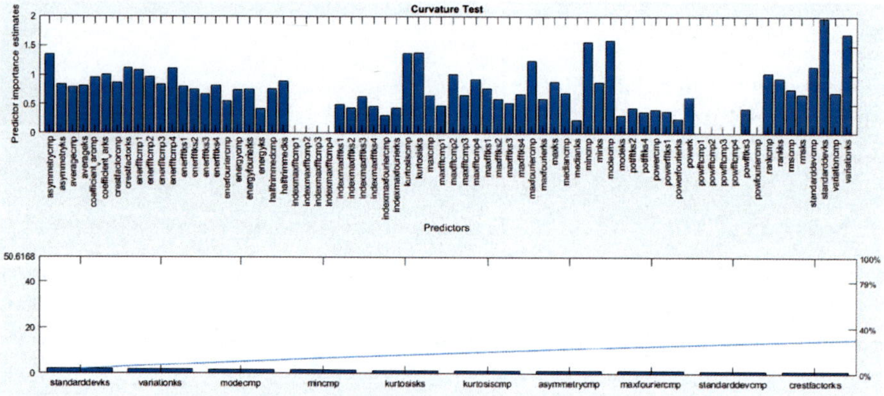

Fig. 4. Predictors with Curvature Test methodology.

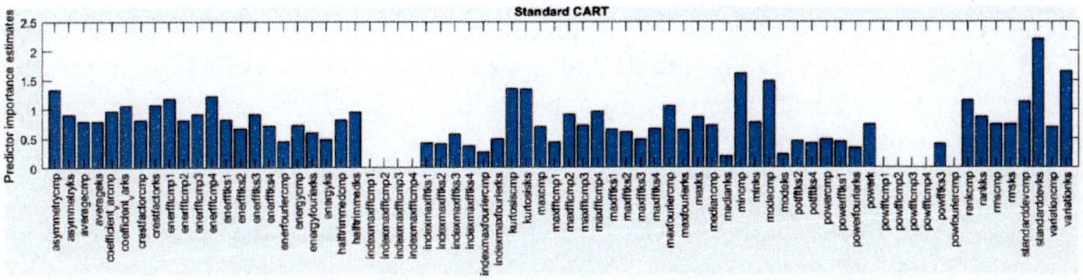

Fig. 5. Predictors with Standard CART methodology.

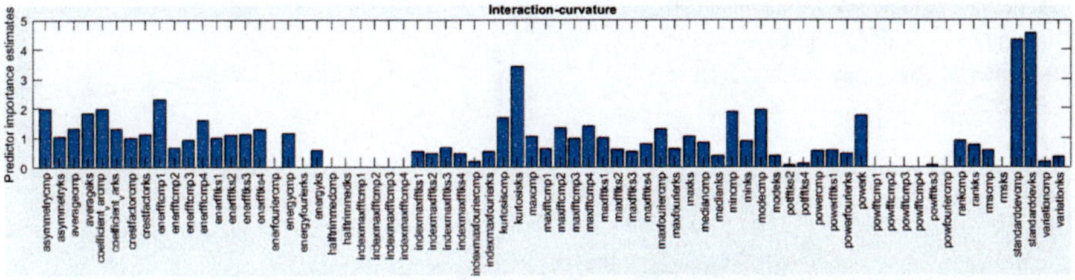

Fig. 6. Predictors with interaction curvature methodology.

Table 3. Attribute matrix

Attribute	N° importance	Value	Accumulated frequency	Pareto value
Standard deviation ks	1	2.10	4.20	–
Variation ks	2	1.83	3.65	7.85
Mode cmp	3	1.59	3.17	11.02
Min cmp	4	1.48	2.95	13.98
Kurtosis ks	5	1.46	2.92	16.90
Asymmetry cmp	6	1.43	2.85	19.75
Kurtosis cmp	7	1.26	2.51	22.26
Enerfft cmp4	8	1.24	2.49	24.75
Range cmp	9	1.17	2.33	27.08
Enerfft cmp1	10	1.10	2.19	29.28
Standard deviation cmp	11	1.09	2.17	31.45

2.8 Neural Network Algorithm

Once the attributes have been selected, a script is developed that allows combining and evaluating different types of: ANN, layers and number of neurons, transfer functions between layers and training functions. The following characteristics such as: Elman backpropagation network type network, with three hidden layers of 20, 40 and 60 neurons between each layer, the transfer functions between each layer were of the tansig and purelin type for the output layer, the % value for el: training, validation and test were 70, 15 and 15 respectively, the number of epochs for training was 2e4; were selected to obtain the least classification error of the faults. In Fig. 7, the training process and characteristics are indicated.

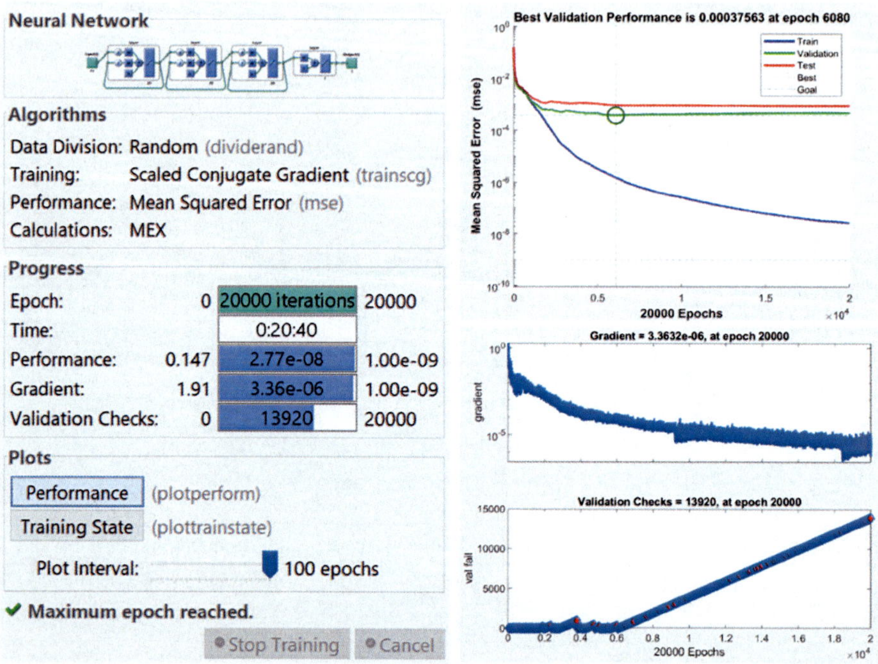

Fig. 7. Characteristics of ANN creation.

2.9 Support Vector Machine Algorithm

For the creation of the SVM, the same attributes are applied as for the ANN, seen in the previous section. The following features identify the parameters and the type of SVM: KFold = 5, Kernel = 0.55, Function = Polynomial, Order = 3, BoxConstrain = 3. These features were selected after different combinations, maintaining the goal of lowering the classification error. of the faults. In Fig. 8, the training process and the characteristics of the confusion matrix and the ROC curve are indicated.

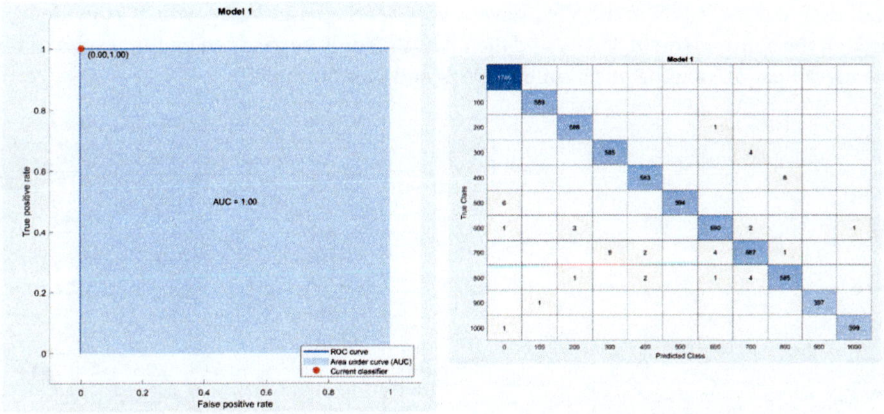

Fig. 8. SVM build features, ROC curve (left), and confusion matrix (right).

3 Results and Discussion

Applying the classification methods for the detection of mechanical failures, various percentages are obtained in the classification accuracy. Figure 9 presents a comparison of the results of the methodologies. For the case of the ANN, the global error is 0.1267% and for the SVM it is 0.0068%, which shows a low dispersion of points for ANN and almost zero for SVM.

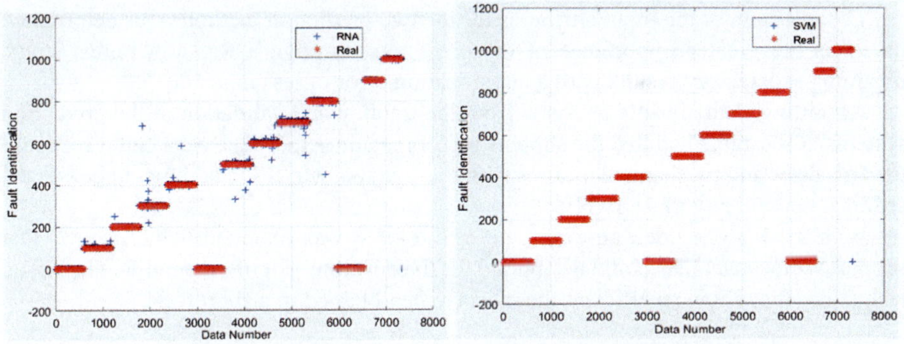

Fig. 9. Failure classification, with ANN (left) and SVM (right)

According to Fig. 9 (right), it can be seen that the SVMs are better suited for the classification of mechanical failures using the attributes of the KS sensor signal, because the maximum value of false positives is 16 in the class 700 (Spark plug failure 3), as evidenced by the confusion matrix in Fig. 8 (left), from which it can be deduced that this classification method obtains an efficiency greater than 95%.

Obtained the results of the different operating conditions of the triggered ignition engine, it can be highlighted that the difference between the responses of both the real condition and the ANN and SVM, have a very marked similarity, which is evidenced through the comparison of Tukey and the ANOVA study carried out on the results of the two classification methods used in this work and seen in Fig. 10.

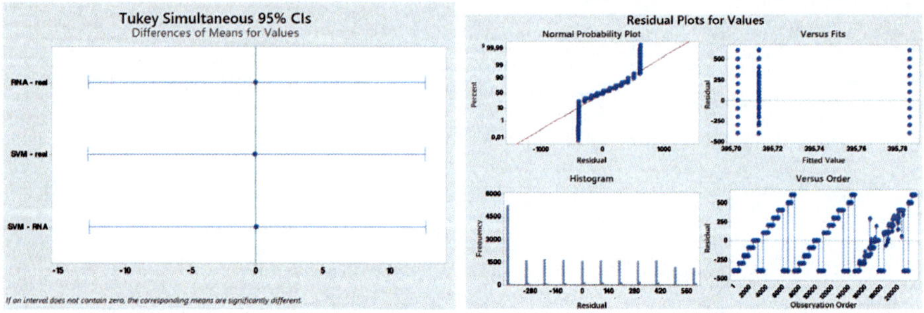

Fig. 10. Tukey comparison (left) and ANOVA study (right) for the results.

4 Conclusions

This article shows the application of two methods based on artificial intelligence: ANN and SVM for the classification of mechanical failures in spark ignition engines, with the advantage of avoiding the disassembly of engine elements and parts with a minimally invasive technique. Simple and efficient.

The selection of the input attributes for the two prediction methods was determined based on the greater importance of the predictors, through a Random Forest study, obtaining satisfactory results with a classification error of less than 1%.

According to the results obtained from the training and validation of the prediction systems, it is established that the support vector machines (SVM) have a better response towards detecting mechanical failures with the signal of Mr. KS, standing out before the artificial neural networks (ANN).

With the SVM model, an error rate of 0.0068% was obtained, which shows that support vector machines could be applied for the development of automotive diagnostic techniques that allow an efficient diagnosis to be obtained in a short time.

References

1. Capata, R.: An artificial neural network-based diagnostic methodology for gas turbine path analysis—part II: case study. Energy Ecol. Environ. **1**(6), 351–359 (2016). https://doi.org/10.1007/s40974-016-0042-7
2. da Silva Junior, E.M., de Sousa, D.R., Marinho, L.C.R.P., Formiga, C.R.B. Matamoros, E.P.: Fault diagnosis in combustion engines using artificial neural networks. SAE International, Warrendale, PA, SAE Technical Paper 2020-36-0076 (2021). https://doi.org/10.4271/2020-36-0076

3. Samanta, B., Al-Balushi, K.R., Al-Araimi, S.A.: Artificial neural networks and support vector machines with genetic algorithm for bearing fault detection. Eng. Appl. Artif. Intell. **7-8**(16), 657–665 (2003). https://doi.org/10.1016/j.engappai.2003.09.006
4. Samanta, B.: Gear fault detection using artificial neural networks and support vector machines with genetic algorithms. Mech. Syst. Sig. Process. **18**(3), 625–644 (2004). https://doi.org/10.1016/S0888-3270(03)00020-7
5. Antory, D.: Fault diagnosis application in an automotive diesel engine using auto-associative neural networks. In: International Conference on Computational Intelligence for Modelling, Control and Automation and International Conference on Intelligent Agents, Web Technologies and Internet Commerce (CIMCA-IAWTIC'06), vol. 2, pp. 109–116 (2005). https://doi.org/10.1109/CIMCA.2005.1631454
6. Widodo, A., Yang, B.-S.: Support vector machine in machine condition monitoring and fault diagnosis. Mech. Syst. Sig. Process. **21**(6), 2560–2574 (2007). https://doi.org/10.1016/j.ymssp.2006.12.007
7. Widodo, A., Yang, B.-S., Han, T.: Combination of independent component analysis and support vector machines for intelligent faults diagnosis of induction motors. Expert Syst. Appl. (2007) https://doi.org/10.1016/j.eswa.2005.11.031
8. Saravanan, N., Kumar Siddabattuni, V.N.S., Ramachandran, K.I.: Fault diagnosis of spur bevel gear box using artificial neural network (ANN), and proximal support vector machine (PSVM). Appl. Soft Comput. https://dl.acm.org/doi/abs/10.1016/j.asoc.2019.08.006
9. Jegadeeshwaran, R., Sugumaran, V.: Fault diagnosis of automobile hydraulic brake system using statistical features and support vector machines. Mech. Syst. Sig. Process. **52** (2015). https://doi.org/10.1016/j.ymssp.2014.08.007
10. Kane, P., Andhare, A.: End of the assembly line gearbox fault inspection using artificial neural network and support vector machines. Int. J. Acoust. Vibr. **24**, 68–84 (2019). https://doi.org/10.20855/ijav.2019.24.11258
11. Sánchez, R.V., Lucero, P., Macancela, J.C., Cerrada, M., Vásquez, R.E., Pacheco, F.:Multi-fault diagnosis of rotating machinery by using feature ranking methods and SVM-based classifiers. https://doi.org/10.1109/SDPC.2017.29
12. Agrawal, P., Jayaswal, P.: Diagnosis and classifications of bearing faults using artificial neural network and support vector machine. J. Inst. Eng. (India): Ser. C **101**(1), 61–72 (2019). https://doi.org/10.1007/s40032-019-00519-9
13. Contreras Urgilés, W., Maldonado Ortega, J., León Japa, R.: Aplicación de una red neuronal feed-forward backpropagation para el diagnóstico de fallas mecánicas en motores de encendido provocado. Ingenius. N.° 21, (enero-junio), pp. 32–40 (2019). https://doi.org/10.17163/ings.n21.2019.03
14. Shin, S., Lee, S., Kim, M., Park, J., Min, K.: Deep learning procedure for knock, performance and emission prediction at steady-state condition of a gasoline engine. Proc. Inst. Mech. Eng. Part D: J. Autom. Eng. **234**(14), 3347–3361 (2020). https://doi.org/10.1177/0954407020932690
15. Hashim, M.A., Nasef, M.H., Kabeel, A.E., Ghazaly, N.M.: Combustion fault detection technique of spark ignition engine based on Wavelet packet transform and artificial neural network. Alex. Eng. J. **59**(5), 3687–3697 (2020). https://doi.org/10.1016/j.aej.2020.06.023
16. Monteiro, R.P., Cerrada, M., Cabrera, D.R., Sánchez, R.V., Bastos-Filho, C.J.:Using a support vector machine based decision stage to improve the fault diagnosis on gearboxes (2019). https://doi.org/10.1155/2019/1383752
17. Zhang, F., et al.: Internal combustion engine fault identification based on FBG vibration sensor and support vector machines algorithm (2019). https://doi.org/10.1155/2019/8469868´
18. Delgado Calle, E.H.: Desarrollo de un algoritmo de diagnóstico para la detección de fallas mecánicas en motores de encendido provocado basados en la transformada Wavelet. http://dspace.ups.edu.ec/handle/123456789/15300

19. Castillo, J., Rojas, V., Martínez, J.: Determinación del Torque y Potencia de un Motor de Combustión Interna a Gasolina Mediante el Uso de Bujía con Sensor de Presión Adaptado y Aplicación de un Modelo Matemático. Revista Politécnica **39**(1), 49–57 (2017)
20. Hou, L., Hao, J., Ma, Y., Bergmann, N.: IWSNs with on-sensor data processing for energy efficient machine fault diagnosis. Int. J. Online Biomed. Eng. (iJOE) **15**, 42 (2019). https://doi.org/10.3991/ijoe.v15i08.10314

Hybrid IIoT & Arduino IoT Cloud Platform for Continuous Industrial Process Analysis

Javier Molina, María Yugcha, and William Montalvo(✉)

Universidad Politécnica Salesiana, 170146 Quito, Ecuador
wmontalvo@ups.edu.ec

Abstract. Hybrid platforms are based on the new technologies of the fourth industrial revolution such as IIoT and cloud computing, which allow local and remote operation. A hybrid system presents a low cost of operation and maintenance. This paper describes the development of a hybrid IIoT (Industrial Internet of Things) & Arduino IoT (Internet of Things) Cloud platform for the analysis of continuous industrial process data in real time through the cloud. This procedure offers a more stable and secure control through the new Cloud technology. The IIoT platform consists of industrial equipment and sensors that are controlled through Delta IIoT devices. An Arduino IoT Cloud with ESP32 module has the functionality of a router that allows sending data to the cloud, thereby achieving the control and monitoring of level and temperature variables of a continuous process. With this type of architecture, the industrial sector has great opportunities to improve its competitiveness by being able to analyze large amounts of data collected from sensors. In addition, the hybrid platform guarantees the visibility of the variables in real time from anywhere in the world where connectivity exists. The results are presented in graphs and are promising as there is no data loss in the cloud and data latency is almost negligible.

Keywords: Industrial Internet of Things (IIoT) · Internet of Things (IoT) · Industry 4.0 · Hybrid IIoT

1 Introduction

Hybrid IIoT & Arduino IoT Cloud platforms enable continuous industrial process analysis [1], which provide solutions for production in a flexible and efficient way to meet the needs of industries, thus achieving an integration between several fields.

Cloud computing is an on-demand resource and one of the leading smart technologies, offering ever-improving developments to expand computing and digital knowledge [2].

The industrial internet of things (IIoT) is a physical network of autonomous devices connected to the internet for industrial studies containing integrated technology for remote sensing and control [3]. It is indeed a generalization of Industry 4.0, which focuses more on the efficiency of industrial processes, which also includes all aspects of industrial operations [4].

V. Robles-Bykbaev et al. (Eds.): CITIS 2022, LNNS 607, pp. 173–181, 2023.
https://doi.org/10.1007/978-3-031-24327-1_15

In response to this, a hybrid IIoT & Arduino IoT Cloud platform was developed for the analysis of data from a continuous process training module. When monitoring temperature and level, sensors are used to send the information via Wi-Fi connectivity to the cloud, since if there were failures in the monitoring of these data, there could be a malfunction of the plant [5].

This technology is based on learning, analyzing and integrating data from all intelligent sensors into a single workspace, so that wiring costs are reduced and manufacturing and production processes are improved, as long as the equipment is used correctly within the industry; that is why the objective of this work is to develop Industrial Cloud Computing through a hybrid platform for the analysis of data from continuous industrial processes [6].

Next, the document is distributed in the following sections that are: Methodology, Results, Discussion and Conclusions.

2 Methodology

2.1 Platform Architecture

For the development of Delta's IIoT platform, the elements reflected in the diagram in Fig. 1 were used; thus, an industrial sensor (PT100) was used to obtain the temperature data of the main tank. The actuator used is a 1500 W nickeline, which will work at 60% of its capacity. On the other hand, a Rosemount level transmitter is also used for level control, while a servo valve is used as actuator, which is controlled through a Fisher transducer.

The system developed implements a topology in parallel to the Delta cloud (DIACloud) with the Arduino cloud (Arduino IoT Cloud), which is in demo phase or phase of test. However, the data received by the cloud from the Arduino is analyzed and saved through the Node-Network platform. This topology shows, in general terms, the existing connectivity in the continuous processes module from the PLC to the Delta equipment and to the cloud, which is provided by the same brand DIACloud and the free remote access it has through static devices (PC) or mobile devices such as tablets or cell phones with the API.

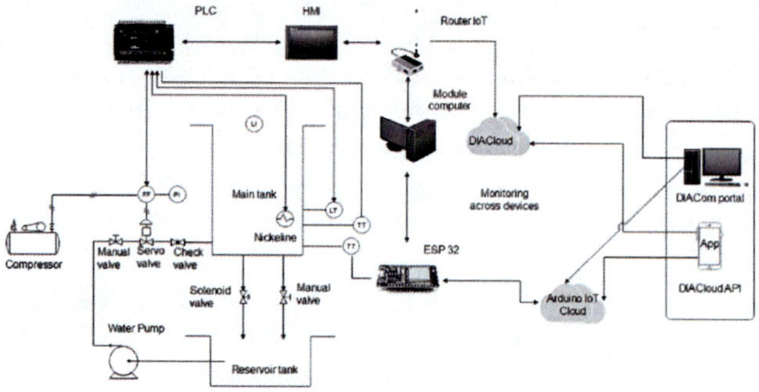

Fig. 1. Platform architecture.

2.2 High-Cost IoT Communication Peripherals

The high cost IoT communication peripheral selected is a Delta Router, model DX-2100RW-WW, which allows to monitor the different equipments remotely through the internet connection, available for web browsers, iOS and Android platforms. It is a 3G industrial equipment that has wireless communication and contains multiple interfaces so it has a highly secure and reliable data transmission; in addition, the technical details of the equipment can be found in the web page [7].

The components that were chosen for the high-cost platform correspond to a series of factors, one of them is the ability to send and save data in the cloud, data that has minimal latency and that in turn can be monitored in real time and from anywhere in the world that has internet access.

2.3 Low-Cost IoT Communication Peripherals

The low-cost communication peripheral is an Arduino ESP32 board, which has a free software and hardware, and uses an Atmel microcontroller, which is why technical details are shown on the web page [8]. In addition, it has a single 2.4 GHz Wi-Fi and Bluetooth combo chip and uses a Tensilica Xtensa LX6 microprocessor [8]. Therefore, it works perfectly for monitoring and sending data to the cloud, and it can also communicate with the NODE-Red platform, which saves the data to be analyzed later.

The components that were chosen for the low-cost platform were chosen because they have similar characteristics of data transmission, just as in the high-cost platform, the data that is rescued from the processes can also be observed in real time, for On the other hand, small sensors must be used, which at too high temperatures will emit false data.

2.4 Design of Screens

The DOPSoft software is specialized for the programming of Delta HMI Screens and was developed by the supplier to be used by programmers without the need for a subscription, since it is free software. From it, a window is created as shown in Fig. 2.

Fig. 2. Main program window.

In the secondary window of the program there is a menu with 3 options, as shown in Fig. 3:

- ACQUISITION: With this option, data can be extracted directly from the HMI by connecting a USB memory stick.
- TEMPERATURE CONTROL: Allows to enter the following window in which a Set Point can be entered to turn on the nickel plating and the water temperature starts to rise.
- LEVEL CONTROL: Finally, there is the option to enter the window as in the temperature control window. Here you can also enter a Set Point so that the module starts to work with the level control.

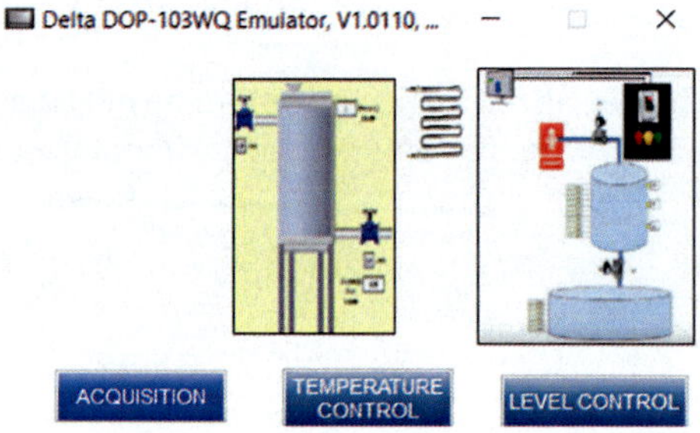

Fig. 3. Program subwindow.

2.5 Low Cost Cloud Configuration

The low-cost environment that was used is Arduino, which has created a platform to connect directly to the Internet or to the cloud through various devices that now include Wi-Fi modules, one of which is the ESP32 [9], the first step is to create a password and a user that will later be used to enter the Arduino IoT environment, then select the device, which as already mentioned, is an ESP32 [10], to access the internet, the password of any internet is entered, then the programming is done in the Arduino sketch [11], where the analog channels are activated. The variable "temp" is also declared, which allows to obtain the temperature data, as shown in Fig. 4.

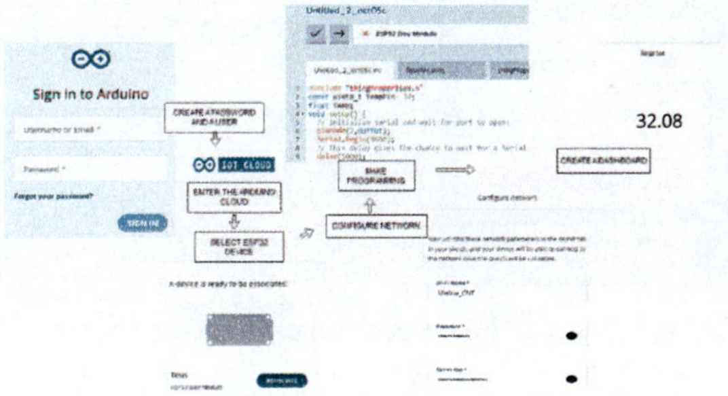

Fig. 4. Flowchart of the steps to set up the low-cost cloud.

2.6 Low-Cost Cloud Communication Configuration

Node-RED is one of the tools that have been appearing with the purpose of supervising, controlling or collecting data through different devices through programming [12]. Since it is a free software in this project, it is very useful and obviously has no cost, in order to work on the Node-RED platform, the steps shown in Fig. 5 must be followed.

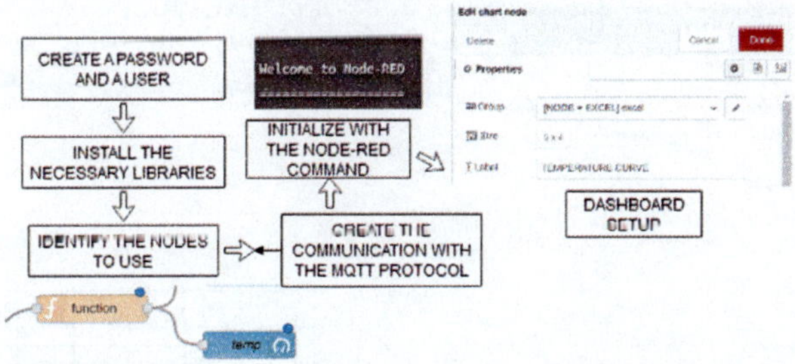

Fig. 5. Flowchart of the steps to set up low-cost cloud communication.

2.7 High Cost Cloud Configuration

The Delta company provides a platform called DIACloud. In order to work on this platform, it is necessary to follow the steps that can be seen in Fig. 6. The first step is to create a user and a password, then a secure tunnel is generated, which provides a secure space for the transmission of data between the platform and the router, finally the records are configured, here is where the data sent to the cloud through the router can be observed.

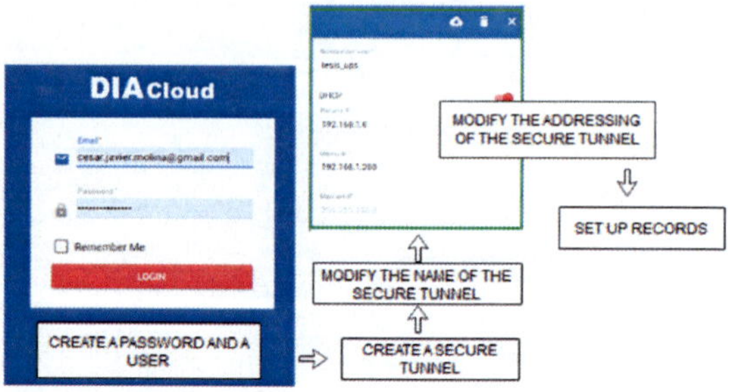

Fig. 6. High-cost cloud configuration.

3 Results

3.1 Comparison of Low-Cost and High-Cost Clouds

The graph in Fig. 7 shows the results obtained through the DIACloud platform. The set point that was entered in the first instance was 28 °C, as can be seen the control is stable, and another set point value was entered, in this case 30 °C, and it also remained stable at the temperature that was entered.

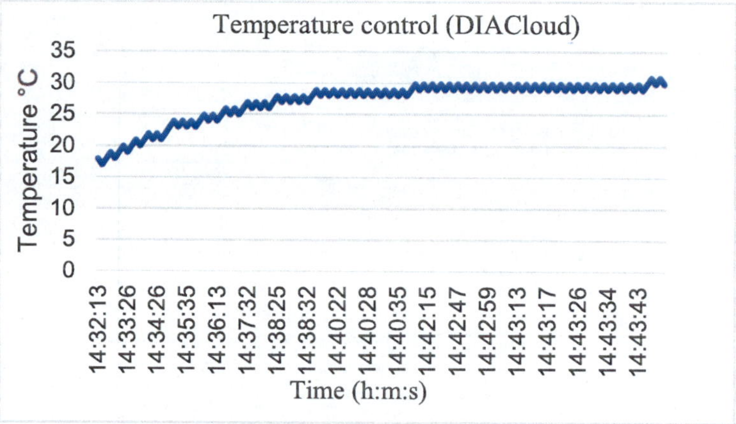

Fig. 7. Temperature control chart in the high-cost cloud.

Figure 8 shows the results obtained through the Node-RED platform. It can be seen that the temperature values from one platform to another vary. This is because the sensors that were used are different. For the DIACloud platform the sensor corresponds to an industrial PT 100, with a 0 to 20 mA transducer, unlike the PT 100 used to work with the Arduino ESP 32, which works with the resistance difference, and for this reason there is this variation in the readings.

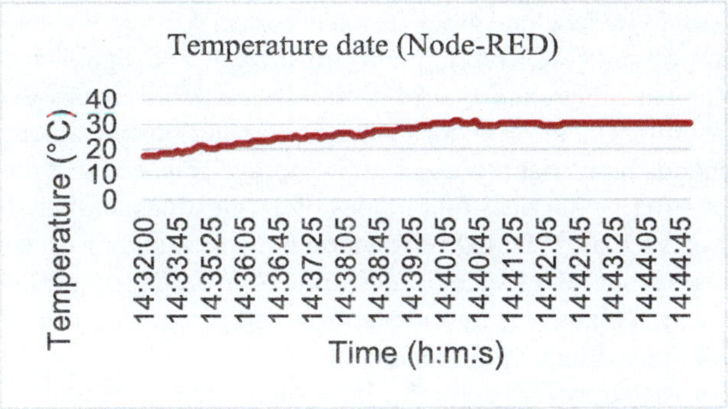

Fig. 8. Low-cost cloud temperature control charting.

Figure 9 shows the results obtained from the level control; in this case 2 tests were performed, one with the set point of 10 cm in height, to which the control response is immediate and the level rises to the desired point, then let the level down completely to 0 cm, to then perform the second test with a value of 34 cm. Likewise, the control response is immediate and reaches the desired value without major problems, i.e., results that have been obtained by performing control tests both temperature and level are very promising and reliable.

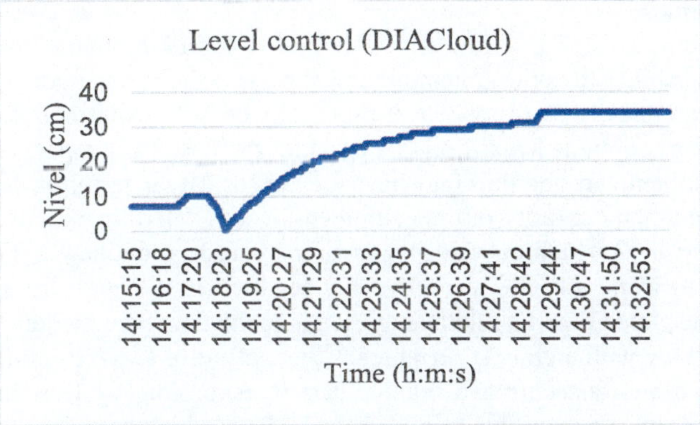

Fig. 9. Graph of level control in the high cost cloud.

4 Discussion

Unlike [1], who used MQTT technologies facilitate communication protocols by allowing the entry of 50 users and implementation making IIoT increasingly transparent is therefore, that the platform offered by the company Delta's main objective is to allow creating an unlimited number of accounts in addition to 10 MB of cloud storage, moreover, the data that is stored in the cloud are in an excel format.

According to [6] and [14], Hybrid platforms present as a significant advantage the ease of technological integration thanks to the diversity of devices that can be used as; sensors, transmitters, actuators, open source computer boards, among many others; therefore, continuous industrial processes allow storage and access of data so they have the advantage of offering services to all types of applications and having several programming options such as: PID, FUZZY, neural networks, among others, which allow working with new technologies such as industry 4.0. In addition, when making a comparison with industrial plants, the hybrid platform will guarantee us a real experience in the control of continuous industrial processes.

The study conducted by [13] and [15], By making use of free tools for the development of hybrid platforms, it is possible to integrate, merge and interoperate data from different sensors, which is a concrete advance in the interoperability of information where they can be implemented in different industrial environments, so a hybrid IIoT platform has the advantage of working through Android mobile devices, IoS and computers with internet access. That is why in the future it is intended to incorporate new functionalities and technologies that help to complement the initial proposal of the project.

5 Conclusions

A hybrid IIoT and Arduino IoT Cloud platform was developed for the analysis of continuous industrial processes, which meets the necessary requirements to control temperature and level variables.

By developing an IIoT platform and Arduino IoT Cloud through a specialized software, it has the ability to control, monitor and supervise the parameters of the level and temperature variables in real time where there may be failures in their security, but to make the settings on the IoT platform that provides the DELTA company, the user must identify and authenticate and thus may have greater security in terms of data collection.

To provide a better quality of the equipment, it was determined that the electronic card Arduino ESP32 which provides applications based on the new IoT technologies which allows to work with the respective devices used in the platform. In addition to this, it was considered that the implement used is in the domestic market.

When working with high-cost peripherals, these allow us to have multiple interfaces and also have a highly secure and reliable data transmission, while when selecting a low-cost peripheral, it is possible to obtain the ESP32, which has free software, and is Therefore, for monitoring and sending data to the cloud it works perfectly, therefore the benefits that the two peripherals offer us are extremely important.

The great advantage of data management in the cloud can be evidenced, demonstrating that it does not require its own infrastructure for data storage because the necessary structure of a provider can be used, as in this project where the Arduino platform was used.

References

1. Solano, N., Rey, J., Hernández, J., Duarte, T.: DISEÑO, DESARROLLO E IMPLEMENTACIÓN DE UNA PLATAFORMA IIoT PARA FORMACIÓN DE PROFESIONALES EN TECNOLOGÍAS DE LA CUARTA REVOLUCIÓN INDUSTRIAL (2021)
2. Bahrami, M.: Cloud computing for emerging mobile cloud apps. In: Proceedings - 2015 3rd IEEE International Conference on Mobile Cloud Computing, Services, and Engineering, MobileCloud 2015, pp. 4–5 (2015). https://doi.org/10.1109/MobileCloud.2015.40
3. Valdeolmillos, C.: El sector de IoT se complicará aún más en 2021 por la diversidad de opciones de conectividad (2020). https://www.muycomputerpro.com/2020/11/06/sector-iot-complicara-2021-opciones-conectividad
4. Del Val Roman, J.L.: Industria 4.0: La transformación digital de la industria (2016)
5. Pillajo, C.: Wireless Network Control Systems (2018)
6. Camargo, M., Osma, G., Ortega, H.: Plataforma de servicios basada en IoT para el monitoreo y control de un sistema de generación fotovoltaica (FV) integrado con vegetación IoT-based services platform for the monitoring and control of a photovoltaic (PV) generation system integrated with ve (2017)
7. DELTA: DX-2100RW-WW. https://www.deltaww.com/en-US/products/0603080301/3325
8. ESP: ESP32 series datasheet. Espr. Syst. 1–65 (2021). https://www.espressif.com/sites/default/files/documentation/esp32_datasheet_en.pdf
9. Novillo, O., Hernandez, J., Mazon, D., Molina, B., Cardenas, J.: Arduino y el Internet de las cosas (2018)
10. Moura, R., Ceotto, L., Gonzalez, A., Toledo, R.: Industrial internet of things (IIoT) platforms - an evaluation model. In: Proceedings - 2018 International Conference on Computational Science and Computational Intelligence CSCI 2018, pp. 1002–1009 (2018). https://doi.org/10.1109/CSCI46756.2018.00194
11. Biswas, A.R., Giaffreda, R.: IoT and and challenges, pp. 375–376 (2014)
12. Ferencz, K., József, D.: Using Node-RED platform in an industrial environment, p. 13 (2020)
13. Contreras Alvarez, J.L.: DISEÑO DE UN MODELO PARA MANTENIMIENTO PREDICTIVO EN MOTORES DE INDUCCIÓN UTILIZANDO TÉCNICAS DE LA INDUSTRIA 4.0. Universidad Tecnológica del Perú, Lima (2020)
14. Ortiz, M.: Implementación y evaluación de plataformas en la nube para servicios de IoT, pp. 13–15 (2019). https://riunet.upv.es/
15. Moreno Hernández, R.: Desarrollo de una aplicación IoT para la gestión de un hogar inteligente mediante el protocolo MQTT y Sistemas en chip (SoC) ESP32. Univ. Politècnica València 1–73 (2020). https://riunet.upv.es/bitstream/handle/10251/152402/Moreno - Desarrollo de una aplicación IoT para la gestión de un hogar inteligente mediante el pro....pdf?sequence=1&isAllowed=y

Check for updates

Discrete Loop Filter for Time-Synchronization: An Approach from Design to Real Implementation with RTL-SDR Systems

Andres Ortega[1](✉), Juan Inga[2], and Diego Lamadrid[1]

[1] Universidad ECOTEC, Samborondón, Ecuador
aortegao@uecotec.edu.ec
[2] Universidad Politécnica Salesiana, Cuenca, Ecuador

Abstract. Nowadays the toolchain for innovating wireless systems technology has a high growth rate with the inclusion of prototyping radio systems. The trend is to migrate toward Software Defined Radio (SDR) devices due to low complexity design, implementation, and speed of data transmission of radio resources. In this context, we highlight Discrete Loop Filter design features for time synchronization Phase Locked Loops (PLL) based on coefficient calculation. The digital filter Type 2 is adapted to the PLL scheme to improve the Time to Achieve Lock and reduce the steady-state error. In addition, we evaluate the filter design into PLL scheme with an RTL-SDR device to demonstrate the performance noise in the receive signal; consequently, this is adapted to other high-performance FPGA technologies for radio signal processing.

Keywords: Loop Filter · RTL-SDR · Steady state · PLL · 5G/6G

1 Introduction

The digital communications systems evolution comes from a sound stage of design and optimization methods. The performance of these systems maintains an exciting compromise between low cost and complexity to build digital signal processing over Radio Frequency Resources with System-on-chip (SoC).

Actuality, the complex architecture of analog systems are combined with modern reconfigurable DSP techniques and development by Xilinx technology through Radio Frequency System on Chip (RFSoC's) [1]. The Zynq architecture integrates Processing System, Programmable Logic (FPGA), and RF-DAC/ADC encapsulated on a single chip to carry out the innovation in future 5G/6G wireless technology [2].

The synchronization process in the receiver is one of the recurring problems in communication systems due to clocks, especially when using high frequencies. All-digital Phase Locked Loop (ADPLL) [3] has an advantage over these problems and, the same way, ease of integration into modern digital SoC.

V. Robles-Bykbaev et al. (Eds.): CITIS 2022, LNNS 607, pp. 182–190, 2023.
https://doi.org/10.1007/978-3-031-24327-1_16

Most filter design strategies are now very well established and integrated into DSP software systems such as CMOS technology that enable the high speed and low noise frequency synthesizers for radar applications [4]. The digital filters are one of the main elements of DSP and are performed using only Multiply-add operations. However, digital filter design programs allow many degrees of freedom.

The digital PLL presents challenges in managing the trade-offs between bandwidth, jitter and lock time [5]. In this context, the digital loop filter coefficients are used to optimize the loop dynamics PLL behavioral.

In [6] an All-Digital PLL for clock generation is designed without synthesis procedures where the Digitally Controlled Oscillator (DCO) is used instead of VCO for fast-locking. This intuitive method describes a more detailed loop filter with an extended dynamic range and high resolution.

Many filter design tools can be implemented for the Digital Phase-Locked Loop, such as MATLAB code [7], Simulink (e.g. System Generator), Xilinx with IP cores (e.g. HLx) or LabVIEW (without knowledge in VHDL code); and that eventually all these technologies can coexist with each other due to Embedded Systems [8].

In [9] the CORDIC algorithm is proposed on FPGA to remove harmonics from error signals, its central role is to reduce the high frequency locking time. In [10] a Polyphase Filter Bank is proposed using Vivado High-Level Synthesis (HLS) with C++ code on Xilinx RFSoC using PYNQ board which facilitates the testing through Python code.

PLL's have been studied for a long time already, however, the design and implementation of digital filters on radio systems such as RTL-SDR are part of the new concept of radio systems.

A Loop Filter Type II design is shown in Fig. 1, which is proposed in Simulink environment of MATLAB for theorical analysis, and finally is placed into to PLL scheme for real-time using the RTL-SDR 2832U [11]. This device includes a FPGA non-programmable, however, the design made in Simulink could be compatible with HDL code generation in order to implement in another Xilinx-FPGA boards.

In addition, we study the problem of tunning and synchronization for analog communications systems, which is independent of the modulation format. For this objective, we design a Loop Filter Type 2 into a PLL scheme, verifying the correct frequency-phase synchronization to achieve the shortest Time Lock.

Furthermore, the loop filter coefficients are calculated based on steady-state behavior from quiescent frequency. Finally, the Receive Spectral Signal is analyzed using Software Defined Radio with RTL device to show the correct performance of the filter design.

The procedure and parameters settings are presented in the algorithm described in Sect. 2. The theoretical and practical results via Simulink and finally with RTL-SDR implementation are demonstrated in the next section. Finally, the conclusion has been shown the performance system of the filter design.

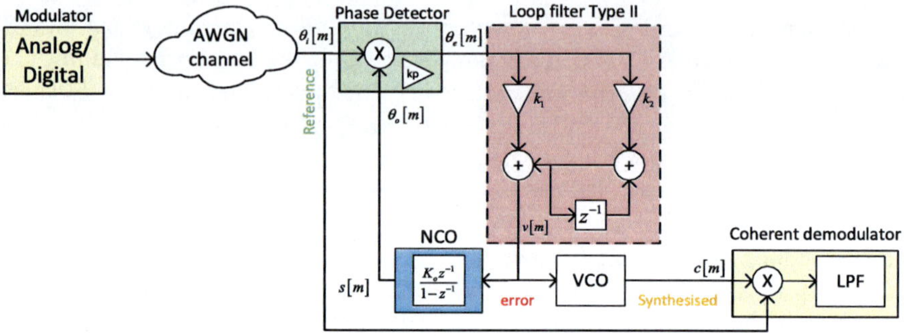

Fig. 1. Schematic of PLL design.

2 Methodology: PLL, Loop Filter and NCO Design

PLL are essential building blocks for synchronization and comprises only three components: Phase Detector, Controllable Oscillator and Loop Filter.

To design the Discrete Loop Filter Type 2, which will be coupled to the PLL for the asynchronous detection of analog and digital signals. For this analysis, we used a high-frequency signal as input to Phase Detector, where the Loop Filter works the process of discretization of the signal.

In Fig. 1, the Phase Detector in the green block, whose primary role is to generate a signal that varies in proportion to the difference in phase between the Numerically Control Oscillator's output signal (NCO), the reference signal or input to the Phase Detector.

Then, the output signal of Phase Detector is defined as:

$$\theta_e[m] = K_p(\theta_i[m] - \theta_o[m]) \tag{1}$$

Then, K_p is the phase detector gain, $\theta_i[m]$ and $\theta_o[m]$ are the estimated phases at sample m of a reference signal and numerically oscillator output signal respectively.

NCO produces the cosine and sine wave, respectively, and it is represented by $c[m] = cos\,cos(2\pi f_o mT + \theta_o[m])$ and $s[m] = sin\,sin(2\pi f_o mT + \theta_o[m])$. The following equation obtains the NCO transfer function in the discrete domain $\frac{K_0 z^{-1}}{1-z^{-1}}$. However, the NCO from Simulink is configured as follows:

- Quiescent frequency (Hz) is a constant defined by $\frac{f_s}{6}$.
- Input Sensitivity (Hz/V): is a dependent variable of the oscillation gain K_0 and sampling frequency f_s. Is defined as $K_0 f_s$.

A PLL attains 'phase lock' when the sine wave generated by the PLL converges to the same frequency as the input reference signal. For this reason, the PLL design is strictly based on the systems gains, i.e., K_1 and K_2 (in the loop filter), K_p (phase detector) and K_o (NCO gain). Together all the systems are achieved locking.

There are some optimization factors for the PLL design, we make use of the **steady-state error**, which is influenced by the Loop Filter design. As is shown in the red block of Fig. 1, the Loop Filter has two gains constants which are sufficient to achieve a **zero phase error**.

The features of Loop Filter Type 2 are:

- Can adapt the correct frequency and phase in the presence of a different initial frequency and phase.
- Can adapt a correct frequency in the presence of a dynamic frequency offset, but leave a residual error phase.

In conclusion, the Type of Loop Filter is very important due to it determines whether it can adapt to the zero-phase error point or have a residual steady-state error.

The damping ratio is another factor that affects the behavior pattern of the PLL. Is denoted by ς, where it relates to the transient behavior it achieves phase lock. It is recommended to handle small damping ratios to better track changes, generally in the range $0.5 < \varsigma < 2$ [12], which achieves a good balance of PLL behavior. For communications applications with good tracking capability $\varsigma = 0.707$.

The PLL bandwidth is another important aspect of the design because this value decides the time to achieve lock. In other words, a fast PLL adaptation is achieved with poor noise performance while a broad bandwidth is enabled; else, a slow PLL adaptation is achieved with narrow bandwidth.

The damping ratio and the PLL noise band is an implicit PLL parameter. In this context, the noise bandwidth is incorporated into the equations from the values of the loop filter coefficients, as shown in the following equations:

$$K_1 = \frac{4\varsigma\left(\frac{B_nT}{\varsigma+\frac{1}{4\varsigma}}\right)}{K_oK_p\left(1+2\varsigma\left(\frac{B_nT}{\varsigma+\frac{1}{4\varsigma}}\right)+\left(\frac{B_nT}{\varsigma+\frac{1}{4\varsigma}}\right)^2\right)} \tag{2}$$

$$K_2 = \frac{4\varsigma\left(\frac{B_nT}{\varsigma+\frac{1}{4\varsigma}}\right)^2}{K_oK_p\left(1+2\varsigma\left(\frac{B_nT}{\varsigma+\frac{1}{4\varsigma}}\right)+\left(\frac{B_nT}{\varsigma+\frac{1}{4\varsigma}}\right)^2\right)} \tag{3}$$

where:

$$\eta = \frac{B_nT}{\varsigma+\frac{1}{4\varsigma}}$$

Then, the Eqs. (2) and (3) can be replace by:

$$K_1 = \frac{4\varsigma B_n T}{K_o K_p\left(\varsigma + \frac{1}{4\varsigma}\right)} \quad (4)$$

$$K_2 = \frac{4(B_n T)^2}{K_o K_p\left(\varsigma + \frac{1}{4\varsigma}\right)^2} \quad (5)$$

Algorithm 1 Design of Loop Filter Type 2

Step 1	Calculate the sampling period with audio signal properties.
Step 2	Set the Phase Detector gain, K_p.
Step 3	Set the NCO gain, K_o
Step 4	Choose the Damping Ratio $0.5 < \varsigma < 2$
Step 5	Set the noise bandwidth B_n .
Step 6	Calculate the Loop Filter Gain Coefficients K_1, K_2

3 Results and Discuss

Thus, to validate the numerical results, the input parameters of the Algorithm 1 are set up in Table 1.

Table 1. Simulation parameter for Simulink PLL design.

Parameters	Name	Value
Sampling period	T	$\frac{1}{f_s}$
NCO gain	K_o	0.5
Phase detector gain	K_p	1
Damping ratio	ς	0.707
Noise bandwidth	B_n	5 Hz

In Fig. 2, the frequency domain result of the input signal, the Phase Detector (Referenced), with the output signal of PLL (Synthesized) is compared. Note that the Synthesized signal has a significant attenuation at high frequency. This effect is due to the cosine component of VCO, thus reducing the amplitude after the loop filter.

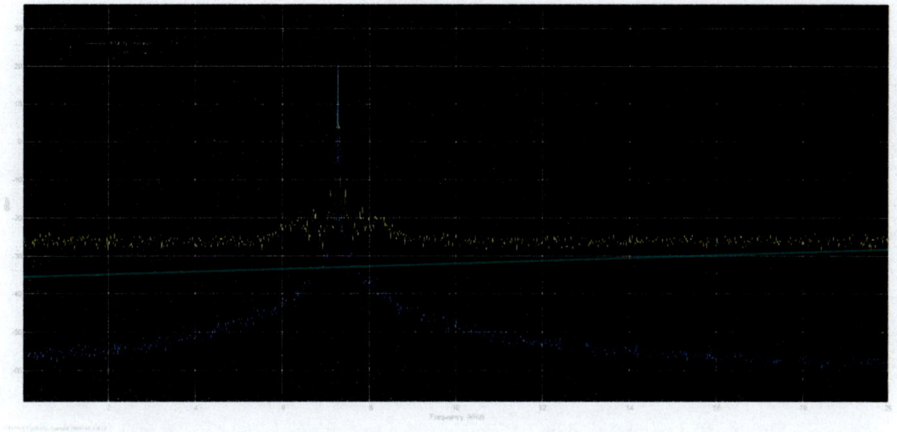

Fig. 2. Spectral analysis of loop filter output vs. referenced signal.

Figure 3, shows the Time Scope of transmitted signal vs. recovered signal with filtering and decimation process. The signal audio is sent to the AWGN channel to demonstrate the effectiveness of the synchronization problem in the PLL design.

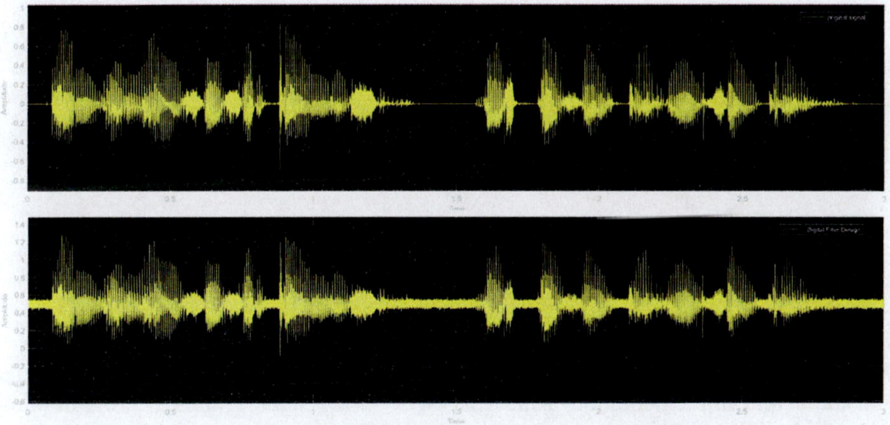

Fig. 3. Audio recovery signal (Output vs. Input).

Therefore, to determine the excellent performance of the Loop Filter, a track of the synthesized signal is made through a ramp or step signal to the input PLL. When the zero-phase error is achieved, then right away, the PLL is adapted.

The time to achieve lock is closed in 125 μs, as is shown in Fig. 4. This performance is due to the deviation of the input reference signal from the quiescent frequency of the NCO. In addition, other factors such as damping ratio ς and bandwidth are influenced by time to achieve lock. In Fig. 5 the damping ratio increases at 1.5 which means that the time of attaining certainty increases in the same way at 500 μs.

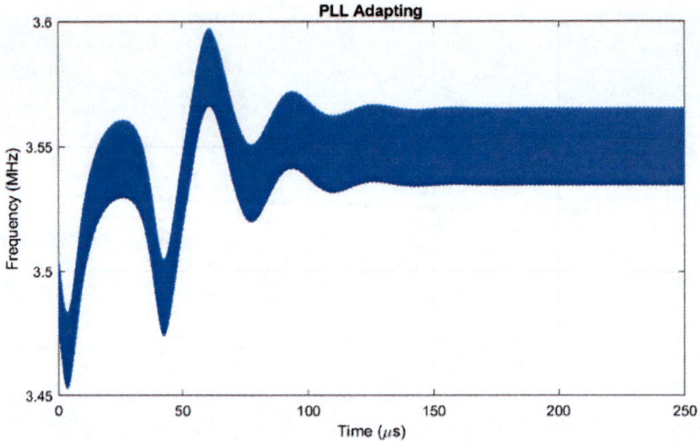

Fig. 4. PLL adaptation at 3.55 MHz (time to achieve lock), $\varsigma = 0.707$

In both graphs, the maximum peak represents the effect of steady-state behavior to achieve zero-phase error and consequently synchronize the system in the shortest possible time.

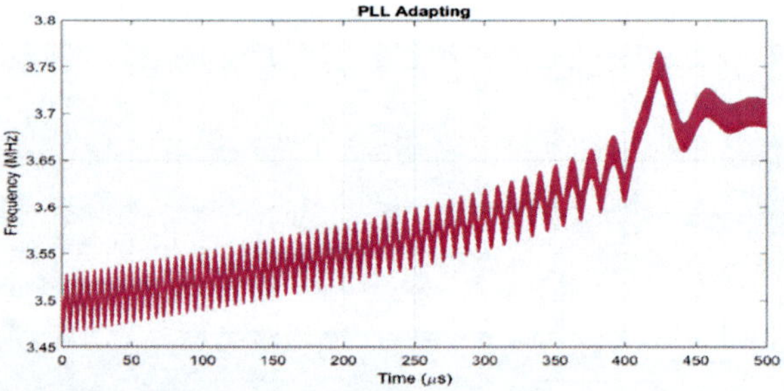

Fig. 5. PLL adaptation at 3.7 MHz (time to achieve lock), $\varsigma = 1.5$

Finally, we have compared the results from the theorical simulation in Simulink, with the actual implementation using RLT-SDR in the same graphical programming environment (Simulink), as shown in Fig. 6. To validate the correct operability of the PLL design. The document shows the recovery signal spectrum with a decimation filtering process at Fs = 48 kHz, as shown in Fig. 7.

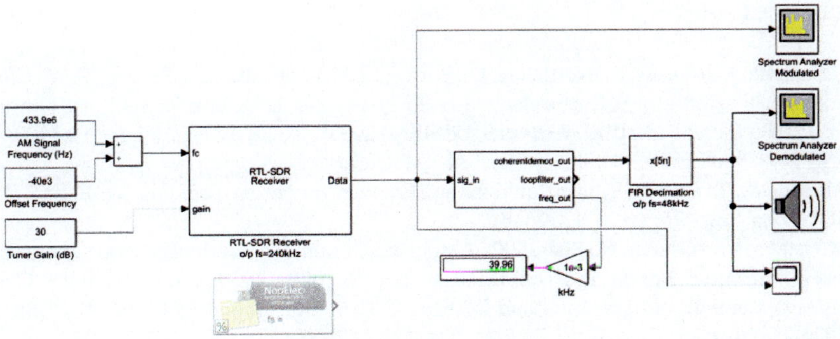

Fig. 6. RTL-SDR testing.

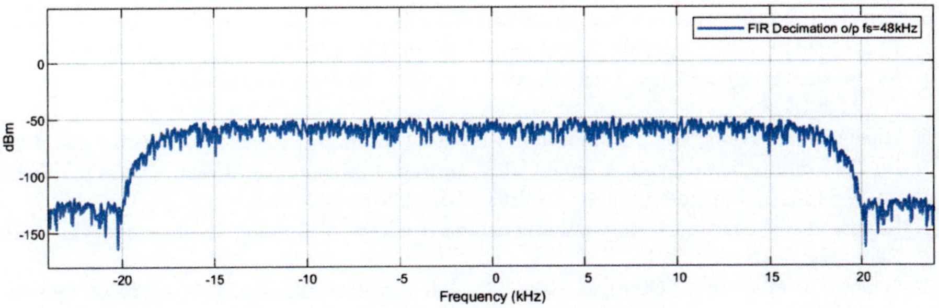

Fig. 7. Spectrum demodulated recovery with RTL-SDR at $f_s = 48$ KHz.

4 Conclusions

The proposed system has two phases previously described in the methodological framework: design and implementation.

We used the MATLAB tool Simulink, which allows us to incorporate the creation of a Loop Filter Type II to incorporate into the PLL. Finally, the design is implemented in an RTL-SDR device for the reception of the signals.

The Type of the PLL dictates the nature of its steady-state behavior. The time to achieve lock is reduced to 125 μs to achieve PLL adapting using a damping factor of 0.707 as the best result for achieving the time lock, which is highly dependent on the NCO parameters such as the quiescent frequency. These parameters are implicit in the calculus of coefficients for the filter design.

These applications can work well on more robust FPGA systems; since Simulink designs can be easily generated in HDL code to use more robust equipment such as RFSoCs for wireless digital processing. The RF developer community is invited to make more contributions with this hardware to provide solutions for developing the framework for the new 5G/6G technologies.

References

1. Goldsmith, J., Ramsay, C., Northcote, D., Barlee, K.W., Crockett, L.H., Stewart, R.W.: Control and visualisation of a software defined radio system on the Xilinx RFSoC platform using the PYNQ framework. IEEE Access **8**, 129012–129031 (2020). https://doi.org/10.1109/ACCESS.2020.3008954
2. Ahmadi, S.: Toward 5G Xilinx Solutions and Enablers for Next-Generation Wireless Systems 5G Vision, vol. 476, pp. 1–32 (2016)
3. Moustafa, A.S., Omran, H., Sharaf, K.: Automatic all-digital phase-locked loop system design and optimization tool. In: Proceedings of the ICCES 2020 - 2020 15th International Conference on Computer Engineering and Systems (2020). https://doi.org/10.1109/ICCES51560.2020.9334613
4. Donno, A., Amico, S.D., Nonis, R., Thurner, P.: Low noise active loop filter for radar PLL applications. In: 2018 International Conference on IC Design and Technology, pp. 77–80 (2018)
5. Safwat, S., Lotfy, A., Ghoneima, M., Ismail, Y.: A 5–10 GHz low power bang-bang all digital PLL based on programmable digital loop filter. In: ISCAS 2012 - 2012 IEEE International Symposium on Circuits and Systems, pp. 1371–1374 (2012). https://doi.org/10.1109/ISCAS.2012.6271498
6. Huang, M.L., Hung, C.C.: Full-custom all-digital phase locked loop for clock generation. In: 2015 International Symposium on VLSI Design, Automation and Test, VLSI-DAT 2015, pp. 0–3 (2015). https://doi.org/10.1109/VLSI-DAT.2015.7114567
7. Mathworks, C., Drive, A.H.: Communications Toolbox™ Support Package for RTL-SDR Radio User's Guide (2021)
8. Gener, A., Valverde, J., Otero, J., Harris, P.: A fast prototyping workflow for reconfigurable SDR applications, vol. 3, pp. 1–9 (2019)
9. Andrich, C., Bauer, J., Grose, P., Ihlow, A., Del Galdo, G.: A fast and stable time locked loop for network time synchronization with parallel FLL and PLL. In: IEEE International Symposium on Precision Clock Synchronization for Measurement, Control, and Communication, ISPCS, vol. 2018-September, pp. 1–4 (2018). https://doi.org/10.1109/ISPCS.2018.8543066
10. Smith, J.P., et al.: A high-throughput oversampled polyphase filter bank using Vivado HLS and PYNQ on a RFSoC. IEEE Open J. Circuits Syst. **2**(October 2020), 241–252 (2021). https://doi.org/10.1109/ojcas.2020.3041208
11. Saber, M., Aroussi, H.K., El Rharras, A., Saadane, R.: Raspberry Pi and RTL-SDR for spectrum sensing based on FM real signals. In: International Conference on Multimedia Computing and Systems - Proceedings, vol. 2018-May, pp. 1–6 (2018). https://doi.org/10.1109/ICMCS.2018.8525867
12. Costamilan, E., Löw, A.M., de Freitas Awruch, M.D., Amico, S.C., Gomes, H.M.: Experimental damping ratio evaluation using Hilbert transform in filament-wound composite plates. Polym. Polym. Compos. **29**(9_suppl), S1578–S1587 (2021). https://doi.org/10.1177/09673911211002063

Simulation by CFD of the Internal Baffles of the Condenser in a 300 L Heat Pump for Residential Water Heating with Refrigerants R717 and R744

Stalyn Oleas, Páez Kevin, Fernando Toapanta-Ramos(✉), and William Quitiaquez

Universidad Politécnica Salesiana, Quito, Ecuador
{ltoapanta,wquitiaquez}@ups.edu.ec

Abstract. This document compares the best alternatives regarding two working fluids and three operating conditions for a helical baffle in a shell and tube heat exchanger. It begins by investigating issues related to the study problem, validated by scientific documents, the most relevant properties of the working fluids and the material used (INOX 304) for the study are presented. The deductive analytical method used with the characteristics obtained shows how the heat transfer area increases as the angle of the baffle increases, the boundary conditions of the water are also detailed as: 10 °C at the entrance with a velocity of 0.8 m/s at a pressure of 25 psi evaluated in the Ansys Fluent software, determining that the best option for residential water heating is the use of refrigerant R744 with a 25° inclination for the helical baffle. The results show that under these parameters a better heat transfer is obtained reaching a temperature of 31.5 °C.

Keywords: Ammonia · Helical baffles · Carbon dioxide · Heat exchanger

1 Introduction

The refrigeration and heating industry is one of the main causes of the deterioration of the ozone layer, especially due to the use of refrigerant gases of the CFC and HCFC type, which have good thermodynamic properties, however, the emissions of hydrofluorocarbon gases (HFC) that also they favor the greenhouse effect, being thousands of times more harmful than CO2 [1]. It is essential to control this circumstance and mitigate the rise in temperature on Earth below 2 °C and even restrict this data to a range of 1.5 °C or less [2]. The search for refrigerant fluids that are respectful of the environment is crucial to achieve a significant reduction in the negative effects and risks caused by the increase in global temperature.

The use of hydrofluorocarbons (HFC), despite being an aid as a substitute for the refrigerants, has an excessive global warming potential (GWP), representing an increasing threat to the 10 to 15% in greenhouse gas (GHG) emissions each year, because of

V. Robles-Bykbaev et al. (Eds.): CITIS 2022, LNNS 607, pp. 191–203, 2023.
https://doi.org/10.1007/978-3-031-24327-1_17

this, in 2016 the agreement is reformulated to also rule out the use of HFC refrigerants worldwide.

Refrigeration is that process in which there is a movement of heat through one place to another with the intervention of a refrigerant in liquid or gaseous phase in a closed cycle [3].

Heat exchangers can commonly be seen in industrial processes where the heat energy that is conceived is used to preheat air and/or water for multiple industrial and commercial applications. According to Mangrulkar et al. [4] within the classes of heat exchangers, the most used are the tubular ones because they are relatively optimal for their design and their maintenance cost is low, this classification of exchangers can be subdivided into heat exchangers. Heat from concentric tubes, tube and shell exchangers, plate exchangers, among others.

Most of this equipment involve two currents, which implies heat exchange through metallic surfaces. A fluid is usually transported through tubular elements, forming a tube bundle. When a tube bundle is housed within a cylindrical shell, the total assembly is called a shell-and-tube heat exchanger [5].

The technology and design of heat exchangers with helical baffles was first developed [6] in Czechoslovakia in 2007. Also called "helixchanger", this design focuses on minimizing the deficiencies that have been manifested in shell and tube heat exchangers with segmented baffles.

A study of exchangers with helical baffles was carried out by Stehlík et al. [7] in which they compared the pressure drop and heat transfer correction factors in a segmented baffle exchanger and one with helical baffles; For design purposes, these correction factors in the helical baffles were analyzed as a function of the angle of inclination of the baffle.

For an efficient heat transfer case, the heat exchanger (IC) must have low pressure drop, high shell-side mass flow, high heat transfer coefficient, and no or very low fouling, and so on. Heat transmission also depends on the turbulence created on the shell side, this condition can be created using baffles [8].

Hu et al. [9] state that the heat transfer capacity for these heat exchangers can be significantly improved with the use of baffles which can be segmented or helical.

Numerical simulations carried out by El Maakoul et al. [10] indicate an improvement regarding both thermal and hydraulic performance in a double tube heat exchanger with helical baffles. In its analysis, a comparative study is carried out between three different exchangers, the first being of segmented baffles commonly found in the industry, the second exchanger of tubes and shell with helical baffles and the third consists of tubes with fins and helical baffles. Their results indicated that when using baffles, a higher thermo-hydraulic performance is obtained, for the baffles that have clover-shaped holes they showed a higher performance in terms of heat transfer, with a notable increase in pressure compared to the first exchanger. With segmented baffles.

The comparative study carried out by Hu et al. [9] on heat exchangers with geometric structures other than ladder-type helical baffles with various inclinations for the angles and their length with the segmented baffles. The results obtained by this analysis demonstrate the impacts on the flow patterns of the different arrangements of the baffles. On the shell side, the heat transfer coefficient is higher in the helical baffles compared to the segmented baffles, concluding that the angle of inclination and the helical pitch intervene directly in the efficiency of the exchanger. Pressure drops and heat transfer coefficient decrease in longer exchangers compared to shorter exchangers. Unlike the device with segmented baffles, all exchangers show lower pressure drops.

According to the investigations of Peng et al. [11] the design of helical baffles shows that heat exchangers that use helical baffles increase at least 10% in their heat transfer coefficient compared to an exchanger that uses segmented baffles at a given pressure drop.

The research carried out by Wu et al. [12] affirms that natural refrigerants such as R744 are a good choice when outlet temperatures not higher than 90 °C are required, while for temperatures below 100 °C it is more suitable the R717. It is intended to meet the demand for a comfort temperature for the water of 45 °C at the outlet [13], the turbulence generated by the helical baffle will be evaluated analytically and with the help of software. Computational to check if this variable is adequate to achieve this objective and to verify which is the best angle as an operating condition for the baffle element. In addition to illustrating which of the working fluids is the one with the best thermodynamic properties to carry out the work of heating water at a residential level.

The refrigerant R744 (carbon dioxide) according to Ara [14] has become an excellent alternative within the refrigeration and air conditioning sector, commercial refrigeration being one of the areas in which it is most used. The simulation carried out by Diaby et al. [15] maintains that CO_2 as a refrigerant can broaden the framework of applications for multifunction heat pumps.

A relevant dimension that they mention in their research Li et al. [16] is about a single negative point that the refrigerant R744 presents, it is its high working pressure, and it is expected to find new methods to correct this problem.

Another possible alternative in the inorganic category for the replacement of refrigerants such as R22 and R502 is the refrigerant R717 (ammonia). Its commercialization on a world scale reaches approximately 120 million tons, where only 5% of this total is destined for the refrigeration sector [17]. According to Li et al. [18] it has a higher resonance frequency than other refrigerants such as R600, R1234yf and R134a under the same operating conditions and in turn also requires a higher energy input which allows it to provide high performance in terms of cooling.

The main objective of this research is to evaluate the appropriate angle of the helical baffles inside a shell and tube heat exchanger, this under the ANSYS CFD numerical simulation approach.

2 Methodology

2.1 Analysis Description

The research is mainly developed using ANSYS Fluent simulation software of residential water heating through a helical baffle heat exchanger which is provided with 23 tubes and made of stainless steel as illustrated in Fig. 1.

In the design you can see the assembly of the heat exchanger provided with a shell, two heads which contain the respective inlets and outlets of the fluids, the bundle of tubes and the helical baffle arranged for heat transmission.

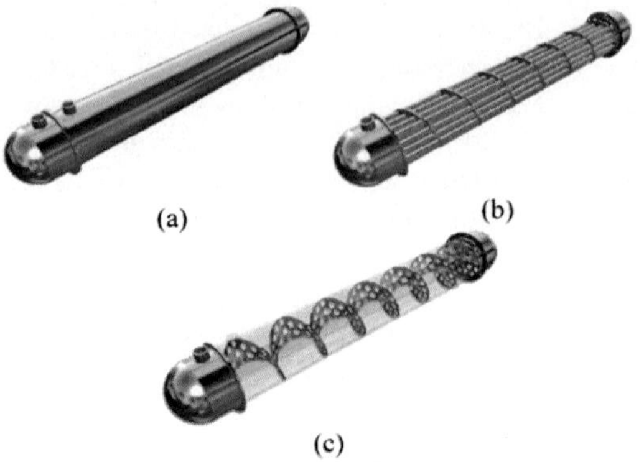

Fig. 1. Heat Exchanger design: (a) complete, (b) tubes with baffle and headers (c) baffle and headers.

The study consists of varying the angle of the baffle helix (25°, 30° and 35°) as illustrated in Fig. 2, generating different operating conditions, in addition to the use of refrigerant fluids R717 and R744 for each case study.

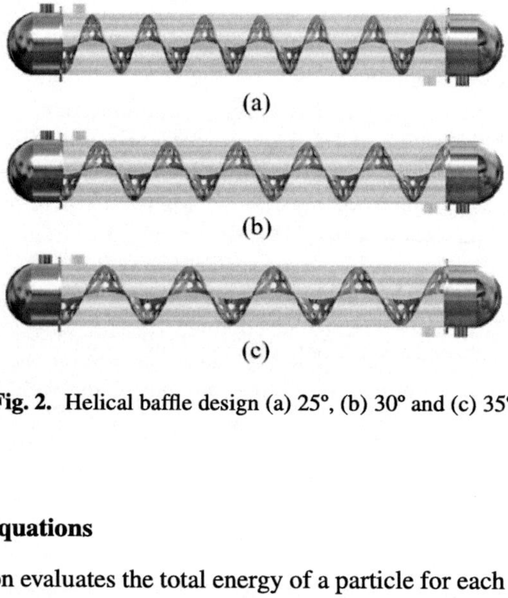

Fig. 2. Helical baffle design (a) 25°, (b) 30° and (c) 35°.

2.2 Governing Equations

The energy equation evaluates the total energy of a particle for each unit of volume, the equations of mass and momentum are involved, and this equality is considered the last one to enable the resolution of flow variable problems. If a flow with minimum velocity is considered ignoring the dissipation of viscosity, the energy equation can be written in terms of enthalpy specific.

$$\frac{\partial}{\partial t}(\rho E) + \nabla \cdot \left[\vec{v}(\rho E + p)\right] = \nabla \cdot \left[K_{eff} \nabla T - \sum_{j} h_j \vec{J_j} + \left(\bar{\bar{\tau}}_{\cdot eff} * \vec{v}\right)\right] + s_h \quad (1)$$

This equation, also known as the mass conservation equation, is based on the principle of conservation of matter, describing the conservation of charge, mass, quantum number, energy, and momentum that occurs in the heat exchange process. It can be written in the following form:

$$\frac{\partial}{\partial t}(\alpha_q \rho_q) + \nabla(\alpha_q \rho_q \overrightarrow{v_q}) = \sum\nolimits_{p=1}^{n} (\dot{m}_{pq} + \dot{m}_{qp}) + S_q \quad (2)$$

The standard k-Ɛ turbulence model is considered standard and has its origin in the equations of turbulent kinetic energy and its average dissipation in a section, the k model starts from the original equation while the ε model is achieved through distant mathematical comparisons and studies. Physical. ANSYS Fluent considers the following equations:

$$\frac{\partial}{\partial t}(\rho k) + \frac{\partial}{\partial x_i}(\rho k u_i) = \frac{\partial}{\partial x_j}\left[\left(\mu + \frac{\mu_t}{\sigma_k}\right)\frac{\partial k}{\partial x_j}\right] + G_k + G_b - \rho\varepsilon - Y_M + S_k \quad (3)$$

$$\frac{\partial}{\partial t}(\rho\varepsilon) + \frac{\partial}{\partial x_i}(\rho\varepsilon u_i) = \frac{\partial}{\partial x_j}\left[\left(\mu + \frac{\mu_t}{\sigma_\varepsilon}\right)\frac{\partial \varepsilon}{\partial x_j}\right] + C_{1\varepsilon}\frac{\varepsilon}{k}(G_k + C_{3\varepsilon}G_b) - C_{2\varepsilon}\rho\frac{\varepsilon^2}{k} + S_\varepsilon \quad (4)$$

The k-Ɛ RNG turbulence model is the statistical technique of renormalization group theory that gives rise to this model, providing it with some improvements such as:

It is ideal for rapidly forced flows since it includes one more term, allowing it to be more exact for the study of these cases. In this model, the turning effect of turbulent fluids is included, increasing the accuracy in stirred flows. It provides an equation used in Prandtl numbers in cases of turbulent flow, whereas the standard specifies these values as constants for the user.

Unlike the standard model that is suitable for high magnitudes of Reynolds number, the RNG model provides a differential equation in cases of effective viscosity that considers the consequences of obtaining a Reynolds number of low magnitude.

It is important to emphasize that to use this model, it must be considered to adequately treat the section closest to the wall.

2.3 System Meshing

Once the parametric geometry of the 3D model has been built in the CAD section of Ansys (Design Modeler), the next step is to mesh the design. The meshing tool that this software has is quite useful since it allows evaluating the quality that the system acquires, delimiting also that it has several functions and meshing methods that can be used in the process to refine any design to achieve the best possible quality. Additionally, boundary conditions that are necessary to enter preset values for subsequent simulation can be placed. Figure 3 illustrates each of the meshes belonging to the most relevant parts involved in the heat exchanger to be evaluated.

Once the mesh is executed and completed, its quality is verified by means of the oblique metric graphic parameter "Skewness" to visualize the qualification of the mesh as far as quality is concerned. Thanks to the large number of projects oriented to the type

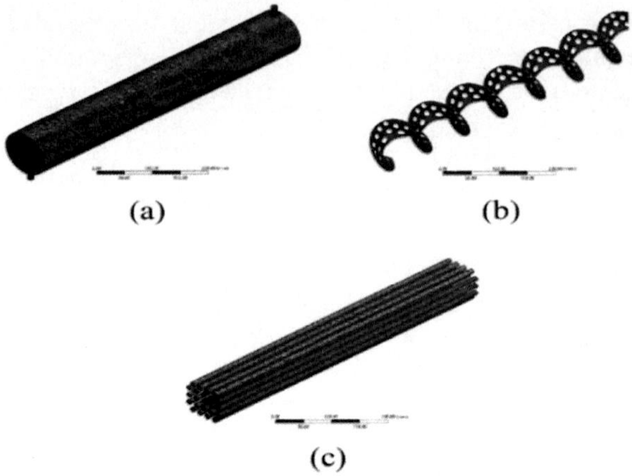

Fig. 3. Mesh: (a) complete system, (b) baffle, and (c) tubes with refrigerant.

of CFD study in Ansys, it is revealed that the standard quality for the result of the meshing process can vary between 0–0.25 units for the mesh parameter with the objective that the results are as timely as possible and significantly resemble reality (Fig. 4).

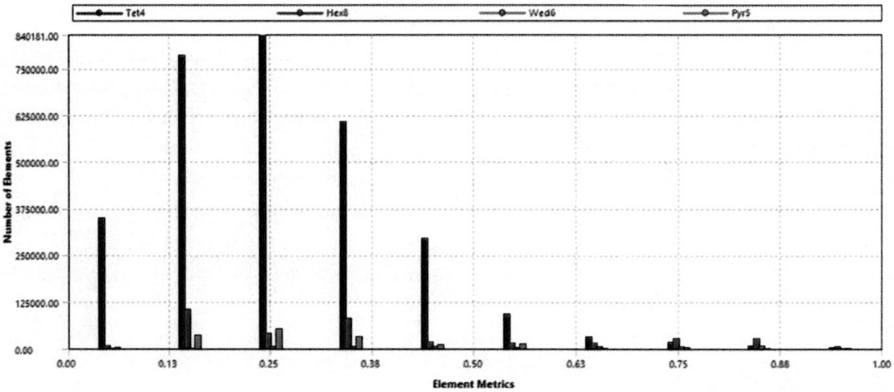

Fig. 4. Mesh convergence of the heat exchanger system with full helical baffles.

3 Results and Discussion

3.1 Analysis of the Heat Exchanger with Refrigerant R717

Figure 5 illustrates the temperature behavior curves obtained through the results section of the ANSYS software, indicating the trend that the water fluid acquires inside the shell of the heat exchanger working with the R717 refrigerant for each operating condition.

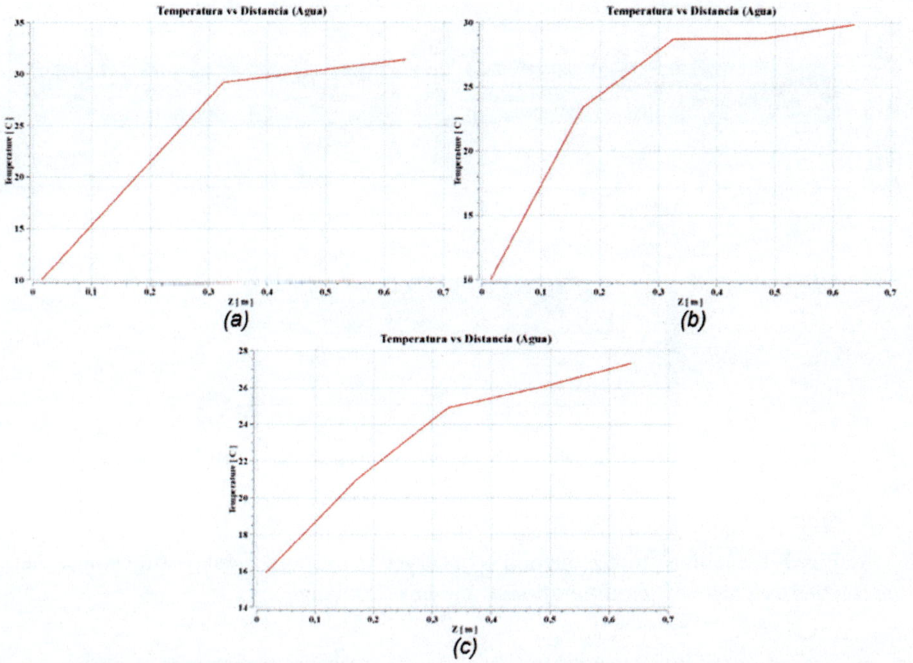

Fig. 5. Thermal behavior of the water through the inclined baffle (a): 25°; (b): 30° and (c): 35° with R717.

The graph shows the increase in temperature of the fluid to be heated from the 10 °C until reaching approximately 31.5 °C in section (a), while in section (b) it approaches 30 °C and in (c) it approaches 27.5 °C, during the lapse of 5 s established in the respective simulation.

For a more notable appreciation of the temperature, Fig. 6 shows the outline of this variable throughout the entire unit, where the respective contrast to the transfer phenomenon that occurs between the water, the tubes and the refrigerant is observed.

Each section of Fig. 6 shows the heating of the water inside the shell of the IC from 10 °C with blue coloration starting from 0 to 650 mm that the shell has, while it advances through the system the color tends to be visualized clearer, it acquires a more yellowish color indicating that the water is heated until it reaches the temperatures detailed in the previous section. On the other hand, the same thing happens with the coloration of the refrigerant, which adopts a reddish color at the inlet, being at 100 °C at its maximum temperature and a range between 75 to 76 °C for the lowest temperature at the outlet, revealing the cooling of the coolant. It is important to clarify that the greenish color that can be perceived surrounding the tubes is where the heat transfer phenomenon is located.

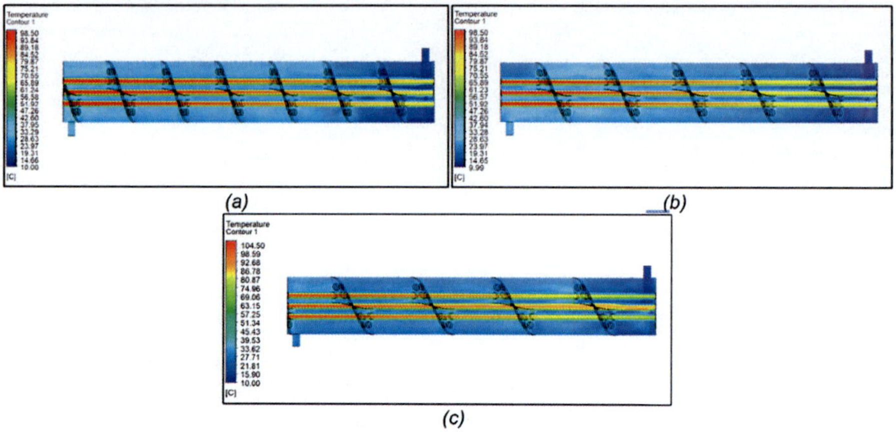

Fig. 6. Inclined baffle heat exchanger temperature contours (a): 25°; (b): 30° and (c): 35° with R717.

In Fig. 7, the temperature reduction is observed in the horizontal section where there are 5 tubes that have the configuration of the heat exchanger.

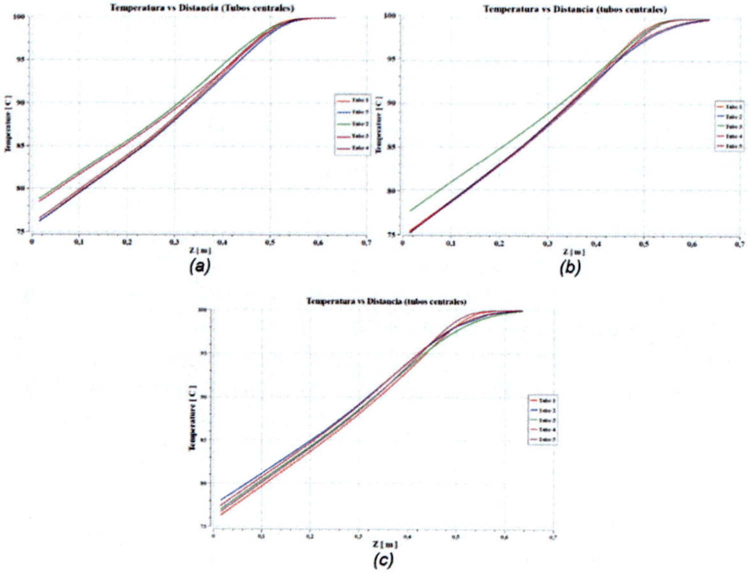

Fig. 7. Thermal behavior of R717 through the five central tubes of the IC, baffle tilt (a): 25°; (b): 30° and (c): 35°

The temperature decreases in each of the tubes and it is observed that the tubes in the central part of the array have a slight difference in magnitude with respect to the tubes that are in the peripheries of the bundle, the variation is approximately 2.5 °C This is due to the fact that in the peripheries, as there is a smaller contour of the cold fluid that surrounds these tubes, they tend to raise their temperature, unlike those found in the regions closer to the center, where there is a greater amount of fluid.

Figure 8 shows the velocity of the same arrangement of 5 horizontal tubes previously mentioned, which correspond to the ammonia gas section.

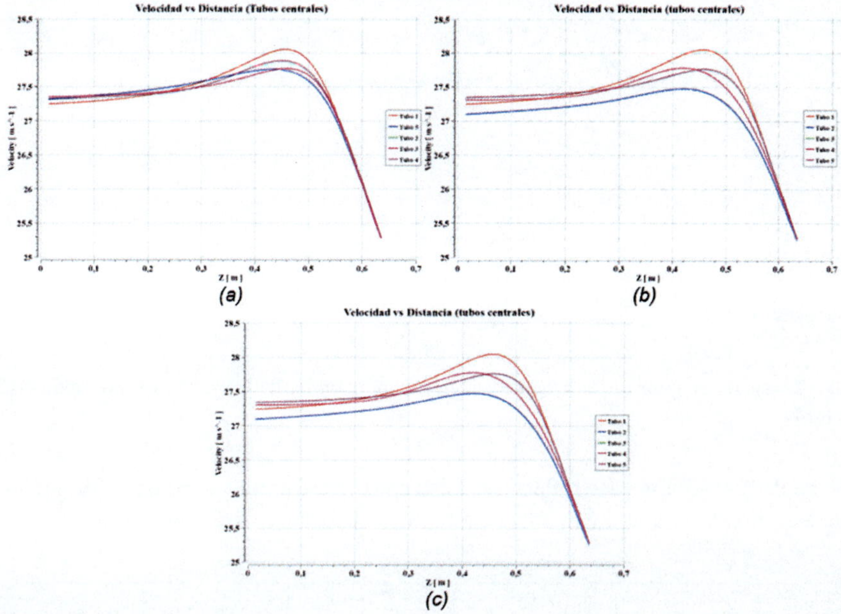

Fig. 8. Behavior of R717 through the five central IC tubes with baffle tilt (a): 25°; (b): 30° and (c): 35°.

In the graph the speed of each of the tubes behaves similarly to each other in the three illustrations, remembering that the system is in counterflow, it follows that the speed of the refrigerant in the section of the tubes increases between 2.1 to 2.4 m/s.

3.2 Analysis of the Heat Exchanger with Refrigerant R744

Figure 9 graphically illustrates the results obtained through the results section of the ANSYS software, indicating the trend that the water fluid acquires inside the heat exchanger with helical baffles working with the R744 refrigerant.

This figure again shows the increase in temperature of the fluid to be heated from 10 °C to approximately 33 °C in section (a), while in section (b) it approaches 31 °C and in (c) approaches 27.2 °C, for 5 s established in the respective simulation.

Figure 10 shows the heating of the water immersed in the shell of the IC from10 °C represented with blue color starting from 0 to 650 mm that the shell has where it reaches the previously described temperatures, the colors become lighter due to heat gain and loss that occurs between the cold and hot fluid, respectively. The same event is evidenced

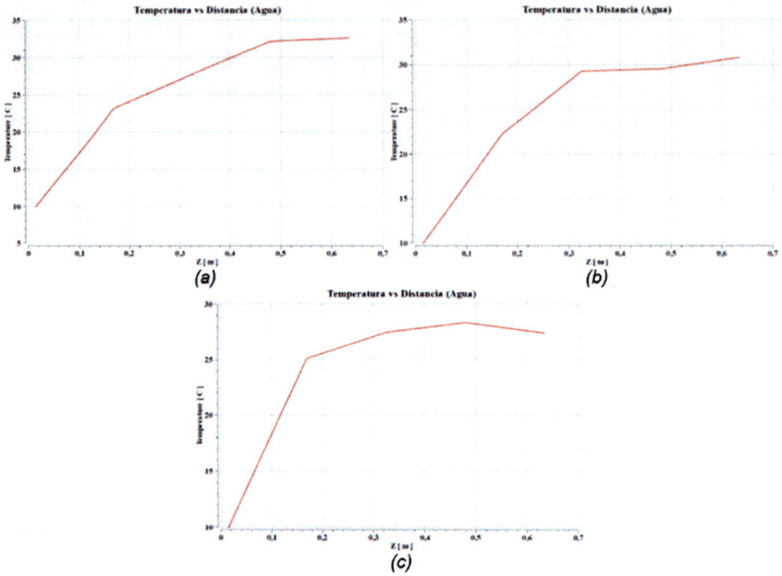

Fig. 9. Thermal behavior of the water through the inclined baffle (a): 25°; (b): 30° and (c): 35° with R744.

in the coolant; It continues to reflect a reddish color at the entrance, being at 88 and 68 °C at the exit, implying that the loss of heat in this substance is being generated.

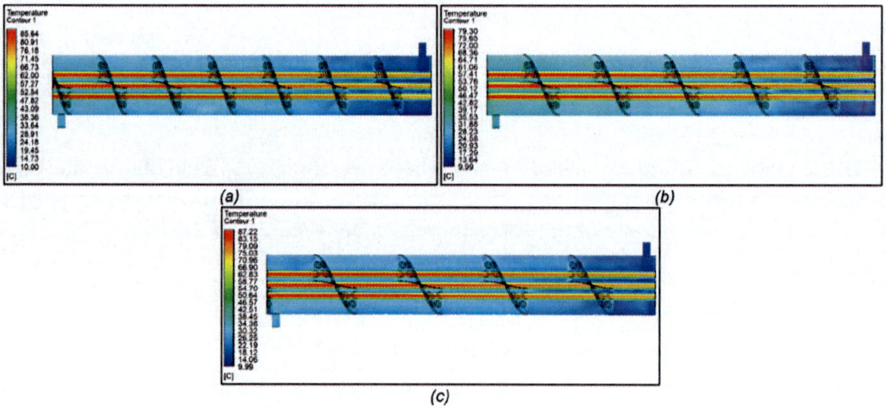

Fig. 10. Inclined Baffle Heat Exchanger Temperature Contours (a): 25°; (b): 30° and (c): 35° with R744.

In Fig. 11, you can see the temperature reduction in the horizontal section where there are 5 tubes of the total tube bundle that the heat exchanger configuration has.

In Fig. 12 the speed of the same arrangement of 5 horizontal tubes for the baffle with 35° of inclination is illustrated.

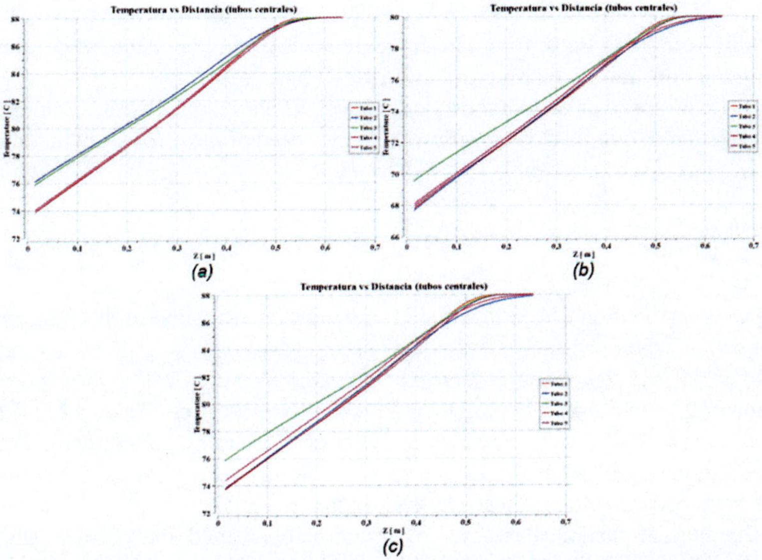

Fig. 11. Thermal behavior of R744 through the five central tubes of the IC, baffle tilt (a): 25°; (b): 30° and (c): 35°.

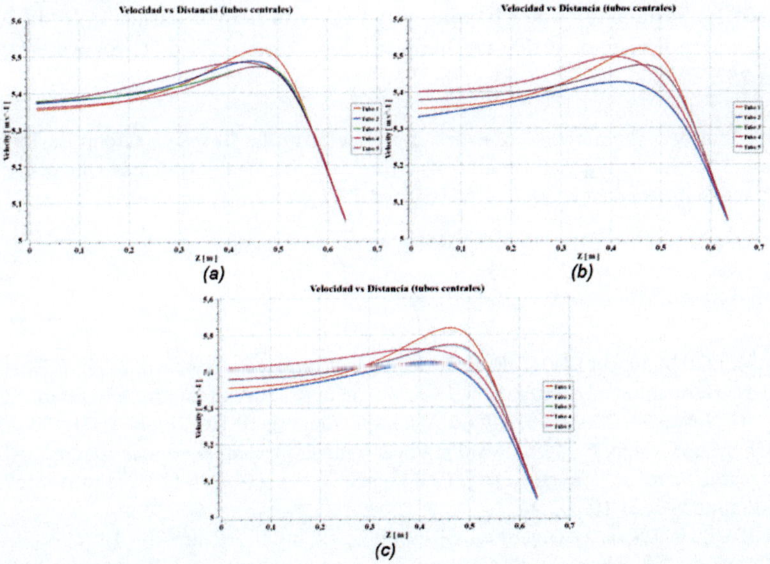

Fig. 12. Behavior of R744 through the five central tubes of the IC with inclination of (a): 25°; (b): 30° and (c): 35°.

The same event regarding the decrease in temperature also occurs in each of the tubes since the array does not have or undergoes any type of change, showing once again the slight dissimilarity of 2 °C that the tubes located on the peripheries of the tube have bundle, with respect to those found in the regions closer to the center where there is a greater amount of fluid circulating in the surroundings.

In Fig. 12 the speed of each of the tubes behaves similarly to each other, remembering that the system is in counterflow, it follows that the speed of the refrigerant in the section of the tubes increases between 0, 33 to 0.41 m/s.

4 Conclusions

The angles of inclination 25°, 30° and 35° of the baffle evaluated in this study present different results when working with both R717 and R744 obtaining that when selecting the angle of 30° the temperature decreases its value by 5 and 7% with respect to the inclination of 25°, while with the angle of 35° the temperature decreases 12 and 17% in relation to the angle of 25° its value at the outlet of the heat exchanger using ammonia and carbon dioxide, respectively. In this way it follows that the helical propeller pitch of 25° is the most suitable for these working fluids.

For the thermal design of the system, it can be confirmed analytically that the IC that works with carbon dioxide needs less heat transfer area compared to ammonia with a difference of 1.57; 0.1 and 1.02% for the angles of 25°, 30° and 35° respectively, reaching a theoretical total heat transfer rate of 2.9175 kW.

Through the ANSYS Fluent simulation tool, it is confirmed that the best refrigerant to be used for residential water heating in an IC with helical baffles is R744 (carbon dioxide), where it is evident that the highest temperature is 31.5 °C for water with the angle of 25°.

Acknowledgment. The authors are especially grateful to the Research Group on Renewable Energies and Mechanical Implementation of SMEs of the Salesian Polytechnic University, Quito-Ecuador, and to the Student Branch of ASHRAE UPS Quito.

References

1. Franco, I.G., Marcucci Pico, D.F., Dall'Onder dos Santos, D., Bandarra Filho, E.P.: A review on the performance and environmental assessment of R-410A alternative refrigerants. J. Build. Eng. **47**(November 2021), 103847 (2021). https://doi.org/10.1016/j.jobe.2021.103847
2. Lv, H., Ma, H., Mao, N., He, T.: Boiling heat transfer mechanism of environmental-friendly refrigerants: a review. Int. J. Refrig. **133**(August 2021), 214–225 (2022). https://doi.org/10.1016/j.ijrefrig.2021.10.007
3. ASHRAE, Handbook-Refrigeration SI Edition. SI Edition, Atlanta (2014)
4. Mangrulkar, C.K., Dhoble, A.S., Chamoli, S., Gupta, A., Gawande, V.B.: Recent advancement in heat transfer and fluid flow characteristics in cross flow heat exchangers. Renew. Sustain. Energy Rev. **113**(November 2018), 109220 (2019). https://doi.org/10.1016/j.rser.2019.06.027
5. Master, B.I., Chunangad, K.S., Boxma, A.J., Kral, D., Stehlík, P.: Most frequently used heat exchangers from pioneering research to worldwide applications. Heat Transf. Eng. **27**(6), 4–11 (2006). https://doi.org/10.1080/01457630600671960

6. Jafari Nasr, M.R., Shafeghat, A.: Fluid flow analysis and extension of rapid design algorithm for helical baffle heat exchangers. Appl. Therm. Eng. **28**(11–12), 1324–1332 (2008). https://doi.org/10.1016/j.applthermaleng.2007.10.021
7. Stehlík, P., Němčanský, J., Kral, D., Swanson, L.W.: Comparison of correction factors for shell-and-tube heat exchangers with segmental or helical baffles. Heat Transf. Eng. **15**(1), 55–65 (1994). https://doi.org/10.1080/01457639408939818
8. Bichkar, P., Dandgaval, O., Dalvi, P., Godase, R., Dey, T.: Study of shell and tube heat exchanger with the effect of types of baffles. Proc. Manuf. **20**, 195–200 (2018). https://doi.org/10.1016/j.promfg.2018.02.028
9. Hu, H., Yang, G., Ding, G., Chen, J., Yang, W., Hu, S.: Heat transfer characteristics of mixed hydrocarbon refrigerant flow condensation in shell side of helically baffled shell-and-tube heat exchanger. Appl. Therm. Eng. **133**(August 2017), 785–796 (2018). https://doi.org/10.1016/j.applthermaleng.2018.01.083
10. El Maakoul, A., Laknizi, A., Saadeddine, S., Ben Abdellah, A., Meziane, M., El Metoui, M.: Numerical design and investigation of heat transfer enhancement and performance for an annulus with continuous helical baffles in a double-pipe heat exchanger. Energy Convers. Manag. **133**, 76–86 (2017). https://doi.org/10.1016/j.enconman.2016.12.002
11. Peng, B., et al.: An experimental study of shell-and-tube heat exchangers with continuous helical baffles. J. Heat Transfer **129**(10), 1425–1431 (2007). https://doi.org/10.1115/1.2754878
12. Wu, D., Hu, B., Wang, R.Z.: Vapor compression heat pumps with pure Low-GWP refrigerants. Renew. Sustain. Energy Rev. **138**(April 2020), 110571 (2021). https://doi.org/10.1016/j.rser.2020.110571
13. Hervás-Blasco, E., Navarro-Peris, E., Corberán, J.M.: Closing the residential energy loop: grey-water heat recovery system for domestic hot water production based on heat pumps. Energy Build. **216**, 109962 (2020). https://doi.org/10.1016/j.enbuild.2020.109962
14. Ara, S.R.: Análisis y cálculo de sistemas de producción de ACS y calefacción con bomba de calor de CO2 (2015)
15. Diaby, A.T., Byrne, P., Maré, T.: Simulation of heat pumps for simultaneous heating and cooling using CO2. Int. J. Refrig. **106**, 616–627 (2019). https://doi.org/10.1016/j.ijrefrig.2019.03.010
16. Li, S., Li, S., Zhang, X.: Simulation research of a hybrid heat source heat pump using R134a, R744 instead of R22 for domestic water heating in residential buildings. Energy Build. **91**, 57–64 (2015). https://doi.org/10.1016/j.enbuild.2015.01.019
17. Lorentzen, G.: Ammonia: an excellent alternative. Int. J. Refrig. **11**(4), 248–252 (1988). https://doi.org/10.1016/0140-7007(88)90082-5
18. Li, Z., Jiang, H., Chen, X., Liang, K.: Comparative study on energy efficiency of low GWP refrigerants in domestic refrigerators with capacity modulation. Energy Build. **192**, 93–100 (2019). https://doi.org/10.1016/j.enbuild.2019.03.031

Leveraging Modal Analysis for Multi-scenario Power System

Gabriel Rodríguez(✉) and José Oscullo

Department of Electric Energy, National Polytechnic School, Quito, Ecuador
gabriel.rodriguezp2@epn.edu.ec

Abstract. In this paper the small-signal stability of the New England 39-Bus Test System, which was modeled by PowerFactory of DIgSILENT, is studied for which a review of the state of the art of modal analysis as tool to obtain the oscillatory modes of an electrical power system. For its study and relying on a Python script, the engine mode of PowerFactory is used to determine the graphic evolution of the trajectory of the electromechanical oscillation modes, paying special attention to the critical modes whose frequency is found between 0.1 to 2 Hz. To obtain the critical modes for various scenarios, a script was developed in DIgSILENT Programming Language (DPL) by PowerFactory is proposed, which considering the variation of the generation offer, for this new operating condition, the modes of the system are determined by means of the modal analysis tool available in PowerFactory. Finally, and in order to visualize the effect produced by the displacement of the critical oscillation modes, through a short circuit in a transmission line is sought observe the RMS simulation of generator variables that vary its potency.

Keywords: Stability of electrical power systems · Small signal stability · Modal analysis · Oscillatory modes · Damping level

1 Introduction

The angular stability is the ability of the power system to maintain the synchronism of the operation of the different generation units while disturbances are present, in normal or quasi-steady operating conditions [1].

When small disturbances appear due to the operation system, it must have the capacity to maintain the generation-demand balance, through the change of state variables and parameters of the production elements and control system. Usually this situation normally generates electromechanical oscillations positively damped, that is, the system reaches a new stable operating point. However, in some particular situation it is possible that oscillations occur with insufficient damping which are characterized by they are low frequency and can lead to loss of system synchronization [2].

Electromechanical oscillations occur in a frequency range of 0.1 to 3 Hz corresponding to the exchange of kinetic energy between the rotating masses of the synchronous machines and the energy stored in the rotating masses present in the system. This type of oscillation occurs in a period of time from milliseconds to one second [3].

V. Robles-Bykbaev et al. (Eds.): CITIS 2022, LNNS 607, pp. 204–213, 2023.
https://doi.org/10.1007/978-3-031-24327-1_18

The exchange of energy through electromechanical oscillations is classified into: [1] local mode, when a generator oscillates differently from the rest of the generators in the system, this kind of oscillation corresponds to a frequency range between 1 and 2 Hz. Intra-plant mode, when a generator or a group of generators of a plant oscillate with each other at a frequency between 2 to 3 Hz; and inter-area mode, in which a group of generators from one area oscillate with respect to a group of generators from another area, these oscillations are characterized by frequencies between 0.1 and 1 Hz [2].

The most critical low-frequency electromechanical oscillations are those that are originated from inter-area oscillation modes given that a group of generators in one area oscillate differently from generators in another area. To analyze this phenomenon, the mathematical tool of modal analysis and linearization techniques of the system model are used around a point of balance, corresponding to a steady-state operating condition of the system, with which it is possible to determine the damping and frequency of the oscillations based on the eigenvalues of the system, which are known as modes of the system. By means of these variables, it is possible to establish control measures to provide the appropriate damping level to the system [4].

Currently, power systems are undergoing a transformation where the preponderant presence of classic synchronous machines is being replaced by generation technology that uses power electronics elements to connect to the network, a situation that is establishing less inertia in the power system. Naturally, the masses of the synchronous machines provide inertia to the system such that they give it the capacity to damp oscillations when disturbances occur. However, under this new reality, the system is leading to a greater presence of low-frequency oscillation modes, which reduce the power transmission capacity in the lines, improper activation of protection equipment that causes lines, transformers and generators to go out of service [5].

These situations have been analyzed considering the modal analysis and the simulation of power systems, as it is presented in [6] where the demand is considered as the parameterization parameter to obtain the modal analysis. In [7], a small-signal stability analysis and transient analysis are carried out for the scenario of a severe disturbance when considering the income of wind generation in which it establishes different all methods to improve its system transient performance. In [8] it presents the use of local signals in conjunction with the frequency to eliminate inter-area electromechanical oscillations through a new type of control and after optimization, modal analysis is used to determine the oscillation modes of the system without control elements and then with the parameters of the proposed control. Thus, although it indicates the extremes of the trajectory of the oscillation modes, it does not allow knowing the trajectory of the dynamics of the oscillation modes.

These situations lead to the power system counting with regulations that establish limits that allow monitor and take control actions against oscillations electromechanical for different scenarios, known as network code. Usually, the system operator considers the point of operation of the maximum demand of the system for determine the limits, however, given the characteristic of that the power system presents high non-linearities, makes it necessary to consider the cases of operating scenarios that present low-frequency electromechanical oscillations and within them identify the critical scenarios, which respond to the dynamics of the oscillation modes given by the operating conditions obtained through its modal analysis.

Therefore, having a tool that allows obtaining the modal analysis of the system in order to determine the dynamics of the trajectory of these oscillation modes before different scenarios of system operation, it is a necessity for the system operator. This can be achieved through the use of the DIgSILENT programming language (DPL) and the Python engine mode provided in the DIgSILENT PowerFactory engineering software. With this information determine the level of stability of each scenario of operation, for example, before the private office of a generation unit.

The purpose of this paper is to develop a methodology of modal analysis that allows to determine the evolution of the oscillatory modes under different supply scenarios using DIgSILENT's PowerFactory.

To achieve this purpose, this work is organized as follows: Sect. 2 presents the modal analysis tool of an electrical system. Section 3 contains the proposal methodology that allows to determine the trajectory of the dynamics of oscillation modes according to dispatch selected. Section 4 presents the characteristics of the study system and the main results. In the Sect. 5 shows an analysis of the results. Finally, Sect. 6 presents the conclusions of the work.

2 Modal Analysis of Electrical Power Systems

The electrical power system is a large-scale system nonlinear whose dynamic response depends on the parameters of the elements that store and transfer energy, as well as of the magnitude of the disturbance. This dynamic, in the case of electrical systems is represented by nonlinear differentials algebraic equations. The oscillations present in the normal operation of the system due to the variation of power in the generators to supply the change of the demand [9].

A condition of operation of the electrical power system is to have a damping level, for a minimum whatever this is, which allows the system to have a small signal stability level which determines the possibility of power exchange between the different system zones. This dynamic establishes trajectories in the different variables of the electrical system, which present oscillatory or monotone behavior. Thus, it is said that the system is stable, if these trajectories are limited and converge to an acceptable operating point, that is, the system presents a quasi-stationary operating point situation that allows the linearization of the nonlinear equations around the point of operation.

The linearization of the system for an operating condition is established by means of the following system formed by Eqs. (1) and (2):

$$\Delta \dot{x} = A\Delta x + B\Delta u \tag{1}$$

$$\Delta y = C\Delta x + D\Delta u \tag{2}$$

where:

$x \in \mathbb{R}^n$: It is the state vector of the system.

$u \in \mathbb{R}^r$: It is the vector of the inputs.
$y \in \mathbb{R}^m$: It is the vector of the outputs.
A_{nxn}: It is the system state array.
B_{nxr}: It is the input or controllability matrix.
C_{mxn}: It is the output or observability matrix, and;
D_{mxr}: It is the matrix that defines the proportion of input that appears directly in the output.

Through the linear model of a nonlinear system for an operating point, the key element of the modal analysis study is constituted, to determine the level of angular stability that the system under study presents. Since the stability of a linearized system is evaluated using the eigenvalues λ, the state matrix A of the system, by the determination of the roots of the characteristic equation of A as its mathematical form is shown in Eq. (3) [10].

$$det(\lambda * I - A) = 0 \tag{3}$$

where $\lambda \in \mathbb{R}^n$ is the i-th eigenvalue of the A_{nxn} matrix, through which it is possible to determine the stability of the system.

The system is stable if the real part (σ) of all the eigenvalues are located in the second and third quadrants of the complex plane, its value represents the measure of the damping of each oscillation mode, and the imaginary part (γ) sets the natural frequency of the oscillation mode. The characteristic value of the damping level (ζ) of the oscillation mode is established by Eq. (3) for a system electrical the specialized literature considers it sufficient to have with a damping level greater than or equal to 5% of all the eigenvalues of the system [9].

$$\zeta = \frac{-\sigma}{\sqrt{\sigma^2 + \omega^2}} \tag{4}$$

To determine the eigenvectors of the system, it is established with the equations, for a given eigenvalue λ_i.

$$(\lambda_i I - A)v_i = 0 \tag{5}$$

$$\psi_i^T(\lambda_i I - A) = 0 \tag{6}$$

where v_i and φ_i are the right and left eigenvectors respectively associated with the i-th eigenvalue λ of the state matrix A itself that depends on the parameters of the system, if any of these is modified it makes the matrix A changes and consequently the eigenvalues λ. This situation enables an eigenvalue to be displaced potentially to a site where the damping level and/or frequency transform it into a critical oscillation mode, if you move towards the positive direction of the real axis the level oscillation mode damping would be reduced from such that the eigenvalue can even cross the axis imaginary making the system unstable.

In general, the i-th element of v_i measures activity and the observability of the state variable x_i with respect to i-th mode, while for the case of the i-th element of φ_i measures the contribution and controllability of the variable of state in the i-th oscillation mode [1, 9].

3 Methodology

For determination of the evolution of oscillatory modes, a methodology has been developed using PowerFactory's modal analysis tool by programming in DIgSILENT Language Programming (DPL) obtain the critical oscillation modes for different scenarios and using the engine mode to use Python to determine the locus of the dynamic trajectory of the oscillation modes of interest selected by the user, as shown in Fig. 1.

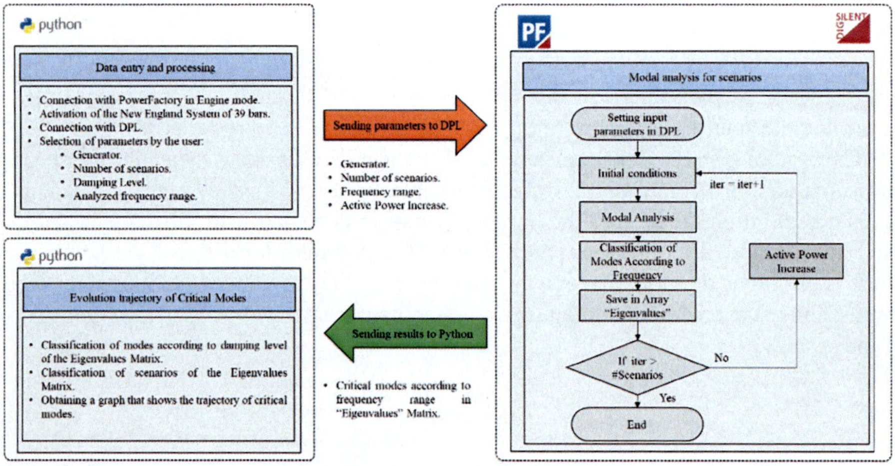

Fig. 1. Methodology implemented to determine the trajectory of the critical modes of the electrical system.

In step 1 the user, through the Python script, selects the generator that modifies its active power, the frequency and damping level classifying the oscillatory mode.

In step 2 the DPL is executed to determine the matrix of eigenvalues for each scenario as presented in Fig. 2.

In step 3 with the data obtained from the eigenvalue matrix, is transferred via engine mode to Python to determine the dynamics of the trajectory of the analyzed oscillatory mode.

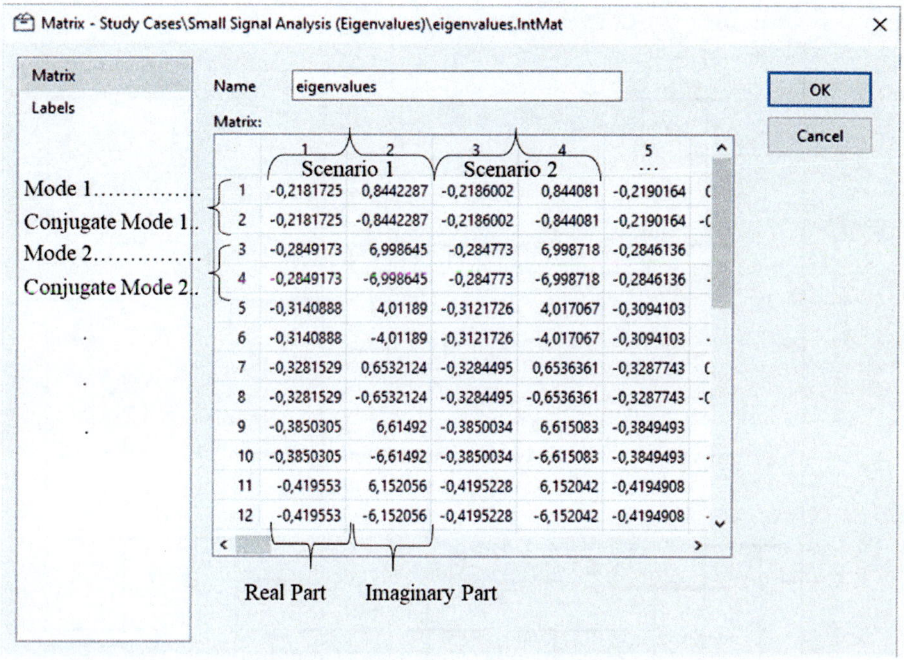

	1	2	3	4	5
1	-0,2181725	0,8442287	-0,2186002	0,844081	-0,2190164
2	-0,2181725	-0,8442287	-0,2186002	-0,844081	-0,2190164
3	-0,2849173	6,998645	-0,284773	6,998718	-0,2846136
4	-0,2849173	-6,998645	-0,284773	-6,998718	-0,2846136
5	-0,3140888	4,01189	-0,3121726	4,017067	-0,3094103
6	-0,3140888	-4,01189	-0,3121726	-4,017067	-0,3094103
7	-0,3281529	0,6532124	-0,3284495	0,6536361	-0,3287743
8	-0,3281529	-0,6532124	-0,3284495	-0,6536361	-0,3287743
9	-0,3850305	6,61492	-0,3850034	6,615083	-0,3849493
10	-0,3850305	-6,61492	-0,3850034	-6,615083	-0,3849493
11	-0,419553	6,152056	-0,4195228	6,152042	-0,4194908
12	-0,419553	-6,152056	-0,4195228	-6,152042	-0,4194908

Fig. 2. Eigenvalue matrix for an analysis case.

4 Study System and Simulations

The proposed methodology was applied in the New England 39 bus test system, this system is used for small-signal stability analysis in various studies of the specialized literature. The system is constituted for 39 bars, 10 generators. The generator G01 represents the equivalent of the electric system of New York and the loads are modeled as constant power [11].

The test system is represented in Fig. 3, where the generators are distributed in: G01 is the slack bus, except the G10 which is hydraulic, all the generators are thermal of coal and nuclear technology.

In the base scenario, it is considering the oscillation modes with a damping level less than 10%, in such a way that they obtain nine modes of which two are interarea and the rest are local as it is presented in Table 1.

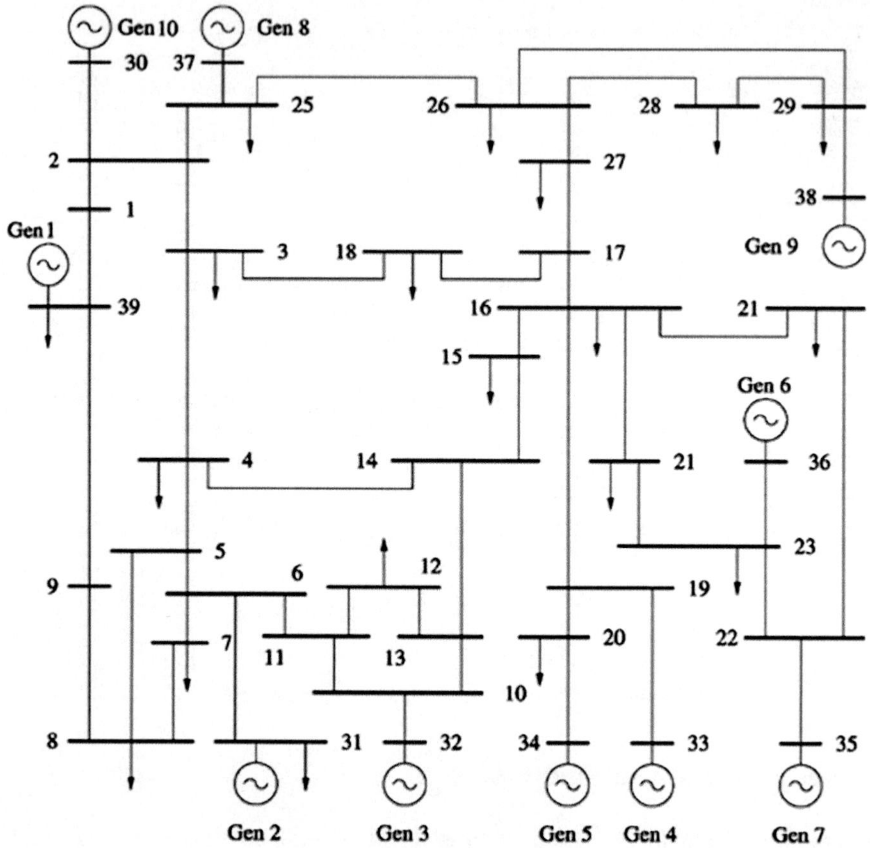

Fig. 3. One-line diagram New England 39-bar test system.

When the variation of the active power for the case of generator G01 in slack bus, the thermal generator G04 and the hydraulic generator G10 are considered, as it can be seen in Fig. 4 that the oscillatory modes 1, 2, 3, 4 and 5 present a shift to the left that corresponds to the stable zone, while oscillatory mode 1 shows an offset that increases the frequency of the mode passing from 1.11 to 1.16 Hz, for the scenario in which it varies it is powered by the G10 hydraulic generator.

The shift of at least one mode to the right by increasing its frequency and in the extreme case decreasing its damping level affects the oscillations in time real. As it is indicated in Fig. 5, a 3-phase short circuit is simulated on the line 21–23, and as the mode 1 corresponds to an interarea mode, it is observed that the speed of the rotor of the generator G10 presents modifications when the active power is incremented.

Table 1. Oscillation modes of the study system – base case.

Scenario	Mode	Eigenvalue	Frequency [Hz]	Mode type	Damping level
Base	1	−0.285 ± 6.99i	1.11	Local	4.07
	2	−0.385 ± 6.61i	1.05	Local	5.81
	3	−0.464 ± 7.52i	1.19	Local	6.16
	4	−0.419 ± 6.15i	0.98	Interarea	6.80
	5	−0.576 ± 7.47i	1.19	Local	7.69
	6	−0.696 ± 8.94i	1.42	Local	7.76
	7	−0.314 ± 4.01i	0.64	Interarea	7.81
	8	−0.739 ± 8.87i	1.41	Local	8.30
	9	−0.820 ± 9.11i	1.45	Local	8.97

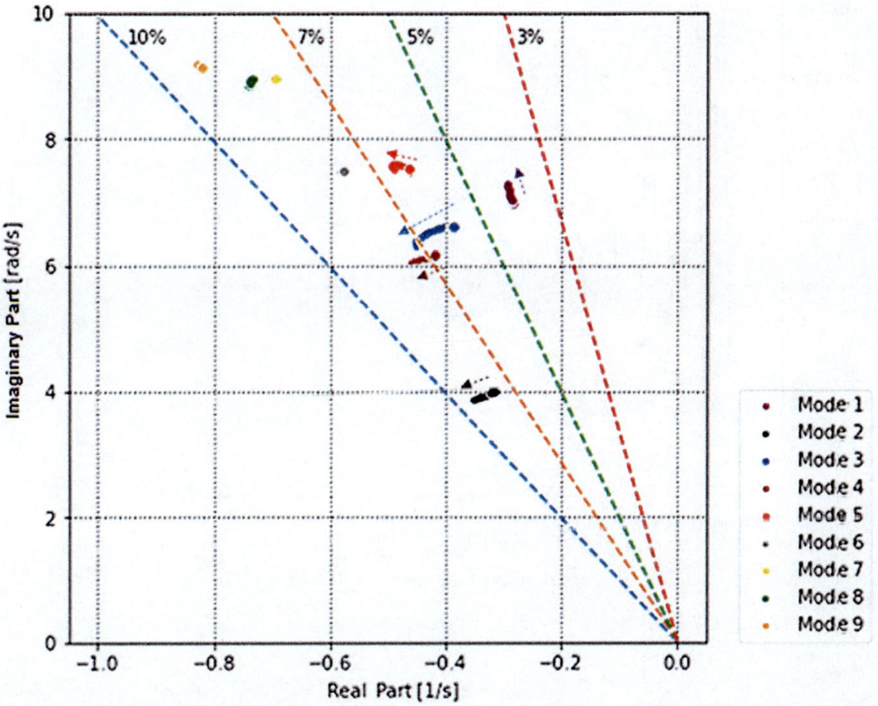

Fig. 4. Trajectory of the oscillation modes for the scenarios before the variation of G01, G04 and G10.

5 Discussion

The studies in this paper can be extended when it is considered the effect produced by including control elements such as power systems stabilizers or voltage regulators since they can modify the damping level of the oscillatory modes.

In addition, it is important to bear in mind that in this paper only the small signal stability study is carried out when the active power of the generators is varied, therefore, studies can be carried out that modify the topology by including wind generation units or that modify the demand. It is important to note that developing tools that allow the operator to visualize the evolution of the oscillatory modes provide information on the stability of the system's small signal and its evolution towards areas of greater or lesser stability.

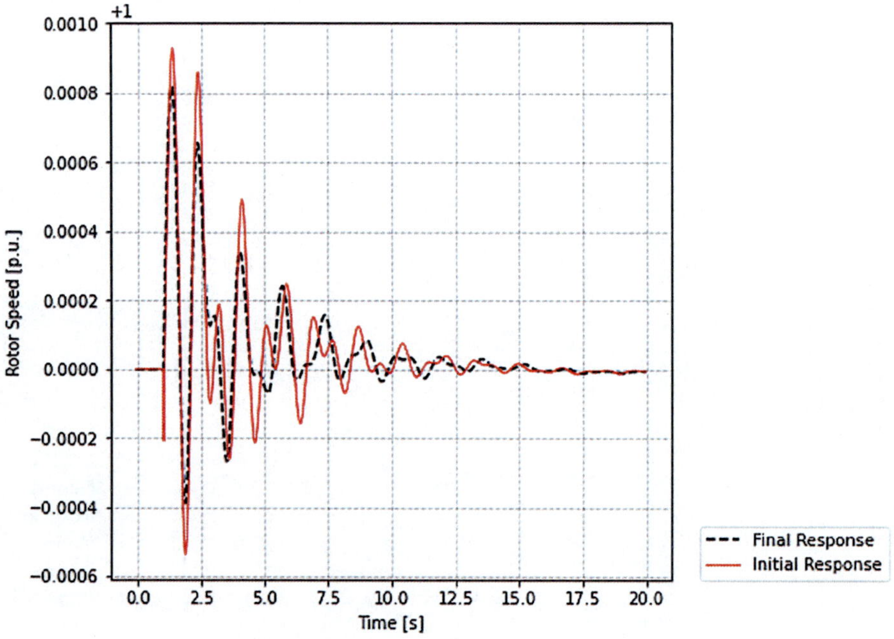

Fig. 5. Simulation of the rotor angle of the G10.

6 Conclusions

In this paper, a tool is developed to determine the dynamic trajectory of critical oscillation modes that occur in an electrical system. The implementation of the modal analysis considering scenarios is relatively simple. However, having the need to classify and select the relevant critical modes from a large volume of modes presented by the electrical system allow have a valuable visual analysis tool for the system operator.

The small-signal stability analysis carried out in the New England 39-Bus Test System allowed to show that the dynamics of the modes can affect the level of damping of the oscillations. In this case, the tool shows positive results that justify the study carried out and constitute the starting point to continue with the analysis to consider the impact of changes in demand or include generating units that consider renewable energy.

References

1. Mondal, A.C., Sengupta, A.: Power System Small Signal Stability and Control, 2nd edn. Elsevier Academic Press, London (2020)
2. Padiyar, K.: Power System Dynamics Stability and Control, 2nd edn. B.S. Publications, India (2022)
3. Rafique, Z., Khalid, H., Muyeen, S., Kamwa, I.: Bibliographic review on power system oscillations damping: an era of conventional grids and renewable energy integration. Int. J. Electr. Power Energy Syst. **136**, 107556 (2022)
4. Machowsky, J., Lubosny, Z., Bialek, W., Bumby, R.: Power System Dynamics: Stability and Control, 3rd edn. Wiley, London (2020)
5. Golpira, H., Messina, A., Brevani, H.: Renewable Integrated Power System Stability and Control, 1st edn. Wiley, New York (2021)
6. Pizarro, S.: Análisis Modal del Sistema Interconectado del Norte Grande ante Escenarios de Oferta-Demanda. Tesis Universidad Técninca Federico Santa María, Chile (2014)
7. Berruti, F.: Análisis modal y transitorio de sistemas eléctricos con incorporación de energía eólica a gran escala. Tesis de Maestría Universidad de la República, Uruguay (2012)
8. Macedo, P.: Improving interarea oscillations damping of power systems through cooperative active power control of distributed energy resources. Master thesis. The University of Tennessee, USA (2021)
9. Dussaud, F.: An application of modal analysis in electric power systems to study inter-area oscillations. Master thesis. KTH Royal Institute of Technology, Sweden (2015)
10. Silva-Saravia, H.: Análisis de la Ubicación de Volantes de Inercia Considerando el Desempeño Dinámico del Sistema de Potencia. Tesis Universidad Técnica Federico Santa María, Chile (2014)
11. Cañizares, C., et al.: Benchmark models for the analysis and control of small-signal oscillatory dynamics in power systems. IEEE Trans. Power Syst. **32**, 715–722 (2017)

Development of a System Based on an Artificial Vision to Determine the Length of Billets in the Cutting Process in Steel Mills

Richard-Xavier Sánchez-Pozo(✉) and Víctor Larco

Grupo de Investigación en Micro-Nanotecnología y Energía Nuclear, (NANOTECH), Universidad Politécnica Salesiana, C. Robles 107 y Chambers, 090108 Guayaquil, Ecuador
r.sanchez.rxsp@gmail.com, vlarco@ups.edu.ec

Abstract. In the steel industry, it is not possible to measure billets' length at the end of the cutting process of a continuous casting machine, due to the high temperatures. To have this measure the billets have to have natural cooling for 24 h. The system developed captures an image of the billets and gets its length in real-time using the AdaBoost algorithm, this algorithm is an artificial vision application through the color segmentation method in HSV space. The system allows obtaining a calculated length near to 99% of the real.

Keywords: Artificial vision · Segmentation method · HSV space · Cutting process · Casting machine · Length of billets

1 Introduction

The role of the steel industry is fundamental in the development of civilization, currently, structural optimization studies reduce the optimal section size of steel beams. The studies show the automatic determination of an optimal length of cast steel blocks, obtaining the minimum loss of cast material of the rolling process [1, 1].

The steel industry's continuous casting machines perform a continuously controlled metallurgical process, transforming the states of steel from liquid to solid, the Fig. 1 shows the schematic process [3].

At the end of the process, the billets have a constant section measurement, and the lengths fluctuate between a minimum and maximum value, the process is through a flame cutting system, and the length to generate an order in the cuts is with an encode, the direct contact with rollers allows for extracting the billet, rollers and billet are in constant hydraulic pressure to prevent displacements [4, 4, 4].

The development of a system based on artificial vision and using the Adaptive Boosting algorithm allows measuring the billet's length in real-time and feedback this measure to the cutting system to keep it constant according to the established set point [7, 8, 9].

The structure of this article is organized as follows. Section 2 presents the methodology. Section 3 presents the results obtained. Section 4 presents the Discussion. Finally, Sect. 5 presents the conclusions.

V. Robles-Bykbaev et al. (Eds.): CITIS 2022, LNNS 607, pp. 214–224, 2023.
https://doi.org/10.1007/978-3-031-24327-1_19

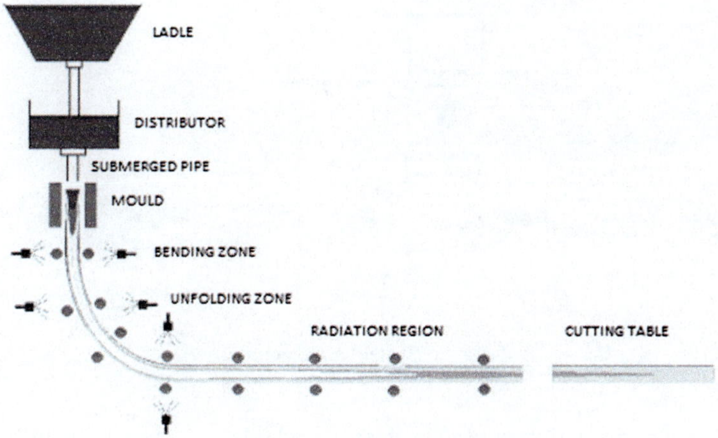

Fig. 1. Continuous casting process diagram.

2 Methodology

2.1 Steel Industry

The review of information from digital repositories showed that artificial vision has been implemented in the steel industry for the inspection of surfaces based on an artificial vision to determine defects such as cracks and corners in steel billets of continuous casting machines [10], it has also been implemented in the development of an automatic marking identification system based on visual recognition of two stages of billets for metal rolling [11].

2.2 System Architecture

Figure 2 shows the system architecture where it is shown how the machine vision system interacts with the automation systems of each process line.

Hardware. The development of the system was carried out using a computer in which the Ubuntu 20.04 LTS operating system was installed.

Making use of this computer for the development, the acquisition of the images was performed through an 8 MP ultra-wide-angle 119° camera, focal aperture f/2.2, focal length 13 mm (lens), sensor size 0.25 inches, pixel size 1. 12 μm pixel, 30 FPS, 720 × 1280 px resolution for cut order generation and an 8 MP ultra-wide-angle 119° camera, focal aperture f/2.2, focal length 13 mm (lens), sensor size 0.25 inch, pixel size 1.12 μm pixel, 30 FPS, 1920 × 1080 px resolution to perform length calculation and billet detection.

Software. For the development of the algorithm, the Python V3.8.10 programming language was used, which comes preinstalled in the Ubuntu operating system that was used and is also multiplatform open-source, currently widely used in web applications, for data analysis, automating operations, creating applications, etc.

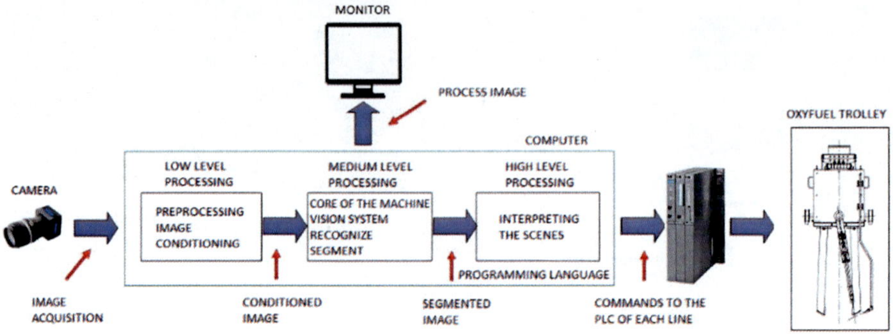

Fig. 2. System architecture

The artificial vision library used is OpenCV, which is an open-source library originally developed by Intel for image and video analysis [12]. Its best virtue is that it is multiplatform and can be run in different programming languages and operating systems.

2.3 System Validation

Figure 3 shows the stages of the algorithm to perform the validation of the system by calculating the length of each billet and performing the recognition and/or detection by the shape of the billets using the AdaBoost artificial intelligence algorithm.

Acquire image. Is the first stage of the algorithm and in general of artificial vision where the images are captured through the camera with which a 3D scene is obtained and converted into a digitized 2D image.

Calibrate reference measurements. In this stage of the algorithm, the measurements that could serve as references of the scene environment are recorded, these data were used to obtain the ratio of pixels concerning the reference measurements preserving the camera position.

Conversion of color space from BGR to HSV. In this stage of the algorithm, the conversion of the BGR (Red, Green, Blue) color space to HSV (Hue, Saturation, Brightness) color space is performed to avoid the variation of the illumination affecting the next stage, which is the segmentation of the billets.

Segmentation. In this stage of the algorithm, the segmentation by color is performed, when segmenting an image the separation of the region or object of interest of the scene is performed, in the developed algorithm the region or object that corresponds to the color of the billets previously selected in the HSV color space of the previous stage is separated.

Billets area greater than the threshold. In this stage of the algorithm, the output of the segmentation process is compared which is a data image containing the region of interest for the analysis, and when this region of interest is greater than the area value that was considered in the algorithm, the next stage of the algorithm is taken to the next stage of the algorithm.

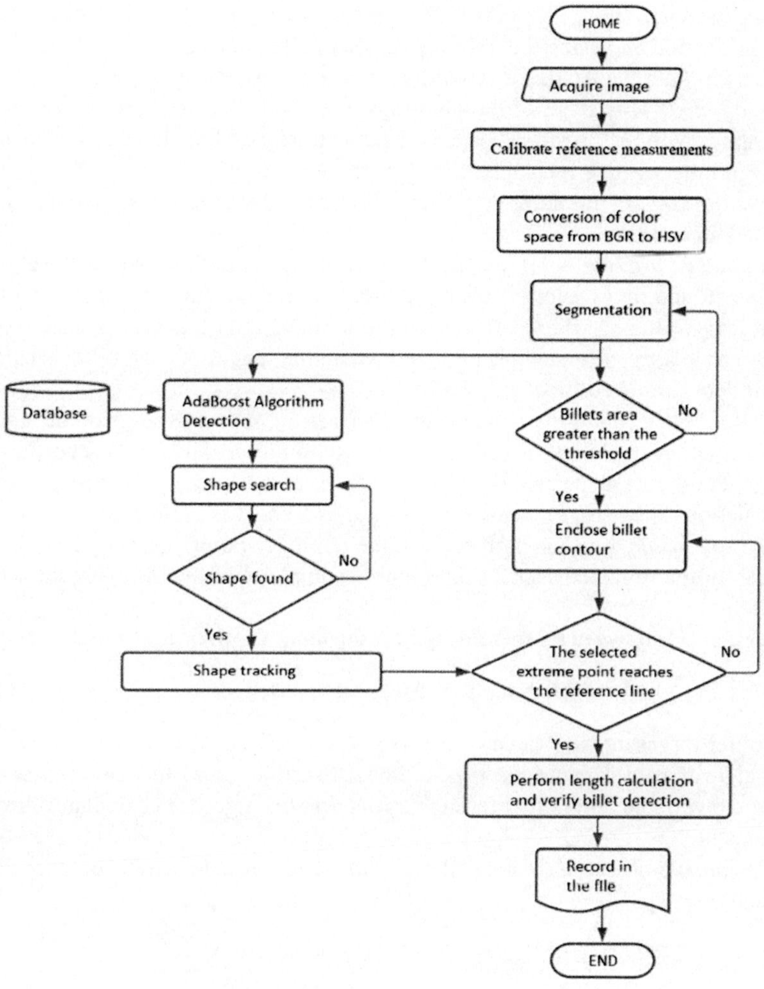

Fig. 3. Process for billet length calculation.

Enclose billet contour. In this stage of the algorithm, the contour of the segmented area is enclosed when it meets the condition of the previous stage by being larger than the region of interest to the area that was considered in the algorithm.

AdaBoost Algorithm Detection. In this stage of the algorithm, artificial intelligence is used with the AdaBoost algorithm for the detection of billets using feature extraction or also called HAAR type feature descriptors.

Database. In this stage of the algorithm, we make use of the training that was performed with a database with positive images and negative images to ensure the correct detection of each billet, making use of the cascade classifier function of OpenCV, at the end of the tests for the detection of the billets the database was completed with 105 positive images of billets and 186 negative images of the scene environment.

Shape search. In this stage of the algorithm we make use of the OpenCV multiscale detection function and using the training we perform the detection of the billets with the characteristics that the AdaBoost algorithm has been previously trained.

Shape found. This stage of the algorithm was evaluated when a shape was found that complies with the previously trained characteristics of the algorithm to enclose the recognized billets with a rectangle.

Shape tracking. In this stage of the algorithm, the billets recognized by the algorithm are tracked in real-time.

The selected extreme point reaches the reference line. In this stage of the algorithm, the most extreme point selected from the billet contour is obtained according to each line and it is evaluated when it reaches the reference point that was considered in the algorithm so that the next stage is executed where the length of each billet is calculated and their detection is verified.

Perform length calculation and verify billet detection. In this stage of the algorithm, the calculation of the length measurement of each billet is performed according to the equation of each line of the process where distances of the scene are related to pixels, and the calculation of the length measurement of lines 1 and 3 is performed at the reference point of 1050 pixels, line 2 is performed at the reference point of 1034 pixels, being the lower part of the digitized scene, each billet is completely visualized and the cutting is finished.

Equation (1) shows in a general way how the total length of each billet is calculated:

$$\text{Med\_total} = \text{Med\_ref} + \text{Med\_calc} \tag{1}$$

Med_ref represents the value of the fixed-length obtained from the reference point for the algorithm to execute the measurement function, and Med_calc represents the value of the variable length that through a function the algorithm calculates this length measurement.

Equation (2) shows in a general way how the variable length of each billet is calculated:

$$\text{Med\_calc} = \text{cdp}(\text{cref}(\text{ne} + 1) + \text{delta} * (\text{n} + \sum\nolimits_{i=1}^{ne} (i))) \tag{2}$$

where, cdp, represents the value of the delta pixel variation from a fixed point to the variable point to be calculated, cref, represents the ratio of centimeters over the delta pixel of the last fixed area, and ne, represents the integer sections of the ratio between the distance of the point of the length to be calculated over the delta pixels or cdp, delta, represents the difference in centimeters over delta pixel of the last fixed area with the variable area to be calculated, n, represents the decimal places of the ratio between the distance of the point of the length to be calculated over the delta pixels or cdp.

In this stage of the algorithm, the detection of the billets is also checked when the reference point is reached for the calculation of the length measurement of the billets according to each line.

Record in the file. In this stage of the algorithm a record was made in a file of the data that were considered for each line such as the length measurement of each billet, the number of billets that were counted, the status of the detection if it is true or false, the date and time of each event.

Figure 4 shows the three lines during the running of the algorithm where it can be seen that the billet on line 1 is being measured and is also being recognized.

Fig. 4. Billet detection on line 1

3 Results

To verify the correct generation of the cutting order by the camera, the length of each billet was recorded, and the measurements of the billets determined by this system were also recorded as shown in Table 1, the nominal value of the length of each billet was 421 cm.

Table 1. Billets dimensions (421 cm).

	Real billet sizes		Measurements calculated by the system	
Sample	Line 1	Line 3	Line 1	Line 3
1	420.9	420.7	420.44	421.19
2	420.3	421	420.5	420.93
3	420.3	421.2	420.5	421
4	420.2	420.9	420.49	421.07

(continued)

It was of great importance to obtain data from each production line, where the system developed in real-time calculates the length of each billet through mathematical calculations based on the equation defined for each line, With these results, a statistical

Table 1. *(continued)*

	Real billet sizes		Measurements calculated by the system	
Sample	Line 1	Line 3	Line 1	Line 3
5	420.4	420.9	420.46	421.07
6	420.9	420.7	420.5	421.16
7	420.2	420.6	420.46	421.06
8	420.7	421.3	420.62	421.19
9	420.4	420.9	420.67	421.16
10	420.2	421.2	420.94	420.9
11	420.6	420.9	420.46	421.14
12	420.2	421.1	420.59	421.17
13	420.5	421	420.48	420.93
14	420.3	420.8	420.57	420.93
15	420	421.4	420.42	421.02

comparison of the data of line 1 as shown in Table 2 and of line 3 as shown in Table 3, where the results obtained for the lengths of the billets such as the minimum, maximum, average, variation or delta, variance and standard deviation of the data grouped according to each line are compared.

Table 2. Comparison of data obtained from the measurement system on line 1 (cm).

Measure	Minimum	Maximum	Average	Delta	Variance	Std. deviation
Real	420	420.9	420.41	0.9	0.07	0.27
Calculated	420.42	420.94	420.54	0.52	0.02	0.13

Table 3. Comparison of data obtained from the measurement system on line 3 (cm).

Measure	Minimum	Maximum	Average	Delta	Variance	Std. deviation
Real	420.6	421.4	420.97	0.8	0.05	0.23
Calculated	420.9	421.19	421.06	0.29	0.01	0.1

Figure 5 shows the normal distribution of the data of the billet lengths of line 1, where the dispersion of the data obtained can be seen graphically.

Figure 6 shows the normal distribution of the data of the billet lengths of line 3, where the dispersion of the data obtained can be seen graphically.

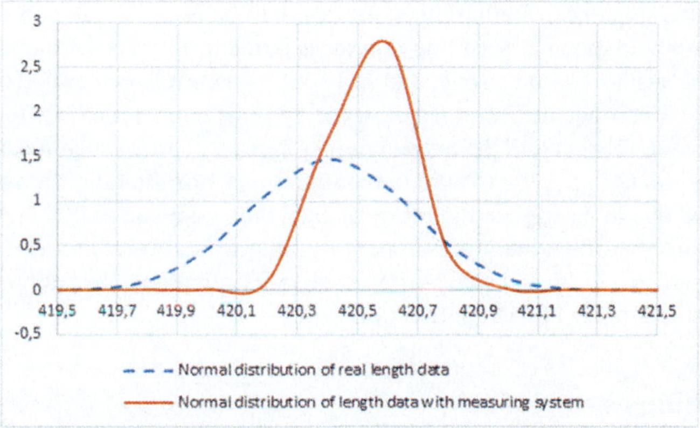

Fig. 5. Normal distribution of the data of the billet lengths of line 1.

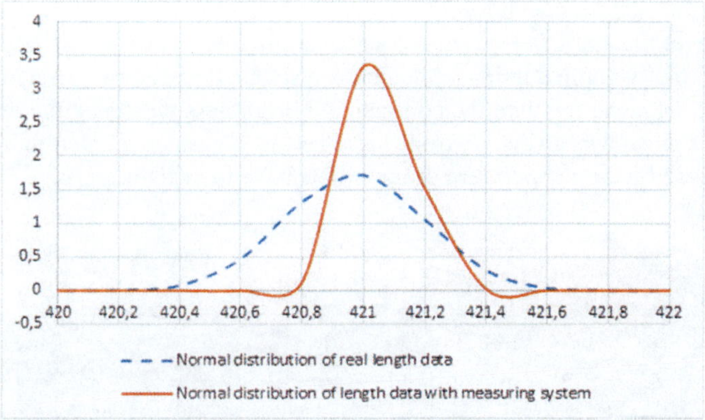

Fig. 6. Normal distribution of the data of the billet lengths of line 3.

This section of the results only shows data from lines 1 and line 3, this is because at the time the tests were carried out to obtain the data to later contrast the results, line 2 was in reserve.

4 Discussion

The length of the billets is of great interest for the continuous casting process because it always seeks to meet the quality parameters established among these is to maintain constant measures because this minimizes waste and can increase productivity, for example in the casting of 25 tons approximately 45 billets of 421 cm in length are obtained with a thickness of 13 cm per side and in this process sequences of approximately 20 castings are made, being this the case in a sequence of 20 castings approximately 900 billets are produced, taking this example if the length of each billet would be increased by 0. 5 cm, at the end of the sequence of 900 billets, there would be an accumulation of 450 cm in length, which would mean at least one billet less in the production of the sequence.

The cutting systems based on encoders only can measure the generation of the cutting order of each billet, once the billet is cut, these systems cannot measure the billets and be able to provide feedback to the automation system to contrast such measurement.

Manually it is not possible to take the measurements of the billets at the end of the cutting process due to the high temperature of each billet, with an artificial vision system we can measure the billets with the process running in real-time and each production line can be managed as an independent process in its references, equations, thresholds and other characteristics of each production line.

5 Conclusions

Much importance should be given to the location and tilt of the camera once the scene is calibrated, thus preserving the relationships of the scene reference measurements to the pixels of the camera resolution.

In this development it was not necessary to apply filters for the preprocessing of the images as normally applied in machine vision systems, for example, applying a filter to equalize the images to regulate the contrast and brightness worsened the segmentation of the images as shown in Fig. 7, from these results it was concluded that the images being processed by the temperature of the billets have their light.

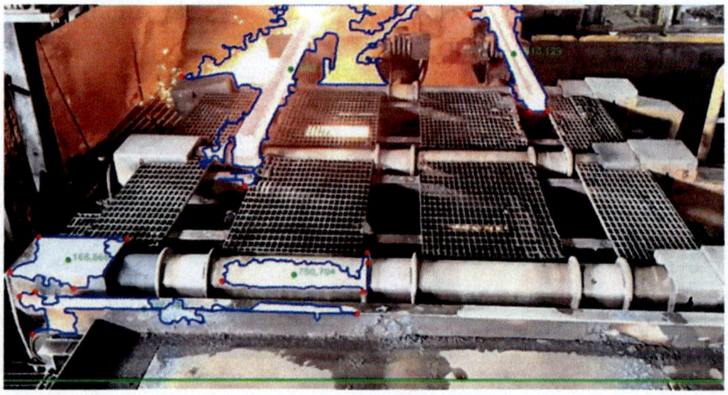

Fig. 7. Equalizing filter application

In this development, a color segmentation was performed because a characteristic of this continuous casting process is that the billet is always processed at an elevated temperature with which the color calibration was performed.

The segmentation was not performed in the RGB color space because the lighting variation can affect it negatively, it was performed in the HSV/HSI color space to avoid being affected by the lighting variation and not require an external lighting system.

For billet detection, artificial intelligence was used by applying the AdaBoost algorithm, which is a supervised learning algorithm based on several weak classifiers to obtain a fast and robust final classifier. The choice of this algorithm was made based on [13], which compares the results of Haar-like feature extractors, Local Binary Patterns (LBP), and Histograms of Oriented Gradients (HOG), which are the three most commonly used detection descriptors.

To obtain the results of line 1, the length for cutting was set at 421 cm, and with the statistical results that were calculated in the results section, it can be concluded that the average lengths of the billets are very close, being 420.41 cm the actual measurements and 420.54 cm the measurements that the developed system calculated, giving a difference of 0.13 cm. 54 cm the measures that the developed system calculated giving a difference of 0.13 cm, it is also concluded that the variations in the lengths of the billets present a delta in the real measures of 0.9 cm, this means that the cutting system presents a high precision in the length of the billets.

To obtain the results of line 3, the length for cutting was set at 421 cm, and with the statistical results that were calculated in the results section, it can be concluded that the average lengths of the billets are very close, being 420.97 cm the actual measurements and 421. 06 cm the measures that the developed system calculated giving a difference of 0.09 cm, it is also concluded that the variations in the lengths of the billets present a delta in the real measures of 0.8 cm, this means that the cutting system presents a high precision in the length of the billets.

References

1. Hao, Z., Tielin, L., Hong, L., Zheng, W.: Optimization design for beam and column of steel structure residence. In: Proceedings - 8th International Conference on Intelligent Computation Technology and Automation, ICICTA 2015, pp. 609–612 (2016). https://doi.org/10.1109/ICICTA.2015.155
2. Clitan, I., Muresan, V., Abrudean, M., Monica, S.D.: Automatic determination of optimal length of casting steel blocks in the context of an imprecise manufacturing. In: Proceedings - 2017 21st International Conference on Control Systems and Computer, CSCS 2017, pp. 99–105 (2017). https://doi.org/10.1109/CSCS.2017.20
3. Zhao, L., Dou, R., Yin, J., Yao, Y.: Intelligent prediction method of quality for continuous casting process. In: Proceedings of 2016 IEEE Advanced Information Management, Communicates, Electronic and Automation Control Conference, IMCEC 2016, pp. 1761–1764 (2017). https://doi.org/10.1109/IMCEC.2016.7867521
4. Santos, J.L., Santos, J., Ferreira, M.J., Alves, N., Guevara, M.: Application of the two-stage one-dimensional cutting stock problem in the steel industry. In: IEEE International Symposium on Industrial Electronics, vol. 2018-June, pp. 683–690 (2018). https://doi.org/10.1109/ISIE.2018.8433734
5. Anuchin, A., Dianov, A., Briz, F.: Synchronous constant elapsed time speed estimation using incremental encoders. IEEE/ASME Trans. Mechatron. **24**(4), 1893–1901 (2019). https://doi.org/10.1109/tmech.2019.2928950
6. Isaev, D.: Implementation of electric drive of withdrawal-roll set in horizontal continuous casting machine for the production of small billets. In: 2016 9th International Conference on Power Drives Systems, ICPDS 2016 - Conference Proceedings, pp. 1–4 (2016). https://doi.org/10.1109/ICPDS.2016.7756723
7. Zhang, X., Xu, S.: Research on image processing technology of computer vision algorithm. In: Proceedings - 2020 International Conference on Computer Vision, Image and Deep Learning,

CVIDL 2020, no. Cvidl, pp. 122–124 (2020). https://doi.org/10.1109/CVIDL51233.2020.00030
8. Viola, P., Jones, M.: Rapid object detection using a boosted cascade of simple features (2001)
9. Barbona, I., Beltrán, C.: Aplicación del algoritmo Boosting Adaptativo (ADABOOST) a un problema de clasificación automática de textos Application of the Adaptive Boosting Algorithm (ADABOOST) to a problem of automatic text classification (2018)
10. Hsu, C.Y., Ho, B.S., Kang, L.W., Weng, M.F., Lin, C.Y.: Fast vision-based surface inspection of defects for steel billets. In: 2016 IEEE International Conference on Consumer Electronics-Asia, ICCE-Asia 2016, vol. 0267, pp. 1–2 (2017). https://doi.org/10.1109/ICCE-Asia.2016.7804804
11. Astafiev, A.V., Orlov, A.A., Provotorov, A.V., Privezentsev, D.G.: Development of the automatic marking identification system based on two-stage visual recognition of metal-rolling billets. In: 2016 Dynamics of Systems, Mechanisms and Machines, Dynamics 2016, pp. 1–4 (2017). https://doi.org/10.1109/Dynamics.2016.7818970
12. Culjak, I., Abram, D., Pribanic, T., Dzapo, H., Cifrek, M.: A brief introduction to OpenCV (2012)
13. Cruz, J.E.C., Shiguemori, E.H., Guimarães, L.N.F.: A comparison of Haar-like, LBP and HOG approaches to concrete and asphalt runway detection in high resolution imagery. J. Comp. Int. Sci **6**(3), 3–20 (2015). https://doi.org/10.6062/jcis.2015.06.01.0101

Evaluation of the Surface Quality Obtained in AISI 201 Stainless Steel by Fiber Laser Cutting in 2 mm Sheet

Bryan Calderón(✉), Kevin Suquillo, Luis Juiña, and Xavier Vaca

Universidad Politécnica Salesiana, Quito, Ecuador
bcalderons@est.ups.edu.ec

Abstract. Fiber laser cutting technology on materials such as stainless steels is one of the most effective compared to other technologies due to its accuracy and speed of cutting. Stainless steel 201 has a lower nickel and chromium content defining it as a non-magnetic austenitic material resistant to corrosion. Its low nickel content makes it inexpensive, and the similar characteristics to AISI 304 make it reliable a substitute. For the development of this study, the effects of cutting process parameters such as: laser power, gas pressure and cutting speed on surface quality were investigated through the surface roughness with the response surface methodology (RSM). In addition to the use of Taguchi methodology and ANOVA variance analysis to determine the influence of the studied factors. The cutting speed parameter had the most significant influence on the surface quality obtained, varying the roughness from 2.112 to 4.663 μm, and from 30 to 100 mm/s with a remarkable percentage difference of 69.43% with respect to the cutting parameters studied, and an influence of the combined parameters of the gas pressure with the laser power of 15.57% .

Keywords: Stainless steel 201 · Anova · Laser cutting · Response surface methodology · Roughness · Fiber laser

1 Introduction

Laser cutting technology is one of the most widely used technologies in the industry due to its precision and cutting speed compared to other methods. Several investigations have been carried out on the influence of cutting on different materials such as stainless steels due to the important applications of these materials in the industry. The most economical stainless steel is AISI 201 which can replace AISI 304 steel in industrial applications such as railway construction, shipbuilding, power generation, nuclear and oil industry [1]. Since stainless steel 201 shows similar characteristics in its composition to stainless steel 304, the most important difference being the reduction of Ni to 3.13% and the presence of Mn at 7.38% [2].

V. Robles-Bykbaev et al. (Eds.): CITIS 2022, LNNS 607, pp. 225–235, 2023.
https://doi.org/10.1007/978-3-031-24327-1_20

Stainless steel used in medical components according to Sun et al. [3] cardiovascular endoprosthesis uses a tube with a diameter of 2.6 mm of stainless steel 316 L, is cut with fiber laser, taking into account the internal surface finish, for this the ideal parameters are defined as: cutting speed at 7 mm/s, laser power at 105 W and frequency at 8000 Hz as a result of these defined parameters a better distribution of the burr of the surface finish was carried.

Ming Li [4] evaluated the effects of temperature and roughness on the cut surface of 1mm thick 2205 duplex stainless steel. 2205 duplex was subjected to fiber laser cutting process by varying the parameters of focal length, laser power and cutting speed, in this study it was determined that the most influential parameters are: laser power and cutting speed; by increasing the laser power from 700 to 1100 W, the temperature rose from 180 to 500 °C, while, by increasing the cutting speed parameter from 4 to 5.6 m/min, the temperature decreased reaching approximately 50 °C increasing the values of the surface roughness between 1.16 and 4.61 μm.

M. Boujelbene [5] investigated the influence of the variation of two cutting parameters in the CO2 laser cutting process on S235 low carbon steel. For the experiment, the speed and power parameters were considered at 3 different levels with a total of 9 experiments. The values used in the experiment were 600, 1400 and 2200 mm/min for speed and values of 1, 2 and 3 kW were used for power. The roughness variation obtained in the experiment varied from 7.49 to 16.65 μm. The mean square roughness method was considered for the analysis of the cut surface. In the experiment it could be observed that there was a slight decrease in roughness with the increase of power varying between 2.78, 1.41 and 0.68 μm. While it had a greater change in roughness with varying speed obtaining values of 8.48, 7.67 and 6.38 μm.

Sharma y Kumar [6] indicate that, for fiber laser cutting on A653 galvanized steel material, the cutting parameters increase. As for the pressure, from 6 to 10 kgf/cm2, for the power from 80 to 100 W and for the frequency from 50 to 70 Hz, this affects to the value of the Material Removal Rate (MRR). By using the desirability function criterion ranging from 0 to 1, they obtain a value of 0.97 considered as acceptable and excellent, where the optimum settings for the width adjustment for minimum and a maximum material removal rate is used gas pressure of 10 kg/cm^2, power 80 W, laser frequency 90 Hz and a feed rate of 150 mm/min.

From the implementation of the fiber laser cutting technology, where the cutting is performed on different steels taking as a priority the stainless steel AISI 201 with a thickness of 2 mm, being this an austenitic steel, economic with similar characteristics to the stainless steel AISI 304 and used in different industrial sectors. The surface quality is evaluated considering the cutting parameters that interact the most in the laser cutting process, in order to obtain the cutting parameters where the surface roughness is minimal.

2 Methods and Materials

2.1 Stainless Steel 201

In the present investigation, the material to be considered is austenitic stainless-steel grade AISI 201. It is characterized as a non-magnetic austenitic steel, having remarkable toughness, high mechanical strength, excellent ductility at low temperatures (cryogenic) and good heat resistance [7]. Austenitic stainless steels grade 201 are used in nuclear and chemical reactors, domestic use in households and reinforcements in structural elements [8–10]. The chemical composition of this steel is given in Table 1.

Table 1. Chemical composition of stainless steel AISI 201.

Element	(%)
C	0.05
Si	0.32
Mn	7.4
Mo	0.05
Cu	0.23
Ni	3.7
Cr	17.48
N	0.2

2.2 Laser Cutting Machine

BODOR A series is a laser cutting machine that uses optic fiber technology for high speed and high precision cutting. Features of this machine include economical series, easy operation, active collision function, high speed and precision cutting. Ideal for cutting carbon steel, stainless steel, aluminum, brass, alloy metals [11]. Table 2 shows the technical specifications of the machine.

Table 2. Technical specifications bodor serie A3 [13].

Model	Bodor Series A3
Workspace	3048 × 1524
Power	1000 W/1500 W/2000 W
x/y axis max line speed	100 m/min

2.3 Rugometer Mitutoyo SJ-210

The roughness meter is a measuring instrument, which illustrates the surface by means of a probe, records the vertical movements of the needle exposing it as the average roughness on the screen of the device [14, 15]. For the research, the Mitutoyo rough meter, model SJ-210, is used, with a total sensor course between 200 μm to +160 μm, a measurement speed of 0. 25 to 0. 75 mm/s and a sampling length up to 2. 5 mm [16].

3 Experiment Design

The experiment was performed on the BODOR machine with A3 series fiber laser cutting technology, maintaining the 1.5 mm nozzle of the laser for all cuts made and considering a compensation of 0.15 mm for the cut. For fiber laser cutting in austenitic stainless steels in thickness of 2 mm, Parmar et al. [18] and Amaral et al. [7], observed that the factors that have an impact on the surface quality of the cut are: gas pressure, laser power and cutting speed. With these factors the levels are defined, according to A.S. Mohamed [17] in his experiment he defines an initial orthogonal arrangement L9 of 3 levels for 3 factors. The values for the 3 levels were taken as a reference, as a manufacturer's recommendation for a power of 1000 W a minimum speed of 117 mm/s is recommended and for a power of 1500 W a maximum speed of 217 mm/s is recommended, on the other hand, the pressure values were used the minimum and maximum values of 7 and 9 bar according to [19]. A rectangular geometry of 40 $\times$ 40 mm is used [20].

When performing the experiment with these established values the laser could not pass through the material, so it was not possible to measure the roughness in any of the specimens. When analyzing the specimens, it was observed that, in the experiments with lower speeds, the laser reached to make the cut, through experimentation the values of the speed were found for the minimum range of 30 mm/s and a maximum of 100 mm/s establishing the cutting parameters for the experiment that are observed in Table 3.

Table 3. Parameters of the experiment.

Factor	Parameter	Units	Level 1	Level 2	Level 3
A	Gas pressure	Bar	7	8	9
B	Cutting speed	mm/s	30	65	100
C	Laser power	W	1000	1250	1500

With these values, the orthogonal arrangement L27 of Taguchi was performed for three levels and three factors, a total of 27 experiments were performed with the objective of increasing the significance of the parameters in the analysis of the values to be obtained; the combinations and values used in each test tube can be seen in Table 4.

Table 4. Parameter of the experiments.

Exp.	Gas pressure (Bar)	Cutting speed (mm/s)	Laser power (W)	Roughness (μm)
1	9	100	1500	2,151
2	8	100	1500	2,112
3	9	65	1500	3,17
4	7	100	1250	2,374
5	7	65	1250	2,702
6	8	100	1000	2,12
7	7	30	1250	3,328
8	8	100	1250	2,213
9	8	65	1250	2,594
10	8	65	1500	3,827
11	9	30	1000	3,971
12	7	100	1500	2,561
13	7	65	1000	3,228
14	8	30	1500	4,101
15	9	65	1000	3,757
16	9	30	1250	4,012
17	8	30	1000	3,951
18	9	100	1250	2,676
19	9	65	1250	3,701
20	7	65	1500	4,101
21	7	30	1500	4,663
22	7	30	1000	3,298
23	9	100	1000	2,894
24	8	30	1250	3,234
25	7	100	1000	2,243
26	8	65	1000	2,894
27	9	30	1500	3,361

4 Analysis and Results

4.1 Taguchi Method Analysis

Table 9 contains the results extracted from ANOVA considering that for surface roughness a certainty level of 95% ($\alpha = 0.05$) was taken. In which it is observed that the cutting speed and the interaction of gas pressure with laser power have a greater effect on surface roughness.

Table 5. ANOVA by Taguchi method.

Fountain	GL	SC Sec.	SC Ajust.	MC Ajust.	F Value	%Contri
Gas pressure	2	3,942	3,942	1,971	5,24	3,55%
Cutting speed	2	77,157	77,157	38,5784	102,56	69,43%
Laser power	2	3,104	3,104	1,5518	4,13	2,79%
Gas pressure*Cutting speed	4	1,841	1,841	0,4602	1,22	1,66%
Gas pressure*Laser power	4	17,3	17,3	4,325	11,5	15,57%
Cutting speed*Gas power	4	4,774	4,774	1,1934	3,17	4,30%
Residual error	8	3,009	3,009	0,3761		2,71%
Total	26	111,126				100,00%
S (adjusted)	R-cuad	R-cuad.ajus				
0,2684	95,90%	86,66%				

The regression equation extracted from the methodology includes linear terms, the present Eq. 1 indicates the relationship between surface roughness and initial parameters of laser cutting of grade 201 stainless steel.

$$roughness = 3,45 + 0,066A - 0,01996B + 0,000376C \quad (1)$$

4.2 Selection of Optimum Conditions for Taguchi Surface Roughness

The optimum parameters obtained from the Taguchi methodology for the lowest Ra surface roughness results in a pressure of 8 bar, speed of 100 mm/s and laser power of 1250 W. The graph shows effects of the parameters on the surface roughness (Fig. 1).

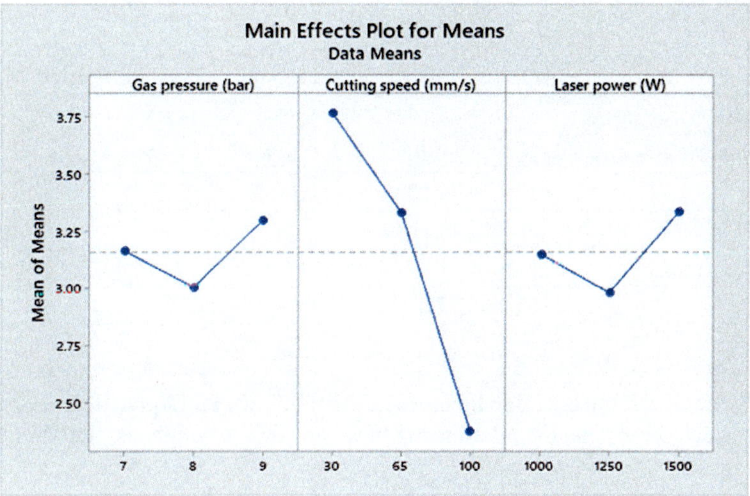

Fig. 1. Main Effects Plot for Means of surface roughness.

4.3 Response Surface Method

From the response surface method in the analysis of variance, for the surface roughness a certainty level of 95% ($\alpha = 0.05$) is carried out. In this it is confirmed that the most influential parameters on roughness is the cutting speed and the interaction of laser power with gas pressure.

Table 6. Anova by response surface method.

Fountain	GL	SC Ajust.	MC Ajust.	F-value	P-value	%Contri
Model	9	12,0096	1,3344	11,12	0	85,48%
Linear	3	9,0232	3,00774	25,07	0	64,23%
Gas pressure	1	0,0793	0,07933	0,66	0,427	0,56%
Cutting speed	1	8,785	8,78503	73,23	0	62,53%
Laser power	1	0,1589	0,15886	1,32	0,266	1,13%
Square	3	1,1328	0,3776	3,15	0,052	8,06%
Gas pressure*Gas pressure	1	0,3111	0,31114	2,59	0,126	2,21%
Cutting speed*Cutting speed	1	0,4065	0,40647	3,39	0,083	2,89%
Laser Power*Laser Power	1	0,4152	0,41519	3,46	0,08	2,96%
Interaction of 2 factors	3	1,8535	0,61784	5,15	0,01	13,19%
Gas pressure*Cutting speed	1	0,0198	0,01985	0,17	0,689	0,14%

(continued)

Table 6. *(continued)*

Fountain	GL	SC Ajust.	MC Ajust.	F-value	P-value	%Contri
Gas pressure*Laser power	1	1,6845	1,6845	14,04	0,002	11,99%
Cutting speed*Laser power	1	0,1492	0,14919	1,24	0,28	1,06%
Error	17	2,0395	0,11997			14,52%
Total	26	14,049				100,00%
S (adjusted)	R-cuad	R-cuad.ajus				
0,3463	85,48%	77,80%				

The MSR regression equation includes linear, nonlinear and squared terms, as shown in Eq. 2, which represents the relationship between surface roughness and laser cutting parameters.

$$\begin{aligned} Roughness =\, & 8,1 - 1,78 \cdot A + 0,0143 \cdot B + 0,00267 \cdot C + 0,228 \cdot A \cdot A \\ & - 0,000212 \cdot B \cdot B + 0,000004 \cdot C \cdot C + 0,00116 \cdot A \cdot B - 0,001499 \cdot A \cdot C \\ & - 0,000013 \cdot B \cdot C \end{aligned} \quad (2)$$

The response surface plot shows the influence of gas pressure vs. cutting speed, the surface roughness decreases when the speed reaches its maximum value of 100 mm/s cutting varying between 2.0 and 3.5 µm, as for the gas pressure, values close to 7.5 bar show a lower surface roughness (Fig. 2).

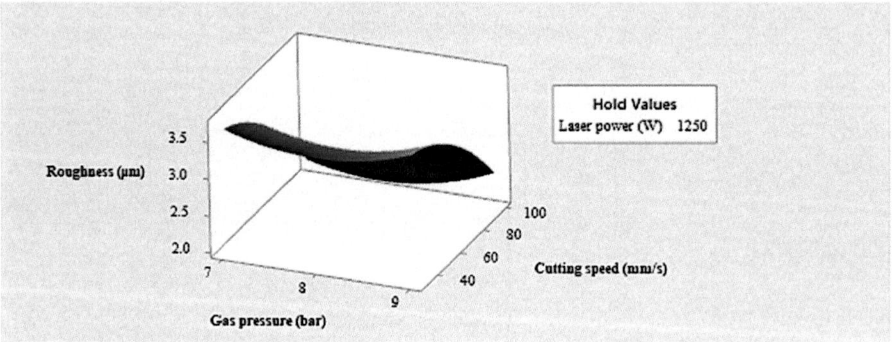

Fig. 2. Gas pressure roughness response surface graph vs cutting speed.

In the combined factors of gas pressure vs laser power (Fig. 3), the roughness decreases as it approaches the gas pressure of 8 bar and a power of 1250 W from its ends.

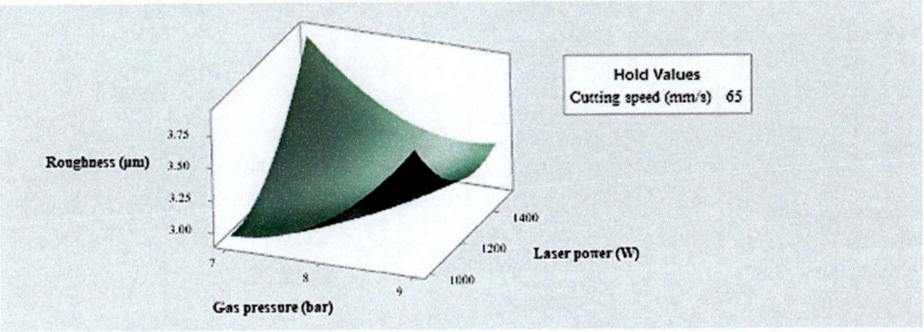

Fig. 3. Surface graph of gas pressure response vs laser power.

In the combined factors of speed vs laser power (Fig. 4), it is analyzed that the increase in speed from 40 to 100 mm/s is directly related to the decrease in surface roughness, while the variation in laser power does not represent a significant change.

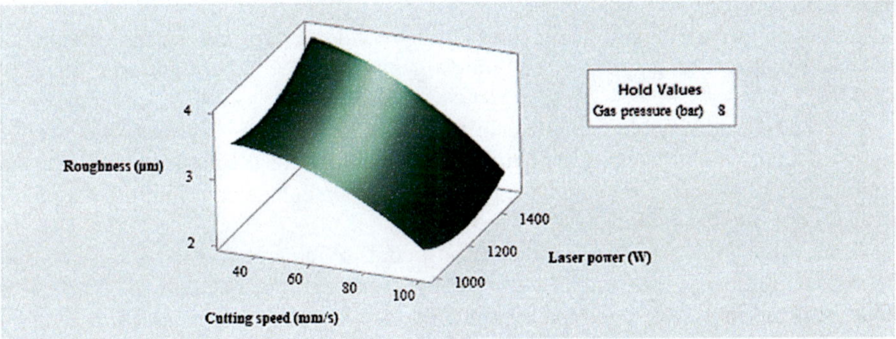

Fig. 4. Surface graph of response of the roughness of the cutting speed vs laser power.

The optimization of roughness with the parameters of pressure, speed, and power (Fig. 5). To achieve the minimum result of the surface roughness (Ra) within the range of minimum and maximum values established for each factor, the laser power parameters are established at 1156.57 W, the speed at 100 mm/s and the gas pressure at 7.4646 bar for a roughness of 2.0207 µm.

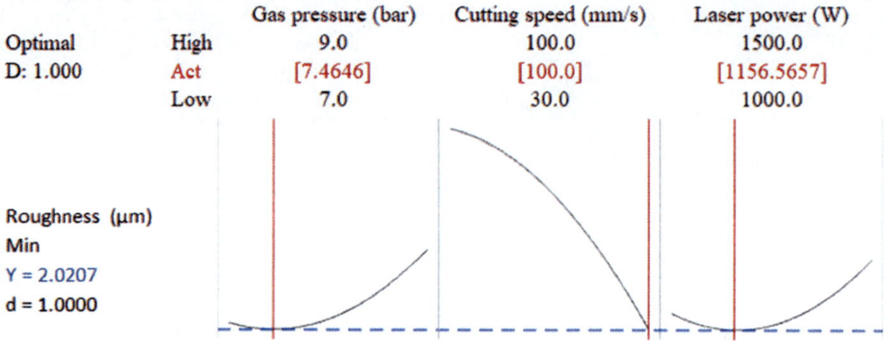

Fig. 5. Optimization minimization chart.

5 Conclusions

Laser speed and gas pressure had a significant effect on the surface quality of the cut material by increasing dross in the cut section for speeds below 100 mm/s and a constant pressure of 8 bar.

The optimum combinations for laser cutting conditions for low surface roughness Ra were set at a cutting speed = 100 mm/s, gas pressure = 8 bar and laser power = 1250 W.

From ANOVA of Table 5 of Taguchi's method, the surface roughness was significantly affected with a percentage contribution of 69.43% for the cutting speed and the combination of gas pressure with laser power with a percentage contribution of 15.57% being this the most significant.

From the ANOVA analysis of the RSM model in Table 6, it was determined that the surface roughness is affected by the parameters of cutting speed in a 62.53%, laser power squared in a 2.96% and the combination of gas pressure and laser power in an 11.99%.

The optimization results of the RSM response surface analysis referring to Fig. 5, considering a minimum surface roughness gives as a result, ideal parameters for cutting speed at 100 mm/s, cutting pressure at 7.46 bar and laser power at 1156.56 W, obtaining from this combination a roughness of 2.0207 µm.

References

1. Chuaiphan, W., Srijaroenpramong, L.: Effect of hydrogen in argon shielding gas for welding stainless steel grade SUS 201 by GTA welding process. J. Adv. Join. Process. **1**, 100016 (2020). https://doi.org/10.1016/j.jajp.2020.100016
2. Chuaiphan, W., Srijaroenpramong, L.: Optimization of gas tungsten arc welding parameters for the dissimilar welding between AISI 304 and AISI 201 stainless steels. Def. Technol. **15**, 170–178 (2019). https://doi.org/10.1016/j.dt.2018.06.007
3. Sun, X., Wei, X., Li, Z., Zhang, J., Wang, Y.: Study on process technology and surface integrity of 316L cardiovascular stents by fiber laser wet cutting. Int. J. Adv. Manuf. Technol. **105**(11), 4517–4526 (2019). https://doi.org/10.1007/s00170-018-03252-2

4. Li, M.: Evaluation of the effect of process parameters on the cut quality in fiber laser cutting of duplex stainless steel using response surface method (RSM). Infrared Phys. Technol. **118**, 103896 (2021)
5. Boujelbene, M.: Influence of the CO2 laser cutting process parameters on the Quadratic Mean Roughness Rq of the low carbon steel. Procedia Manuf. **20**, 259–264 (2018). https://doi.org/10.1016/j.promfg.2018.02.038
6. Sharma, V.K., Kumar, V.: Study on material transfer and surface properties during fiber laser cutting of A653 galvanized steel sheet. J. Braz. Soc. Mech. Sci. Eng. **41**(8), 1–17 (2019). https://doi.org/10.1007/s40430-019-1842-4
7. Tsuge, S.: Recent advances in stainless steel. Encycl. Mater. Met. Alloy. **2**, 200–207 (2022). https://doi.org/10.1016/b978-0-12-819726-4.00023-5
8. Souza Filho, I.R., Almeida Junior, D.R., Gauss, C., Sandim, M.J.R., Suzuki, P.A., Sandim, H.R.Z.: Austenite reversion in AISI 201 austenitic stainless steel evaluated via in situ synchrotron X-ray diffraction during slow continuous annealing. Mater. Sci. Eng.: A **755**, 267–277 (2019). https://doi.org/10.1016/j.msea.2019.04.014
9. Souza Filho, I.R., Zilnyk, K.D., Sandim, M.J.R., Bolmaro, R.E., Sandim, H.R.Z.: Strain partitioning and texture evolution during cold rolling of AISI 201 austenitic stainless steel. Mater. Sci. Eng.: A **702**, 161–172 (2017). https://doi.org/10.1016/j.msea.2017.07.010
10. Ahmed, K.S., Habib, M.A., Asef, M.F.: Flexural response of stainless steel reinforced concrete beam. Structures. **34**, 589–603 (2021). https://doi.org/10.1016/j.istruc.2021.08.019
11. BODOR: Catálogo Bodor. https://www.tecarsa.com/pdf/Bodor%20Laser.pd
12. Sotysiak, R., Wasilewski, P., Sołtysiak, A., Troszyński, A., Maćkowiak, P.: The analysis of fiber and CO2 laser cutting accuracy. Matec Web Conf. **290**, 1–6 (2019). https://doi.org/10.1051/matecconf/201929003016
13. La Fé-Perdomo, I., Ramos-Grez, J., Mujica, R., Rivas, M.: Surface roughness Ra prediction in selective laser melting of 316l stainless steel by means of artificial intelligence inference. J. King Saud Univ. - Eng. Sci. (2021). https://doi.org/10.1016/j.jksues.2021.03.002
14. Toapanta, O.G.: Determinación de los parámetros de corte láser por co2 en el acero ASTM A500 y la aleación HSLA ASTM A1011 grado 50 y su incidencia en la rugosidad superficial, dureza del material y efecto térmico para aplicaciones en la industria metalmecánica. http://repositorio.uta.edu.ec/handle/123456789/28263%0Awww.uta.edu.ec. (2020)
15. Parmar, V., Kumar, A., Prakash, G.V., Datta, S., Kalyanasundaram, D.: Investigation, modelling and validation of material separation mechanism during fiber laser machining of medical grade titanium alloy Ti6Al4V and stainless steel SS316L. Mech. Mater. **137**, 103125 (2019). https://doi.org/10.1016/j.mechmat.2019.103125
16. Amaral, I., Silva, F.J.G., Pinto, G.F.L., Campilho, R.D.S.G., Gouveia, R.M.: Improving the cut surface quality by optimizing parameters in the fibre laser cutting process. Procedia Manuf. **38**, 1111–1120 (2019). https://doi.org/10.1016/j.promfg.2020.01.199
17. Parthiban, A., Chandrasekaran, M., Muthuraman, V., Sathish, S.: Optimization of CO2 laser cutting of stainless steel sheet for curved profile. Mater. Today Proc. **5**, 14531–14538 (2018). https://doi.org/10.1016/j.matpr.2018.03.042
18. García, A.T., Levichev, N., Vorkov, V., Rodrigues, G.C., Cattrysse, D., Duflou, J.R.: Roughness prediction of laser cut edges by image processing and artificial neural networks. Procedia Manuf. **54**, 257–262 (2020). https://doi.org/10.1016/j.promfg.2021.07.040

Design and Construction of a Prototype Backpack-Type Ergonomic Cutting System with Exoskeleton for Oil Palm Fruit

July Ana Masache Apolo[1] and Fabio Eduardo Obando Herrera[2](✉)

[1] Universidad Politécnica Salesiana, Quito, Ecuador
[2] Biomechatronics and Bioengineering Research Group (GIByB), Universidad Politécnica Salesiana, Quito, Ecuador
fobando@ups.edu.ec

Abstract. This research details the design and construction process of a backpack-type ergonomic cutting system for African palm fruit implemented in the San Vicente community. Limoncocha parish, province of Sucumbíos – Ecuador. In the community of San Vicente, the harvesting process is carried out by hand, inspection of the land to be harvested (power lines), observation of the fruit palm by palm, working in a safe area (free from falling fruit), the cutting tool used is a stick, usually made of wood with a sickle at one of its ends, the movement is alternative made by the arms at angles outside the comfort zone and overhead. A passive exoskeleton has been adapted to the cutting system in the upper limbs, which will serve as support for the user so that their work is carried out comfortably, reducing muscle activity in the shoulders and arms, and thus reducing the times cutting and improve harvest yield. The anthropometric study was based on the Ecuadorian technical standard NTE INEN-ISO 7250-1 and on Title III of the Regulations for the safety and health of workers and improvement of the work environment. For the design of the prototype, the mechanical components were designed with the support of the Autodesk Inventor 2021 software. For the electronic and control system, the identification of the plant was implemented with the support of the MATLAB software, the transfer function of the behavior of the system and parameters of the PID controller and the position and orientation of the exoskeleton.

Keywords: Exoskeleton · African palm · Anthropometry

1 Introduction

In eastern Ecuador, especially the community of San Vicente - Limoncocha, the cultivation, harvesting and processing of oil palm fruit has increased markedly, as indicated in INIAP [2] harvesters do not collect all the fruits because there is a discrepancy between getting a clean cut or damaging the plant. The maneuver for the harvest is clearly directed by the harvester in an artisanal way, his tools are a rope to climb the palm, and a knife to cut the branches and fruits, or a sickle at the end of a long stick. The local market shows a harvester with a 1.2 hp 2-stroke engine linked to an amplification mast.

V. Robles-Bykbaev et al. (Eds.): CITIS 2022, LNNS 607, pp. 236–248, 2023.
https://doi.org/10.1007/978-3-031-24327-1_21

As Rosero stated [3], the oil palm agro-industrial chain represents 14.1% of Ecuador's agricultural GDP, throughout the chain, it generates 48,000 direct jobs and 60,000 indirect jobs not only for production but also for all services that are generated around the sector.

As Llaneza states [4], work is qualified as an activity or management determined in its relationship by internal behaviors referring to the state of the worker in relation to the task. There are several institutions that provide support to palm fruit harvesters.

During the palm fruit harvest, the worker is exposed to some risk agents, the main one is given by postures that ergonomically imply danger and that could cause musculoskeletal injuries that vary from mild to severe [5]. Some of these agents are mentioned Gómez et al. [6], load or force applied outside the comfort zone (work on the head), poor distribution of body weight increases the possibility of suffering an injury [7]. In the harvest, for much of the day, body overexertion is observed, especially in the process of cutting the fruit, with the most affected areas being the back and neck, producing muscle contractures, root discomfort due to nervous pressure, paralysis, and muscle atrophy. Two types of injuries occur, some acute-strong and others long-lasting chronic [4].

The cut of the oil palm fruit must be executed by an individual who is in adequate positions [8], since when the individual experiences discomfort, he or she performs the work efficiently (See Fig. 1).

Fig. 1. Fruit harvest using the traditional method and angles of upper limbs [9]

2 Materials and Methods

An observational, correlational study was carried out on 30 people dedicated to the harvest of oil palm fruit, 22 men and 8 women, community members of San Vicente, Limoncocha parish, aged 27.5 ± 4.0 years, with a working time in the palm fruit harvest of 7.3 ± 3.7 years.

When estimating the position of the worker considering the load on him, it is convenient to consider the degree of physical demand inflicted by the assigned work and that it is within the tolerable physiological, biomechanical, and anthropometric limits since the capacity can be exceeded of the person with the consequent health risk [10–12].

2.1 Anthropometric Measurements – Test Individuals

Anthropometric measurements [4, 13] were carried out in a population of 30 workers, who voluntarily and in writing expressed their desire to carry out the practical measurement tests and perform a work routine in the harvest of oil palm fruits [14, 15]. The same ones that are summarized in the Table 1:

Table 1. Anthropometric measurements of harvesters

	Cod.	Gender	Age (years)	Work experience (years)	Height (cm)	Weight (kg)	Length shoulder-elbow (cm)	Length elbow-hand (cm)
1	I1	F	27	6	160	52,3	19,2	24,5
2	I2	F	25	6	162	45,5	19,4	24,8
3	I3	F	34	12	165	46,8	19,8	25,2
4	I4	F	25	5	164	50,9	19,7	25,1
5	I5	F	30	10	170	50,0	20,4	26,0
6	I6	F	30	12	168	47,3	20,2	25,7
7	I7	F	36	18	171	46,4	20,5	26,2
8	I8	F	26	11	164	48,2	19,7	25,1
9	I9	M	28	8	165	54,5	19,8	25,2
10	I10	M	33	12	174	76,4	20,9	26,5
11	I11	M	31	5	168	79,1	20,2	25,6
12	I12	M	23	6	171	75,0	20,5	26,1
13	I13	M	24	4	173	76,4	20,8	26,4
14	I14	M	36	8	166	70,5	19,9	25,3
15	I15	M	23	2	160	72,7	19,2	24,4
16	I16	M	29	5	167	73,6	20,0	25,5
17	I17	M	25	5	170	77,3	20,4	25,9
18	I18	M	24	6	173	74,5	20,8	26,4
19	I19	M	26	8	167	75,5	20,0	25,5
20	I20	M	28	10	168	75,5	20,2	25,6
21	I21	M	29	10	170	70,9	20,4	25,9
22	I22	M	26	8	173	74,5	20,8	26,4
23	I23	M	31	10	175	75,0	21,0	26,7
24	I24	M	24	4	170	75,0	20,4	25,9
25	I25	M	24	2	158	76,8	19,0	24,1
26	I26	M	33	11	160	76,4	19,2	24,4

(continued)

Table 1. *(continued)*

	Cod.	Gender	Age (years)	Work experience (years)	Height (cm)	Weight (kg)	Length shoulder-elbow (cm)	Length elbow-hand (cm)
27	I27	M	24	5	162	74,5	19,4	24,7
28	I28	M	25	5	161	76,4	19,3	24,6
29	I29	M	23	4	172	72,3	20,6	26,2
30	I30	M	22	2	170	80,0	20,4	25,9
Avg			27,5	7,3			20,1	25,5
SD			4,0	3,7			0,6	0,7

The anthropometric measurements of shoulder-elbow length and elbow-hand length, necessary to measure the links of the exoskeleton, were worked with the averages for both women and men, obtaining for women: $19.9 \pm 0.18''$ shoulder to elbow and $10'' \pm 0.23''$ elbow to hand and for men: $8'' \pm 0.25''$ shoulder to elbow and $10'' \pm 0.30''$ elbow to hand.

2.2 Proposed Model

For the proposed prototype, the following were considered: time of the cutting process, construction cost, ergonomics of workloads and posture. It is based on an EksoVest exoskeleton, used at Ford in the automotive parts assembly area, designed to raise and support a worker's arms while performing general tasks.

For the armor of the prototype backpack, ASTM 36 steel will be used due to its good resistance, durability, affordable costs and existence in the local market, low weight per unit of length, as a propellant medium, a 2-stroke internal combustion engine will be used times. The links of the exoskeleton prototype whose main function is to improve ergonomics and balance the workload of the operator, were printed on Dremel 3D Digilab 3D40 FLEX 3D Printer, obtaining an exoskeleton with 2 degrees of freedom with angular movement shoulder - arm and elbow - forearm. The transmission system for the cutting system is by connecting rod – crank, implemented in the gearbox for cutting the African palm fruit. The proposed model of the prototype is shown (see Fig. 2).

Design of Mechanical Elements. Base for the engine: for the design of this element, the most critical element is taken, and it is designed in bending, the load that acts is the weight of the 2-stroke internal combustion engine, Elite brand with a power 1.47 kW/7000 rpm, with a weight of 6 kg = 58.9 N (see Fig. 3).

The maximum bending moment was calculated [16], with the support of the free software MDSolids 4.0, obtaining 120 Nm, with this value and being the selected material structural steel ASTM A-36 [17], with Sy = 248.2 MPa [16, 17] and with an allowable bending stress of 0.6 Sy [18]. We calculate the section modulus $S = 0.4\ cm^3$, using the catalog [19], selecting a round tube Ø = ¾ in, e = 0.04 in.

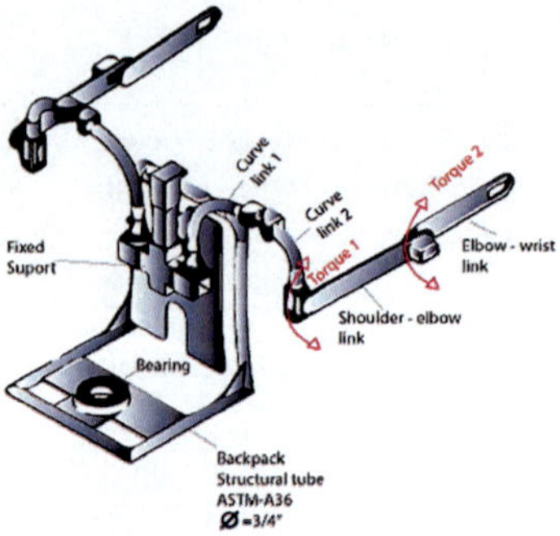

Fig. 2. Structure of the proposed exosqueleton

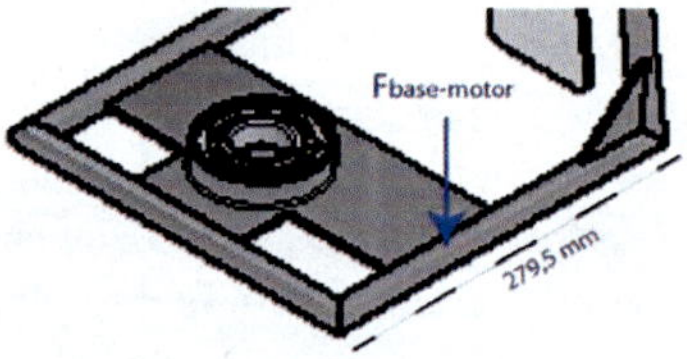

Fig. 3. State of loads - critical element

With the support of Autodesk Inventor 2020 software, the structure of the backpack is simulated, obtaining a maximum deformation of 0.8 mm, which is compared with the admissible deformation [14, 17].

$$v = \frac{L}{350} \tag{1}$$

Obtaining a deformation permissible of 0.79 mm, showing that it works within the range, also in the simulation a safety factor of 15 was obtained in the entire backpack (Fig. 4).

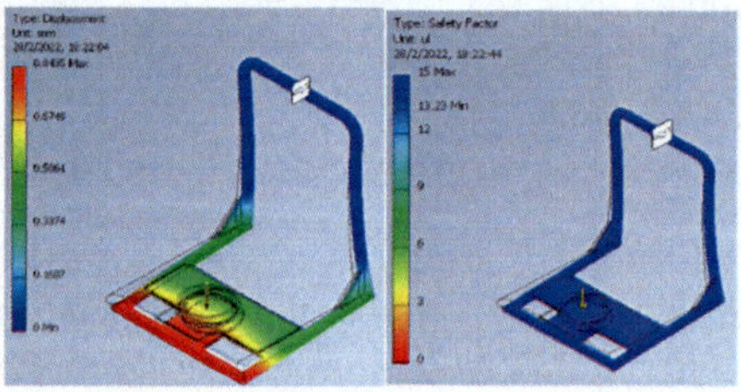

Fig. 4. Deformation and factor of safety backpack obtained in Autodesk Inventor

Sizing of Exoskeleton Links. For the size of the exoskeleton links, dimensions are proposed such that they fall within the range of averages for men and women (See Fig. 5).

Table 2. Averages of anthropometric measurements for men and women - Proposed measurements of the links

	Shoulder – elbow (cm)	Elbow – hand (cm)
Women	19.9	25.2
Men	20.1	25.6
Links	20.0	25.4

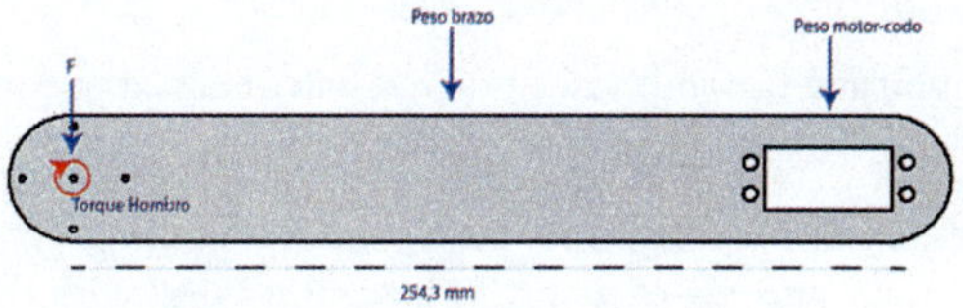

Fig. 5. Shoulder - elbow link

The main purpose of the exoskeleton is to support movement of the entire arm of the worker [20, 21], so each link (shoulder – elbow; elbow – hand) generates a resistance torque, torques that must be overcome by the servomotors. Using the calculation of segmental weights proposed in [22], the following results are obtained for the heaviest worker (80 kg) (Tables 2 and 3).

Table 3. Segmental weights of the heaviest worker

Largest weight (N)	Forearm (N)	Hand (N)	Arm (N)
784.8	12.7	4.9	21.3

For the calculation, the forces acting on the movement of the elbow are considered, the main weight to be lifted is 10 kg where 3 kg corresponds to the cutting arm and this is the force to which the elbow link is subjected. Obtaining the following torques: Elbow torque = 21.77 Nm; Shoulder torque = 56.27 Nm.

With these torques, the appropriate servomotor is selected that meets the parameters and functionality suitable for the passive exoskeleton according to the study based on benefits and limitations of the use of occupational exoskeletons, passive exoskeletons support a percentage criterion with respect to the loading fraction of muscle mass [23].

According to the design parameters, a servomotor is required that meets the necessary torque for its correct operation, for which the MG996R 12 kg servomotor with a torque capacity of 107.9 Nm was implemented.

The curved links were also simulated in Autodesk Inventor, meeting a satisfactory design (Fig. 6).

Regarding the simulation, it was carried out with the Autodesk Inventor 2019 software, before entering the data into the simulator, a previous calculation of the most

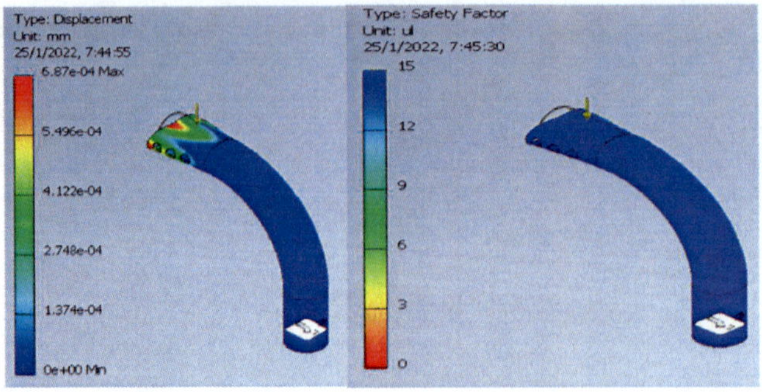

Fig. 6. Deformation and factor of safety curved link obtained in Autodesk Inventor

critical element subjected to bending was made, since it is the predominant effort in the structure of the backpack, with the software It was simulated with the workloads (motor and accessories - approx. 150 N), as it is a simple design, a medium size mesh of element factor 0.1 was used.

Regarding the restrictions in the simulation, rigid supports were used at the ends of the element (opposite to the workload).

Being the model a mechanical system that works with "low loads", the results of the simulation, safety factor 15 (maximum), guarantee us an optimal design and safety in it.

2.3 Electronic Design

For the data acquisition it is done through the connection of the Arduino and the Excel PLX-DAQ_R2, and it is obtained now of operation of the system and creates a database, with this the identification of the system is achieved using MATLAB generating the transfer function.

For the power supply for both the servomotor control module, the use of an external source is necessary, for which a charger with solar cells was implemented that has two 5v outputs and a current of 1A (see Fig. 7).

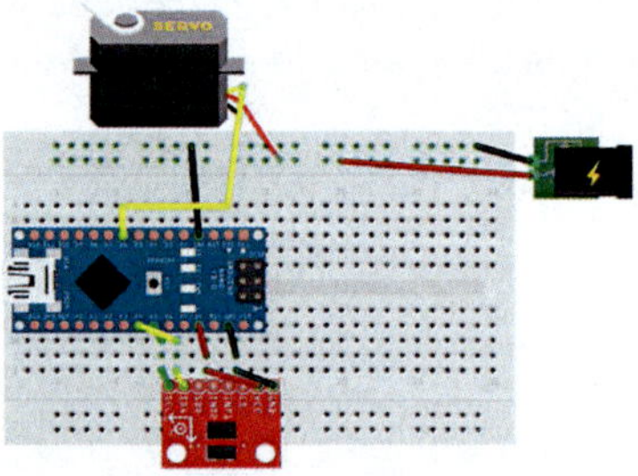

Fig. 7. Phase connection diagram: perception and data processing

The sensors and actuators used are: ADXL345 accelerometer, I2C digital interface; MG996R 12 kg servomotor, PWM and digital outputs. These sensors and actuators are controlled by the microcontroller that is implemented in the designed PCB. The diagrams are defined for each link of the exoskeleton movement, which have the same elements to implement, for each arm two identical systems are designed, one for the shoulder and the other for the elbow (Fig. 8).

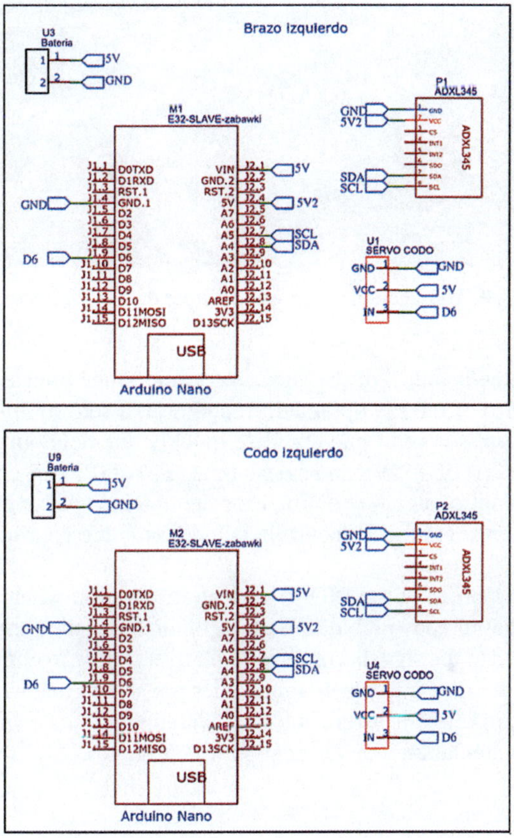

Fig. 8. Connection of sensors and actuators

3 Results and Discussion

3.1 Development of the Control System

The control system for executing the movement of the exoskeleton is developed by means of data acquisition and signal control with MATLAB and Arduino.

For the acquisition of the signals for the movement of the exoskeleton, an individual ADXL345 accelerometer was used for each link of the exoskeleton. Its operation is based on the movement of three independent axes (x, y, z) that detects the acceleration in them and allows obtaining the orientation of the sensor, obtaining signals in relation to the force of gravity, allowing control of the displacement of the arm and forearm links (See Fig. 9).

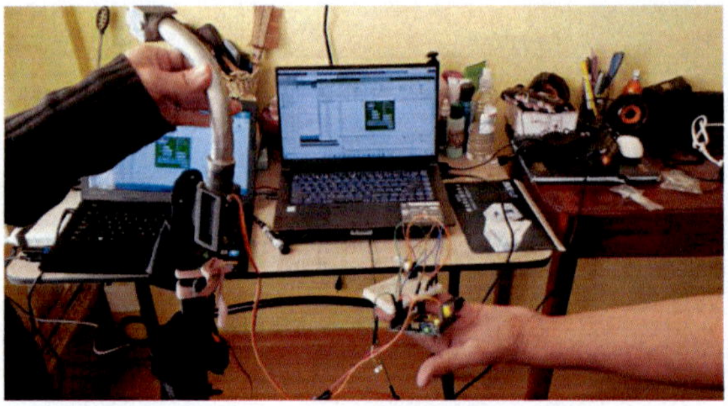

Fig. 9. Acquisition of input and output data of the system

The results obtained control of the knee, according to the transient analysis, a maximum overshoot (Mp) of 0.01 is obtained. In addition, a settling time (Ts) of 0.594 s, which makes the system control and stabilize quickly, the delay time (Td) of 0.120 s, likewise, a rise time (Tr) of 0.239 s and a peak time (Tp) of 0.734 s, concluding that with these variables the control signal is optimal for the obtained parameters of Kd, Kp and Ki that will be implemented for the controller, and finally the controlled signal is stable over time.

For the implementation of the PID, it is done in Arduino where the PID library is implemented, which is already included in the Arduino tools, the variables of the parameters are programmed to the signals corresponding to each link to obtain a more immediate establishment time. The implementation codes are shown below. With these values, the control of the exoskeleton system is implemented where its respective operation is verified in the implementation and verification (See Fig. 10).

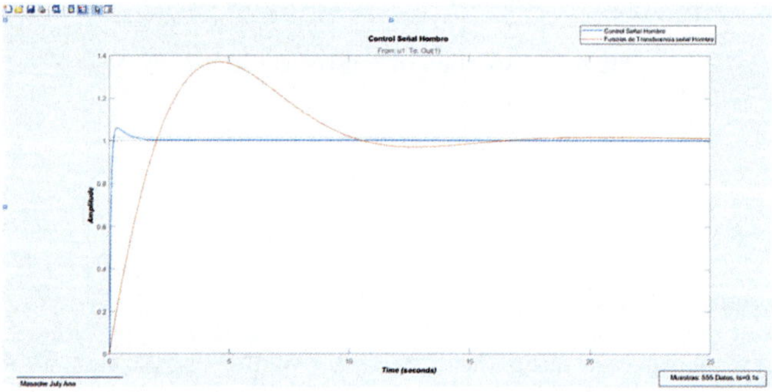

Fig. 10. Controlled signal graph of the shoulder

3.2 Range of Movements

The measurement [26] of the angles [27] of movement of the upper limb was carried out using the Protractor – KHTSXR application for Android, obtaining the results shown in the Table 4:

Table 4. Measurements of angles of movement of the exoskeleton

Optimal range	Flexion elbow	Optimal range	Flexion shoulder	Optimal range	Extension elbow	Optimal range	Extension shoulder
0–110	109.54	0–120	118.48	110–0	0.40	120–0	0.80
	110.34		120.34		0.30		0.90
	110.53		119.90		0.50		0.20
	110.20		119.89		0.00		0.70
	109.89		120.20		0.30		0.20
Av	110.10	Av	119.76	Av	0.30	Av	0.56
SD	0.39	SD	0.74	SD	0.19	SD	0.34

3.3 User Satisfaction Survey

Calibrations carried out, both positions, speeds, torques. Field tests were carried out with the prototype with the entire population of workers - fruit harvesters, due to the degree of technical instruction of the same, it was decided to carry out a survey with simple questions, so as not to have problems with the understanding of the same and that these are done individually.

The survey was carried out after a harvest day, prior to the use of the prototype, a training was carried out with all the attendees (See Fig. 11).

The evaluation of user satisfaction was made by all the workers - harvesters, through a personal and written survey, with an evaluation criterion from 1 to 10 (1 being difficult, expensive, slow) and privately expressed proposals verbally, the following parameters were evaluated in the survey: cost of the cutter, ease of operation, aesthetics, ease of transportation, working time, obtaining the following results (average values of each category).

Expressed in the graph, the qualifications are around 9 points, which indicates total satisfaction of the workers, except in the question about the cost, which has a qualification of 5.7 on average, which indicates that there is no satisfaction, because in the area where its inhabitants (workers) were tested is an area of limited economic development (See Fig. 12).

Fig. 11. Exoskeleton tests

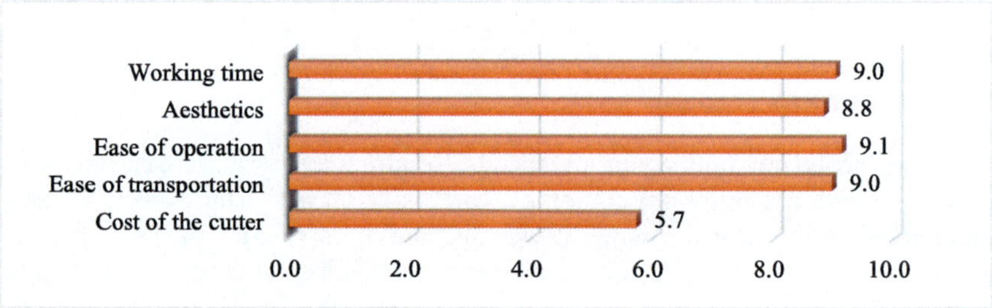

Fig. 12. User satisfaction survey results

4 Conclusions

An exoskeleton was designed for the upper extremity capable of maintaining the weight of the arm with a minimum energy expenditure of the worker, the structure was based on an anthropometric study of 30 people who are dedicated to the harvest of palm fruit, the exoskeleton is ergonomic and comfortable, it improves the user's posture and makes it easier to cut the fruit.

It is concluded that the precision elbow flexion-extension movement is ±0.3472° of reliability that represents 99.12% and in the flexo-extension movement of the shoulder the precision is ±0.76° which represents 99.57% reliability. In addition, the cutting cycle of the average African palm fruit for plantations of 1–3 years is 2.786 s per stem, and for plantations of 3–5 years it is 3.71 s per stem, detailing the path of 20 mm per stem. Each stroke of the blade. In addition to this, the exoskeleton can move with two degrees of freedom, the first in flexo-extension movement in the elbow link in a range of 0–110° and the second in the shoulder link from 0–120°.

References

1. Loor, C.: Investigation of force multiplier mechanisms for the design and construction of a robotic upper limb exoskeleton for lifting loads. ESPE (2018)
2. INIAP: Oil palm cultivation manual. Santo Domingo, Ecuador (2015)
3. Rosero, J.: Characterization of the oil palm sector in Ecuador. Pan American Agricultural School, Zamorano, Honduras (2010)
4. Llaneza, J.: Ergonomics and applied psychology. Manual for the formation of the specialist. Valladolid, Spain, Lex Nova (2010)
5. Sánchez, C., Rosero, C., Galleguillos, R., Portero, E.: Evaluation of the musculo-skeletal risk factors in the assembly area footwear. UNEMI Science Magazine, pp. 69–80 (2017)
6. Gómez, L., Rivera, S., Granado, M., Suárez, O.: Most common injuries in sport fishermen and the application of natural and traditional medicine. PODIUM Revsita de Ciencia y Tecnología en la Physical Culture, pp. 242–249 (2017)
7. COE: https://www.gestionderiesgos.gob.ec/coe-nacional/. Accessed: 15 Apr 2020
8. Estrada, J.: Basic Ergonomics. Bogotá, Colombia, Editorial de la U (2015)
9. Parra, L., Quntero, J., Maradei, F.: Design of a cutting tool for oil palm bunches. UIS Eng. Mag. **17**(1), 59–68 (2018)
10. Montiel, M., et al.: Assessment of postural load and musculoskeletal risk in workers of a metal-mechanic company. Workers' Health (2006)
11. González Cabanach, R.: Communication, stress and accident rates. Three current factors, La Coruña, Spain: Editorial University of La Coruña, Publications Service (1998)
12. OSHA: https://osha.europa.eu/publications/fact-sheet-73-hazards-and-risk-associated-manual-handling-loads-workplace. Accessed 12 Oct 2021
13. Obando, F., Maldonado, C.: Ergonomic diagnosis of postural changes and ergonomic risk assessment of a left-handed operator in handling a pedestal drill, using the REBA, RULA and OCRA checklist methods. Ind. Data Mag. **22**(2), 157–172 (2019)
14. Bernal, J.: Vigas, Buenos Aires, Argentina: Bibliográfika (2005)
15. Preciado, L., Aranda, C., Pando, M.: Work stress and burnout síndrome. In: Factores psicosocieles y mental health at work. Guadalajara - Mexico: Editorial Universitaria de Guadalajara (2006)
16. Hibbeler, R.: Mechanics of Materials. Pearson Education, Mexico (2006)
17. Avallone, E., Baumeister, T.: Manual of Mechanical Engineer. McGraw Hill, Mexico (2012)
18. Mott, R.: Resistance of Materials. Pearson Education, Mexico (2009)
19. https://www.importacareos.com/ecuador-quito/tubo-redondo-mecánico. Accessed 12 Jan 2021
20. Chávez, M., Rodríguez, F., Baradica, A.: Exoskeletons to enhance human capabilities and support rehabilitation. Biomed. Eng. J. **4**(7), 63–73 (2010)
21. Colín, J.: Exoskeletons, they are already here. Institutional Repository Universidad Iberoamericana Puebla **11**(1), 1–5 (2016)
22. Monterrosa, A.: Biomechanical analysis of the shonal chigüí technique in black belts of the martial art Taekwondo. Revista Digital **12**(108), 1–6 (2007). http://www.efdeportes.com
23. Amada, D.: Benefits and limitations of the use of occupational exoskeletons for the prevention of musculoskeletal disorders (2020). http://uisek.edu.ec/
24. WHO: Mental disorders and behavior. Diagnostic Research Criteria. World Health Organization (1997)

25. Sauceda, I., Miranda, C., Cine, G., Susarrey, O., Corona, A.: Analysis of steel pipes A-35 gr. B exposed to geothermal fluids. Científica, pp. 11–16 (2011)
26. Quitiaquez, W., et al.: Numerical analysis of flow pattern maps of refrigerant R600A in a collector/evaporator of a solar-assisted heat pump. INGENIUS **1**(26), 111–121 (2021)
27. Quitiaquez, P., Cocha, J., Quitiaquez, W., Vaca, X.: Investigation of geometric parameters with HSS tools in machining polyamide 6 using Taguchi method. Mater. Today: Proc. **1**(49), 181–187 (2022)

Roughness Analysis on Concave Surface for Martensitic Steel AISI 420 Obtained in the 3D Machining Process

Luis Christian Juiña(✉), David Gavilanez, Julio Campos, and Emilio Dávalos

Universidad Politécnica Salesiana, Quito, Ecuador
ljuina@ups.edu.ec

Abstract. The objective of this project is to study the surface roughness and the percentage influence of the factors involved in a machining process on a convex surface on the AISI 420 steel, using a variety of parameters on cutting speed, tool feed and machining strategy. These parameters were applied on a software CAD-CAM to generate the different machining strategies, resulting in 27 test which were carried out on a CNC machine.

Once the machining process was completed, the Taguchi methodology and ANOVA statistical analysis were applied to analyze the roughness obtained on the surface under study, showing that the machining strategy has an influence on the roughness value of 78.931%, followed by the cutting speed with 9.812% and finally the tool feed with 0.981%. In addition, this analysis indicates that the optimum machining parameters for this type of surface on the AISI 420 are the use of a Radial Zigzag strategy with a cutting speed of 100 m/min and a tool feed of 0.15 mm/tooth.

It should be mentioned that by applying the ANOVA statistical analysis, it was possible to obtain an equation that allows to predict the average value of roughness on this surface. Showing that it can predict the values with an error of less than 20% with respect to the values obtained experimentally, thus allowing validation of the equation for its use.

Keywords: Steel · CNC milling · Milling cutting parameters · Optimum roughness · Taguchi and ANOVA statistical method · Analysis of variance

1 Introduction

The purpose of this study is the quantitative and statistical analysis of the roughness of AISI 420 Steel, achieved through the 3-axis milling machining operation.

Steel is one of the most used materials in the production of dies for plastic products, also in the manufacture of surgical products, as well as motor drivers in centrifugal pumps and various mechanical elements. The steel under study, has characteristics that make it stand out from other materials such as excellent machinability with high hardness, good dimensional stability and considerable mechanical resistance due to stresses subjected to compression and surface deterioration of the material.

V. Robles-Bykbaev et al. (Eds.): CITIS 2022, LNNS 607, pp. 249–261, 2023.
https://doi.org/10.1007/978-3-031-24327-1_22

Due to the manufacturing process that the material undergoes to obtain the final form, it's necessary to take into consideration the surface roughness, being the most specified customer requirement and the major indicator of surface quality on machined parts.

There is various simple surface roughness amplitude parameters used in industry, such as roughness average (Ra), root-mean-square (RMS) roughness (Rq), and maximum peak-to-valley roughness (Ry or Rmax), etc. The average roughness (Ra) is the area between the roughness profile and its mean line, or the integral of the absolute value of the roughness profile height over the evaluation [1].

The surface roughness is mainly a result of various controllable or uncontrollable process parameters, and it's harder to attain and track than physical dimensions are. A considerable number of studies have researched the effects of the cutting speed, feed, depth of cut, and other factors on the surface roughness [2].

López de Lacalle [3], mentioned in their paper that nowadays it's essential to achieve a good surface quality after the machining process without the need to apply additional work. To achieve this, it's necessary to take into consideration the material to be machined and the tool, because these will give the necessary machining parameters to obtain a better surface roughness. (Improving the high-speed finishing of forming tools for advanced high-strength steels).

In their study, Tapie L [4], mention that it's difficult to select a machining strategy based on the shape of the surface, because in the machining process there are decreases in the feed rate of the tool, resulting in an investigation where they determined the impact of the tool path on a complex surface. (Machining strategy choice Performance viewer).

Finally, Laurence Colares M. and Espindola Ferreira J. [5], in their study machined hardened H13 steel, using different tool paths for both roughing and finishing processes. Thus, demonstrating that it's possible to obtain up to 25% of improvement in machining times between trajectories and that the surface roughness is significantly affected by the tool feed and the trajectory of the tool on the surface. (Assessment of tool path strategies for milling complex surfaces in hardened H13 steel).

To analyze the data obtained from the machining process and carry out the study, it was necessary to use the Taguchi methodology and an ANOVA statistical analysis, which is a statistics-based tool that measures the significance of the process parameters of the experimental design on the quality characteristic at a specific confidence level, and is performed for means and S/N ratios [6].

While the Taguchi method provides a more simplified way to set up the combination of experimental parameters and ANOVA was employed to determine statistically significant parameters [7].

The main objective of this paper is to study the roughness on a convex surface of martensitic steel AISI 420 obtained in a 3d machining process, giving to the industry in general data that allow to obtain the best possible roughness when machining this type of surfaces. In addition, to present an equation capable of predicting the roughness to be obtained on the surface from the parameters established like feed rate of tool, cutting speed and machining path, using a round mechanical milling cutter as a cutting tool.

2 Material and Machining Parameters

2.1 AISI 420 Stainless Steel

The AISI 420 steel is a very particular material, it's a martensitic steel with chromium in its structure, giving the material a great tenacity and resistance to wear [8]. Usually, this material is intended for the manufacture of dies for plastic products, due to a good polishing ability on the surface and a great resistance to corrosion, but this material can be used to produce accessories for pumps, valves for the chemical industries and tools such as scalpels or tweezers for medical use [9]. Table 1 shown below, details the chemical composition of the study material [10].

Table 1. Chemical composition of AISI 420 martensitic stainless steel.

C%	Si%	Mn%	Mo%	Ni%	Cr%
0,27	0,30	0,65	1,00	0,85	14,50

As can be seen in Table 1, this material has a 14,5% of chromium in its structure, which explains why the material has good machining properties. In this study, 9 test pieces mad of this material were used, with dimensions of 30 × 25 × 25 mm, which can be seen in the Fig. 1 shown below.

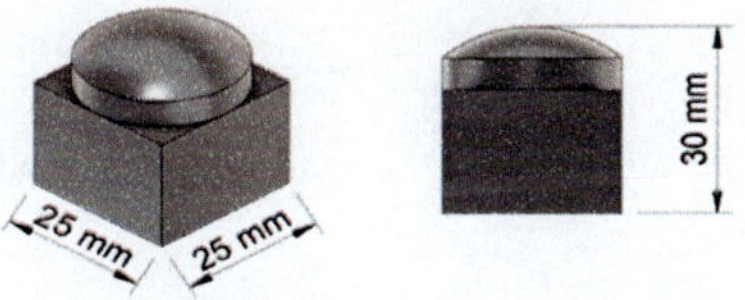

Fig. 1. Representation of the test pieces with their respective dimensions.

2.2 Machining Parameters of Study Parts

The milling parameters required for the study of surface roughness in AISI 420 steel and in any other material, are generally the tool used for the machining process, the feed of the tool, the cutting speed and the machining strategy on the surface. For the specific case of the study material, the Table 2 indicates the range of recommended parameters necessary to machine the material and ensure an adequate surface roughness.

This table indicates only a part of the parameters necessary to machine this material, because the machining strategies are selected according to the surface. Taking this into consideration, a cutting speed range of 70, 85 and 100 m/min was selected, with a tool feed of 0.05, 0.10 and 0.15 mm/tooth. It should be mentioned that this study does not take into consideration the depth of cut, since all machining operations were carried out at the same depth.

Table 2. Recommended values as a guide for the milling of AISI 420 steel [11]

Cutting parameters	Milling with carbide tools	
	Rough milling	Fine milling
Cutting speed (vc) [m/min]	80–150	70–100
Feed rate (fz) [mm/tooth]	0.2–0.4	0,01–0,2

2.3 Machining Strategies for Convex Surfaces

Using a 3D model of the test pieces in a CAD-CAM software [12], it was noted that this kind of surface presents problems with most of the commonly used roughing and finishing machining strategies. Only 3 strategies were properly attached to the surface, which are Zigzag, Concentric Zigzag and Radial Zigzag. Figure 2 indicates how these strategies were coupled on the surface under study, taking into consideration that the space between the tool passes is exaggerated to appreciate how the strategies react on the surface.

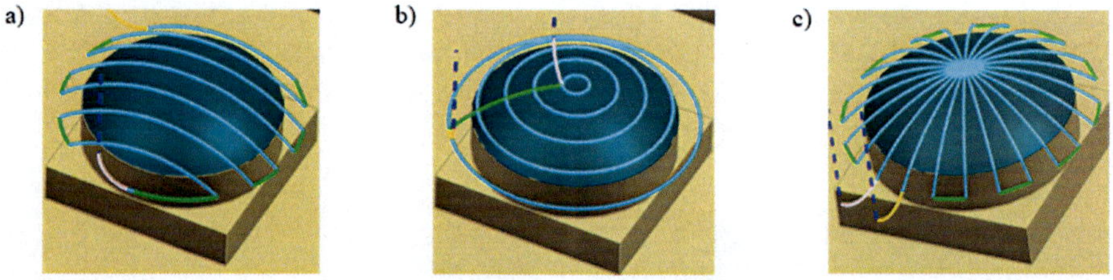

Fig. 2. Machining strategies used, where a) represent Zigzag, b) Concentric Zigzag and c) Radial Zigzag.

2.4 Machining Tool

The tool used for the study, was mainly a ball point tool who had a diameter of 6 mm, which was selected according to the recommendations of the material manufacturer. As a particularity of this tool, it has has 2 cutting helixes and is manufactured based on a special type of carbide, called "UWC" which has an aluminum titanium nitride coating (AlTiN - MS), which allows working at high revolutions, and offers greater resistance to wear, thus preventing the tool from sticking on the surface [11].

3 Roughness Results Over the Study Surface

In the machining process, the study pieces were organized according to the cutting speed and tool feed rate, resulting in the pattern shown in Fig. 3, in order to facilitate roughness data collection.

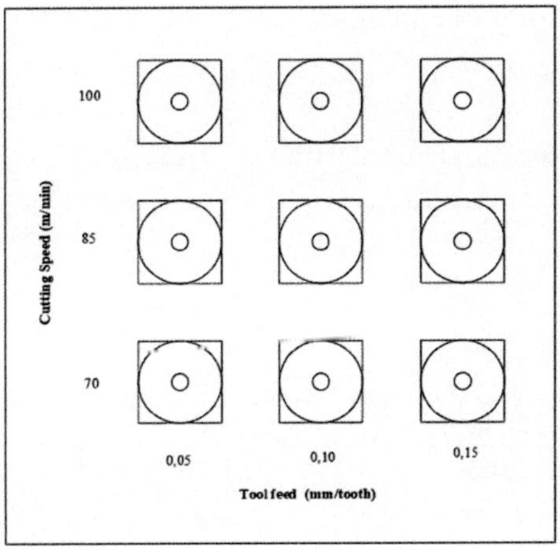

Fig. 3. Distribution of test pieces based on cutting speed and tool feed rate.

For roughness data collection, a roughness meter was used, which was a Surftest series device, model SJ-210; capable of determining the roughness of the samples, through transverse and longitudinal measurements [13]. For this work the measurements were obtained with the following parameters: measuring range of 4 mm, cutting length of 0.8 mm and a measuring speed of 0.5 mm/s.

The roughness measurement was performed tangentially to the convex surface using the previously induced roughness meter. It's necessary to mention that 27 machining processes were carried out, where 3 roughness measurements were made to obtain an average roughness per parameter. Figure 4 shown below, indicates how these values were obtained.

Fig. 4. Process of taking roughness in the study surface.

Once the machining process was completed, the roughness results were tabulated according to each strategy used, resulting in the data shown in Tables 3, 4, and 5. Table 3, indicates the roughness obtained by applying a Zigzag strategy on the surface, where an average roughness value of 2,375 μm was obtained by applying a cutting speed of 100 m/min with a feed rate of 0,10 mm/tooth, being the best possible value for this type of strategy.

Table 3. Roughness values obtained by applying a Zigzag strategy.

Strategy:	Zigzag		
Vc	fz		
	0,05	0,10	0,15
100	2,536	2,404	2,150
	2,561	2,348	2,500
	2,057	2,374	2,650
85	4,480	4,509	3,938
	4,502	4,787	3,486
	4,278	4,688	4,149
70	4,907	4,989	4,752
	4,938	4,202	4,413
	4,741	4,741	4,386

In the case of the Concentric Zigzag strategy, a worsening of the roughness values was evidenced. This strategy yields an average roughness of 4,542 μm at a cutting speed of 70 m/min with a tool feed rate of 0,15 mm/tooth, which represents a decrease in quality of 52,28% compared to the Zigzag strategy. Table 4 shows the complete values obtained by applying this strategy.

Table 4. Roughness values obtained by applying a Concentric Zigzag strategy.

Strategy:	Concentric Zigzag		
Vc	fz		
	0,05	0,10	0,15
100	4,940	4,673	4,706
	4,807	4,639	4,652
	4,215	4,751	4,587
85	4,991	4,644	5,108
	4,199	5,113	4,937
	4,697	4,690	4,775
70	4,829	5,759	4,877

(*continued*)

Table 4. (*continued*)

Strategy:	Concentric Zigzag		
Vc	fz		
	0,05	0,10	0,15
	4,873	5,523	4,466
	4,814	5,001	4,282

Finally, applying the same process, Table 5 shows that the radial Zigzag strategy shown an improvement of 63,1% compared to the Concentric Zigzag strategy, with an average roughness value of 1,237 μm at a cutting speed of 100 m/min with a tool feed rate of 0,10 mm/tooth.

Table 5. Roughness values obtained by applying a Radial Zigzag strategy.

Strategy:	Radial Zigzag		
Vc	fz		
	0,05	0,10	0,15
100	1,817	1,374	1,590
	1,598	1,122	1,290
	1,901	1,215	1,588
85	2,160	3,157	1,767
	2,814	1,601	1,356
	1,761	1,668	1,428
70	3,188	3,093	1,746
	1,679	1,789	1,183
	1,602	1,738	1,551

It should be mentioned that after performing these machining operations, the best parameters for roughing this type of surface are found in the Table 6 below. As a little resume of the machining process.

Table 6. Best roughness value obtained after machining.

Strategy	Vc [m/min]	Fz [mm/z]	Average Ra [μm]
Radial Zigzag	100	0.10	1,237

3.1 ANOVA Statistical Analysis

The information obtained, was optimized by applying a statistical software, through a statistical ANOVA analysis, which yield a linear regression equation capable of predicting the roughness values. In addition to indicating the percentage of influence on the machining of each parameter mentioned earlier in this study. These data are tabulated in Table 7, which is shown below.

Table 7. Results obtained from the ANOVA analysis.

Source	DF	S	V	F value	%P
Strategy (A)	2,000	41,725	20,863	69,133	78,931
Vc m/min (B)	2,000	5,187	2,594	8,594	9,812
Fz mm/tooth (C)	2,000	0,519	0,259	0,859	0,981
Error	18,000	6,900	0,383		12,476
ST (total)	26,000	55,304			100,000

The data shown in Table 7, which were obtained through the application of an ANOVA analysis mentioned above, allows us to appreciate that the machining strategy was the most influential factor with a 78,931%, followed by the cutting speed with 9.812% of influence on the machining process.

3.2 Analysis of Variance

In Taguchi's method it's important to study the variance or the variation of the response where the signal/noise or response action is used [14]. The optimal arrangement of the control parameters is emphasized, which makes the procedure great in the face of the variation caused by said noise factors. In Table 8 the values of the variance in the study carried out are shown.

Table 8. Results of the variance analysis of the levels according to their factors.

	Factor		
Level	A	B	C
1	4,798	3,853	3,551
2	3,794	3,692	3,577
3	1,807	2,854	3,271
Variation	2,992	1,000	0,280
Rank	1,000	2,000	3,000
Average	3,466		

As can be seen, factor A (Strategy) has a greater variance, so it is located in range number 1, so it's shown that it's the factor with the most influence in the study, factor C (Fz) has a very small variation, therefore it is located in range 3. And it is shown that it is the factor that has the least influence.

3.3 Optimum Machining Levels

According to the analysis of variance, the appropriate parameters for finishing milling are obtained with lower average roughness values on the surface. The best average surface roughness Ra is obtained with the Zigzag radial machining strategy which belongs to factor A, cutting speed of 100 m/min corresponding to factor B and finally with a feed per tooth of 0.15 mm/z corresponding to factor C. Figure 4 shows the influence of the parameters on the quality of the surface finish.

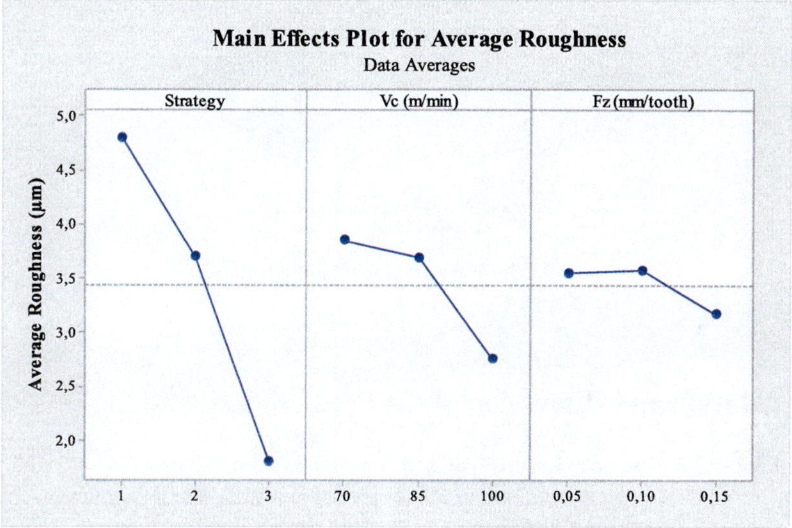

Fig. 5. Influence of finish milling parameters.

3.4 Roughness Prediction

For the prediction of the average Ra roughness, a regression equation is used using the statistical software that predicts the Ra roughness, said equation is obtained based on the results of the iterations of the most influential factors in the process, which in the study carried out in this project are the strategy and the cutting speed so that only factors A and B are considered in the regression equation.

The predicted roughness according to the calculated predictive model is detailed in the equation below.

$$R_a = 9,21 - 1,21A - 0,0305B - 0,003AB \tag{1}$$

where:

A = Represent the strategy (1: Concentric Zigzag, 2: Zigzag, 3: Radial Zigzag).

B = Represent the cutting speed (70, 80, 100 [m/min]).

The equation shown Eq. (1), which allows to predict the roughness values on this type of surface, responds to the graph shown in Fig. 5, where all the possible values that can be obtained by means of this equation can be observed.

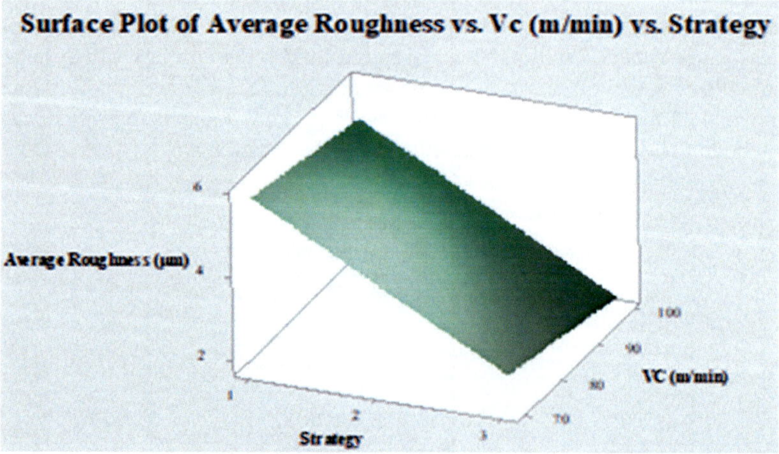

Fig. 6. Surface plot of machining factors.

4 Validation and Discussion of the Predicting Equation

For validation, it was necessary to carry out a new machining process with the optimal machining parameters obtained in the statistical analysis. This optimum machining parameters and the expected surface roughness Ra to be obtained are tabulated in Table 9.

Table 9. Optimum machining parameters.

Strategy	Vc [m/min]	Fz [mm/z]	Ra theoretical [μm]
Radial Zigzag	100	0,15	1,36

To compare this theoretical roughness value, a machining and roughness measurements were performed, obtaining a roughness value of 1,707 μm. This value can be seen in Fig. 6 shown below, which correspond to the graph generated by the roughness meter previously mentioned (Fig. 7).

This roughness value taken on the surface, differs with the value calculated by means of the prediction equation, giving an error of 20,32% in relation to the real roughness

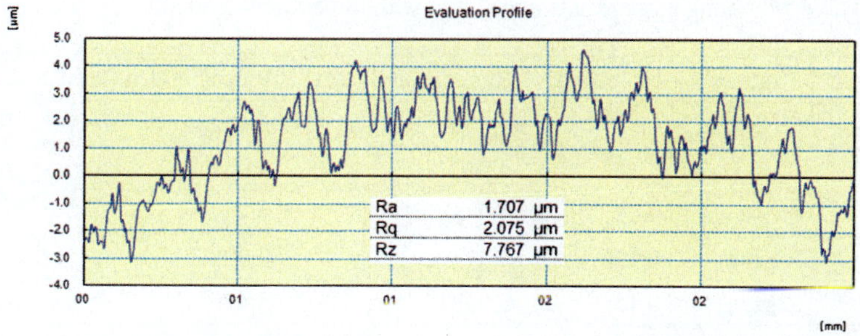

Fig. 7. Roughness measurement with optimal machining parameters.

value obtained by experimentation. To mitigate this error, it's necessary to calculate a confidence interval, which is given by Eq. (2) shown below:

$$C_I = \pm\sqrt{F(1, DF_e) * V_e * \left[\frac{1}{neff} + \frac{1}{R_p}\right]} \quad (2)$$

where:

DF_e: The degrees of freedom of the error (For this case 4,414).

V_e: Represent the error variance (In this study the value is 0,302).

neff: Represents the effective number of replications.

R_p: Represents the number of roughness measurements.

To calculate the value of *neff*, it's necessary to apply a new Eq. (3) indicated below:

$$neff = \frac{N}{1 + V_t} \quad (3)$$

where:

N: Represents the total number of tests.

V_t: Represents the number of degrees of freedom of the most influential factors on the process.

Once the *neff* is calculated, all the data are replaced in the Eq. (3), resulting in a value of C_I of ±0.77. To complete the process of validation of the Eq. (1), It's necessary to calculate the error percentage, which is determined based on new roughness measurements on the study surface; in addition, the roughness calculation of the tests is performed with the prediction equation, which contains the appropriate advance of 0.15 mm/z. These values are tabulated in the Table 10 shown below.

As the results of the errors obtained on the roughness prediction formula can be seen, which fluctuate between 0.02 and 19.34%, it is considered a relatively minimal error, therefore, the equation obtained complies with what was foreseen in terms of the optimal conditions for the machining. Then the validity of the study is ensured, the results are in the range of the optimal values obtained from the process according to the values and calculations obtained from the calculated predictive model.

Table 10. Roughness results calculated with the optimal feed of 0.15 mm/z.

Vc [m/min]	Strategy	Fz [mm/z]	Ra measured [μm]	Ra calculated [μm]	% error
70	C. Zigzag	0,15	4,542	5,442	19,8
85	C. Zigzag	0,15	4,940	4,941	0,02
100	C. Zigzag	0,15	4,648	4,44	4,48
70	Zigzag	0,15	4,517	3,992	11,62
85	Zigzag	0,15	3,858	3,446	10,67
100	Zigzag	0,15	2,43	2,956	19,34
70	R. Zigzag	0,15	1,493	1,789	19,19
85	R. Zigzag	0,15	1,517	1,767	16,0
100	R. Zigzag	0,15	1,489	1,36	8,66

5 Conclusions

According to the ANOVA and Taguchi statistical analysis, the parameters that significantly influence the variability of the surface roughness are the machining strategy and the cutting speed with 78.93% and 9.81%, respectively.

It was shown that that the best milling strategy for lower roughness values is the radial zigzag strategy with a cutting speed of 100 (m/min) and a feed per tooth of 0.15 (mm/z). These roughness values of the radial zigzag strategy represent an improvement of 51.2% with respect to the concentric zigzag and 63.1% with respect to the normal zigzag strategy.

In addition, the study revealed that the errors in the roughness values obtained by the application of the prediction equation do not exceed 20% of error, due to the confidence interval that helps to mitigate them. The value obtained experimentally with the best strategy, cutting speed and advance was 1,707 μm, this value is within the values of the confidence interval calculated by the regression equation ranging from 1,697 μm to 3,237 μm, thus confirming the validity of the statistical study carried out.

As we know, the surface roughness is an important parameter in industry in general, it intervenes directly in industrial processes that require the transformation of even more refined products with an improvement in the structure of their material. The statistical regression equation obtained allows to obtain a prediction of optimal roughness values to efficiently machine a convex surface of M303 steel.

References

1. Moghaddam, M.A., Kolahan, F.: Improvement of surface finish when EDM AISI 2312 hot worked steel using Taguchi approach and genetic algorithm. Int. J. Eng. **27**(3), 417–424 (2014). https://doi.org/10.5829/idosi.ije.2014.27.03c.09
2. Azadi Moghaddam, M., Kolahan, F.: Application of orthogonal array technique and particle swarm optimization approach in surface roughness modification when face milling

AISI1045 steel parts. J. Ind. Eng. Int. **12**(2), 199–209 (2015). https://doi.org/10.1007/s40092-015-0137-3

3. López de Lacalle, L.N., Lamikiz, A., Muñoa, J., Salgado, M.A., Sánchez, J.A.: Improving the high-speed finishing of forming tools for advanced high-strength steels (AHSS). Int. J. Adv. Manuf. Technol. **29**(1–2), 49–63 (2006). https://doi.org/10.1007/s00170-004-2482-z
4. Tapie, L., Mawussi, K.B., Anselmetti, B.: Machining strategy choice: performance viewer. In: Advances in Integrated Design and Manufacturing in Mechanical Engineering II, pp. 343–356. Springer, Dordrecht (2007). https://doi.org/10.1007/978-1-4020-6761-7_23
5. Magalhães, L.C., Ferreira, J.C.E.: Assessment of tool path strategies for milling complex surfaces in hardened H13 steel. Proc. Inst. Mech. Eng. Part B J. Eng. Manuf. **233**(3), 834–849 (2019). https://doi.org/10.1177/0954405418755824
6. Pinar, A.M., Filiz, S., Ünlü, B.S.: A comparison of cooling methods in the pocket milling of AA5083-H36 alloy via Taguchi method. Int. J. Adv. Manuf. Technol. **83**(9–12), 1431–1440 (2015). https://doi.org/10.1007/s00170-015-7666-1
7. Radune, M., Lugovskoy, S., Knop, Y., Yankelevitch, A.: Use of Taguchi method for high energy ball milling of CaCO3. Int. J. Mech. Mater. Eng. **17**(1), 1–9 (2022). https://doi.org/10.1186/s40712-021-00140-8
8. Juiña, L.C., Dávalos, E.J., Landazurí, D.S., Guaño, S.E., Moreno, N.V.: Roughness analysis of a concave surface as a function of machining parameters and strategies for AISI 420 steel. Mater. Today: Proc. **49**, 85–92 (2022). https://doi.org/10.1016/J.MATPR.2021.07.477
9. Juiña, L.C., Dávalos, E.J., Pilataxi, E., Pillajo, O., Cabrera, V., Moreno, N.V.: Analysis of the surface quality of a convex machining for a martensitic steel AISI P20. Mater. Today: Proc. **49**, 100–106 (2022). https://doi.org/10.1016/J.MATPR.2021.07.479
10. Uysal, A., Demiren, F., Altan, E.: Applying minimum quantity lubrication (MQL) method on milling of martensitic stainless steel by using nano MoS2 reinforced vegetable cutting fluid. Procedia Soc. Behav. Sci. **195**, 2742–2747 (2015). https://doi.org/10.1016/j.sbspro.2015.06.384
11. Mitsubishi Materials: Catalogo herramienta de corte 2020–2021, p. 1816 (2019). http://www.mitsubishicarbide.com/en/download/catalog_top/catalog_sa
12. Juiña, L., Cabrera, V.H., Reina, S.: Aplicación de la teoría de restricciones en la implementación de un Sistema de Manufactura CAD-CAM en la industria Metalmecánica-Plástica. Enfoque ute **8**(3), 56–71 (2017). https://doi.org/10.29019/enfoqueute.v8n3.167
13. Mitutoyo: Portable surface roughness tester SURFTEST SJ-410 series. Bolletin No. 2110, no. 2140, p. 12 (2013)
14. Juiña, L.C., et al.: Determinación de Estrategias de Mecanizado Tres Ejes de Superficies Cóncavas y Convexas en Aluminio EN-AW-7075n. Ingeniería Mecánica, Tecnología y Desarrollo **6**(3), 79–86 (2018)

Environment

Fog Collectors Systems with IoT Sensors in the Andes and Coastal Regions of Ecuador Southamerica and Data Processing

David Vinicio Carrera-Villacrés[1,2](✉), Felipe Benjamin Carrera-Villacrés[3], José Luis Carrera Villacrés[4], and Lucía Vernaza[5]

[1] Departamento de Ciencias de la Tierra y la Construcción y Departamento de Ciencias Exactas, Universidad de las Fuerzas Armadas ESPE, Av. General Rumiñahui s/n, Sangolquí, Ecuador
dvcarrera@espe.edu.ec

[2] Facultad de Ingeniería en Geología Minas Petróleos y Ambiental FIGEMPA, Carrera de Ambiental, Universidad Central del Ecuador, Av. Universitaria, Quito, Ecuador

[3] Facultad de Ciencias de la Vida, Carrera Ingeniería en Ecosistemas, Universidad Regional Amazónica IKIAM, Av. Km 7 vía Alto Tena, Tena, Ecuador

[4] Innovaiyayku, Av. Panzaleo y Av. Natabuela, Quito, Ecuador

[5] Departamento de Gestión Ambiental, Carrera de Ingeniería en Recursos Renovables, Pontificia Universidad Católica del Ecuador (PUCE), Sede Esmeraldas, Av. Subida a Santa Cruz, Santa Cruz, Esmeraldas, Ecuador

Abstract. Worldwide, one in three people do not have access to safe drinking water, and two out of five people do not have a basic facility for washing hands with soap and water, according to goal 6 Sustainable Development of United Nations, clean water and sanitation. Due to different meteorological factors that are happening today altering the water cycle and therefore making its regeneration more difficult. The use of fog and obtaining atmospheric water becomes crucial for low-income communities where their supply is limited for their needs. The objectives of this work were to describe the seven fog collector systems that have been installed in various areas of Ecuador, as well as the implementation of IoT sensors to measure some climatic variables in conjunction with the capacity to capture water from the fog, finally, make correlation and covariance matrices of the data generated by the sensors. The fog collector system that worked best was that of Galte Laime with an average water intake of 1.91 L/m^2/day. The correlation matrix shows strong positive relationships among temperature, environmental light, and battery.

Keywords: Fog collector system · Bamboo · IoT · Sensors

1 Introduction

1.1 Fog Collector Systems

In an age in which freshwater is increasingly scarce around the world, different countries and international organizations are vigorously promoting water conservation as part of their way of life [1].

V. Robles-Bykbaev et al. (Eds.): CITIS 2022, LNNS 607, pp. 265–275, 2023.
https://doi.org/10.1007/978-3-031-24327-1_23

Collecting atmospheric moisture is far from a new idea: evidence shows that the original inhabitants of the Canary Islands dug holes under trees with large foliage to collect the mist water that condenses on the leaves. While the general concept can be found in indigenous heritage and practice, 20th and 21st-century technologies have allowed mist collection to be considered in more systemic tap water supply approaches [2]. The water quality is the main aspect, if it decreases it will pose a risk to people's health.

Modern fog interception technology was first introduced in the mid-20th century, and with major advancements in its collection systems in recent decades. One of the first documented experiments aimed at determining the volume of fog deposition was carried out between 1901 and 1904 in South Africa, to investigate the viability and productivity as a natural complement to limited water resources [3].

One of the alternative water resources is mist collectors, which involve the use of mist collection technology [4] formed with a net that is arranged perpendicular to the prevailing wind direction [5]. Large water collection towers are of commercial interest, they can be used to supply water to a community [6] to take advantage of this type of collector, they can be installed in these different models.

The Two-dimensional fog collector systems consist of a collection surface that is framed between two posts so that it is perpendicular to the wind direction, they differ in terms of dimensions and materials for anchoring to the ground and the fabric fastening system [7].

The Three-dimensional fog collector systems are less common, although some are commercially available, most are associated with pilot projects suggesting innovative solutions to optimize water harvesting performance [7]. These fog catchers require a more detailed examination of their results. Some are designed to be installed in locations with multidirectional fog, thus, providing multiple fronts with fog water catchment surfaces, such as the macro diamond fog catcher and the cylindrical [7].

1.2 IoT Approach for Environmental Monitoring

IoT stands for Internet of Things. IoT aims to provide access and control to devices by using the Internet. These devices are called things. Things are embedded with electronics, software, and sensors which enable data collection without human interaction [8]. Environmental parameters can be monitored and collected by using the IoT approaches. The better is the data collection process, the better is the understanding of the monitored environment.

IoT sensors make it possible to improve the designs of fog collector systems because all the environmental information that influences the capture of water is known. It is possible to generate report anomalies, as well as to respond in real-time. Predominantly, an infrastructure IoT is equipped with a set of rules that it needs to infer to be able to perform efficient detection [9].

Today, a variety of platforms have emerged in response to the rapid growth of wireless devices used to measure and automate industrial and residential processes. IoT platforms allow the recording of data captured by sensors, the routing of data in real-time, and the remote control of devices. In addition, they provide an interface as an access point to their resources through Internet protocols; that allow interaction with high-level languages such as LabView, Matlab, and Python to perform data processing and operate a device remotely [10].

The objectives of this work are to design and build seven fog collector systems in the Andes Mountains and one fog collector system on the coast of Ecuador to determine their capacity to capture water. Likewise, in the fog collector systems, the IoT system was deployed to measure and data process the atmospheric pressure, relative humidity, air temperature, wind speed, wind direction, visibility (fog), the load of the solar panels, and environmental light.

This work is organized as follows, first, the areas of Ecuador where the fog collectors system of different shapes and materials were implemented, second, the installed IoT sensors are explained and how they send the information to the internet. Third, the numerical values of minimum, maximum, and average fog water uptake of each model are presented. Finally, A first run of the data generated by the IoT sensors is carried out to discover some type of correlation or model that explains their behavior.

2 Methodology

2.1 Study Zone

The location map of the seven fog collector systems is presented in Fig. 1.

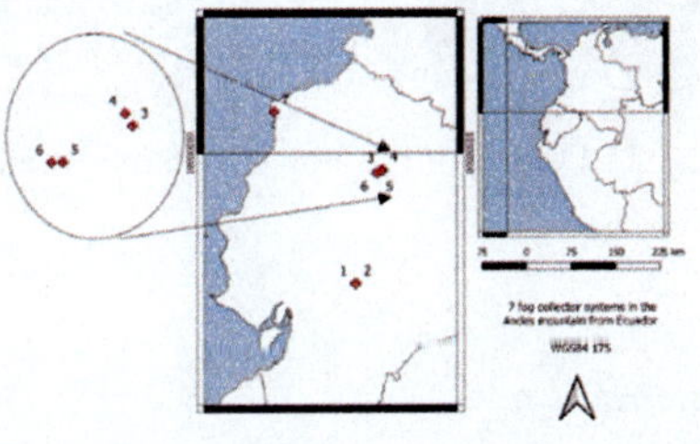

Fig. 1. Location map of the seven fog collector systems

The detailed information on collector systems is in Table 1.

Table 1. Description of the location of the fog collector systems

Id	X (m)	Y (m)	Description	Altitude (m)
1	743,387.1	9,776,062.61	Galte Laime as Warka Water	3,820.00
2	742,992.51	9,776,808.4	Fog collector in two dimensions	3,900.00
3	788,337.96	9,969,224.13	Farm Urku Huayku	2,800.00
4	787,407.45	9,970,848.54	Ilalo mountain	3,100.00
5	779,460.86	9,964,545.15	Conocoto Quinta Girasoles (pyramid)	2,600.00
6	778,023.04	9,964,506.73	Conocoto Parque Metropolitano del Sur	2,724.00
7	606,602.25	10,070,547.39	Bunche	20.00

2.2 Design, Installation, and Construction of the Fog Collector Systems

Three forms of fog collector systems were tested to determine which is the best alternative (Fig. 2). The construction of the fog collector systems has been carried out with financing from Switzerland with the program: Leading House for the Latin American Region, and Germany with the GIZ: 4th Call for the Innovation Fund.

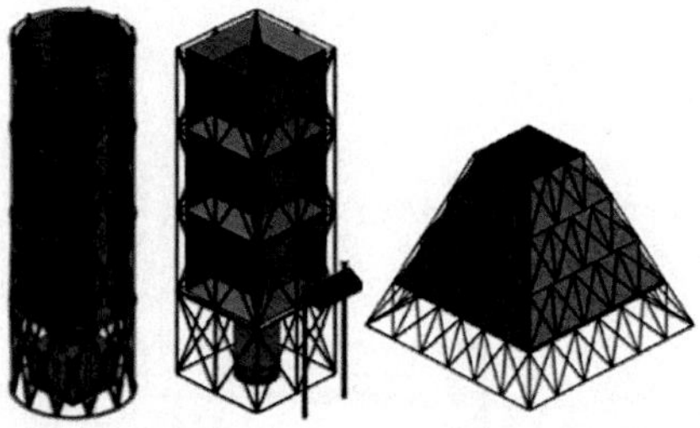

Fig. 2. Design of the Fog collector systems

2.3 IoT System Implementation

The design, implementation, and construction of the towers were carried out by the Universidad de las Fuerzas Armadas ESPE, Department of Earth and Construction Sciences, and the Research Group on Environmental Pollution (GICA) [11].

The measured parameters are transmitted through a gateway to a network server. The network server implements integration with cloud services for data storage, data analysis, and monitoring. One of the goals of data analysis is to determine the relationships among environmental parameters, the fog collector model, and the among of water collected in the storage tanks.

We have implemented a customized IoT weather station. The weather station measures several environmental parameters through embedded sensors. Measurements are wirelessly sent through the network infrastructure to be persisted in a database hosted in the Cloud infrastructure (Table 2).

Table 2. IoT weather station components

Component	Model
Atmospheric pressure sensor	BME280
Relative humidity and temperature sensor	SHT-31
Anemometer	DFR SEN086
Wind vane	DFRSEN086
Visibility (fog) sensor	MiniOFS
Environmental light	ALS-PT19
Liquid level	KIT039
Microcrontoller	Arduino leonardo
Solar power manager	DFR0535
LoRa transceiver	RN2483A

Network Infrastructure (Wireless Platform), Server, Gateways, and Cloud Infrastructure. LoRaWAN technology is well suited for applications that transmit a small amount of data with low bit rates. It is based on the LoRa spread spectrum modulation technique that transmits data over long ranges with low power consumption. Thus, LoRaWAN is the de-facto wireless platform for IoT applications [12]. Thus, in this work, we deployed a LoRaWAN network as wireless infrastructure. The network server enables connectivity, management, and monitoring of devices, gateways, and end-user applications. In this work, we use the TTI LoRaWAN Network Server. This server is also called The Things Stack. Further information can be found in [13]. The gateway serves as a communication device within the network between end devices and network servers. In this work, we deployed the TTOG LoRaWAN outdoor LoRa gateway. Further description of the gateway can be found in [14].

We use The Things Network (TTN) for the LoRa WAN network server hosting. Additionally, we integrate TTN to Amazon Web Services (AWS) to make use of further cloud services such as Dynamo DB data storage, Timestream time series storage, and AWS Grafana for data visualization.

3 Results and Discussion

3.1 Fog Collector Systems Towers in Ecuador Installed by GICA

In the Andean region of Ecuador, the water deficit is a constant problem that Galte communities located at an altitude higher than 3,500.00 m above sea level struggle with, the proposed solution of a two-dimensional fog-catching tower as shown in Fig. 3, with dimensions of 1.0 m long and 1.0 m high; a partial solution to the problem since they only cover the agricultural needs of the area, collecting between 5 to 10 L per day and around 20 L in the rainy season [15].

Fig. 3. Fog collector in two dimensions

Seven models of fog collector systems towers were designed and installed by GICA: 1- Fog collector in two dimensions (Fig. 3), 2- Galte Laime (Fig. 4a), 3- Farm Urku Huayku (Fig. 4b), 4- Ilalo volcano (Fig. 4c), 5- Conocoto Quinta Girasoles (Fig. 4d), 6- Conocoto Parque Metropolitano del Sur, collapsible tower (Fig. 4e), 7- Bunche, collapsible tower (Fig. 4f). Fog collectors in two dimensions caught 12 L/m^2/day, Galte Laime as Warka Water and Farm Urku Huayku fog collector systems managed to capture a volume of water of up to 0.85 and 4.57 L/m^2/day respectively [16]. Conocoto Quinta Girasoles (pyramid) fog collector system caught 3.0 L/m^2/day on his worst day.

Fig. 4. a) Galte Laime, b) Farm Urku Huayku, c) Ilalo volcano, d) Conocoto Quinta Girasoles, e) Conocoto Parque Metropolitano del Sur, collapsible tower, f) Bunche, collapsible tower.

A summary table of the installed fog collector systems is presented with the maximum capacity captured, average, minimum, and contact area with the fog (Table 3).

Table 3. Summary characteristics fog collector systems towers installed in Ecuador

No.	Model	Peak performance	Average performance	Minimum performance	Catchment area	Reference
		L/m^2/day	L/m^2/day	L/m^2/day	m^2	
1	Galte Laime	2.63	1.91	0.65	30.34	[17]
2	Fog collector in two dimensions	1.33	0.87	0.33	15.0	[15]
3	Farm Urku Huayku	4.57	0.56	0.07	23.4	[16, 18]
4	Ilalo mountain	0.85	0.23	0.1	33.9	[16, 18]
5	Conocoto Quinta Girasoles	2.52	0.42	5.0	60.0	[16, 18]
6	Conocoto Parque Metropolitano del Sur	2.20	0.25	0.10	24.0	
7	Bunche	0.80	0.40	0.00	24.0	

Ecuador is a country where some peasant communities are dedicated to the planting, harvesting, treatment, and marketing of bamboo, therefore, the latest design of the fog collector systems was made with this material with metal parts so that it is disassembled and easy to install.

3.2 Fog Collector Systems with IoT Sensors and Data Processing

A mist collection system located in the Ilaló volcano region based on a water condensation tower and Internet of Things (IoT) technology with real-time monitoring is presented. Figure 5 shows the design of the condensation tower deployed [18]. It is built with a combination of inexpensive and easy-to-obtain materials [18], the average water intake was 13,155 L and its maximum collection is 55,049 L [19].

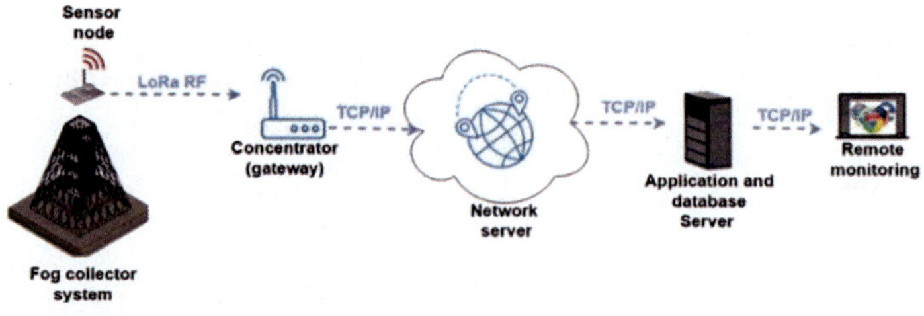

Fig. 5. Schematic of the fog collector systems with IoT sensors

The data is being collected and stored in a cloud-based database. To save battery consumption, the IoT weather station was configured to send sensor information every four hours. Figure 6 shows an overview of the collected environment dataset. For information processing, we used the open-source library "Pandas" for data analysis in "Phyton". We computed the covariance and correlation matrix among the collected environmental variables.

	barPressure	battery	humidity	light	temperature	visibility	windDirection	windSpeedAverage
0	809	4.07	68.92	52	19.84	1.92	270	0.00
1	751	4.07	68.76	52	19.84	3.09	270	0.00
2	751	4.07	69.05	51	19.84	3.08	270	0.00
3	751	4.07	69.08	43	19.84	3.10	270	0.00
4	751	4.06	68.89	50	19.87	3.09	270	0.00
...	...	...	...	...	...	...	...	...
265	468	3.69	96.65	6	12.97	3.97	180	0.00
266	464	4.10	65.75	122	20.25	3.95	270	0.89
267	464	4.17	42.69	121	25.22	3.97	270	0.89
268	464	3.69	60.60	114	20.06	3.97	270	0.00
269	464	4.16	38.57	111	27.49	4.06	270	0.00

270 rows × 8 columns

Fig. 6. Environmental variables dataset overview.

Covariance Matrix. The covariance matrix measures the linear relationship between two variables. Thus, covariance values show the impact of one variable over another. Figure 7 shows the covariance matrix computed from the environmental variables dataset.

	barPressure	battery	humidity	light	temperature	visibility	windDirection	windSpeedAverage
barPressure	12713.766543	-0.327361	436.296572	-406.678067	-46.271234	-7.316431	202.784387	0.586914
battery	-0.327361	0.030960	-2.828243	5.777189	0.730670	-0.008584	1.013569	0.013031
humidity	436.296572	-2.828243	479.394268	-906.696337	-115.291598	1.332150	-151.331437	-2.164348
light	-406.678067	5.777189	-906.696337	2759.085240	224.199211	2.303066	471.785626	3.692911
temperature	-46.271234	0.730670	-115.291598	224.199211	29.721096	-0.303355	50.125130	0.525930
visibility	-7.316431	-0.008584	1.332150	2.303066	-0.303355	0.465480	-6.370322	0.019955
windDirection	202.784387	1.013569	-151.331437	471.785626	50.125130	-6.370322	5998.438662	-6.869083
windSpeedAverage	0.586914	0.013031	-2.164348	3.692911	0.525930	0.019955	-6.869083	0.113371

Fig. 7. Covariance matrix.

Correlation Matrix. The correlation evaluates the strength of a relationship between two variables. Figure 8 shows the correlation matrix computed from the environmental variables dataset. [20] explains that the correlation of wind with other climatic variables presents inversely proportional relationships, it is also visualized that evaporation with wind speed and precipitation register higher R^2 values, classified as moderate-acceptable to good.

Environmental Variables Analysis. The covariance matrix shows a directly proportional relationship between bar pressure and humidity. Nevertheless, the covariance matrix shows that the strength of the atmospheric pressure and humidity relationship

	barPressure	battery	humidity	light	temperature	visibility	windDirection	windSpeedAverage
barPressure	1.000000	-0.017000	0.177000	-0.069000	-0.075000	-0.095000	0.023000	0.015000
battery	-0.017000	1.000000	-0.734000	0.625000	0.762000	-0.072000	0.074000	0.220000
humidity	0.177000	-0.734000	1.000000	-0.788000	-0.966000	0.089000	-0.089000	-0.294000
light	-0.069000	0.625000	-0.788000	1.000000	0.783000	0.064000	0.116000	0.209000
temperature	-0.075000	0.762000	-0.966000	0.783000	1.000000	-0.082000	0.119000	0.287000
visibility	-0.095000	-0.072000	0.089000	0.064000	-0.082000	1.000000	-0.121000	0.087000
windDirection	0.023000	0.074000	-0.089000	0.116000	0.119000	-0.121000	1.000000	-0.263000
windSpeedAverage	0.015000	0.220000	-0.294000	0.209000	0.287000	0.087000	-0.263000	1.000000

Fig. 8. Correlation matrix.

is close to zero, which means this relationship is not strong. As expected, the correlation matrix shows strong positive relationships among temperature, environmental light, and battery. The battery variable evaluates the battery charge level. Since the batteries are charged from solar energy, the battery variable could be used to relate the solar light intensity during a period. Relative humidity above 70% and temperatures below 15 degrees Celsius have proven to be the most important variables in capturing water from fog [18]. IoT sensors to generate data on climatic variables have been shown to improve the quality of life of citizens while managing natural resources in a sustainable way [21].

4 Conclusions

The seven fog collector systems built-in Ecuador continues to operate, capturing the water from the fog in places that are difficult to access and in populations that lack this vital liquid, proving to be a viable alternative. The IoT sensors installed in the fog collector systems sent meteorological information in real-time to the internet and their behavior could be modeled as a first approximation, also demonstrating that information can be generated to improve fog collector systems and obtain meteorological variables to work in various fields such as climate change. The prototype that captured the most water was Galte Laime with an average of 1.91 L/m^2/day. The correlation shows strong relationships between temperature, environmental light, and battery.

References

1. Ghosh, R., Ganguly, R.: Fog harvesting from cooling towers using metal mesh: effects of aerodynamic, deposition, and drainage efficiencies. Proc. Inst. Mech. Eng. Part A J. Power Energy **234**(7), 994–1014 (2020). https://doi.org/10.1177/0957650919890711
2. Qadir, M., Jiménez, G.C., Farnum, R.L., Trautwein, P.: Research history and functional systems of fog water harvesting. Front. Water **3**, 1–11 (2021). https://doi.org/10.3389/frwa.2021.675269
3. Qadir, M., Jiménez, G.C., Farnum, R.L., Dodson, L.L., Smakhtin, V.: Fog water collection: challenges beyond technology. Water **10**(4), 372 (2018). https://doi.org/10.3390/w10040372
4. Tan, F.J., Estanislao, M.A.P., Gregorio, A.M.A., Navea, I.J.D.: The potential of fog harvesting in tropical highlands as an alternative water resource: the case of Atok, Benguet, Philippines. In: E3S Web of Conferences, vol. 117, p. 00007. EDP Sciences (2019). https://doi.org/10.1051/e3sconf/201911700007

5. Ismail, Z., Go, Y.I.: Fog-to-water for water scarcity in climate-change hazards hotspots: pilot study in Southeast Asia. Glob. Chall. **5**(5), 2000036 (2021). https://doi.org/10.1002/gch2.202000036
6. Bhushan, B.: Design of water harvesting towers and projections for water collection from fog and condensation. Philos. Trans. R. Soc. A Math. Phys. Eng. Sci. **378**(2167), 20190440 (2020). https://doi.org/10.1098/rsta.2019.0440
7. Duque Pardo, V., Mora Estupiñan, M.C., Muñoz Cardenas, L.M.: Análisis de viabilidad de captación de agua atmosférica como fuente alterna de abastecimiento para el municipio de Villavicencio, Meta, p. 14 (2020)
8. Lyu, Y., Yin, P.: Internet of Things transmission and network reliability in complex environment. Comput. Commun. **150**, 757–763 (2020). https://doi.org/10.1016/j.comcom.2019.11.054
9. Nasir, M., Muhammad, K., Bellavista, P., Lee, M.Y., Sajjad, M.: Prioritization and alert fusion in distributed IoT sensors using Kademlia based distributed hash tables. IEEE Access **8**, 175194–175204 (2020). https://doi.org/10.1109/ACCESS.2020.3017009
10. Quiñones-Cuenca, M., González-Jaramillo, V., Torres, R., Jumbo, M.: Monitoring system of environmental variables using a wireless sensor network and platforms of internet of things. Enfoque UTE **8**(1), 329–343 (2017). https://doi.org/10.29019/enfoqueute.v8n1.139
11. U.D.L.F.A.E. DECTC: Grupo de Investigación en Contaminación Ambiental – GICA (2017)
12. The Things Industries: "The things network," What are LoRa and LoRaWAN? (2022). https://www.thethingsnetwork.org/docs/lorawan/what-is-lorawan/
13. The Things Industry: "The Things Stack," What is the Things Stack? (2022). https://www.thethingsindustries.com/docs/getting-started/what-is-tts/
14. The Things Industries: "The Things Outdoor Gateway," The things network (2022). https://www.thethingsnetwork.org/docs/gateways/thethingsoutdoor/
15. Carrera-Villacrés, D.V., Robalino, I.C., Rodríguez, F.F., Sandoval, W.R., Hidalgo, D.L., Toulkeridis, T.: An innovative fog catcher system applied in the Andean communities of Ecuador. Trans. ASABE **60**(6), 1917–1923 (2017). https://doi.org/10.13031/trans.12368
16. Gómez García, A.J., Quinteros Carabalí, J.A.: Diseño e implementación de torres atrapanieblas (3D) y ecosistema informático de monitoreo con internet de las cosas y aprendizaje automático. Universidad Central del Ecuador (2020)
17. Cárdenas Espinoza, C.A., Moncayo Basurto, A.D.: Diseño e Implementación de sistemas de captación de agua lluvia y condensación atmosférica tipo "Warka water" en la parroquia Palmira, Provincia de Chimborazo. Universidad de las Fuerzas Armadas ESPE. Carrera de Ingeniería Civil (2017)
18. Villacrés, D.C., Carrera Villacrés, J.L., Braun, T., Zhao, Z., Gómez, J., Carabalí, J.Q.: Fog harvesting and IoT based environment monitoring system at the Ilalo volcano in Ecuador. Int. J. Adv. Sci. Eng. Inf. Technol. **10**(1), 407–412 (2020). https://doi.org/10.18517/ijaseit.10.1.10775
19. Carrera-Villacrés, D.V., et al.: Relationship between the morphometric and physicochemical parameters of the Urkuhuayku microbasin where the Urku Yaku 2.0 fog collector system tower is located. In: LACCEI International Multi-conference for Engineering, Education and Technology, p. 11 (2020). https://doi.org/10.18687/LACCEI2020.1.1.161
20. Núñez-González, G., Velázquez-Pérez, D., Pelayo-Cortés, F.J., Barboza-Jiménez, P.: Analysis of reference evapotranspiration behavior during the rainy season at five weather stations in the Lerma-Chapala basin. Ing. Agrícola Biosist. **11**(2), 147–159 (2019). https://doi.org/10.5154/r.inagbi.2018.06.014
21. Gómez, J.E., Marcillo, F.R., Triana, F.L., Gallo, V.T., Oviedo, B.W., Hernández, V.L.: IoT for environmental variables in urban areas. Procedia Comput. Sci. **109**, 67–74 (2017). https://doi.org/10.1016/j.procs.2017.05.296

An IoT Device for Determining the Level of Chlorine Present in Water, Based on Electrochemical Reactions

Marcelo Flores[1](✉), Miguel Mayancela[1], Angélica Zea[1], and Juan Capella[2]

[1] Salesian Polytechnic University, Calle Vieja 12-30, Cuenca, Ecuador
mfloresv@ups.edu.ec

[2] Polytechnic University of Valencia, Camino de Vera S/N, 46022 Valencia, Spain

Abstract. This work describes the design and construction of an IoT device to determine the level of free residual chlorine present in drinking water.

The device measures the level of free residual chlorine through the correlation of the signals from three sensors (ORP, pH and temperature), the signals obtained by these sensors are processed and stored in the device, later this information is taken to the cloud through a mobile phone network.

The work begins with the characterization of the response of the electrical signal of the ORP and pH sensors to contact with water with different concentrations of chlorine and different temperatures.

The design of the circuit for amplification and filtering of the electrical signals obtained by the ORP and pH sensors is described, as well as the design of the digital and communication circuit, and the printed circuit, considering the immunity to possible electromagnetic interference.

An analysis of the operating limits of the device is carried out by performing contrast tests with conventional measurement systems.

The designed device is part of the research project "Automatic telemonitoring system of the level of chlorine present in drinking water in treatment plants belonging to small rural communities located in Ecuadorian territory", carried out by the GIHP4C research group of the Salesian Polytechnic University.

Keywords: IoT sensors · ORP · pH · Free chlorine · Water quality

1 Introduction

Access to safe drinking water is essential for health, a basic human right and a component of an effective health protection policy. [1].

The quality of drinking water directly influences the health of consumers, for this reason the monitoring of water quality is a priority for the entities responsible for public health in Ecuador [2].

An excessive level of chlorine would increase the level of toxicity of the water and a very low level would be ineffective for the elimination of microorganisms, therefore effective monitoring is necessary for people's health [2].

V. Robles-Bykbaev et al. (Eds.): CITIS 2022, LNNS 607, pp. 276–285, 2023.
https://doi.org/10.1007/978-3-031-24327-1_24

The large treatment plants that supply urban centers have expensive monitoring systems, personnel, and laboratories, in charge of verifying the correct purification process.

Far from urban centers in rural areas of Ecuador, there are small and numerous communities that, due to distances and geographical conditions, cannot be supplied by the same water treatment plants that supply urban centers. These communities are supplied through small water treatment plants located in high areas, called "Juntas de Agua" [2].

In Ecuador, currently, the monitoring of the level of chlorine in the drinking water of the "Juntas de Agua" is carried out manually by the personnel of the Ministry of the Environment, Water and Ecological Transition, who periodically travel to these places and take samples of manually for later analysis in the laboratory. The difficult access, in addition to the number of joints, generates a high operating cost in monitoring, which is why the frequency with which monitoring is carried out is well below what is necessary [3].

The IoT device developed in this work would allow this monitoring to be carried out automatically and remotely, substantially improving the monitoring process currently used.

2 Methods

2.1 Measurement of Free Residual Chlorine

Currently, the staff of the Ministry of the Environment, Water and Ecological Transition take samples manually for later analysis in the laboratory using the DPD colorimetric method.

The colorimetric method is based on the reaction of diethyl-p-phenylene-diamine (DPD). This reaction is a standard analytical approach for the analysis of free residual chlorine and other chlorine oxidants and is based on the formation of colored products with DPD [6].

For the IoT device developed in this work, it is necessary to use an electrochemical method that allows the measurement to be carried out without human intervention.

Amperometric sensors are relatively free from interference. Manganese, iron, nitrate and chromate, substances that interfere with other methods, have little influence on amperometric sensors [7], but the high cost of their components (gold cathodes and silver anodes) [5] cost of the amperometric sensor makes the system more expensive.

To reduce the cost of the measurement system, in this work it has been decided to carry out this measurement through the correlation of the signals of the ORP oxidation-reduction potential sensors, the pH and the temperature of the water.

Oxidation reduction potential is measured as a single voltage in millivolts (mV). [4] Fig. 1 shows the variation curve of the voltage delivered by an ORP sensor as a function of the free residual chlorine level expressed in parts per million (ppm) when the temperature is 20 °C and the pH is 7 (Table 1).

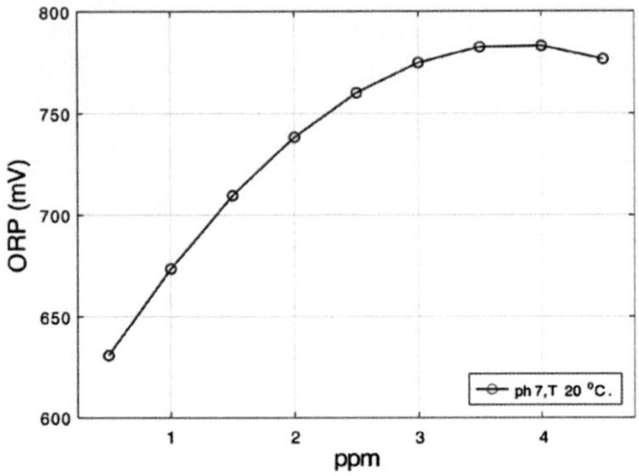

Fig. 1. ORP sensor behavior at a temperature of 20 °C and constant pH.

Table 1. ORP values measured at pH 7 and temperature 20 °C

Concentration (ppm)	Average voltage (mV)
0.5	630.2
1	674.1
1.5	709.7
2	738.5
2.5	761.3
3	774.2
3.5	781
4	784.3

The voltages delivered by the ORP sensor are also a function of pH as shown in Fig. 2 and temperature as shown in Fig. 3

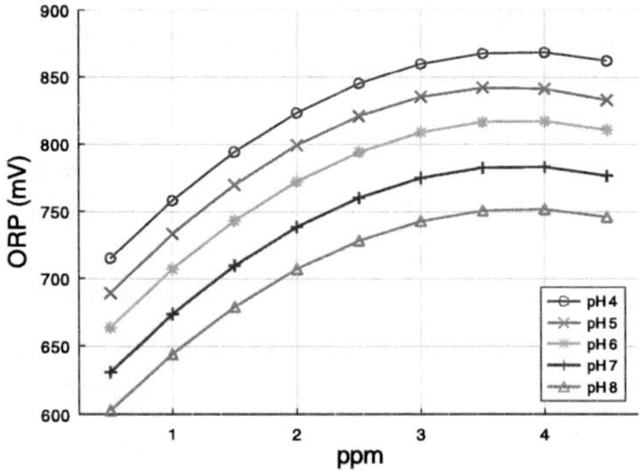

Fig. 2. Behavior of the ORP sensor at a temperature of 20 °C and variable pH.

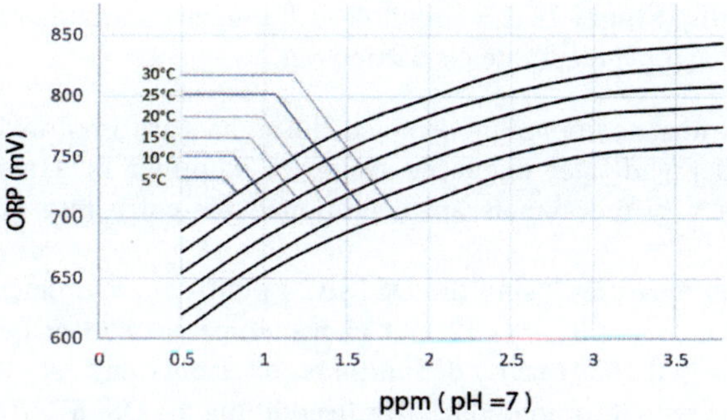

Fig. 3. ORP sensor behavior at constant pH and variable temperature.

The behavior of the ORP sensor changes only in its vertical offset when the pH and/or temperature changes.

2.2 Design and Construction of the Device

Hardware Overview. To determine free chlorine levels, the designed device uses the signals delivered by the ORP, pH and Temperature sensors. The information generated is transmitted to the cloud through a mobile phone network.

Figure 4 shows the different stages that make up the designed circuit.

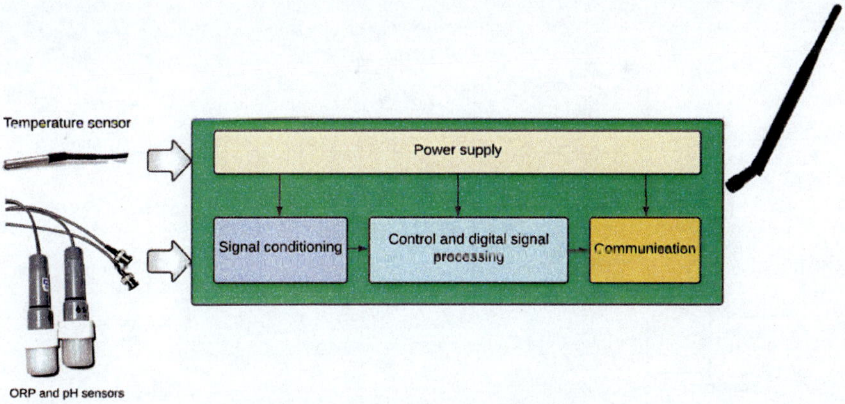

Fig. 4. General diagram of the device

Signal Conditioning Stage. In this circuit, the filtering and amplification of the signals given by the ORP and pH electrodes is carried out.

In order to eliminate common mode noise, the signals are amplified using an instrumentation amplifier and later displaced using an amplifier in a non-inverting configuration, in this way the signals are conditioned for entry into the digital analog converter.

Temperature is measured using the DS18B20 digital thermometer, this is an integrated circuit that provides 9-bit to 12-bit Celsius temperature measurements and communicates over a 1-Wire bus that by definition requires only one data line (and ground) for communication with the microcontroller. In addition, the DS18B20 can derive power directly from the data line ("parasite power"), eliminating the need for an external power supply [8]. The signal delivered by this device does not need to be conditioned, it is connected directly to the microcontroller.

Circuit for ORP Signal Reading. For the reading of the ORP, the circuit shown in Fig. 5 will be implemented in broad strokes, it is composed of an INA128 instrumentation operational amplifier and a TL071 general purpose operational amplifier, this acts as a non-inverting adder for calibration purposes.

Circuit for pH Signal Reading. The ORP and pH sensors, specifically the S651CD models used, have an output impedance of around 300 Megaohms, so it is necessary to use an operational amplifier that meets the minimum conditions for the design, initially multiple operational instrumentation amplifiers can be used, however the structure that is most suitable for reading this signal is shown in Fig. 5.

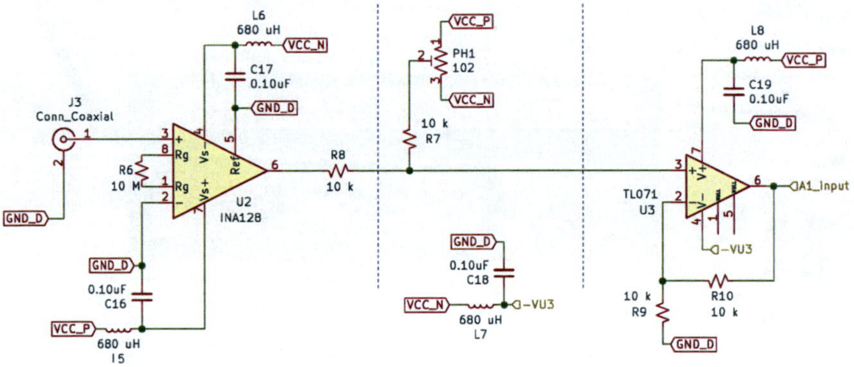

Fig. 5. Circuit for pH reading and calibration

Microcontroller Circuit. The integrated circuit intended to handle the control stage and digital processing of digital signals is the ATmega2560. The schematic design for this purpose is shown in Fig. 6.

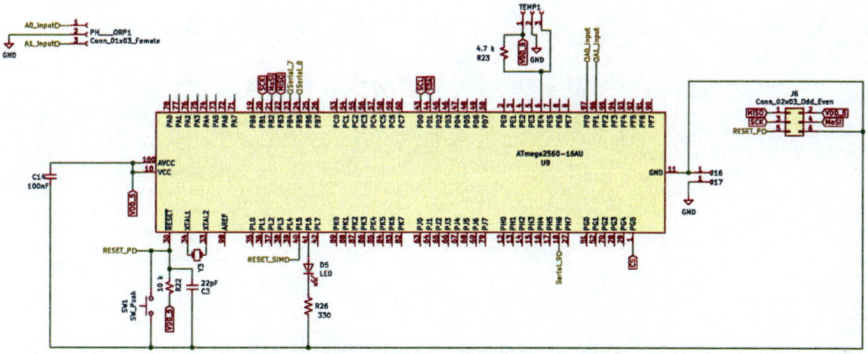

Fig. 6. General microcontroller circuit.

Communication Stage. In this circuit, the data from the ATmega2560 microcontroller is taken for sending to the cloud through a mobile phone network.

The main component of this circuit is the SIM900 communication module which, in addition to a SIM device, requires some external components shown in the circuit of Fig. 7.

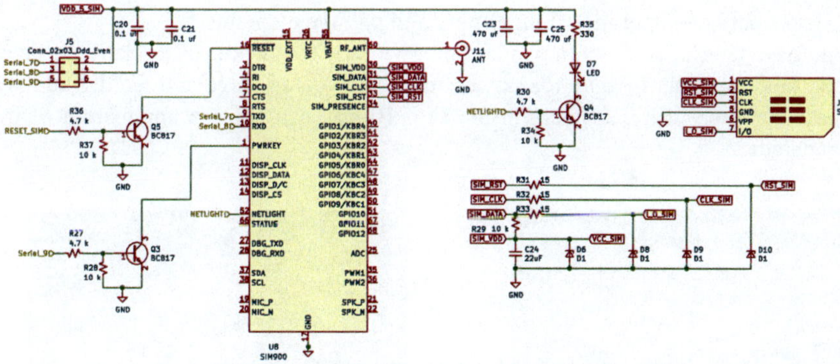

Fig. 7. Schematic for the SIM900 module.

Feed Stage. In this circuit, a symmetrical source is used in order to work with the operational amplifiers both above the reference value of zero volts and below that reference value, this would allow the reading of the ORP signals from both oxidizing samples, as well as reducing, for the pH value it would allow the reading of values above 7 whose reference value is zero volts (Fig. 8).

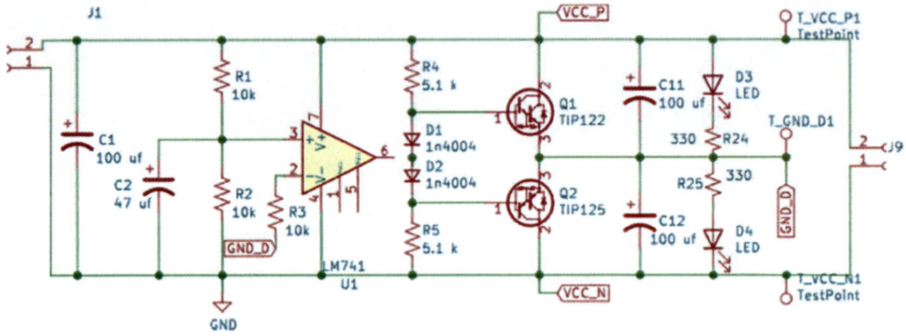

Fig. 8. Double symmetric source diagram.

Printed Circuit Board Design. Some considerations for the layout of the tracks and soldering of the components are described below.

General track width: For this, it must be considered if the design in general is working at high speeds, in the case of signal processing, these cannot cause electromagnetic disturbance because it is not working at high frequencies within the block., so in the design of the tracks both in the analog part and in the digital part corresponding to the IC ATmega2560, there is a certain "freedom" to select a track width, in this case a minimum track width of 1 mm was selected, sufficient to connect to the pads, Fig. 9 shows the 1 mm. Trace between operational amplifiers and passive elements.

Topology: One of the recommendations given for the layout of tracks is to determine the topology, that is, both the power stage, the analog stage and the digital stage must be completely differentiated, in the Fig. 10 you can see the application of this recommendation.

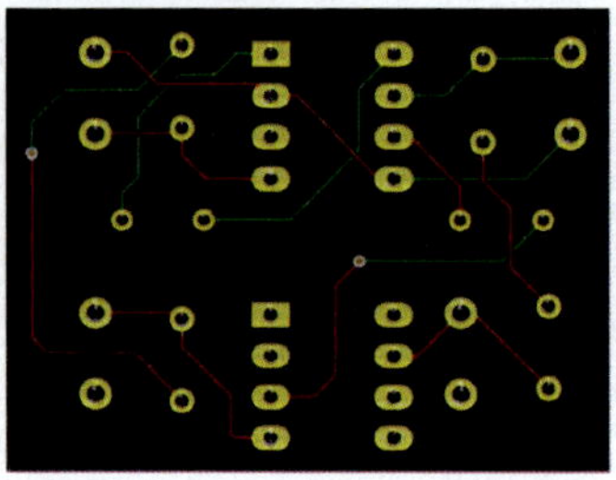

Fig. 9. Track width of 1mm.

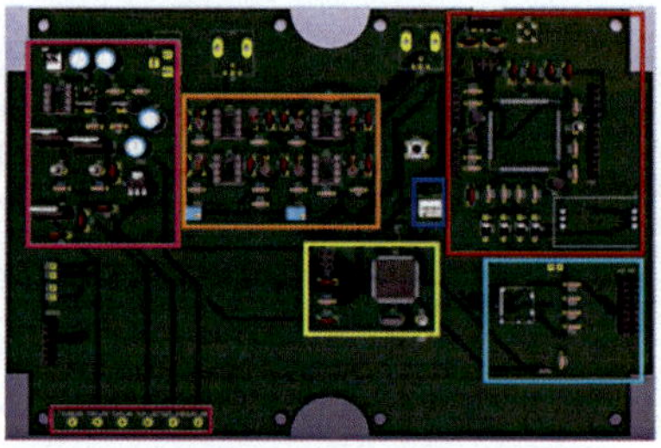

Fig. 10. Front view of the PCB.

Manufactured PCB: Applying the specified design rules, the PCB was created as shown in Fig. 11.

Hardware Ingress Protection Rating. The IEC (International Electrotechnical Commission) has established an enclosure rating system that produces an IP rating. [9] The case used in this device has IP66 (Ingress Protection 66) protection, it is the protection of an electronic device against dust particles and powerful jets of water, the selected

Fig. 11. PCB board manufactured.

Fig. 12. Implementation of IP68 connectors.

case is provided by Hammond Manufacturing, and is manufactured in ABS has a UL 94 HB flammability rating, which means that in the event of possible combustion of the furniture, it will gradually extinguish.

In Fig. 12 the connectors of the redox, pH, temperature and power input sensors are implemented in addition to the antenna connector, all the connectors are of IP68 protection level, which allows to obtain a higher degree of protection in places with high humidity.

3 Results and Discussion

Commercially available electrochemical measurement devices use amperometric sensors to determine the level of residual free chlorine.

Measurement systems based on amperometric sensors offer better results than the method used in this device, but due to the high cost of the sensor components (gold cathodes and silver anodes) [5] the use of these sensors would increase the cost. of each node making its massive implementation economically unfeasible, since the cost rises by almost 500%.

The method used by this device to measure the level of residual free chlorine does not allow chlorine to be discriminated from other oxidizing substances, that is, if an oxidizing solution other than sodium hypochlorite were introduced into the drinking water, the device would continue to deliver level measurements. of residual free chlorine, however this does not reduce the usefulness of the method since in the drinking water in the "Juntas de agua" the only oxidizing substance used is sodium hypochlorite.

Before being used each device requires to be calibrated using standard solutions with known concentrations of residual free chlorine pH.

After the calibration, several tests were carried out where it was possible to observe that the measured value stabilizes after one minute of introducing the sensors in the new solution.

To analyze the repeatability, a battery of tests was carried out using solutions with residual free chlorine levels of 0.2, 1, 1.5, 2, 2.5, 3, 3.5 and 4 ppm, temperatures of 5, 10, 15, 20, 25 and 30 °C, pH of 4, 5, 6, 7 and 8, during the tests it was possible to observe variations of less than 8% between the measured values.

State monitoring entities use the DPD colorimetric method to measure the level of residual free chlorine, in this work contrast tests were carried out between the values measured by the IoT device and those measured by the DPD colorimetric method.

Tests were carried out in solutions with 0.2, 1, 1.5, 2, 2.5, 3, 3.5 and 4 ppm at temperatures of 5, 20 and 30 °C, the differences between the values measured by the developed device and the DPD colorimetric method did not exceed 5%.

4 Conclusions

This work shows the development of an IoT device that allows measuring the level of residual free chlorine present in drinking water of small communities far from urban centers.

The tests carried out show differences of less than 5% between the values measured by the device developed in this work and the DPD colorimetric method used by state entities.

Thanks to the measurement method used, amperometric sensors are not used, this allows obtaining a low-cost node whose massive implementation is economically viable.

The designed device is part of the research project "Automatic telemonitoring system of the level of chlorine present in drinking water in treatment plants belonging to small rural communities located in Ecuadorian territory", carried out by the GIHP4C research group of the Salesian Polytechnic University, in this project has developed a mobile application and a website that, through a mobile phone network, allow access to the IoT device developed in this work.

The electronic circuit and printed circuit board have been designed to resist the problems caused by electromagnetic interference generated in the circuit and externally.

The design has considered the use of hardware (case and connectors) with ingress protection rating IP in accordance with the environmental conditions of the implementation sites in the "Juntas de Agua".

References

1. World Health Organization: Guidelines for Drinking-water Quality. Fourth Edition. World Health Organization, p. xv (2011)
2. Capella, J.V., Flores Vazquez, M.E., Navarro, J.: Proposal of a network for intelligent monitoring of chlorine level in water-drinking water in treatment plants of small rural communities. In: Proposal of a Network for Intelligent Monitoring of Chlorine Level in Water-Drinking Water in Treatment Plants of Small Rural Communities, Editorial UPV, pp. 1–3 (2019)
3. Segovia Cobos, A., Flores Vázquez, M.: Sistema de telemonitoreo del nivel de cloro presente en agua potabilizada. http://dspace.ups.edu.ec/handle/123456789/18450, Accessed 12 Mar 2022
4. Sensorex: Oxidation Reduction Potential (ORP). https://sensorex.com/oxidation-reduction-potential-orp/, Accessed 15 Mar 2022
5. Sensorex: Free Chlorine Amperometric Sensors. https://sensorex.com/wp-content/uploads/2021/09/FCL5-Instructions.pdf, Accessed 15 May 2022
6. Ecociencia: Sensores amperométricos o analizadores DPD. https://ecociencia.com.pe/sensores-amperometricos-o-analizadores-dpd/, Accessed 15 May 2022

7. Rosemount Analytical Inc: Chlorine Measurement by Amperometric Sensor. https://www.emerson.com/documents/automation/application-data-sheet-chlorine-measurement-by-amperometric-sensor-rosemount-en-68454.pdf, Accessed 15 May 2022
8. Maxim Integrated Products, Inc.: DS18B20 Programmable Resolution 1-Wire Digital Thermometer. https://datasheets.maximintegrated.com/en/ds/DS18B20.pdf, Accessed 15 May 2022
9. Hammond Manufacturing Ltd.: Definition of Protection Grades IEC 60529. https://www.hammfg.com/electrical/technical/iec, Accessed 15 May 2022

Quantification of Contaminants Through Infrared Spectrometry in Naphtha Engine Oils

Diego Díaz-Sinche(✉), Jairo Castillo-Calderón, and Rubén Carrión Jaura

Facultad de la Energía, las Industrias y los Recursos Naturales no Renovables, Universidad Nacional de Loja, Ciudad Universitaria Guillermo Falconí, Loja, Ecuador
diego.v.diaz@unl.edu.ec

Abstract. The research focuses on the quantification of contaminants in the lubricating oil of a spark-ignition engine; a controlled percentage of contaminants is induced, and they will be verified with the standard oil sample. The selection of the lubricating oil is developed based on an analytical hierarchy process. The Citroen C-ELYSEE vehicle is the experimental unit of the study, from which the oil samples will be extracted for further analysis; Kendall SAE 20W50 lubricant is used, due to the selection, the contaminants are ECOPAIS fuel with a RON 87 and a refrigerant with an ethylene glycol concentration between 85–90%. The samples are processed on optical equipment using the Fourier transform infrared spectrometry technique, both in terms of Absorbance and Transmittance. The results indicate that small percentages of glycol in the lubricant cause large variations in the chemical properties of the oil. Where the common frequency band extends in a range of about 920 to 1250 cm^{-1}, and the percent absorbance grows with the glycol level. Finally, the carbonaceous matter product of combustion is identified in the wavenumber of 1700 cm^{-1} due to the type of fuel used.

Keywords: Analysis of lubricating oils · Ethylene glycol · Fourier transform infrared spectrometry (FT-IR) · Fuel · Nitrates · Oxides · Sulfates

1 Introduction

The machines are subject to wear and tear for the period of their service life; their performance and efficiency depend to a large extent on the quality of the lubricants. Pollutants such as coolant and gasoline are highly influential in the degradation of engine oil [1]. Some parameters can be used to quantify its condition, such as vibration, noise, wear residues, temperature, and lubricant contamination [2]. Some research is related to the evaluation of the state of the lubricant, such as the work of Borín Alessandra and Poppi Ronei Jesus, who use a multivariate quality control together with the Fourier transform infrared spectrometry (FT-IR) technique, which determines the class and state of the lubricating oil [3]. The spectrometry technique has been used to measure oxidation levels in motor oils, which have characteristic vibration frequencies between 1700 and 1750 cm^{-1}, , in the case of nitration products they have absorption frequencies between 1600 and 1650 cm^{-1}, [4]. The study by Wolak, Krasodomski, and Zakhaz extends the

V. Robles-Bykbaev et al. (Eds.): CITIS 2022, LNNS 607, pp. 286–298, 2023.
https://doi.org/10.1007/978-3-031-24327-1_25

range of combustion-product oxidation processes from 1700 to 2000 cm^{-1}, to 3600 to 3700 cm^{-1} [5].

In the last decade the measurement of the properties of the engine lubricant to establish the functional limit and the change interval, employ direct and indirect measurements, for this, they use sensors and monitoring systems. In the work of Aghayan and Bordatchev, through a statistical optical analysis based on the shape of the object, they show favorable results in contents of 1%, 4%, 7%, and 10% of water, refrigerant, and gasoline, individually [1]. In the presence of some type of fuel in the engine oil, in the case of diesel, the kinematic viscosity at 40 °C presents a reduction above the allowed values, resulting in considerable changes in the properties of the lubricant [6]. A great loss of properties in lubricating oils is the presence of water, its monitoring is important to avoid damage to engine components. The combination of attenuated total reflectance (ATR) with FT-IR are precise, non-invasive technique on the quality of the lubricant that can obtain results of up to 97.7% for the viscosity state in the presence of water [7].

The parameters that can be included in an analysis of lubricating oil in spark ignition engines, to analyze the degradation, are kinematic viscosity, total acidity index (TAN), water content, flash point, soot, wear metals and additive elements [8]. The presence of fuel in the lubricating oil can be a natural thing, due to the hot flue gases that sneak through the gap between the rings and the crankcase. In the work of Díaz, Castillo and Carrión it is established that exhaust gases contain high hydrocarbons associated with an unburned fuel [13]. At low operating temperatures, such as the starting stage, there is an increased condensation of fuel vapor in the oil. It is estimated that between 1–2% of gasoline can pass into oil in the average shift interval of 5000 km. In diesel engines, the fuel passage will be about 4%. Percentages above these values affect engine elements such as fuel injectors and sealing rings. If we add to this the causes caused by driving conditions, engine design, fuel type, lubricant type, we have a variation in kinematic viscosity. The result is an excessively thin lubricating film, as well as the appearance of sediment, increased friction, and wear and, at minimum engine revolutions, the presence of preignition [10]. A gas chromatography analysis indicates that the presence of the fuel in the lubricating oil causes a lag with the standard spectrum measurement, the lag being proportional to the amount of pollutant. However, a parameter established by engine and oil manufacturers to measure their quality is the total base number (TBN), this can be very accurate in engines that are exposed to high temperatures and require effective protection against degradation [11].

The presence of water in the lubricating oil can be a common phenomenon in engines and is something that cannot be avoided, due to the combustion of the mixture itself; especially, CO2 and H2O are typical of gasoline or diesel. The problems are most noticeable in the cold start, and it is regulated when it reaches the operating temperature. During the initial operation of the engine, water vapor is observed in the exhaust duct, which is directed to the environment, and another considerable amount enters the engine crankcase through the piston rings. This causes the steam to condense and water to accumulate in the oil, resulting in emulsions. The presence of coolant in the engine oil can occur due to cracks in the cylinder heads, damaged gaskets, or leaking seals. In this case, glycols and additives react with the oil - these changes can be catastrophic due to loss of fluidity, the presence of insoluble deposits and the appearance of engine problems. Glycol can be identified in an interval between 1040 and 1080 cm^{-1}, but care should be taken with the presence of water [12]. Engine oil should be controlled in relation to the number of pollutants and the decrease of additives, where the samples have been defined in terms of the number of waves and the absorption coefficient, which help to discriminate the different levels of pollutants [9].

This research aims to quantify the degradation of spark ignition engine lubricating oils, which can be caused by the period of use, the type of fuel used, and the presence of contaminants, such as coolant, due to internal leaks in the engine. The presence of impurities identified early can only be evaluated with techniques based on infrared spectrometry. The characterization of pollutants helps to make decisions about the maintenance and a significant decrease in pollutant emissions.

The research contribution to the new knowledge is related to the characterization of failure points, which cannot be determined with a simple oil analysis and require special analysis, using infrared spectrometry techniques, to detect losses in the additive properties of a certain lubricant.

It has been considered a vehicle in normal operating conditions, intended for the transport of passengers. The FT-IR technique is used for the analysis and the lubricating oil samples detailed in Table 1 have been considered, where the type of sample, the lubricating conditions in each analysis, as well as the induction of contaminants in small percentages are described. All this in relation to the kilometers traveled by the vehicle.

Table 1. Characteristics of the analyzed samples using FT-IR.

Sample	Engine oil SAE20W50	Oil Condition	Kilometer traveled (Km)
1	No-pollution	Fresh	0
2	Pollution	Used	5000
3	Gasoline (Ecopais) – 5%	Used	3000
4	Gasoline (Ecopais) – 10%	Used	5000
5	Etilenglycol – 1%	Used	3000
6	Etilenglycol – 2%	Used	5000

The lubricant used is SAE 20W50, Kendall brand. Six samples have been defined: the first corresponds to the base spectrum, the second is spontaneously altered with the travel of the vehicle, exhibiting the presence of oxidation and nitration phenomena. In the rest of the samples, contaminants have been induced, corresponding to the presence of fuel at 5% and 10%, and coolant at 1% and 2% in the new engine oil, in relation to the periods of 3000 and 5000 km of use. It is anticipated that the presence of glycol in the oil would cause a high degradation of the lubricant properties.

2 Materials and Methods

2.1 Equipment and Procedure

The degradation of the lubricating oil in a provoked ignition engine was studied. The Citroen C-ELYSEE vehicle was used as an experimental unit and the technical specifications are described in Table 2. A Thermo Scientific spectrometry equipment, Nicolet iS10 FT-IR model, is used, which allows a signal-to-noise peak-to-peak measurement in 1 min of 50000:1 and a spectral resolution of up to 0.4 cm^{-1} (see Fig. 1); the general characteristics of the equipment are described in Table 3.

A review of the methodologies of the ASTM standards has been carried out, aimed at the quantification of the wear level of engine oil. Each of the oil samples extracted during engine operation are subjected to a protocol established in the ASTM-E2412 standard, which measures the determination of contaminants such as: water, carbonaceous matter, ethylene glycol, fuels, also the degradation of the oil by oxidation and nitration content. The wavenumber ranges for oxidation are centered between 1670 and 1800 cm^{-1}; for nitration at 1600 and 1650 cm^{-1}; for the presence of gasoline at 745 and 755 cm^{-1}; finally, ethylene glycol residues were found between 1030 and 1100 cm^{-1} [14]. The ASTM-D7214 standard measures the level of oxidation using the FT-IR technique, but it does not quantify an absolute value of oxidation and its results do not have a high precision; except for mechanical transmission lubricating oils [15]. Finally, the ASTM-D7414 standard measures the oxidation level of oils in service in machines such as: gear transmissions, internal combustion engines, hydraulic systems, and compressors [16]; which will be used for quantification of results.

Table 2. Engine specifications.

Citroen C-ELYSEE	
Power (CV/rpm)	115/6050
Torque (Nm/rpm)	150/4000
Displacement (cm^3)	1587
Fuel	Nafta
Year of manufacture	2016
Operation	Spark ignition

Each sample is obtained with a vacuum pump located in the duct of the engine oil level dipstick and poured into a 60 ml reservoir. The cells used for the analysis of samples are potassium bromide (KBr), which corresponds to the ASTM-D2412 standard with a spectral range from 350 to 7800 cm^{-1}. Outside the indicated range, the measurements will have a low sensitivity. To quantify the impurity content in the lubricating oil, contaminants related to gasoline and antifreeze have been induced. In addition to incipient failures, the experimental unit was previously checked and prepared to ensure ideal operating conditions, avoiding any mechanical, electrical, or electronic failure.

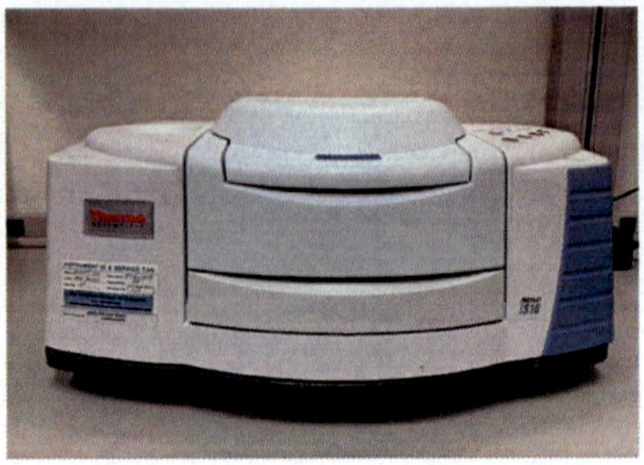

Fig. 1. Thermo scientific nicolet iS10 FT-IR spectrometry equipment.

Table 3. Equipment specifications.

Nicolet iS10 FT-IR	
Spectral range	7850 a 350 cm^{-1}
Spectral resolution	0.4 cm^{-1}
Interface	PC USB 2.0
Maximum speed	40 spectra per second at 16 cm^{-1}

2.2 Selection of Engine Lubricating Oil

Applying a hierarchical approach of multiple criteria for decision-making, the lubricants in greatest demand in the automotive market of the city of Loja, Ecuador, where the research was carried out, have been established. Four categories have been defined for the selection of lubricating oil, of which we have: oil process, a decision factor that includes quality and price, the category that describes the vehicle with the greatest request. Finally, in the SAE grade, the most used 5W30, 10W30 and 20W50 multigrade oils are established.

Of all the conditions and alternatives proposed, available, the most used oil will be selected based on surveys previously defined by premises dedicated to vehicle maintenance and professional experience. The decomposition of the problem into a hierarchical structure has been carried out, which allows us to evaluate the influence of the criteria with the aim of the final decision. The problem is divided into several stages, at the first level the selection of the lubricating oil has been defined as a goal, the object of study, there are seven alternatives and four criteria to be considered. The structure allows to build paired comparison graphs, where each element is faced one by one, first the criteria with each other and then the alternatives depending on each criterion [17]. There is a stage that values the elements, in which a binary comparison is made that determines the relative importance between each of them, this phase involves the synthesis and analysis of the results to achieve a quantitative classification [18]. The level of inconsistency reached is 0.08, below 0.1 of the allowed value, which allows an acceptable state of the selection results (see Fig. 2, Fig. 3) [19].

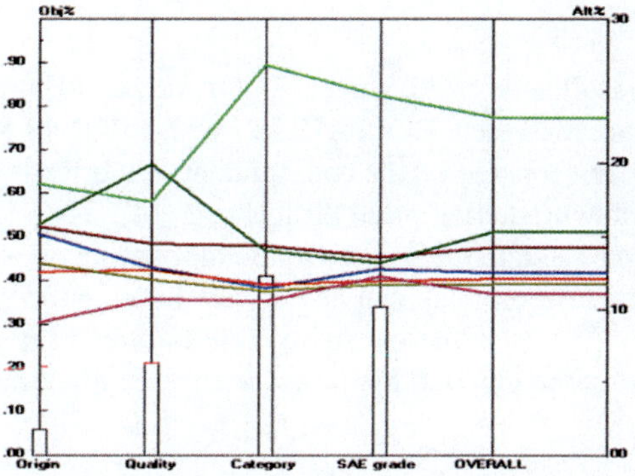

Fig. 2. Sensitivity graph – Engine lubricating oil selection.

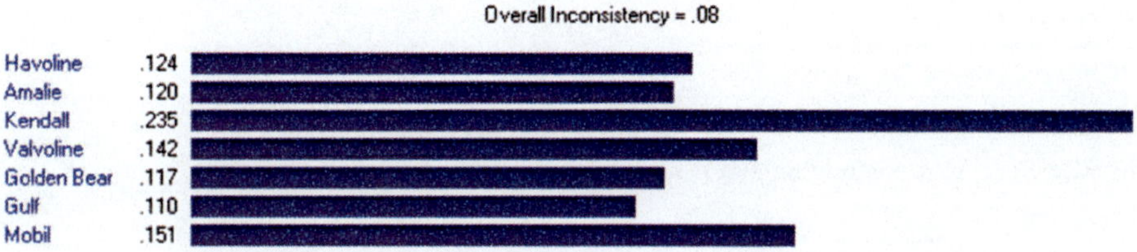

Fig. 3. Synthesis with respect to Engine Lubricating oil selection.

The properties of the selected lubricant, the object of study, have been defined in Table 4. It is based on a semi-synthetic formulation that protects against oil thickening, oxidation, and corrosion, low volatility, protection in cold start stages, high resistance to foaming, smoke reduction in exhaust gases, suitable for the use of engines running on ethanol-containing fuels up to E85 [21].

Table 4. Lubricating oil specifications.

Kendall 20W-50 (High Performance)	
Specific Gravity @ 60 °F	0.881
Density, lbs/gal @60 °F	7.35
Color, ASTM D1500	3.5
Flash Point °C	230
Pour Point °C	−30
Viscosity, Kinematic @ 40 °C (cSt)	181
Viscosity, Kinematic @ 100 °C (cSt)	19.2
Viscosity Index	127
Total Base Number (TBN), ASTM D2896	7.0

2.3 Characteristics of Pollutants

In Ecuador, three types of gasoline are sold: Super, ECOPAIS, and Extra, however, in the city of Loja, the latter is not marketed. ECOPAIS fuel with a RON of 87, established by the NTE INEN 935, has been considered the contaminant of lubricating oil, which is in higher demand due to its low marketing price [20].

Coolant has been defined as a contaminant of engine lubricating oil; concentrations are described in Table 5. An antifreeze leak can be very common in the internal part of the engine, causing loss of viscosity in the oil and engine failures. It is not enough to change the oil, as a small part of the glycol remains in the engine, affecting even the new oil [12].

Table 5. Physical and chemical properties.

Antifreeze/Coolant	
Ethylene Glycol (%)	80–95
2-Ehtyl Hexanoic Acid, Sodium Salt (%)	1–5
Neodecanoic Acid, Sodium Salt (%)	1–5
Diethylene Glycol (%)	0–5
Freezing Point (°C)	−38
Flash Point (°C)	Greater than 100
Relative density	1.07–1.14
pH	8.7–9.2
Boiling Point (ªC)	171

2.4 Optical Technique by FT-IR

The FT-IR technique has been defined by its response speed, ease of testing and access to instrumentation. Electromagnetic radiation is used that conducts a flow of particles and decreases when the radiation travels from one medium to another, so that a beam can vary its intensity relative to photons per unit area [22]. The electromagnetic spectrum includes many wavelengths, and a logarithmic scale is required for the representation of the wavelength range. Infrared spectra, in a zone established between 4000 and 400 cm^{-1}, are used for the study of organic molecules, a range in which spectral analyses will be defined [23]. For the quantification of radiation absorption by infrared spectroscopy, magnitudes are used in terms of transmittance (T) and absorbance (A), both are used to describe the same sample, where the first is directly proportional to the concentration of the elements existing in the sample. The absorbance of a body increases with the amount of light impregnated and with it a smaller amount of light will be transmitted by the body. The mathematical formulation of the study focuses on the Beer-Lambert Law, which expresses how light is absorbed by a body [24].

$$A = \log_{10}\left(\frac{1}{T}\right) = \varepsilon \cdot C \cdot L \tag{1}$$

$$T = \log_{10}\left(\frac{I}{I_0}\right) \tag{2}$$

where, I is the radiation energy transmitted by a sample, I_0 the incident energy, L is the thickness of the cell, the absorptivity constant ε and C the concentration of the substance to be quantified.

3 Results and Discussion

The values of the infrared spectrum correspond to the graphical representation of the wavelength values, against the values of the transmittance and absorbance percentages. As identified in Fig. 5, for a certain wavelength or frequency cm^{-1} there is a drop in the transmittance percentage. In the case of the spectrum in terms of absorbance, such a drop manifests itself as a frequency peak or band (see Fig. 4). Contaminants such as oxidation and nitration are not easily quantifiable compared to the standard sample. According to ASTM-E2412, both pollutants are present in a spectral range comprised between 1800 and 1600 cm^{-1}. When the sample is extracted after 5000 km of travel of the experimental unit, changes in the spectrum are present approximately at high frequencies from 1200 to 850 cm^{-1} and low from 3000 to 3600 cm^{-1}. In the research of Wolak, Krasodomski and Zajac, the use of the FT-IR technique identifies the presence of oxides and nitrates in ranges between 1700 - 2000 cm^{-1} and 3600 to 3700 cm^{-1}. These results have similar characteristics to the research data, but the presence of contaminants with low wavenumber ranges is also observed [5]. It has been defined in Fig. 4, the wavelength characteristics of the pollutants, the degradation product of the lubricant in use, with the presence of oxidation in 1700 cm^{-1}, nitration at 1620 cm^{-1} and sulfation at 1100 cm^{-1}, values very similar to those of other investigations [12]. Finally, there is the presence of water that varies the spectrum to a high wave number that is between 3100 and 3600 cm^{-1}, reaching a value of 0.28% in terms of absorbance.

Figures 6 and 7 show the absorption and transmission coefficients for the two levels of glycol contamination in SAE 20W50 engine oil. An increase in glycol is present throughout the spectral range. The characteristic frequency of this pollutant is shown in the range of 920 to 1250 cm^{-1}, this frequency band can be altered by other pollutants, such as sulfation residues and antiwear additives, typical of lubricating oil, which degrade with use. Similar trends are obtained in the results of other researchers where the absorption coefficient increased with the increase of glycol, reaching a maximum wave number of 2743 cm^{-1}, which is associated with an amount of 500 ppm [14]. At a high number of waves, the water increase is noticeable in the spectrum, relative to the standard sample, reaching a maximum value of 0.45% absorbance.

The presence of contaminant by fuel is shown in the range of 755 to 745 cm^{-1} according to ASTM-E2412. In Figs. 8 and 9, the range is from 820 to 960 cm^{-1}, where an increase in the level of the spectrum is observed due to an increase in the contaminant in the oil sample. The existence of carbonaceous matter, a product of combustion, contributes to the degradation of the oil, but its value is influenced to a greater extent when the octane level contained in the fuel is lower, causing incomplete combustations that favor this phenomenon. Finally, the presence of this contaminant gives rise to a wave number close to 1700 cm^{-1}, where the value in terms of absorbance has an amplitude of 0.13% and a transmittance loss of 2% [25].

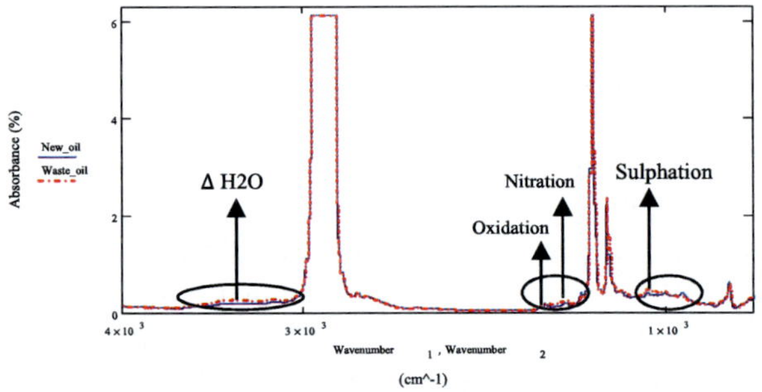

Fig. 4. Result of the measurement of the oxidation and nitration content at 5000 km of travel in terms of absorbance.

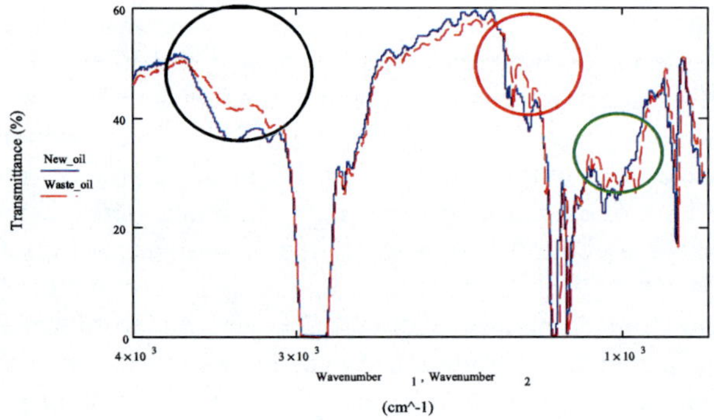

Fig. 5. Result of the measurement of the oxidation and nitration content at 5000 km of travel in terms of transmittance.

Finally, the presence of fuel in a lubricant analysis by TBN can be confirmatory with the variation of kinematic viscosity if we take the value of Table 4; this is 19.2 @100 with the standard sample and varies to 14.987@100 (see Fig. 10). The presence of fuel in the lubricant modifies the degree of viscosity and affects the function of the engine.

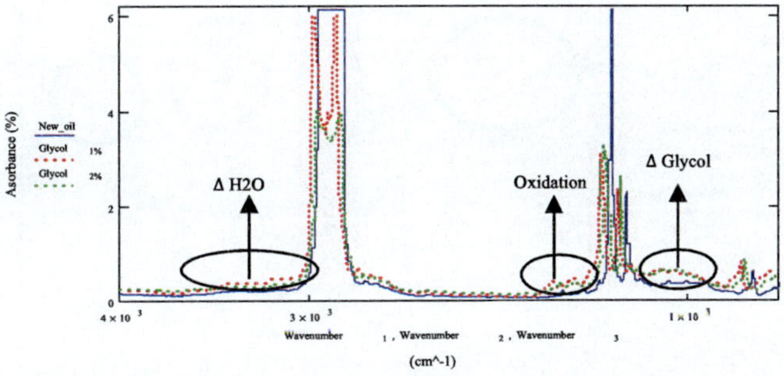

Fig. 6. Result of the measurement of the glycol content at 1% and 2% in terms of absorbance.

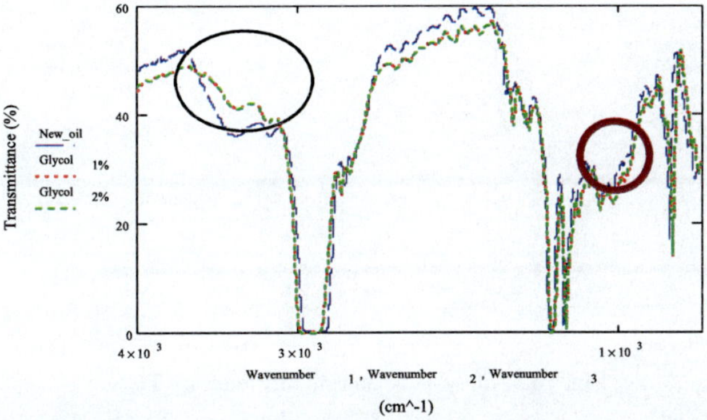

Fig. 7. Result of the measurement of the glycol content at 1% and 2% in terms of transmittance.

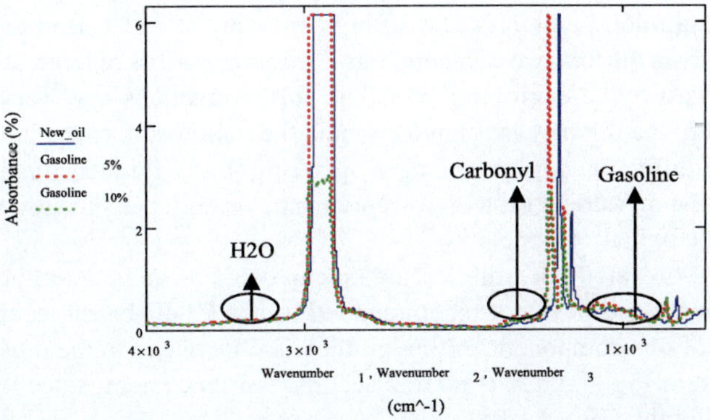

Fig. 8. Result of the measurement of the gasoline content at 5% and 10% in terms of transmittance.

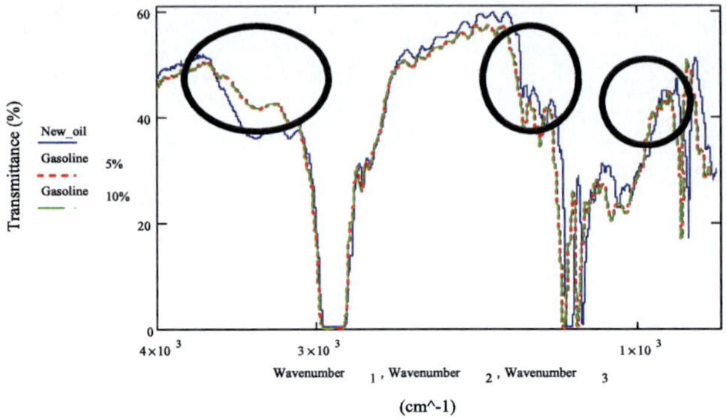

Fig. 9. Result of the measurement of the gasoline content at 5% and 10% in terms of transmittance.

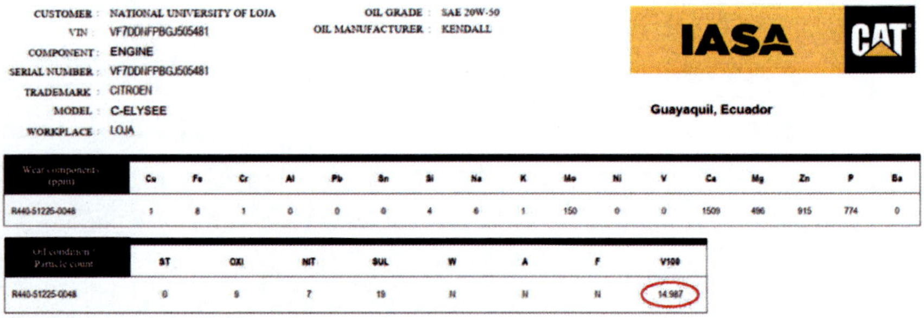

CUSTOMER : NATIONAL UNIVERSITY OF LOJA
VIN : VF7DDNFPBGJ505481
COMPONENT : ENGINE
SERIAL NUMBER : VF7DDNFPBGJ505481
TRADEMARK : CITROEN
MODEL : C-ELYSEE
WORKPLACE : LOJA

OIL GRADE : SAE 20W-50
OIL MANUFACTURER : KENDALL

IASA CAT
Guayaquil, Ecuador

Wear components (ppm)	Cu	Fe	Cr	Al	Pb	Sn	Si	Na	K	Mo	Ni	V	Ca	Mg	Zn	P	Ba
R440-51225-0048	1	8	1	0	0	0	4	6	1	150	0	0	1509	496	915	774	0

Oil condition / Particle count	ST	OXI	NIT	SUL	W	A	F	V100
R440-51225-0048	0	9	7	19	N	N	N	14.987

Fig. 10. Analysis of engine lubricant by TBN.

4 Conclusions

From the results of the analysis of engine lubricating oils; contaminants such as oxidation, nitration, and sulfation in the presence of high amounts of water, lead to measurement errors, especially in the first two contaminants. This means a loss of lubrication efficiency, and with it, catastrophic engine failures. The sulfation values may vary if the glycol concentrations in the oil increase above 2%, and the monitoring of the lubricant may be affected by the different additive packages operating in this frequency range. As can be seen in Fig. 6, the mixture of oil with refrigerant shows high percentages of absorbance, especially in the oxidation zone.

The amount of water that a lubricating oil can contain can be interfered with by the presence of glycol, and it is not recommended to use FT-IR based techniques for the quantification of this component. Although there are increases in the presence of water in the spectrum of Fig. 6, there is no real accuracy of its content, since it also contains the Oxyhydryl group (HO^-).

A small amount of contaminant can affect the chemical properties of the lubricant, even if the oil analysis values (TBN) are above the minimum allowable value. The spectrometry technique helps predictive maintenance and early detection of contaminants.

References

1. Abdulmunem, O.M., Abdul-Munaim, A.M., Aller, M.M., Preu, S., Watson, D.G.: THz-TDS for detecting glycol contamination in engine oil. Appl. Sci. **10** (2020)
2. Yendhe, V., Lawande, V.: A study on effect of added contaminants in lubricants by using wear debris and vibration analysis technique. i-Manager's J. Instrument. Control Eng., 9–17 (2019)
3. Borin, A., Poppi, R.J.: Multivariate quality control of lubricating oils using fourier transform infrared. Sociedade Brasileira de Química **15**, 570–576 (2004)
4. Coates, J.: Interpretation of Infrared Spectra, A Practical Approach. John Wiley & Sons Ltd, Chichester (2000)
5. Hamid, A., Evgueni, B., Jun, Y.: Two methodologies for optical analysis of contaminated engine lubricants. Meas. Sci. Technol. **23**, 1–15 (2012)
6. Bezerra, K.D.S., Neto, T.D.O.Z., Souza, C.L., Gomes, L.O., Filho, N.R.A.: Monitoring of lubricating oils used in diesel engine by biodiesel contamination from fuel dilution. Lubr. Sci. **33**, 432–438 (2021)
7. Chimeno-Trinchet, C., Murru, C., Diaz-Garcia, M.E., Fernandez-Gonzalez, A., Badia-Laino, R.: Artificial Intelligence and fourier-transform infrared spectroscopy for evaluating water-mediated degradation of lubricant oils. Talanta **219** (2020)
8. Hasannuddin, A., et al.: Performance, emissions and lubricant oil analysis of diesel engine running on emulsion fuel. Energy Conv. Manag. **117**, 548–557 (2016)
9. Taylor, R.I.: Fuel-lubricant interactions: critical review of recent work. Lubricants **9** (2021)
10. Sejkorova, M., Sarkan, B., Veselik, P., Hurtova, I.: FTIR spectrometry with PLS regression for rapid TBN determination of worn mineral engine oils. Energies **13** (2020)
11. Honig, V., et al.: Determination of tractor engine oil change interval based on material properties. Materials **13** (2020)
12. Wolak, A., Krasodomski, W., Zajac, G.: FTIR analysis and monitoring of used synthetic oils operated under similar driving conditions. Friction **8**, 995–1006 (2020)
13. Díaz-Sinche, D., Castillo-Calderón, J., Carrión Jaura, R.: Vibrations associated with the quality of combustion derived from induced failures in a spark ignition engine. Commun. Smart Technol. Innov. Soc. **252**, 755–768 (2021)
14. ASTM-D7414. Standard Test Method for Condition Monitoring of Oxidation in In-Service Petroleum and Hydrocarbon Based Lubricants by Trend Analysis Using Fourier Transform Infrared (FT-IR) Spectrometry, USA (2009)
15. ASTM-E2412. Standard Practice for Condition Monitoring of In-Service Lubricants by Trend Analysis Using Fourier Transform Infrared (FT-IR) Spectrometry, USA (2004)
16. ASTM-D7214. Standard Test Method for Determination of the Oxidation of Used Lubricants by FT-IR Using Peak Area Increase Calculation, USA (2007)
17. Ji, P., Jiang, R.: Scale transitivity in the AHP. J. Oper. Res. Soc. **54**, 896–905 (2003)
18. Saaty, T.L.: Relative Measurement and Its Generalization in Decision Making Why Pairwise Comparisons are Central in Mathematics for the Measurement of Intangible Factors The Analytic Hierarchy/Network Process. Revista de la Real Academia de Ciencias Exactas, Fisicas y Naturales, pp. 251–318 (2008)
19. N. I. 935. Productos derivados de petrólero. Gasolina Requisitos. Quito-Ecuador (2016)
20. Castillo-Calderón, J., Díaz Sinche, D., Carrión Jaura, R., Rivera Campoverde, D.: Influence of failures not detected by the on-board diagnostic system on the performance and pollutant emissions of a spark ignition engine. In: Botto-Tobar, M., Cruz, H., Díaz Cadena, A. (eds.) CIT 2020. AISC, vol. 1326, pp. 389–403. Springer, Cham (2021). https://doi.org/10.1007/978-3-030-68080-0_29

21. Kendall. GT-1® High Performance Motor Oil with LiquiTek® (2022). https://kendallmotoroil.com/wp-content/uploads/2020/05/Kendall-GT-1-HP-w-LiquiTek-GF6-ES.pdf, Accessed 10 Feb 2022
22. John, C.: Interpretation of Infrared Spectra: A Practical Approach, pp. 10815–10837. John Wiley & Sons Ltd, Chichester (2000)
23. Hernández, L., González, C.: Introducción al Análisis Instrumental. Ariel Ciencia (2002)
24. Tormos, B.: Diagnóstico de motores diésel mediante el análisis del aceite usado. Reverté, S. A, Barcelona (2012)
25. Notay, R.S., Martin, P., Fox, M.F.: The influence of lubricant degradation on measured piston ring film thickness in a fired gasoline reciprocating engine. Tribol. Int. **129**, 112–123 (2019)

Measurement of Free Residual Chlorine in Drinking Water Using Clean Technologies "Oxidation and Reduction Potential" as an Alternative Technique to the Use of Chemical Reagents

Angélica Geovanna Zea(✉), Gabriela Ordoñez, Boris Quiroz, and Marcelo Flores

Universidad Politécnica Salesiana, Cuenca, Ecuador
azea@ups.edu.ec

Abstract. In the field of health, one of the main aspects is to ensure access to drinking water, within its control, monitoring and analysis there are some important physical chemical parameters, however one of the most relevant at the time of monitoring of water quality to be able to guarantee the disinfection of this is the analysis of free residual chlorine in an on-site manner in the water treatment plants, however this monitoring is not carried out routinely due to the high costs that it implies and the large amount of chemicals that are used and that end up being released from the environment. This article presents a proposal for the use of ORP sensors or Redox sensors, an electrochemical device that, when in contact with a solution, generates an electrical signal whose voltage varies depending on the capacity of the solution to act as an oxidizing or reducing agent. Which would currently be an alternative to the use of chemical reagents in the measurement of residual free chlorine in drinking water plants, and for this work chlorine acts as an oxidizing agent, therefore the sensor will measure its oxidation capacity in units of millivolts and will relate them to the research obtained from the assessment of the analytical technique by the method of the European Standard UNE-EN ISO 7393-1:2000, "Water quality. Determination of free chlorine and total chlorine. Method by titration with N,N-diethyl-p-phenylenediamine", and the results obtained will be correlated with the influence of pH and temperature, in order to find an equivalence in parts per million as a unit of residual free chlorine concentration of According to the standard, to validate the method, the precision, accuracy, sensitivity, selectivity, among others, were calculated and it was determined that the sensor readings are related to the concentration of free chlorine present in the water, as well as pH and temperature, the latter being indirectly proportional factors that influence measurements, so it is crucial to control these parameters; There is no linearity in the checks, however, it will be prolonged that the system is accurate for the measurement of free chlorine since it is within the accepted criteria in the validation of analytical techniques, being an adequate and economical alternative for continuous monitoring in Water treatment plants.

Keywords: Free residual chlorine · ORP sensor · Clean technology · Electrochemical technique · Voltage · Oxidizing agent · Reducing agent · Water quality · Public health

V. Robles-Bykbaev et al. (Eds.): CITIS 2022, LNNS 607, pp. 299–311, 2023.
https://doi.org/10.1007/978-3-031-24327-1_26

1 Introduction

Water is considered an indispensable good for life, and in 2010 the right to its sanitation was recognized by the United Nations Organization, being included in the objectives established by the United Nations in terms of Sustainable Development. Currently, the situation is still very unfortunate, since every year around five million people die from consuming water that has not undergone an adequate disinfection process, according to Oxfam Intermón [1]. Being of great interest a good water quality, which is included in the objectives of the 2030 Agenda approved in 2015, where the importance of ensuring access to water and an optimal disinfection process ensuring its safety is manifested [2, 3].

Ecuador ranks third among the countries of the Americas in terms of contamination by protozoan microorganisms in water, whose effects on the population are intestinal disorders such as diarrhea, especially in children between one and five years of age [4].

The quality of drinking water decreases in the distribution network and collection points due to bacterial growth, affecting its taste and odor, even causing the presence of suspended solids, which can create a bacterial layer in the pipes, causing corrosion, reducing their hydraulic power, and hindering the presence of antiseptic residual substances [5].

In public health it is a priority to ensure access to drinking water for the population, and currently this is not a reality since millions of people in the world do not have this resource in optimal conditions to ensure their health due to several causes such as: the lack of allocation of resources for its infrastructure, treatment and service [6], being one of the functions of local governments it is a priority to ensure access to drinking water in safe conditions for consumption through the analysis and sampling of its physical-chemical parameters as main indicators of drinking water quality, considering the analysis of residual free chlorine one of the main ones since it guarantees the safety of water and its conservation during storage and transport through the distribution network [7].

Free chlorine concentration is usually determined by one of several methods using colorimetric methods such as DPD (color change associated with the reaction of free chlorine with NN-diethyl-p-phenylenediamine (DPD) or iodide, chemiluminescence, detection by liquid chromatography or electrochemical detection. However, most of these techniques require reagents or laboratory techniques that are not the most suitable for low-cost and autonomous continuous monitoring [8], due to the acquisition of chemical reagents and its proper use in treatment plants, since this type of technique requires adequate monitoring and that when using any chemical substance implies an impact on the environment. For this project, the ORP sensor is used, which is an electrochemical device that enters in contact with water and generates an electrical signal in units of voltage, and its value will vary according to the capacity of the solution. Ion to act as an oxidizing agent, in this case the chlorine that comes into contact with the water, and the measurements obtained were related to the concentration of free chlorine in standard samples, previously adequately validating the technique according to the European Standard UNE-EN ISO 7393-1:2000, "Water quality. Determination of free chlorine and total chlorine. Method by titration with N, N-diethyl-p-phenylenediamine This measurement was related to the free chlorine concentration of the standard samples prepared in the laboratory to determine the concentration of free chlorine present in the water capable of disinfecting it, analyzing together with the pH and temperature factors, to determine their influence on the results.

Then it is said that the ORP can measure the total effectiveness of all oxidizing disinfectants that are in solution in the water either as for example: hypochlorous acid, hypochlorite, monochloramine, dichloramine, ozone, peracetic acid, and it is mentioned that there is a correlation between ORP and known disinfection control methods, measurements and bacterial plaque counts, since this measurement provides an accurate picture of residual chlorine disinfection activity reported as an empirical number that is not subject to visual interpretation as is the case with colorimetric techniques, and also states that this technique has been advocated as the best way to judge the residual disinfectant power of chlorinated water by water quality experts since the 1960s [9].

ORP measurements do not directly measure the concentration, they can be correlated with free chlorine levels in ppm, and for this the pH and temperature variables must be controlled, as well as the presence of any interfering chemicals or oxidizing or reducing substances which should be considered and more ideally eliminated [9]. Once these are controlled, ORP can be related to concentrations by laboratory experimentation or by mathematical formulas such as the Nernst Equation, which indicates the connection between the electrode potential of a specific chemical in a solution and its concentration, but this process can often be complex and laborious [9]. ORP measurement is based on the principle of an inert metal electrode such as platinum and gold, which due to its low resistance, will yield electrons to an oxidant or accept electrons from a reductant, developing a potential, due to the accumulated charge, typical accuracy is $\pm$ 5 mV [10].

The Sensorex ORP sensor was used in this work, which has a measurement electrode and a reference electrode, where the measurement cell is made of a noble metal such as platinum or gold which detects changes in the oxidation and reduction potential, while the reference cell provides a stable comparison signal [10–12].

The objective of this project is to present a proposal for the measurement of residual free chlorine in drinking water plants using an electrochemical ORP sensor as a proposal for the substitution of chemical reagents for monitoring and control, and that according to the different tests that were carried out in the laboratory controlling pH and temperature that influence, it is shown that it is possible to measure the concentration of residual free chlorine, and that it was found that there is a directly proportional relationship between the voltage that the solution presents and the concentration of free chlorine present in the solution.

The process of standardizing the chlorine solution that will be used according to the regulations is described, as well as the different validation tests such as accuracy, precision of the method used and that is proposed below (Fig. 1).

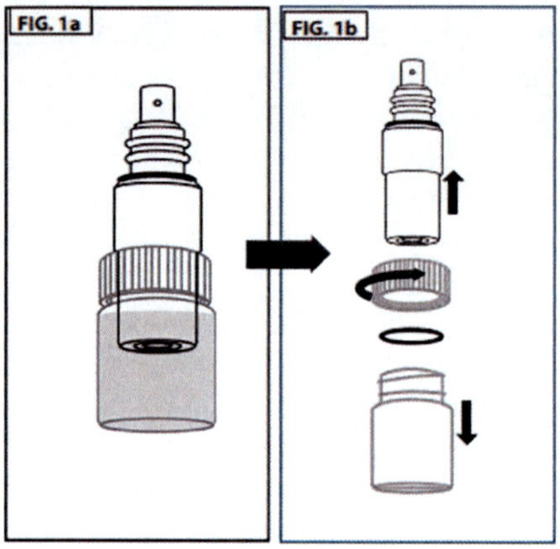

Fig. 1. Sensor ORP-Sensorex

2 Methodology

To achieve the objectives established in this project, to measure the concentration of free residual chlorine by electrochemical techniques, and to check its effectiveness, an ORP sensor was used to measure the oxidation potential of free chlorine present in water. The experimental development consisted first of all in measuring free residual chlorine in solutions prepared with known concentrations of distilled water and total chlorine, which was titrated with sodium thiosulfate; The measurement was performed based on the colorimetric technique with Diethyl P-phenylene diamine DPD reagent of the brand [13–15], which was used as a reference method for the objective measurement of this project, which consists of the measurement of free residual chlorine based on electrochemical techniques, Therefore, SENSOREX ORP sensors were used to measure the oxide reduction potential of the water, which is equivalent to the voltage read in the measurement, which was related jointly with the pH and temperature variables to create a database that allows us to determine the amount of free chlorine present in the water.

2.1 Development of the Free Chlorine Residual Calibration Curve

To perform the calibration curve of free residual chlorine, the HACH Company technique was used [15], whose procedure is equivalent to USEPA and Standard Method 4500-Cl G for drinking water, which is based on the measurement of free chlorine by using the reagent diethyl p-felinen diamine, which when in contact with free chlorine causes a pink coloration in the sample, The technique consists of the application of the reagent contained in an aluminum-coated envelope designed to protect it and ensure its preservation, which is added to a volume of 10 ml of water, then the solution is shaken for 20 s so that it can react and homogenize completely, then the absorbance of the sample is measured in the spectrophotometer at a wavelength of 530 nm. For which a stock solution of 100 ppm was previously prepared from commercial sodium hypochlorite, and then standard solutions of 0.5 to 4 ppm were prepared in intervals of 0.5.

2.2 Measurement of Free Residual Chlorine with ORP Sensors

Once the concentration of free residual chlorine in the standard samples was verified, free chlorine was measured using an ORP sensor manufactured by Sen-sorex. This sensor is based on the measurement of redox reactions, that is, the oxidation and reduction potential, whose measurement reflects the ability of a molecule to oxidize or reduce another molecule:

Since chlorine is an oxidizing substance, the sensor will measure the oxidation potential through the emission of a single voltage in millivolts (mV), with oxidizing substances giving positive values and reducing substances giving a signal of a negative ORP value.

2.3 Voltage Readings

The voltage values of measurements made with ORP sensors can vary depending on the temperature of the solution to be measured, so for this project measurements were made at known concentrations in parts per million of free residual chlorine at different temperature ranges, ranging from low temperatures of 5 °C to higher temperatures of 30 °C, at intervals of 5 °C between the indicated values, the samples were also taken to different pH values comprising values of 6, 6.5 and 7.

The standard samples were previously prepared with known concentrations of free residual chlorine, ranging from 0.5 ppm to 4 ppm, with intervals of 0.5; each of the standard samples was regulated with acid and alkaline substances to obtain the pH required in the solution, 0.1 N sulfuric acid was used as acidifying substance and 0.01 N sodium hydroxide as alkalinizing substance, the pH was controlled with 0.01 N sodium hydroxide. 01 N, the pH was controlled with the help of a Metter Toledo pH meter and the temperature of the samples was controlled with a digital thermometer. All the samples were prepared with distilled water to ensure that there were no inferences with other substances and to avoid contamination of the samples, in order to ensure the accuracy of the results obtained. The redox sensor was connected to the Meter Toledo equipment, the ORP measurement function was determined and for each reading the Read button was pressed and the estimated time until the value stabilized on the screen was 5 to 10 min, once stable, the value read by the equipment was noted down.

2.4 Determination of the Measurement Number to Perform

Because the true value of the magnitude to be measured is unknown, an estimate is made of it and its level of error (Posadas, 2009), in order to obtain a certain statistical validity in the measurements, it is convenient to repeat its measurement several times. Determination: this number is based on the dispersion, obtaining a dispersion percentage of 0.05%, according to the applied calculations, the number of samples to choose is 3, however a greater number of repetitions decreases the random error, and since it is an unstable substance such as free chlorine, n = 5 number of measurements were taken.

2.5 Voltage Readings

The voltage values of measurements made with ORP sensors can vary depending on the degree of temperature in which the solution to be measured is found, which is why for this project measurements were made at known concentrations in parts per million of residual free chlorine. at different temperature ranges, from low temperatures of 5 °C to higher temperatures of 30 °C, at intervals of 5 °C between the indicated values, the samples were also brought to different pH values that included values of 6; 6.5 and 7.

2.6 Statistic Analysis

Precision can be measured either by the degree of reproducibility or repeatability of the analytical method, under normal operating conditions [16]. Therefore, we will use the repeatability to determine the precision of the measurements. The precision was analyzed through the value of the coefficient of variation as established in the Validation of analytical methods (Vázquez, 2008), which being less than 1.5% there is a precise system in the concentration interval of 0.5 to 4 ppm of free chlorine.

3 Results

3.1 Free Chlorine Calibration Curve Values

Table 1 and Fig. 2 show the concentration values with the value of the respective absorbance readings obtained by spectrophotometry, whose correlation coefficient was 0.9985, this value being accepted within the criteria established by the equation of the Law de Beer to meet acceptance. Criterion, the correlation coefficient must be greater than or equal to 0.999[18] (see Fig. 2).

Table 1. Residual free chlorine concentration values, obtained from the absorbance of the Uv-Vis spectrophotometer.

Standard	Concentration (mg/l)	Absorbance
0,5	0,5	0.0340604
1	1	0.150898
1,5	1,5	0.231571
2	2	0.346023
2,5	2,5	0.456486

3.2 Measured Values of Free Chlorine with ORP Sensor

It is worth mentioning that 5 repetitions were carried out on each sample according to the statistical calculation, and that the ORP values obtained were correlated to different ranges of temperature and pH, thus obtaining residual free chlorine values from the correlation of pH, temperature and ORP in millivolts, some of the significant tables obtained are shown, since due to the length of the article, more tables of the results cannot be added.

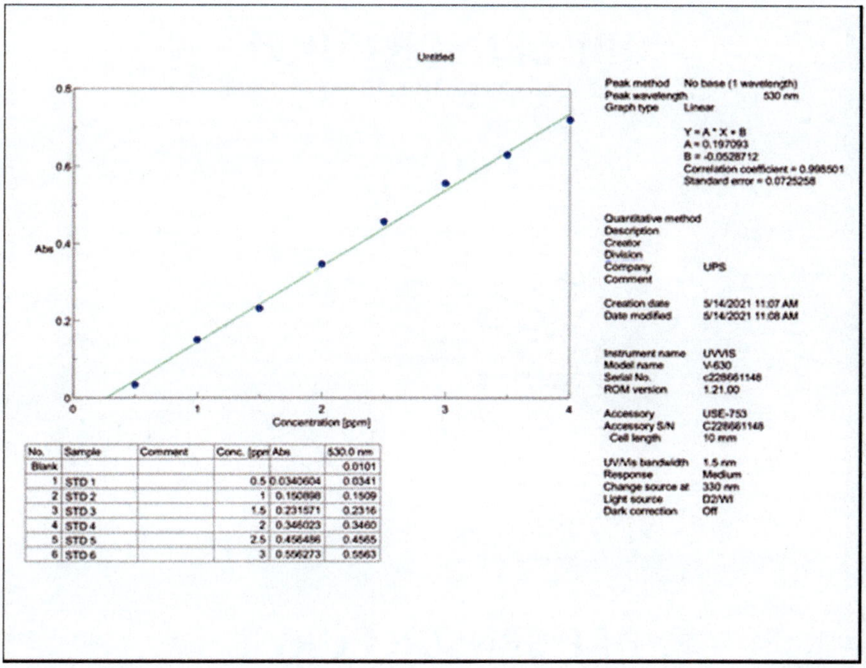

No.	Sample	Comment	Conc. [ppm	Abs	530.0 nm
Blank					0.0101
1	STD 1		0.5	0.0340604	0.0341
2	STD 2		1	0.150898	0.1509
3	STD 3		1.5	0.231571	0.2316
4	STD 4		2	0.346023	0.3460
5	STD 5		2.5	0.456486	0.4565
6	STD 6		3	0.556273	0.5563

Fig. 2. Residual free chlorine concentration values, obtained from the absorbance of the Uv-Vis spectrophotometer

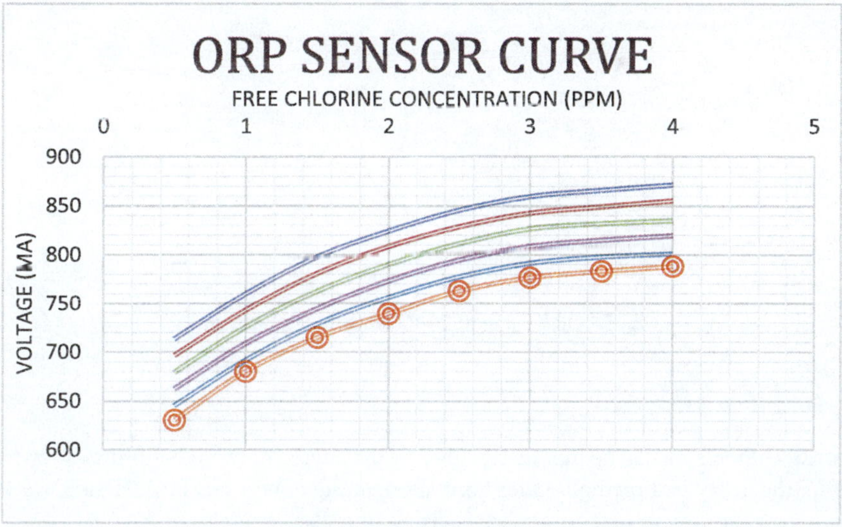

Fig. 3. ORP curve at pH 6

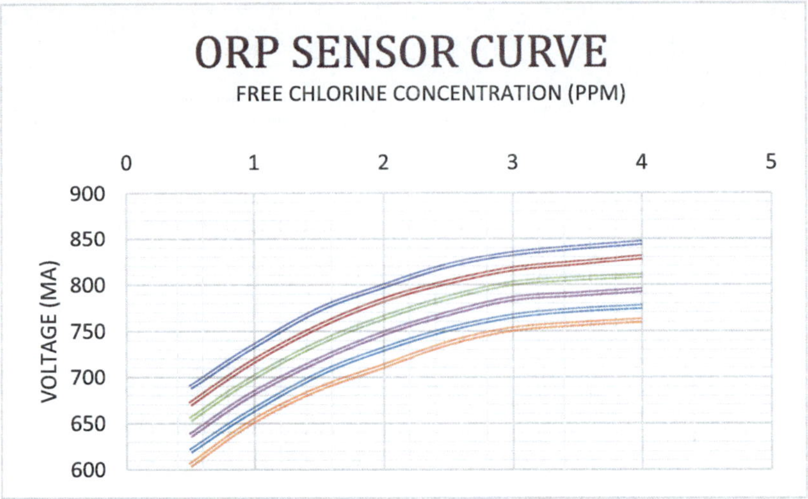

Fig. 4. ORP curve at pH 7

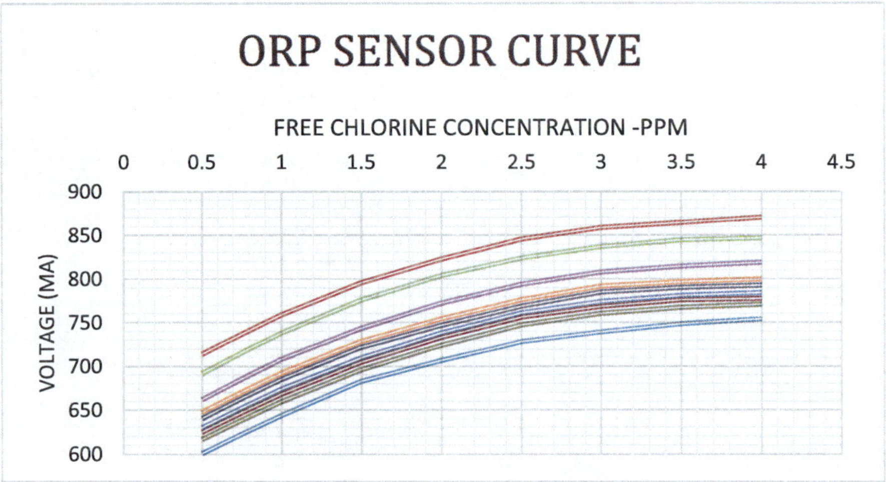

Fig. 5. ORP curve at 20 °C ambient temperature

3.3 Statistical Analysis

Accuracy. Precision can be measured either by the degree of reproducibility or repeatability of the analytical method under normal operating conditions [16]. Hence, we will employ repeatability to determine the precision of measurements.

Accuracy was analyzed using the coefficient of variation value as established in the Validation of Analytical Methods [16], which, being less than 1.5%, is an accurate system in the concentration range of 0.5 to 4 ppm of free chlorine, as shown in Tables 2, 3 and 4.

Table 2. Electrochemical Technique with ORP Sensor (n = 5 each level), at temperature 20 °C, and pH 6

Concentration added (ppm)	Average (Measured concentration ±	Coefficient of variation CV (%)
0,5	662 ± 3.97	0.60
1	708.2 ± 3.88	0.55
1,5	743.8 ± 4.09	0.55
2	772.6 ± 3.70	0.48
2,5	794 ± 4.08	0.51
3	808 ± 3.98	0.49
3,5	814.5 ± 4.10	0.50
4	819 ± 4.02	0.49

Table 3. Coefficient of variation of the samples at 20 °C and pH 7

Concentration added (ppm)	Average (Measured concentration ±	Coefficient of variation CV (%)
0,5	630.2 ± 3.97	0.63
1	674.1 ± 3.88	0.58
1,5	709.7 ± 4.09	0.58
2	738.5 ± 3.70	0.50
2,5	761.3 ± 4.08	0.54
3	774.2 ± 3.98	0.51
3,5	781 ± 4.10	0.52
4	784.3 ± 4.02	0.51

4 Discussion

Within the electrochemical measurement techniques are the Redox or ORP sensors whose capacity is the measurement of the oxide reduction potential by means of an electrode made up of noble metals, being very useful to know the activity of the electrons present in the water or in aqueous systems [19].

As shown in Fig. 2 and 3 the ORP curves at different concentrations, the relationship between the voltage value and the temperature can be seen, being an inverse relationship

Table 4. Coefficient of variation for samples at 20 °C and pH 8

Concentration added (ppm)	Average (Measured concentration ±	Coefficient of variation CV (%)
0,5	600.5 ± 3.97	0.66
1	644.0 ± 3.88	0.60
1,5	682.9 ± 4.09	0.60
2	706.8 ± 3.70	0.52
2,5	728.4 ± 4.08	0.56
3	739.4 ± 3.98	0.54
3,5	748.7 ± 4.10	0.55
4	754.2 ± 4.02	0.53

since as the pH of the solution decreases the voltage increases, as the pH solutions are 6, 6.5 and 7, the highest ORP value is in the solution with pH 7.

ORP sensors are basically made up of two electrochemical half-cells, the measurement electrode in contact with the solution being measured and a reference electrode that is connected to an isolated reservoir of highly concentrated saline solution, when the solution being measures has a high concentration of oxidants, it accepts more electrons than it loses, and the measurement electrode develops an electrical potential greater than the stable potential of the reference electrode [20]. The difference in electrical potential will be displayed in millivolts, a positive reading indicates an oxidizing solution, and a negative reading indicates a reducing solution, and more positive or negative values mean that the oxidizing or reducing agents present are stronger, they are present in concentration. -higher ratios or both [20].

In the present work, the values obtained by means of ORP are correlated with the levels of free chlorine in ppm, also controlling the pH values and the temperature, this correlation was applied through experimentation in the laboratory, but this correlation can also be obtained through formulas mathematics such as the Nernst Equation, which indicates the connection between the potential of the electrode of a specific chemical substance in a solution and its concentration [20].

Regarding chlorine measurements with ORP sensors, the limitations that can occur are the slow response time and the nonlinear response of ORP to free chlorine.

The ORP probe in drinking water treatment plants has a great advantage, since it can be placed directly in continuous flow and it is not necessary to extract or take a water sample to add reagents or perform a spectrophotometric analysis on the water sample, as is the case of N, N-diethyl-p phenylendiamia (DPD).

Figure 5 shows the ORP and free chlorine curve at 20 °C, however it is worth mentioning that in the laboratory it was carried out at different temperatures, showing similar results as shown in Fig. 4, therefore it can be seen that there is a relationship between the temperature at which the solution is found and the voltage value, being an inversely proportional relationship, since if the temperature value increases the voltage decreases, and in the opposite direction.

Indeed, it is corroborated that there is a relationship between the voltage value and the concentration of free chlorine present in the samples, and it was determined that it is a directly proportional relationship, since as evidenced, as the concentration of free chlorine in the samples increases, the voltage value increases, however, this relationship is not linear, at a concentration of approximately 2.5 ppm the graph tends to stabilize, becoming relatively more linear.

The voltage measurements obtained and plotted are not linear, as for the precision calculated by repeatability, no coefficient of variation exceeds the value of 1.5%, so it is said to be within the acceptance criteria, i.e. the system is accurate in terms of residual free chlorine measurements.

5 Conclusions

With the preparation of this article it is concluded that it is possible to carry out the measurement of one of the main parameters established in the water quality regulations, which is the concentration of residual free chlorine, based on electrochemical techniques specifically with the use of ORP sensors, since it was found that there is a directly proportional relationship between the voltage presented by the solution and the concentration of free chlorine present in the solution, However, the measurements are linked to factors such as pH and temperature, which are two determining factors in the voltage of the solution, so it is crucial to consider them for future measurements, and it is also important to control oxidation-reducing substances present in the water, since the sensor is not selective to free chlorine and may be susceptible to interference by these substances.

ORP sensors are an excellent tool for measuring free chlorine on a large or small scale, both for large cities and small towns, the latter being administered by drinking water boards, as they do not require the use of chemicals which are toxic to both health and the environment as they can affect the development of aquatic life, whose waste is considered hazardous and must be properly managed by regulatory bodies for proper disposal.

ORP measurements, unlike conventional techniques, require relatively less time, since chlorine, being an unstable substance, requires more time to stabilize the reading, with a minimum time of 5 min and a maximum of approximately 10 to 15 min. In addition, one of the advantages of this technique is that it does not require the use of laboratories or the mobilization of water samples, even avoiding the demand for technical personnel. By means of a statistical analysis it was proved that the ORP sensor is an accurate tool, and with it a continuous and real time monitoring of free residual chlorine and water quality can be carried out prefer the use of American English.

Although there are some articles where they talk about ORP and its use to measure the level of water disinfection based on its millivolt result, this paper demonstrates the connection between ORP, pH and temperature to inform with these three variables free chlorine residual in water in parts per million (ppm), since according to the different articles there are two techniques, one is mathematical and the other is in the laboratory, therefore the latter has been demonstrated following all the analytical chemical steps and statistics to obtain different tables that will be used as a reference when it comes to measuring ORP values and comparing them with the concentration of residual chlorine in the water in parts per million, in addition to representing an environmentally friendly proposal since chlorine monitoring residual free by this technique does not require the use of chemical reagents, including sampling and transfer to the laboratory for analysis, and because it can be done on-site and continuously, which is what is required to guarantee water disinfection, continuous monitoring and control of residual free chlorine in water.

References

1. May, P., Ehrlich, H.-C., Steinke, T.: ZIB structure prediction pipeline: composing a complex biological workflow through web services. In: Nagel, W.E., Walter, W.V., Lehner, W. (eds.) Euro-Par 2006. LNCS, vol. 4128, pp. 1148–1158. Springer, Heidelberg (2006). https://doi.org/10.1007/11823285_121
2. Burgues, M.: DÍA MUNDIAL DEL AGUA 2021: LA IMPORTANCIA DEL AGUA QUE NO SABEMOS. Boletin Electronico; Ok Diario (2021). https://okdiario.com/salud/dia-mundial-del-agua-2021-importancia-del-agua-que-no-sabemos-6988152
3. Czajkowski, K., Fitzgerald, S., Foster, I., Kesselman, C.: Grid information services for distributed resource sharing. In: 10th IEEE International Symposium on High Performance Distributed Computing, pp. 181–184. IEEE Press, New York (2001). https://doi.org/10.1109/HPDC.2001.945188
4. Cepeda Granda, J.G.: DISMINUCIÓN DE LA CALIDAD DEL AGUA POR VARIACIONES DEL CLORO LIBRE RESIDUAL ENTRE LA RED DE DISTRIBUCIÓN Y LA RED INTERNA DE LOS USUARIOS EN URBANIZACIÓN CIUDAD VERDE DE LA CIUDAD DE MACHALA, PROVINCIA DE EL ORO. Trabajo de Titulación Ingeniería Civil **151**(1), 10–17 (2015). http://repositorio.utmachala.edu.ec/handle/48000/2956, National Center for Biotechnology Information. http://www.ncbi.nlm.nih.gov
5. Carrasco, Z., Renato Iván Morales Samperio, H.: Importancia de la cloración del agua : sitios de abastecimiento con presencia de bacterias patógenas Importance of water chlorination : pathogenic bacteria in supply sites Resumen Introducción Desinfectantes y colonización bacteriana Efectos del material. Enf. Inf. Microbiol. **39**(3), 86–92 (2019). https://www.medigraphic.com/cgi-bin/new/resumen.cgi?IDARTICULO=92252
6. OMS. 1 de cada 3 personas en el mundo no tiene acceso al agua potable, según UNICEF y la OMS. Comunicado de Prensa, 3 (2019). https://www.unicef.org/es/comunicados-prensa/1-de-cada-3-personas-en-el-mundo-no-tiene-acceso-a-agua-potable%0Ahttps://www.who.int/es/news/item/18-06-2019-1-in-3-people-globally-do-not-have-access-to-safe-drinking-water-unicef-who%0Ahttps://www.who.int/
7. OMS. Guías para la calidad del agua de consumo humano. Organización Mundial de La Salud, 4, 608(2011). https://apps.who.int/iris/bitstream/handle/10665/272403/9789243549958-spa.pdf?sequence=1&isAllowed=y
8. Hoque, E., Hsu, L.H.H., Aryasomayajula, A., Selvaganapathy, P.R., Kruse, P.: Pencil-drawn chemiresistive sensor for free chlorine in water. IEEE Sensors Letters **1**(4), 1–4 (2017). https://doi.org/10.1109/lsens.2017.2722958
9. Spahl, R.J.: FC E TM : Groundbreaking Measurement of Free Chlorine Disinfecting Power in a Handheld Instrument Myron L Company ' s Free Chlorine Equivalent (FC E TM) (2012). http://www.myronl.com/PDF/fcetr.pdf
10. Sensorex.. Flat Surface CPVC and PVDF pH and ORP Sensors (2011)
11. Sensorex. Flat-Surface, Self-Cleaning pH and ORP Electrodes (2017)
12. Sensorex. ¿Qué es ORP? (2021)
13. HACH. Hoja De Datos De Seguridad. 2012, 1–4 (2011)

14. HACH Company. (n.d.). Método USEPA DPD1. Método 8021
15. HACH Company. Chlorine , Free (2014)
16. González Hinostroza, C.M.: Aplicación del cálculo de incertidumbre combinada ala validación de una metodología analítica por hplc, en un producto que contiene citalopram bromhidrato. Universidad Austral de Chile, 66 (2006)
17. Vázquez, A.: Instituto Politecnico Nacional Unidad Profesional Interdisciplinaria De Biotecnología Ingeniero Farmacéutico Presenta 47 (2008). http://tesis.ipn.mx/bitstream/handle/123456789/14634/ayded%0Avazquezmoreno.pdf?sequence=1
18. Martinez Osorio, F.S., Pérez Epinoza, I., de los, A.: Calibración de un espectrofotómetro UV-Visible y evaluación de la incertidumbre (2009). http://riul.unanleon.edu.ni:8080/jspui/retrieve/2738
19. Schettino Sanhueza, E.A.: Desarrollo de sistema de control y mantenimiento de calidad de agua dulce en grandes concentraciones mediante IoT [Universidad Andres Bello] (2019). http://repositorio.unab.cl/xmlui/bitstream/handle/ria/17973/a131451_Schettino_E_Desarrollo_de_sistemas_de_control_2019_tesis.pdf?sequence=1&isAllowed=y
20. Secretaria del Agua.. Instructivo Para Conformación Y Legalización De Juntas Administradoras De Agua Potable Y Saneamiento ; Juntas Administradoras De Agua Potable Y Saneamiento Regional ; Y , Juntas De Segundo Y Tercer Grado , Y El " Instructivo Para Conformación Y Legaliz (2016)
21. Asociación Española de Normalización y Certificación. UNE-EN ISO 7393- 1. Calidad del agua. Determinación de cloro libre y de cloro total. Parte 1: Método por valoración con N,N-dietil-1,4-fenilendiamina (2000)

Design and Implementation of a Photovoltaic System for the Simón Bolívar Basic Education School in the Community La Masa 2, Puna Gulf Island

Luis Alcívar, Esteban Villegas, and Gary Ampuño(✉)

Universidad Politécnica Salesiana, Guayaquil, Ecuador
gampuno@est.ups.edu.ec

Abstract. The objective of this project is to deliver electricity and raise awareness of the importance of providing renewable energy in places where there is a lack of basic services due to the lack of resources for these areas, since due to their geographical characteristics they are difficult to access or very remote. The project was implemented in the Simón Bolívar Basic Education School located in the community of La Masa 2 in the province of Guayas, where the main problem is the lack of electricity to provide classes to children in the initial stage; at the same time, by implementing this project, tools are provided that allow learning with technologies such as ICTs and with the lighting of their classrooms allow more hours of study, more concentration and less reading effort.

This work also describes everything related to the community La Masa 2, providing data on its population, number of houses, activities, religion, and how they generate energy in order to establish the needs of the community. This information was obtained through bibliographic and field research to generate new works that contribute to the development of the community.

Finally, a 400 W isolated functional photovoltaic system was delivered, through the study of load and demand necessary for the sizing of the generator, accumulator, charge controller and inverter, which was conducted both analytically and simulated by software; the hours of use of the system per day (6 h) were determined. A computer was also provided by the Salesian Polytechnic University, which will be of help to facilitate the study of its students.

Keywords: Renewable energies · Off-grid systems · Photovoltaic energy · Isolated communities

1 Introduction

In 2015, the member states of the Organization of American States made a call to end poverty, protect the planet, and ensure peace and prosperity for all people by 2030. This call is called Sustainable Development Goals (SDGs) or Global Goals.

There are 17 SDGs that are integrated to balance environmental, economic, and social sustainability. This has committed countries to accelerate promises such as zero

V. Robles-Bykbaev et al. (Eds.): CITIS 2022, LNNS 607, pp. 312–321, 2023.
https://doi.org/10.1007/978-3-031-24327-1_27

poverty, zero hunger, zero AIDS and zero discrimination against women and girls. It takes creativity, knowledge, technology and financial resources from across society to achieve the SDGs [1].

Through the present project, it is envisaged to cover the goals:

Goal 7: Affordable and Clean Energy and Goal 13: Climate Action.

In 2018, the initiative "Don Bosco Green Alliance" was born, which is a global Salesian platform that includes young people of the Salesian Family committed to environmental protection [2].

It has about 274 members and is present in fifty-six countries. And it works on three priority areas in which they are:

Fighting pollution, reducing global warming, and eliminating disposable plastic.

In the framework of the Annual Operating Plan 2021 in the Strategic Plan Intervention D.1.3.1.1.3. "Implement in the UPS the principles of the Don Bosco Green Alliance (DBGA)1 initiative" where the careers propose initiatives of the environmental sustainability plan, are executed within the Don Bosco Green Alliance initiative [3].

The proposal states: Elaborate the environmental sustainability plan at the UPS sites considering sustainable mobility, care of the natural heritage, reduction of the ecological footprint, adequate waste management, and eco-efficient building.

Currently, electric energy is considered an important and necessary factor for the development of human beings [4]. Electricity allows the enjoyment of basic services such as drinking water, electric power, sewerage, telephone, among others. Social development has been affected by the lack of electricity lines and in accordance with the provisions of Article 314 of the Constitution of the Republic of Ecuador, which states that every citizen has the right to this service.

However, there are sectors and communities that lack these privileges because they are in remote and difficult to access areas, preventing the construction of an electrical distribution network involving substations, transformer elements or public lighting [5, 6].

With the above mentioned, public and private companies and higher education institutions are making efforts to supply electric energy to sectors that lack electric energy. In works such as Tinajero and Mayorga present a study to verify which type of photovoltaic system works (centralized or decentralized systems) for the Mata de Cacao community [7]. In the Gulf of Guayaquil there are communities isolated from the electrical grid due to their geographic location, giving way to investigations such as the one proposed by Panchana and Valverde, for the design of the autonomous energization system of the Bellavista Island, located in the Gulf of Guayaquil [8]. Another locality intervened in the same sector is the Cerrito de los Morreños Island where Peralta Jaramillo et al. show the evaluation of the commune's power systems to optimize the use of energy [9]. The locality with more inhabitants on the shores of the Gulf of Guayaquil is the Puerto Roma commune of Mondragon Island, where Vicente and Ruiz implement a solar energy system [10].

A similar work, but with a greater impact due to the amount of equipment installed, is the one conducted in the community of La Masa 2, where load-gauge studies were carried out. The result was the implementation to improve the quality of life of the inhabitants [11–16].

La Masa 2 commune is one of the communities that lacks basic services such as electricity, sewage, solid waste collection, telephone networks and internet. The inhabitants of the commune dedicate their only activities to crab catching, fishing and animal husbandry [17, 18]. The lack of lighting in and around the commune is one of the main problems faced by its inhabitants, as the generators from which they obtain their main energy are rented by the hour at a cost of US$25 per week [19].

La Masa 2 is a community located in the Gulf of Guayaquil south of the city (Latitude: −2.38; −79.86) (Fig. 1), which can be accessed by sea [20]. Starting at the Caraguay Pier, the trip takes approximately 45 min by boat and by land through the access road to the Las Exclusas electrical substation, entering through the Songa Shrimp Farm with the authorization of its directors. The overland trip is fifteen kilometers from the control point and takes 20 to 25 min.

This paper is structured as follows. After the introduction, Sect. 2 presents the methodology and analysis prior to system installation. Section 3 shows the implementation of the PV system and the results obtained. Finally, the conclusions of the paper are presented.

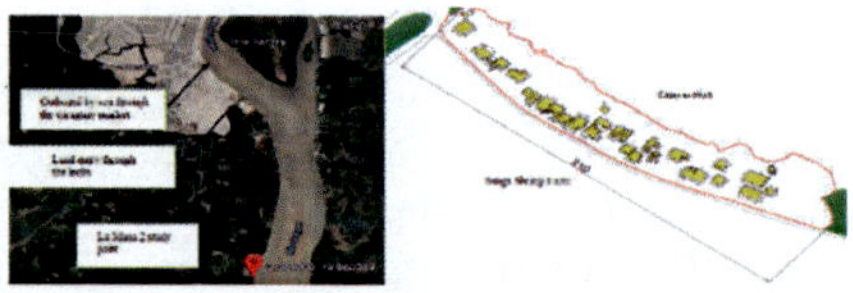

Fig. 1. a) GPS location of the commune with departure points (by sea and land), obtained from Google Maps. b) Perimeter and longitude of the commune La Masa 2

The commune has a population of sixty-six people, 19 families and 17 houses, most of which are made of cane and only 3 of cement. The population is 40 men, of which 20 are adults (between 18 and 60 years old), two adolescents (13 and 17) and 18 children (between 1.5 and 10 years old), and 26 women, 15 are adults (between 18 and 60 years old), 4 adolescents (between 12 and 16 years old) and 7 girls (between 1 and 9 years old) [21, 22]. Within the community is the Simón Bolívar Study Center located in the Guayaquil canton, Puna, Gulf Island – La Masa 2 precinct (Fig. 2), which began in January 2012, through a fiscal system and face-to-face modality, under the direction of Jenny Carol Villalba Cabezas, Licentiate in Initial Education. This school consists of two classrooms without ventilation and lighting, where it receives twenty-two students (ten girls and twelve boys) with an age range between seven and seventeen years old. The general director teacher: Yesenia Anastacio Quinde is the person who helps us with the access to the commune.

In 2019, the technical degree projects on the design and implementation of photovoltaic systems with solar panels to electrify and provide peripheral lighting to the Simon Bolivar School - Commune La Masa 2 began simultaneously. At the same time, groups of extensionists visited the sector and through the contribution of the Salesian Polytechnic University, it was possible to conduct part of the assembly of the photovoltaic system and the delivery of a computer for the school. For peripheral lighting, three lighting poles with photovoltaic energy are placed at the entrance of the sector.

Fig. 2. Simón Bolívar School La Masa 2, Gulf Island – Puna [22]

The project was delivered into the hands of the users of the educational unit, an institution that now has lighting that allows the community members a place where they can hold meetings in the evenings.

2 Methods

In any design of an electrical installation, it is necessary to obtain information that helps us to perform the calculations for the sizing of all equipment such as the orientation and inclination of solar panels, calculation of optimum inclination, information of the beneficiaries, current architectural and electrical plans of the homes and one-line diagram of the photovoltaic electrical system.

2.1 Beneficiary Information

The families that participated in the study have diverse types of electrical appliances such as household appliances that require different power consumption. Figure 3 shows the architectural and electrical plans of a house that serves as a reference for the calculations prior to obtaining the photovoltaic system. Table 1 shows the consumption of each electrical circuit of a house.

Table 1. Electrical loads and consumptions

Quantity	Equipment	Indvidual power	Accumulated power	Time	Energy
#	Description	[W]	[W]	[hours]	[Wh]
1	Fan	60	60	4	240
1	PC	55	55	4	220
2	Spotlights	15	30	4	120
1	Miscellaneous	100	100	4	400
TOTALS			245		980

2.2 Photovoltaic System Calculation Software

The PVGIS software is a web application that helps the user to obtain data on solar radiation and energy production of a photovoltaic system, from anywhere in the world,

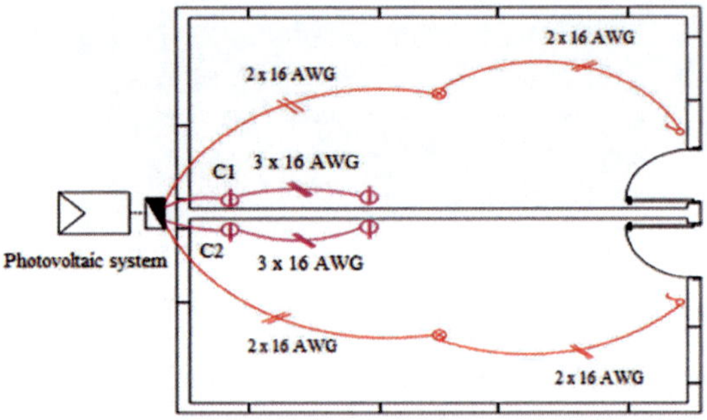

Fig. 3. Electrical diagram [22]

it can be used to obtain a series of different calculations. In Fig. 4 we can see the results of the performance of an off-grid PV system that was placed in the commune La Masa 2.

Fig. 4. Performance of an off-grid PV system in the commune La Masa 2

Another widely used program for sizing photovoltaic systems is PVSYST. PVSYST is a program created so that a user can accurately examine different configurations and estimate the results for the optimal possible solution.

The software provided the design of the system by means of the desired power, entering data such as number of elements, unit powers and operating hours.

It is also possible to position the photovoltaic modules in orientation and angles that have been previously calculated and the season (winter or summer). In this case, being a building without obstacles around it helps a lot in terms of electricity generation.

The system asks for a geographical point to establish the amount of radiation, temperature, global, diffuse, and normal radiation automatically through the METEONORM program to start the simulation.

With respect to the load regulator, it helps us to see the number of regulators in series or parallel that can be used with the required frequency either 60 Hz or 50 Hz.

The software has been very satisfactory to compare the simulated and calculated results which have been similar and give as a conclusion that the calculations have been developed correctly. In Fig. 5 we can observe the one-line diagram of the resulting photovoltaic system for a home and Table 2 Calculation results and software usage.

Table 2. Calculation results and software usage

Value	Symbol	Result
Optimum tilt	β	5.34°
Losses due to orientation and tilt	P_{oi}	3%
Irradiance	I	1000 W.m^{-2}
Cell working temperature	T_c	25 °C
Temperature efficiency	η_{Temp}	87.86%
Maximum power consumption	E_{CM}	1176 Wh
Nominal power	P_n	366.79 W
Peak power	P_p	440.15 W
Number of photovoltaic modules	No_{Mod}	2
Inverter power	P_{inv}	560 W
PV modules in series	N_S	1
PV modules in parallel	N_P	2
Battery system capacity	C_{SB}	420 Ah
Number of batteries in series with equation	N_{SB}	1
Number of batteries in Parallel	N_{pB}	4
Regulator input current	Iin	23 A
Regulator output current	I_{out}	46.29 A
Performance ratio	PR	65.70%

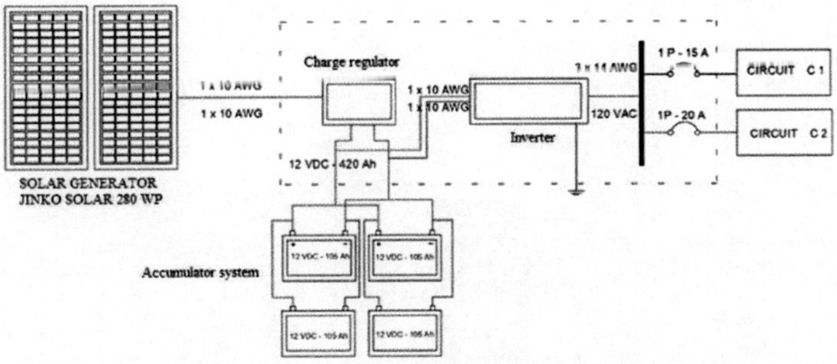

Fig. 5. One-line diagram of the stand-alone photovoltaic system [22]

3 Implementation of the Solar Photovoltaic System

The following is a photographic compilation of the elements that are part of the photovoltaic system and the work conducted for the implementation.

Figure 6 a) shows the photovoltaic panels and their installation on the roof of the houses. The structure to be used to support the solar panels is made of stainless steel to prevent corrosion. Figure 6 b) shows the connection of the batteries inside the tank.

a) Battery system, regulator, and inverter b) Mounting of photovoltaic panels [22]

Fig. 6. Installation of photovoltaic panels

The batteries used in this project are protected by a charge controller and are in a place where dust and rain do not affect them. They must also be ventilated to avoid inhaling the pollutant gases emanating from the batteries. To prevent a voltage drop it is necessary that they are placed as close together as possible and placed on a wooden base so that they do not discharge quickly.

3.1 Results and Discussion

This section shows the results of the PV system described above. To conduct this presentation, different measurements have been made during the year 2021 such as the daily demand projection of the users, the charging and uncharging of the batteries, losses in the PV system, the resulting energy delivered to the user.

Figure 7 shows the consumption characteristics of the users. The highest consumption demand is from 7 to 10 a.m. when the school is in operation. At night there is a lower consumption due to the use of luminaries because the school is used as a meeting place for the residents.

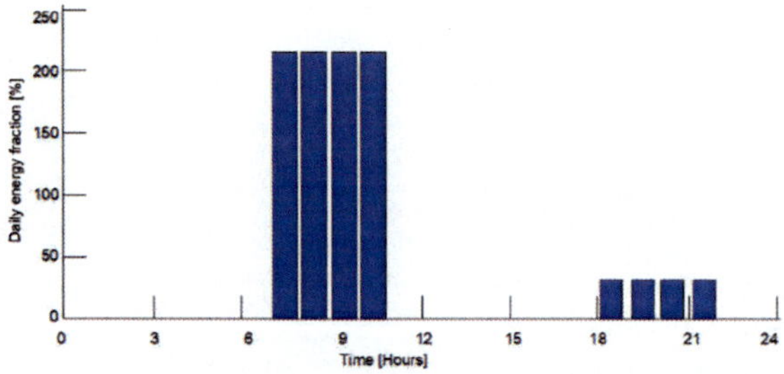

Fig. 7. Timetable profile for use at school

Figure 8 shows the specific annual yield (kWh/kWp) of the installed PV system. The graph also presents the unused energy (when the battery is full), the system and battery load losses, and at the end the energy supplied to the user. As can be seen in the bar graphs, the system and battery losses compared to the battery energy are able to supply the necessary energy to the users. Demonstrating that the system has a high efficiency. The total annual power delivered by the system is 256 kWh/year and its characteristic graph is shown in Fig. 9.

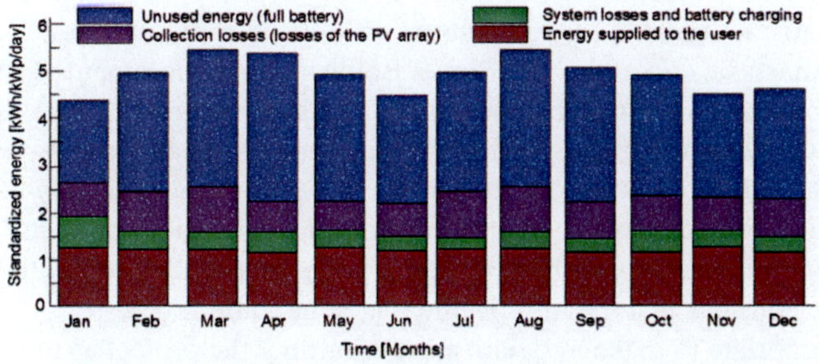

Fig. 8. Normalized production per installed kw peak

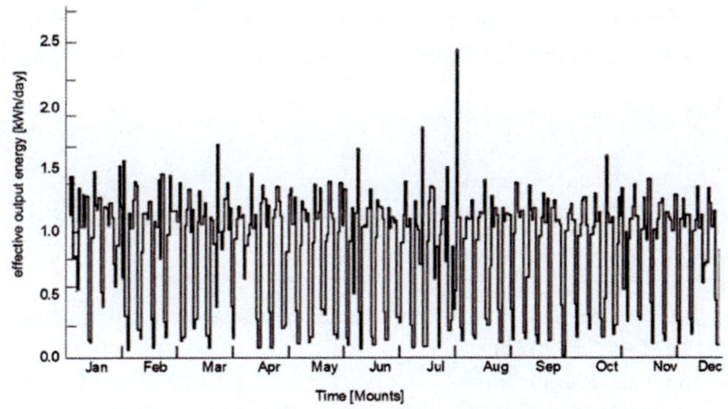

Fig. 9. Installation of photovoltaic panels

4 Conclusions

Prior to implementation, a two-part survey was developed. The first eight questions give us information on the way of life, lifestyle, and primary needs of the community members, such as the number of people, their ages, how many study in the educational unit located in the Masa 2 commune, the number of adults and children that make up the inhabitants of the site.

In the commune there are seventeen houses that currently have electricity supply, but through their own generators that require the burning of fuel for its operation. The generators supply their basic loads such as light bulbs for internal lighting and momentary loads for mobile devices. One of the main problems mentioned to us was the cost of operation, which is around $120 per month for each home.

In the second part of the survey, which corresponds from question 9 to the end, shows us their need for energy supply improvement. We conducted an analysis considering several aspects such as economic, social, security and environmental. This analysis, being far from a conventional network, we opted for a self-sustainable system with solar panels, so we were able to contribute to the environmental, social and safety aspects of the implementation project dedicated to the lighting of the main road of the Commune, showing the satisfaction of the inhabitants as announced in an interview.

"Gisella Anastasio, director of the Simon Bolivar School, commented: "I am grateful to the guys from the Universidad Politécnica Salesiana, because it is the first time they carry out this type of project here in the community, I have been working here for seven years and both for me and for the parents of the institution, we are happy that university students come to leave a mark, we hope they continue with more projects."

In the economic part, since the inhabitants do not have an external lighting for the main road, we consider that the improvement is remarkable.

It was appropriate to make a socialization regarding the project as much as a before making emphasis to the improvement that they would have as it during this way they would have knowledge of the procedure that we made for the installations of the equipment, at the same time to have an appropriate training more to the leaders of the commune for the teaching of the operation of the equipment and maintenance of these since they have not used until the present day equipment equal to the ones that were installed.

References

1. ONU. Los Objetivos de Desarrollo Sostenible, pp. 1–3 (2018). (http://www.undp.org/content/undp/es/home/sustainable-development-goals.html)
2. Don Bosco Green Alliance. Where Green Heartprints A (2021).
3. Universidad Politécnica Salesiana. Plan Operativo Anual de La Universidad Politécnica Salesiana pp. 2021–2022 (2021)
4. Consejo Nacional de Planificación. Plan Nacional Del Buen Vivir **1**, 2017–2021. Quito: Secretaría Nacional de Planificación y Desarrollo, Senplades (2017)
5. Carrion, L., Mendoza, K.. Diseño Optimo de Sistemas Aislados Fotovoltaicos Para La Iluminación de La Vía Principal de La Comuna Masa 2 Golfo de Guayaquil. Universidad Politécnica Salesiana (2021)
6. Gustavo, G., Kevin, C.: Sistema de Alumbrado Público Aplicado Mediante Energía Renovable Para La Comuna Masa 2, Golfo de Guayaquil. Universidad Politécnica Salesiana (2021)
7. Tinajero, J., Mayorga, W., Ampuño, G.: Simulation of centralized and decentralized photovoltaic generation systems for isolated sectors of the electricity grid. In: Rocha, Á., López-López, P.C., Salgado-Guerrero, J.P. (eds.) Communication, Smart Technologies and Innovation for Society. SIST, vol. 252, pp. 85–95. Springer, Singapore (2022). https://doi.org/10.1007/978-981-16-4126-8_9
8. Panchana, G., Valverde, R.: Diseño del Sistema de Energización Autónomo de la Isla Bellavista, localizada en el Golfo de Guayaquil. Escuela Superior Politécnica del Litoral (2018)

9. Jaramillo, J.P., Plaza, E.D., Benitez, A., Neira, A., Ulloa, G.A.: Evaluación de los sistemas de energización rural de la Isla de Cerrito de los Morreños en el Golfo de Guayaquil -Ecuador (2016). https://doi.org/10.18687/laccei2016.1.1.276
10. Andrade, G.V.S., Ruiz Buchelli, W.:Universidad de Posgrados, Estudio, Diseño e implementacion de un sistema de energía solar en la comuna Puerto Roma de la Isla Mondragon del Golfo de Guayaquil, Provincia del Guayas. Rev. EIA, ISSN 1794–1237, vol. 17, pp. 1–10, (2015). http://dspace.ups.edu.ec/handle/123456789/10070
11. Maldonado, M., Velastegui, D.: Implementación de un sistema fotovoltaico para residencias, en la comuna masa 2. Universidad Politécnica Salesiana (2022)
12. Villao, O., Consuegra, E.: Dimensionamiento óptimo de un sistema fotovoltaico a través de herramientas computacionales en zonas aisladas (2021)
13. Méndez, S., Manuel, J., González, T., Santiago, D.: Estudio e implementación de un sistema de luminaria de alumbrado público de 150w mediante sistema fotovoltaicos autónomo en la Comuna Masa 2, Isla del Golfo de Guayaquil (2021)
14. Suarez, M., Vallejo, J.: Implementación de un sistema fotovoltaico para sistemas aislados de la red pública, con una carga de 200 W. Universidad Politécnica Salesiana (2022)
15. Reinoso, W., Sánchez, D.: Diseño e implementación de electrificación en un sector rural del golfo de guayaquil mediante sistema fotovoltaico aislado. Universidad Politécnica Salesiana (2019)
16. Chiluiza, V.: Implementación de un sistema fotovoltaico para abastecer de energía a un sector rural del golfo de Guayaquil mediante el análisis de carga y simulación por software. Universidad Politécnica Salesiana (2022)
17. Mestanzan, J.: Guayaquil Insular Busca Tener Mejores Días, a Través Del Turismo. vol. 1, pp. 1–9
18. Duarte, K., Suarez, G.: Análisis del sector turístico de la Isla Puná del cantón de Guayaquil, período 2019. Universidad de Guayaquil (2021)
19. Diario el Universo. Salesianos Incentivan El Uso Energía Limpia En Comuna Del Golfo de Guayaquil. December 24, 1–14.Galarza, Rivera, Luis Antonio, Ordoñez Garzón, and Carlos Josué. 2021. Diseño de Un Arreglo Fotovoltaico Para La Iluminación Interior de Viviendas de La Comuna MASA 2 – Golfo De Guayaquil. Universidad Politécnica Salesiana (2020)
20. Moisés, P.: El Reflejo Del Golfo de Guayaquil, 16 December 2012, pp. 1–3 (2012)
21. Borbor, W.: Sistema Fotovoltaico Para Casas Individuales En La Comunidad Masa 2 – Golfo De Guayaquil. Universidad Politécnica Salesiana (2021)
22. Villegas, E., Alcivar, L.: Diseño de un sistema fotovoltaico para la escuela de educación básica Simón Bolívar en la comunidad Masa 2, Golfo de Guayaquil. Universidad Politécnica Salesiana (2020)

Photovoltaic Energy for Mobility and Building Operation in Unconditioned Equatorial Households: A Case of Study in Quito

Santiago Vaca-Jiménez, Anthony López-Vega, Carlos Naranjo-Mendoza, and Freddy Ordóñez(✉)

Department of Mechanical Engineering, Escuela Politécnica Nacional, Ladrón de Guevara E11-253, Quito, Ecuador
freddy.ordonez@epn.edu.ec

Abstract. Currently, humanity is transforming toward resilient and sustainable cities (RSC). Cities are composed of people who live and work in the built environment and, consequently, move from one place to another to perform their activities. Both buildings and transport are energy-intensive, and the energy source is generally not sustainable. Thus, a RSC must reduce the demand of these un- sustainable energy sources to operate its buildings and provide transportation. This paper evaluates the possibility of using a photovoltaic (PV) system to pro- vide the electricity required for the building operation and the urban private mobility of an average middle-class household in Quito, Ecuador. For this, we surveyed different households in the city to estimate the average consumption. Then, we modelled the electricity output of the PV system using the NREL's SAM software. Results show that it is feasible to fulfill up to the 100% of the household electricity requirement, including transportation (using a plug-in hybrid vehicle), by implementing a medium-size grid-connected PV system. This is possible since buildings in high altitude equatorial regions are generally unconditioned, and the cities are usually compact. These results suggest a path for high altitude equatorial regions to become RSC.

Keywords: Urban private mobility · Building operation · Photovoltaic system · Unconditioned equatorial buildings · Resilient · Sustainable cities

1 Introduction

Cities need to transform towards a resilient and sustainable framework to face the challenges linked to Climate Change. There are two ways to deal with this problem: Adaptation or Mitigation. On the one hand, cities can adapt to the consequences of Climate Change by building better and more resilient infrastructure that could provide shelter and support to extreme weather events. Adaptation goes hand in hand with resiliency. On the other hand, cities can mitigate Climate Change by introducing renewable energies to fulfill their energy needs. Mitigation relates to sustainability. Thus, the RSC framework is related to the actions and strategies of adaptation and mitigation of cities [1].

V. Robles-Bykbaev et al. (Eds.): CITIS 2022, LNNS 607, pp. 322–331, 2023.
https://doi.org/10.1007/978-3-031-24327-1_28

Mitigation-wise, it is imperative to know and assess the critical energy consumers so that they can be aimed toward a transition to renewable sources of energy, and buildings' operation and urban mobility are the largest energy consumers in cities.

Worldwide, the building sector consumes a large amount of primary energy [2]. For instance, in 2018, the residential sector of OECD countries demanded the 20% of the final energy worldwide [3]. Most of this consumption is used in active heating and cooling to maintain thermal comfort in buildings [4]. However, for high altitude equatorial regions, since the ambient temperature is mostly mild, most buildings are unconditioned [5]. Thus, the operation of buildings in these regions requires much less energy compared to other latitudes.

Similarly, passenger transport is responsible for approximately the 21% of share of final energy uses for 2018 [3]. In cities, urban mobility implies many quotidian movements, such as commuting for working, shopping, ludic activities, etc. Public transportation covers most of these daily trips [6]. However, a significant amount of them is made in private cars, especially for middle- and high-class citizens [7, 8].

In this context, any energy strategy that could reduce the energy consumption from the grid and increase the renewable energy share in a city is imperative. This paper argues that implementing small, grid-connected, PV systems in buildings could provide a feasible alternative to supply the electricity required for the building's operation and charge an electric car. A grid-connected PV installation is a system that uses the electric grid to balance the production of electricity from the PV panels [9, 10]. The PV system provides electricity to the household when there is enough solar radiation. If there is not enough PV output to cover the demand, the system takes the electricity required from the grid [11].

The objective of this study is to assess the feasibility of providing the electricity demand of a middle-class household in the city of Quito, by the means of such PV system. The paper has the following structure: Sect. 2 describes the methods used, Sect. 3 presents the results of our analysis, Sect. 4 discus these findings to contextualize the results obtained, and finally, Sect. 5 shows the conclusions of the study.

2 Methods

Buildings in equatorial regions are generally unconditioned. It means that the energy used for its operation is due mainly for appliances, cooking and domestic hot water (DHW). To explore the feasibility to cover a part of the energy needs for the building operation and for the mobility of its occupants with a small PV system the following methods were used. First, the energy needs for the building operation and for the mobility using a private car were investigated. Next, a small PV system was designed to cover the energy needed for the building operation and for charging a plug-in-hybrid electric vehicle (PHEV). For this part, the Ecuadorian regulation (ARCERNNR-001/2021, [12]) was considered. This regulation permits the injection of the PV electricity to the grid and recover it when necessary. This avoids the use of batteries or any other storage system. In the next subsections, the parts of these methods are described.

2.1 Household Description

The case of study consists in a residential household to be built in Quito, Ecuador. It was designed to be occupied by a family composed by four people, which is representative of a middle-class average family, according to the National Socioeconomic Stratification Survey [13]. Hence, the family is energy intensive consumers since it has all appliances for a modern style life. The parameters used for the Ecuadorian Government to classify a family as middle-class can be found in [13]. Some of the characteristics are the use of wood or similar as floor material, two bathrooms in the house, the use of several appliances: washing machine, TV, stove with oven, fridge, etc., the family has at least one car, internet access, the family head has superior education and works as technician or professional, among others.

The house has living surface of 120 m^2 and an available roof surface of 60 m^2, which may be used to implement the PV panels. This roof is tilted approximately 16° and has an azimuth of 100°. Even though the tilt is not the best for the Ecuador's latitude (0°), it may help avoid dust accumulation and self-cleaning. In Fig. 1 a 3D schematic of the house is presented.

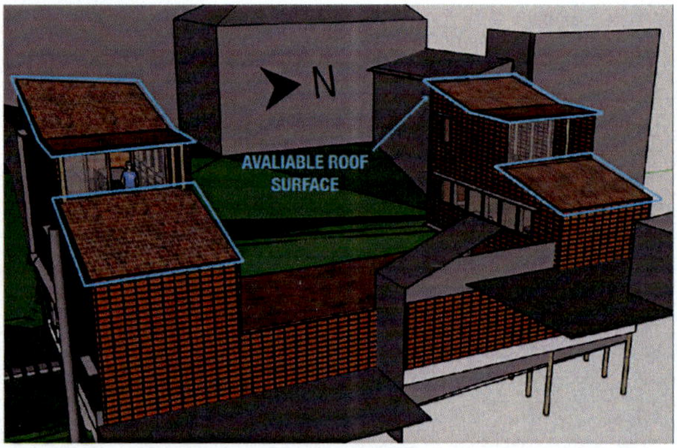

Fig. 1. 3D schematic of the studied household.

2.2 Energy Used in a Residential Household

In Ecuador, there is not a complete understanding how the final energy is used inside of a standard household. In the residential sector, the main energy carrier used is liquefied petroleum gas (LPG) with 52%, followed by electricity with 38%, firewood with 9.7%, and natural gas with the remainder [14]. In general, electricity is used for appliances, while LPG is mostly used for cooking and domestic hot water (DHW). It is worth mentioning that, in the last decade, the Ecuadorian government has promoted a transition from LPG to electricity: from LPG-fired to induction stoves, and LPG-fired to electric DHW systems. Part of the reason behind this transition is to take advantage of the extra electricity production by the new hydro-power plants in operation [15].

In this context, to have a preliminary idea of how the energy is used in a household, a survey of twenty-one middle-class families was conducted. First, the electricity consumption of these families from the city electricity company (Empresa Eléctrica Quito, EEQ) was found. Next, the survey was intended to classify the energy carrier used for the three main activities: appliances' consumption, cooking and DHW systems.

2.3 Energy for Urban Private Mobility

In Ecuador, transportation is the most demanding energy sector historically [16], reaching up to 38% in 2020 [14]. In the same way of the useful energy in the residential sector, there is not a complete understanding of the amount of energy consumed for the mobility of citizens inside cities.

Quito's municipality has conducted some studies to understand the different components of the city mobility [17]. Quito, as many Latin American cities, has been built around a hyper-center, which concentrates the most of city equipment, such as hospitals, public services, educational institutions, financial institutions, among others. Around the hyper-center, the macro-center, with less services than hyper-center, concentrates the residential buildings. Moreover, out of these two zones, some sites have become residential neighborhoods from which people mobilize every day to their activities in the hyper or macro center [17]. In Fig. 2, a scheme of the previously described zones in the city and the daily trips in private cars distribution is shown.

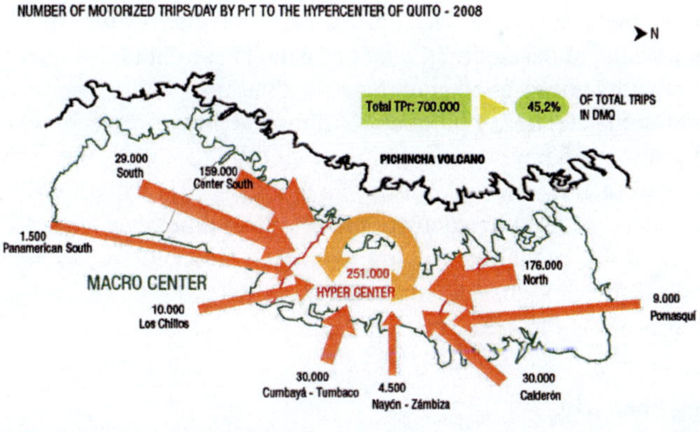

Fig. 2. Schematic of the daily trips distribution in private car to the hyper-center for Quito. Figure taken from [17].

According to the Master Plan of Mobility for the Metropolitan Area of Quito [17], there are 1 548 000 trips every day in Quito, from which 45.2% (700 000) are made in private car. The average time for a trip in private car is 23 min. From the data of the daily trips in Quito for the different sectors of the city, the average distance run by a private car was estimated.

For the present study, a plug-in hybrid electric vehicle (PHEV) to be charged by the PV system was selected. The characteristics considered for the study were: the electric autonomy of the PHEV, the battery capacity and the gasoline consumption in mixed conditions. In Table 1 are shown these characteristics.

Table 1. Technical characteristics of the plug-in hybrid electric vehicle.

Electric autonomy	50 km
Battery capacity	8.8 kWh
Gasoline consumption in mixed conditions	1.3 L/100 km

2.4 PV Production Model

The PV production modeling was made using the software System Advisor Modeling (SAM), developed by the National Renewable Energy Laboratory (NREL) of United States. This software enables to model dynamically the electricity production of a PV plant, taken as inputs, the solar resource of the site (global horizontal irradiation, GHI), the tilt and azimuth of the panels and the technical characteristics of the elements (PV panels, inverter, etc.).

For the model, the GHI was taken from the National Solar Data Base (NSDB) [18]. This data base estimates the solar irradiation by using satellite pictures in five different spectra in the visible and near infrared zones and produces a Typical Meteorological Year (TMY) of hourly meteorological data. In this case, for the location of the case study, the TMY was calculated based on 60 years of data These data have been compared to ground measurements, given good confidence of their use [19].

For computing the electricity production of the PV array, SAM firstly computes the plane of array (POA) irradiation with the usual theory [20]. Next, the software uses an extended data base of PV panels and inverters that exist in the market. The data of the technical data sheets are used to calculate the PV panel efficiency by the CEC model.

More details of this model can be found in [21]. Other losses, such as dirt losses, shadows, wire transmission losses, mismatch losses, etc. are taken into account in the performance ratio (PR) term.

2.5 Modeling Scenarios

Three scenarios of energy consumption were considered in this work. The first scenario includes the electricity consumption by home appliances, domestic hot water using an instantaneous electric water heater, cooking using an induction stove and the urban private mobility using one PHEV. The second scenario includes the appliances, cooking and the urban private mobility. Finally, the third scenario includes the appliances and the urban private mobility only.

The PV system considered for the study is a grid-connected system. The electricity provided by the PV and the grid is used for the electricity appliances, induction stove and the electric water heater (depending on the scenario). Figure 3 shows the diagram of the PV system.

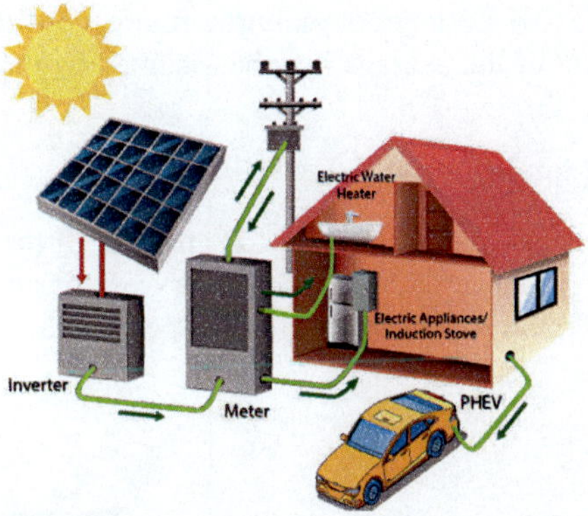

Fig. 3. Schematics of the designed PV system.

3 Results

3.1 Energy Required for Mobility

As described in the method section, to estimate the energy required for mobility, the average distances travelled of a private car in the City must be calculated. Table 2 shows the average distances to the hyper-center from different sectors of the city, and its corrected distance based on the number of trips from one sector to the other. It shows that the daily corrected-average distance travelled by a private car (one way) is 9.68 km.

Table 2. Average and corrected distances, per trip, travelled by a private car in Quito.

Trip	Distances [km]	
	Average	Corrected
North to hyper-center	10.62	2.67
South to hyper-center	10.62	0.44
Center-South to hyper-center	10.62	2.41
Panamerican to hyper-center	23.92	0.05
Los Chillos to hyper-center	12.61	0.18
Cumbayá-Tumbaco to hyper-center	12.61	0.54
Nayón-Zambiza to hyper-center	12.61	0.08
Calderón to hyper-center	12.61	0.54
Pomasqui to hyper-center	23.92	0.31
Within hyper-center	6.86	2.46
Total		9.68

Thus, considering 50 km as the expected autonomy of the PHEV, an average middle-class user of the car will need approximately to charge 40% of the battery per day to accomplish his daily trip (3.4 kWh per day). Furthermore, considering that it is a hybrid vehicle, other journey out of the average may be assumed by the petrol-engine.

3.2 Energy Required for Household Activities

Based on the analysis of the energy bill and the answers of the surveys made for this study, the average energy consumption, per category of consumption, of a household in the city was found. Figure 4 shows the average, per capita, energy demand of three consumption categories.

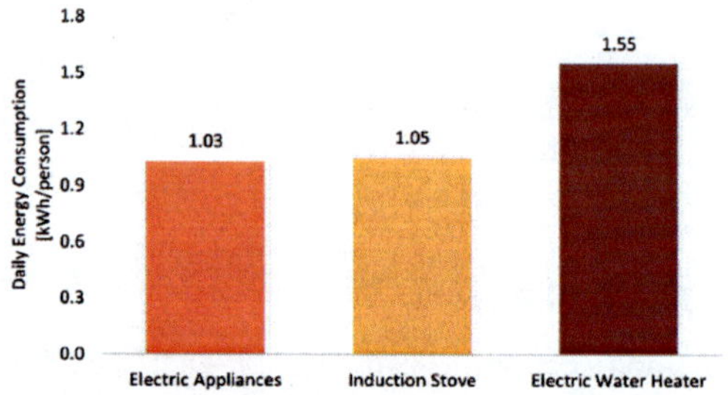

Fig. 4. Average energy demand (per capita) of three consumption categories: electric appliances, induction stove and instantaneous electric water heater.

3.3 Energy Requirement of the Household for the Three Scenarios

Considering an average of four-occupants, and one-private car household in Quito, Figure 5 shows the daily energy requirement of the household for the three proposed scenarios. It shows that, for the best-case scenario of this average household, the minimum energy required to be provided by the PV system is approximately 7.53 kWh per day, while in the worst-case scenario, the PV system must provide 17.95 kWh per day. Furthermore, it is worth mentioning that the largest energy consumer in these scenarios is the electric instantaneous water heater, while the smallest is the PHEV. The energy required for the PHEV is almost 3.7 times smaller than the electric water heater.

3.4 PV System

The PV module used in the three scenarios is a 345 Wp SunPower panel. Scenario 1 used a SMA inverter of 3800 Wac maximum output power, Scenario 2 a Growatt inverter of 3000 Wac, and Scenario 3 a Growatt inverter of 2000 Wac.

Figure 6 shows the annual electricity demand, the simulated PV production, and the area required for the installation of the PV modules for the three studied scenarios. In this case, it is shown that the PV system´s design is accurate for all the scenarios, as the electricity production is larger than the demand, but the difference is not big enough to treat as an overestimation of the PV system (and higher investment costs).

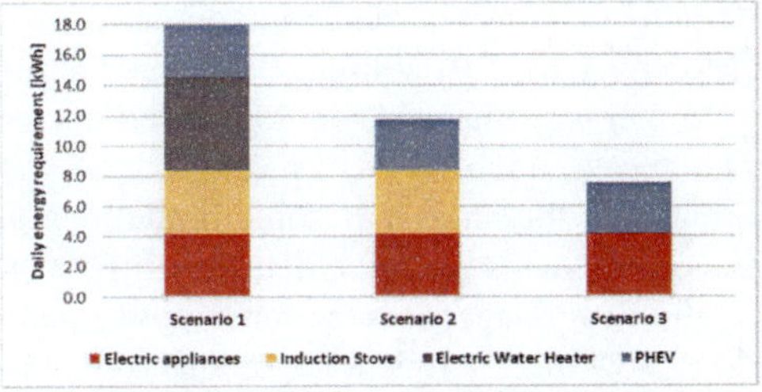

Fig. 5. Total daily energy requirement of an average household in Quito, considering three scenarios of electricity consumption (based on mobility, appliances, cooking and hot water systems).

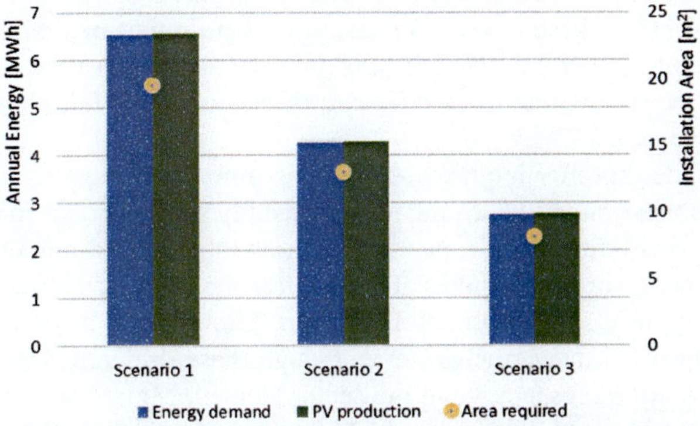

Fig. 6. Annual electricity demand, PV production and the area required for the installation of the PV modules.

In terms of the required area for the PV system, Scenario 1 has the largest requirement with 19.6 m2 (12 PV panels), and Scenario 3 has the smallest with 8.2 m2 (5 PV panels). The PV system for any scenario could fit in the building considering the 60 m2 of available area in the roofs.

4 Discussion

Results show that for unconditioned buildings, the size of the PV system necessary to cover the energy consumption of the building operation and mobility may be small in comparison to other cases or studies. This is because the locations of this study do not require active cooling or heating, reducing the energy requirement for these purposes, and that the city is very compact in its size. Hence, the covered distances are not extremely large, even when commuting from outside of the city to the hyper-center.

It is imperative to indicate that the proposed PV systems are grid-connected. Even though, the PV systems provide more than enough energy throughout the year, this does not mean that they will provide enough electricity all the time through the year. The system does not include batteries and the peaks in power requirement were not included.

There might be days (cloudy or rainy) in which the output is not enough to fulfill the requirement instantly. Moreover, the PHEV might be charged during the night, in which the solar output is zero. Either way, the grid should be used to cover the demand when the solar array does not.

It is worth mentioning that the study has some limitations that affect the results obtained. The main limitation comes from the databases used. In terms of the energy consumption of the household, this paper uses the consumption patterns of 21 households of the studied sector. We know that there is a large range of social realities in the city that will affect largely the results obtained, and 21 households is only a small fraction of it. However, considering that there are no available and validated databases of this, this approach is the only possible, as it uses real data from households in the city, and not assumption or estimations. We are confident that this approach does not introduce a significant uncertainty in our results, as the extreme cases of energy consumption in the targeted household are covered. For instance, there might be households that use other devices to fulfill their DHW needs (as shower head heaters), or to cook (as electric stoves). Nevertheless, despite being different, these cases will fall between the range studied by the scenarios.

In terms of the mobility requirements, we assume an average case, in which the user of the private car moves an average distance daily. This average considers that the car will need to cover different distances throughout the year. Despite Quito's mobility problems, there are no updated studies of trips within the city made by private cars. Here we had to use the available information and projections made by Municipality more than a decade ago [17]. Nevertheless, even though these difficulties in available data, the results found are representative to the sector studied. Moreover, the results should be large different if the mobility of the whole family is considered. For instance, by the introduction of additional PHEVs, or other vehicles like scooters.

5 Conclusions

The results obtained in this study indicate that an on-grid PV system is a feasible alternative to provide the electricity required for household activities and the mobility of a middle-class, four members, family in the city of Quito. However, these results should not be extrapolated for other cities with different geographical and climatic characteristics.

We used three scenarios to portrait the range of electricity demand inside the household. We found that relatively small PV systems (4 to 12 PV panels) may provide enough electricity to cover the needs of the household. This is possible due to the characteristics of Quito. The city is compact enough, so the distances travelled are not large, and the household does not require active cooling or heating systems because of the climatic conditions of the city. Thus, the electricity requirement is small in comparison to other cities in higher and lower latitudes.

Finally, this study is relevant for the discussion of the energy transition (electrification) of the household energy demand, and the mobility needs. Considering that the charging infrastructure of the city is almost non-existent at the moment, any type of introduction of electric mobility must be charged at home. Thus, it is imperative to analyze the possibilities of the households to incorporate systems that may provide that extra electricity required to charge their vehicles. Future works should focus on more detailed analysis, and in to include a larger and more updated scenarios of mobility.

References

1. Mendizabal, M., Heidrich, O., Feliu, E., García-Blanco, G., Mendizabal, A.: Stimulating urban transition and transformation to achieve sustainable and resilient cities. Renew. Sustain. Energy Rev. **94**(May), 410–418 (2018)
2. IEA, World energy balances (Edn. 2021), Paris (2021)
3. IEA, Key World Energy Statistics. Int. Energy Agency, vol. 33(August), p. 4649 (2020). https://www.iea.org/reports/key-world-energy-statistics-2020
4. IEA, Shares of residential energy consumption by end use in selected IEA countries, IEA Data (2018)
5. Ordóñez, F., Jácome, F., Castro, P., Naranjo-Mendoza, C.: Sensitivity analysis of the variables affecting indoor thermal conditions on unconditioned dwellings in equatorial high-altitude regions from an experimentally validated model. Adv. Build. Energy Res. **2549** (2019)
6. UITP, Urban public transport in the 21st century, Brussels (2017)
7. Gates, S., Gogescu, F., Grollman, C., Cooper, E., Khambhaita, P.: Transport and inequality: an evidence review for the department for transport, London (2019)
8. Hernández, D.: Public transport, well-being and inequality: coverage and affordability in the city of Montevideo. CEPAL Rev. **2017**(122), 151–169 (2018)
9. Pradhan, J.D., Hadpe, S.S., Shriwastava, R.G.: Analysis and design of overcurrent protection for grid-connected microgrid with PV generation. Glob. Transitions Proc. **3**, 349–358 (2022)
10. Mokhtara, C., Negrou, B., Settou, N., Bouferrouk, A., Yao, Y.: Optimal design of grid-connected rooftop PV systems: an overview and a new approach with application to educational buildings in arid climates. Sustain. Energy Technol. Assessments **47**, 101468 (2021)
11. Zhang, Z., Ma, M., Wang, H., Yun, P.: An optimal grid-connected strategy for improving the DC voltage utilization rate by increasing the number of PV modules connected in series in the string. Sol. Energy **236**, 687–702 (2022)
12. ARCERNNR, Resolución ARCERNNR-001/2021 (2021). https://www.controlrecursosyenergia.gob.ec/wp-content/uploads/downloads/2021/03/Resolucion-ARCERNNR-001-2021.pdf
13. INEC, Encuesta de Estratificación del Nivel Socioeconomico, Inec, p. 37 (2011). http://www.ecuadorencifras.gob.ec//documentos/web-inec/Estadisticas_Sociales/Encuesta_Estratificacion_Nivel_Socioeconomico/111220_NSE_Presentacion.pdf
14. MERNNR, Balance Energetico Nacional 2020, Quito (2021)
15. Vaca-Jiménez, S., Gerbens-Leenes, P.W., Nonhebel, S.: Water-electricity nexus in Ecuador: the dynamics of the electricity's blue water footprint. Sci. Total Environ. **696**, 133959 (2019)
16. IIGE, Balance Energético Nacional (2018). https://www.recursosyenergia.gob.ec/wp-content/uploads/2020/03/Balance-Energético-Nacional-2018.pdf
17. DMQ, Plan Maestro de Movilidad para el Distrito Metropolitano de Quito, 2009–2030 (2009)
18. NSRDB, National Solar Radiation Database (2017). https://nsrdb.nrel.gov/nsrdb-viewer
19. Ordóñez, F., Vaca-Revelo, D., López-Villada, J.: Assessment of the solar resource in Andean regions by comparison between satellite estimation and ground measurements: study case of Ecuador. J. Sustain. Dev. **12**(4), 62 (2019)
20. Duffie, J., Beckman, W.: Solar Engineering of Thermal Processes, 4th edn. Wiley, Hoboken (2006)
21. Gilman, P.: SAM photovoltaic model technical reference. Natl. Renew. Energy Lab. **63** (2015)

Mobility

Estimation of Areas with the Highest Accident Rate on the Cuenca-Loja Road According to the Driving Maneuvers

Néstor Rivera Campoverde(✉), Paúl Molina Campoverde, and Juan Molina Campoverde

Universidad Politécnica Salesiana, Cuenca, Ecuador
nrivera@ups.edu.ec

Abstract. Cuenca-Loja Road is a great reflection of the national reality, in the last five years, there have been 13 accidents of public transport vehicles, which left as consequences 159 injuries and 18 deaths. According to data from the National Traffic Agency (ANT) the main cause of accidents is the driver inexperience being 65.56% due to speeding, 23.33% due to invasion of the lane in an overtaking maneuver, and 11.11% for lost control of the vehicle. This research determines the areas with the highest accident rates along the Cuenca - Loja Road based on the physical characteristics of the road and the driving style. Fort the data acquisition a data logger device connected to the OBD II port was used that stores information such as speed, acceleration, and GPS, among others. Then, using automatic learning tools, the correlation between the different characteristics of the road (long slopes, road curves, etc.) and the driving style was determined in order to establish driving parameters appropriate to the geography and road surface.

Keywords: Traffic accidents · Driving behavior · Machine learning · Driving style

1 Introduction

Traffic accidents produce a large number of fatalities, and it is estimated by 2030 it will be the seventh cause of death in the world [1], several factors such as driving style, vehicle performance, traffic, weather conditions, and road infrastructure affect directly driver safety and air quality [2, 3]. As can be seen, the cause of traffic accidents is not influenced by a single reason, in contrast is the result of many factors [4].

Ecuador it is ranked at seventh position in the world with the number of deaths per habitant due to traffic accidents, according to statistics reported by the "Agencia Nacional de Tránsito", it has been estimated that the main cause of deaths is speeding with 65.56%, followed by the invasion of the lane during the overtaking maneuver with 23.33% and finally the lost control of the vehicle with 11.11% considering only the driving style in their statistics [5]. Traffic accidents according to vehicle classification occurred in vehicles used for private use (86.34%), public service vehicles (8.87%),

V. Robles-Bykbaev et al. (Eds.): CITIS 2022, LNNS 607, pp. 335–345, 2023.
https://doi.org/10.1007/978-3-031-24327-1_29

commercial vehicles (3.15%) and 1.64% in vehicles that provide other services. Driving style is one of the most studied parameters, especially with people who drive for long periods of time, such as professional drivers, which suffer 27% more accidents than non-professional drivers for reasons such as: stress and fatigue [6]. In Ecuador, vehicles with more than eight seats, without count driver's seat and total weight greater than 5000 kg, belongs to M3 category [7]. This type of vehicles during 2019 year produced 1076 traffic accidents with a balance of 852 people injured and 52 dead [8].

In Ecuador there are few studies about road accidents, but several authors propose different solutions for this problem. Algora et al. [9], Using information about traffic accidents extracted from the database of the National Traffic Agency since January to December of 2016, describe statistically the main epidemiological reasons for death in traffic accidents occurred in Ecuador. León et al. [5], applying data mining and decision trees techniques to classify traffic accidents based on causes and link distracted driving with side collision and drunk driving with the generation of a single vehicle crash. Yu et al. [10], propose a Deep Spatio-Temporal Graph Convolutional Network to predict traffic accidents, to evaluate their model they use an extensive database of traffic accident records, GPS taxi, meteorological observations, etc. Lin et al. [11], develop a model to predict vehicle traffic after a traffic accident with crowdsourcing data obtained from social networks through a hierarchical scheme, the authors identify four levels and the duration of these events. Alkheder et al. [12], performs three data mining analyzes to extract risk factors related to traffic accidents, showing that the position where the passenger was injured is the most important variable in the study. Ali et al. [13], present a model based on social network analysis using an algorithm that transforms unstructured data into structured data, automatically labeling each sentence to extract the exact information of a traffic accident with the aim of understanding the condition which cause traffic vehicle. Tang et al. [14], they use a numerical traffic model that considers disruption effects such as traffic accidents and bus stations.

This research develops a model for estimating the areas with the highest accident rates in the Cuenca Loja Road using multivariate statistical methods and machine learning algorithms, in order to identify the areas where a vehicle of category M3 has more probability of suffer an accident. It should be noticed that the province of Azuay is the fourth state with most traffic accidents. The Cuenca - Loja is the main road in the south of the country where serious accidents it had been occurring in past years, therefore identifying conflict areas could save lives and prevent accidents in the future.

In the next section the description of the model used for estimating the trajectory of the vehicles through the Cuenca Loja road using the coordinates obtained by GPS, after that the data is processed and compared according to certain indices to finally verify the validity of the models.

2 Material and Methods

Most of the studies use large databases of governments, public entities or information from digital media, but as previously described, the cause of traffic accidents is not due to a single factor instead is the sum of different scenarios. The current study tries to determine the places most prone to accidents along the E35 road that connects the cities of Cuenca and Loja in the south of the country. First, statistical information it was collected of traffic accidents and the areas where they have occurred, considering information from 2015 to 2019. Later, data is post-processed to determine the areas with high traffic accidents by means an algorithm that determines the curvature and radio on the road with the use of a data logger device connected to the OBD II port of the public transport units, to determine car skidding and rollover speed data in order to relate this information with "Ley de Transporte Terrestre, Tráfico y Seguridad Vial (RLTTTSV)", to determine the places in which a vehicle category M3 is more accidents-prone. The process carried out is described below.

2.1 Experimental Overview

Characteristics associated with driving style is acquired through the OBD II communications protocol to obtain information from sensors such as: IAT, TPS, RPM, VSS, GPS, etc. [15, 16]. Data logger devices does not adhere mass to test conditions and stores data under normal driving conditions unlike Portable emissions measurement system (PEM'S) [17]. The device used as a data logger was "Freematics ONE +" with free access, which records all the information in a plain text file. The variables used for this study are shown in Table 1.

Table 1. Stored variables [18]

Variables	Nomenclature	Units
Longitudinal acceleration	Acceleration	m/s^2
Vehicle speed sensor	VSS	m/s
Engine speed	RPM	rpm
Throttle position sensor	TPS	%

2.2 Plot Route

Using the developed algorithm, the data collected by the GPS such as latitude, longitude and altitude are analyzed to plot the route Cuenca - Loja observed in Fig. 1a, but in order to perform the different calculations during this study, it was necessary to transform the GPS information into UTM coordinates (Universal Transverse Mercator) to be able to work with the International System units.

2.3 Calculation of Curvature at Points on the Route

To find the different curvature values at each point of the route (Cuenca - Loja), Eq. 1 is used because is possible to associate every element of the route in terms of the radio.

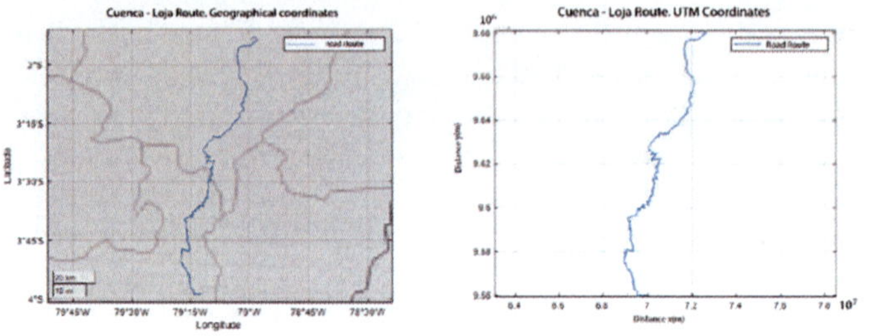

Fig. 1. Plot route Cuenca - Loja. a) Geographical coordinates b) Coordinates UTM

This parameter can be shown throughout the route and in the areas specified by the user [19] (Fig. 2).

$$k = \frac{\left| f'(t) \cdot g''(t) - g'(t) \cdot f''(t) \right|}{\left[(f'(t))^2 + (g'(t))^2 \right]^{\frac{3}{2}}} \tag{1}$$

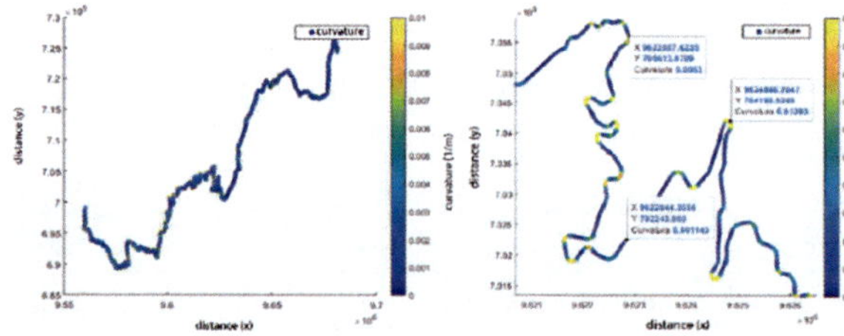

Fig. 2. a) Curvature values on the road; b) Specific zone with curvature values.

Calculation of the Radius of Curvature. For the calculation of all the radius of curvature for each point of the route, the inverse of the curvature is used, obtaining the results shown in Fig. 3.

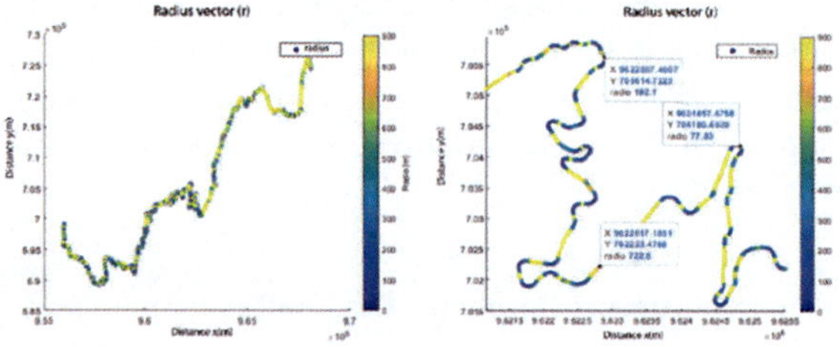

Fig. 3. Radius of Curvature; a) The whole route b) Specific places

2.4 Testing Radius of Curvature

The radius of curvature is the initial condition of the problem to determine the skidding and rollover speed of public transport units. The values obtained by the algorithm are compared with the radius found by the Google Earth Pro application, it should be noticed that there are small differences between the algorithm and application due to the type of sampling performed by Google. Some of the data are presented in Table 2 and Fig. 4 and Fig. 5.

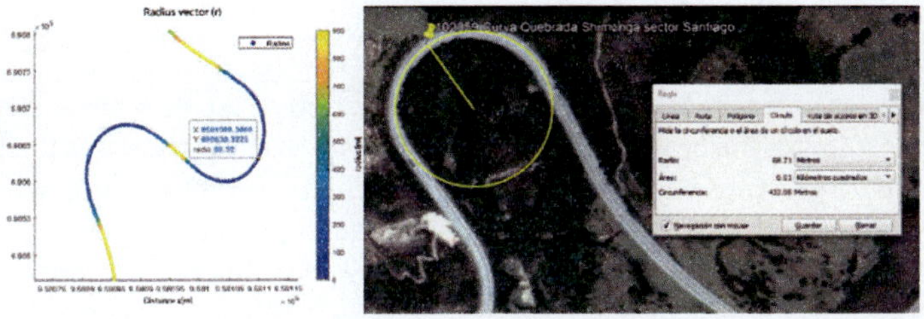

Fig. 4. Comparison for Santiago location, Algorithm vs Google Earth

Table 2. Checking radius of curvature algorithm vs Google Earth

Point	Vector location number	North coordinate	East coordinate	Algorithm, radius (m)	Google Earth Pro, radio (m)
1	102059	9581098.3007	690630.3224	68.52	68.71
2	46368	9624509.2012	701624.3790	78.42	78.29
3	16351	9663357.9709	717332.7068	34.34	31.5
4	73696	9600998.1028	700765.9106	85	85.28

2.5 Calculation of the Skidding Speed

This parameter allows to determine the limit speed of skidding in a curve, thus the limit speed at which the car must enter in a curve without slip. For this, Eq. 2 is used. The variables involved in the equation are gravity (g), radio (R), coefficient of pneumatic grip on the ground μ_{ymax} and the cant of the road in degrees ($\tan\zeta$). The algorithm can describe the speed along the route or at a specific part of the route

$$V_{ld} = \sqrt{g \cdot R \cdot \frac{\mu_{ymax} + tg\zeta}{1 - \mu_{ymax} * tg\zeta}} \tag{2}$$

2.6 Calculation of the Rollover Speed

Equation 3 identifies the areas where the traffic accident is imminent, the variables are the same as those used in the past calculation of the skid speed with the difference that the term B represents the vehicle's track width and h, the height of the center of gravity.

$$V_{lv} = \sqrt{g \cdot R \cdot \frac{B}{2h}} \tag{3}$$

2.7 Braking Distance Calculation

For areas where the transport unit needs the application of the braking system, Eq. 4 is used, which estimates the braking distance and determines how risky it is to drive at high speeds. The equation presents variables such as: vehicle weight P, rotating mass coefficient γ_f, coefficient of resistance C, coefficient of aerodynamic drag cx, vehicle frontal area Af, vehicle speed in m/s, V_1^2, friction force F_f, coefficient of rolling resistance f_r and gravity g.

$$S_p = \frac{P \cdot \gamma_f}{2 \cdot g \cdot C} \cdot ln\left[1 + \frac{C \cdot V_1^2}{\eta_f \cdot \mu \cdot P + P \cdot sen\theta + P \cdot f_r}\right] \tag{4}$$

2.8 Classification of Roads: Straight Road, Medium Curves, and Dangerous Curves in Urban and Rural Areas.

Route E35 is made up of curves with various radius and slopes given the geography of the southern part of the country and is classified as can be seen in Table 3.

Table 3. Types of curves in urban and rural areas

Type of curve (urban area)	Radio	Identifier
Straight road	>200 m	0
Medium curve	>95 m y < 199 m	2
Dangerous curve	<94 m	4
Type of curve (rural area)	Radio	Identifier
Straight road	>200 m	1
Medium curve	>95 m y < 199 m	3
Dangerous curve	<94 m	5

3 Results and Discussion

Article 191 of "RLTTTSV" specify the limits and ranges of circulation established for public vehicles (see Table 4), with this information the route taken by a bus is compared with the speed limits established by the article. The analysis of speeds of the public unit is used to find the places with the highest accident rates on the Cuenca-Loja route and to know the driver behavior.

Table 4. Speed limits for public transport vehicles.

Type of road	Maximum speed limit	Moderate range (Art. 142.g)	Out of range (Art. 145.e)
Urban	40 km/h	>40 km/h–150 km/h	>50 km/h
Beltway	70 km/h	>70 km/h–100 km/h	>100 km/h
Straight road	90 km/h	>90 km/h–115 km/h	>115 km/h
Curve road	50 km/h	>50 km/h–65 km/h	>65 km/h

3.1 Comparison of the Driver's Speed and the Limited Speed to Determine the Accident Zones

The algorithm compares the speed of the driver against the speed imposed by the traffic law, in this comparison the points in which the driver exceeds the speed limits are obtained, this will generate a possible risk in the area where the vehicle is transiting. Figure 5 shows the actual speeds of the vehicle in the route and compares with the speed limited by the algorithm to determine the most dangerous areas. As can be seen, the driver reaches a maximum speed of 105 km/h when the law allows 90 km/h in this area.

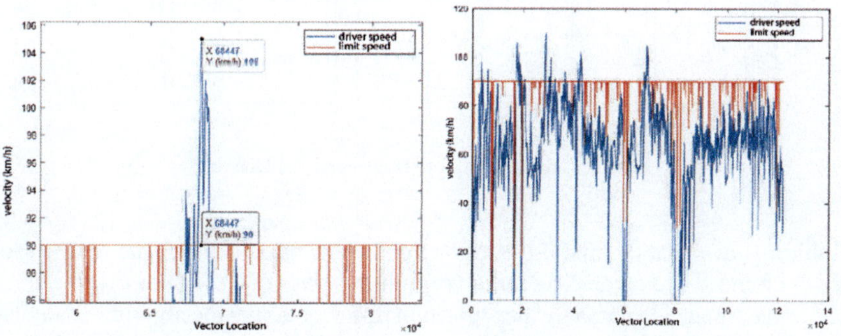

Fig. 5. Comparison of actual driver speed and limited speed

3.2 Driver Behavior Classification

Driver behavior allows knowing the driving style of the driver on the road, with this information a safety index is created as shown in Table 6 (Table 5).

Table 5. Classification of driving behavior pilot

Index	Driver behavior
0	Moderate
0.5	Medium
1	Aggressive
>1	Reckless driving

3.3 Measures of Central Tendency Applied to the Driver Based on the Safety Index

The measures of central tendency used to analyze the driver are mean, median and mode, these summarize the set of values in one variable. The algorithm calculates its own trend and compares respect to the measures of central tendency. For this reason, is taking in account the behavior of the two drivers as it can be seen in Fig. 6. After the analysis, the algorithm evaluates the behavior of two drivers to identify the appropriate safety values as can been seen in Table 7.

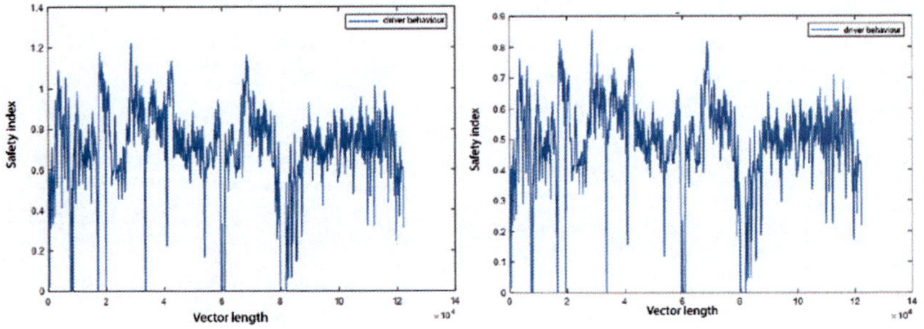

Fig. 6. Driver behavior. a) Driver 1, b) Driver 2

Table 7 shows that the first driver has a trend value of 0.7087235 and for the second driver the trend value created by the algorithm is 0.4961064. To sum up, the first and second value trend calculated by the algorithm differs, making the algorithm calculation more accurate and reliable since the outliers do not influence. Another way to determine the behavior of the driver is through statistical analysis of the standard deviation, allowing to know how dispersed the data are from each other. Table 7 shows the analysis of the two drivers.

Table 6. Measures of central tendency

Tendency	Driver 1	Driver 2
Algorithm	0.7087235	0.4961064
Mean	0.7087262	0.4961083
Median	0.7333333	0.5133333
Mode	0.7111111	0.4977777

Table 7. Standard deviation for drivers

Driver	Standard deviation values
1	0.2073581
2	0.1451507

As can been seen in Table 7, the first driver has a variable driving behavior, because tend to change from low speeds to high speeds aggressively, generating a bad road behavior with a maximum velocity of 105 km/h with an index safety of 1.2 compared with the second driver with a 0.85 index safety resulting in a moderate behavior without modifying the velocity instantly as can been seen with the factor of 0.145 of standard deviation.

4 Conclusions and Future Works

Traffic speed limits are established by the "RLTTTSV" unfortunately the transport laws are breaking according to the data obtained in this study due to the imprudence of the evaluated drivers. As an example, Cumbe sector is taken where it can be seen that the speed of circulation that the algorithm record is 90 km/h generating a risk situation if it is compared with the maximum speed of circulation which is 50 km/h for this type of area. Some of the areas where the driver has this behavior is: Highway Cuenca, Tarqui, Cumbe, entance to Nabón, sector "Parque Natural Silván", Hacienda Aventure, La Paz, Progreso, "Estación de Servicios S. F. Oña", straight road Oña, Hacienda el Prado (Potrerillos), sector "Centro de Retiros Carboncillo", San Lucas, y Carigán.

In this study it could be established a recommendation for every radius of curvature along the road that the driver should adopt, unfortunately, according to the data obtained in this study, the drivers do not respect speed limits, for example, in the Gañadel sector, the public transport unit enters with a speed of 69 km/h greater than the maximum limit established by "RLTTTSV". (Curves on road 50 km/h).

The measures of central tendency allow us to know how the behavior of the driver of the transport unit is and with the analysis carried out on each driver it is possible to maintain safe driving without having the risk of suffering traffic accidents. Through the mean, median, mode and the trend created by the algorithm, a guide is provided for public bus drivers in order to create a responsible culture driving. In future works, it would be pertinent to generate a road safety education plan so that professional drivers can distinguish the highest risk areas and can reduce the high probability of an accident that exists in the country.

References

1. Ramírez, F., Valencia, C.: Spatiotemporal correlation study of traffic accidents with fatalities and injuries in Bogota (Colombia). **149**(November 2020), 1–18 (2021). https://doi.org/10.1016/j.aap.2020.105848
2. Vipin, N., Rahul, T.: Road traffic accident mortality analysis based on time of occurrence: evidence from Kerala, India. Clin. Epidemiol. Glob. Health **11**(11), 100745 (2021). https://doi.org/10.1016/j.cegh.2021.100745
3. Kmeť, R., Kvet, M.: Traffic accident monitoring information system of the selected region. Transp. Res. Proc. **55**, 1452–1459 (2021). https://doi.org/10.1016/j.trpro.2021.07.132
4. Bucsuházy, K., et al.: Human factors contributing to the road traffic accident occurrence, no. 2019 (2020). https://doi.org/10.1016/j.trpro.2020.03.057
5. León-Pluas, E., Pumares-Romero, A., Guaña-Moya, J., Loayza-Valarezo, P., Naranjo-Villota, D., Salgado-Reyes, N.: Analysis of causes of traffic accidents in ecuador using data mining | Análisis de causas de accidentes de tránsito en el ecuador utilizando minería de datos. STI – Rev. Iberica Sist. Tecnol. Inform. **2020**(E26), 540–547 (2020)
6. Han, W., Zhao, J.: Driver behaviour and traffic accident involvement among professional urban bus drivers in China. Transp. Res. Part F: Psychol. Behav. **74**, 184–197 (2020). https://doi.org/10.1016/j.trf.2020.08.007
7. de Normalización, S.E.: ECUATORIANA NTE INEN 2656, Clasificación vehicular (2016)
8. Agencia Nacional de transito: Estadísticas de siniestros de tránsito. https://www.ant.gob.ec/?page_id=2670. Accessed 22 July 2021
9. Af, A.-B., Pr, S.-B., Ar, G.-G.: Epidemiological study of fatal road traffic accidents in Ecuador Epidemiological study of fatal road traffic accidents in Ecuador Research Group on Working Conditions and Occupational Health. International University SEK, Ecuador What this study adds, no. April (2017). https://doi.org/10.21767/AMJ.2017.2951
10. Yu, L., Du, B., Hu, X., Sun, L., Han, L., Lv, W.: Neurocomputing deep spatio-temporal graph convolutional network for traffic accident prediction. Neurocomputing **423**, 135–147 (2021). https://doi.org/10.1016/j.neucom.2020.09.043
11. Lin, Y., Li, R.: Real-time traffic accidents post-impact prediction: based on crowdsourcing data. **145**(March), 1–11 (2020). https://doi.org/10.1016/j.aap.2020.105696
12. Alkheder, S., Alrukaibi, F., Aiash, A.: Risk analysis of traffic accidents' severities: an application of three data mining models. ISA Trans. **106**, 213–220 (2020). https://doi.org/10.1016/j.isatra.2020.06.018
13. Ali, F., Ali, A., Imran, M., Ali, R., Hameed, M., Kwak, K.: Traffic accident detection and condition analysis based on social networking data. Accid. Anal. Prevent. **151**(September 2020), 105973 (2021). https://doi.org/10.1016/j.aap.2021.105973
14. Tang, T., Shi, W., Huang, H., Wu, W., Song, Z.: A route-based traffic flow model accounting for interruption factors. Phys. A **514**, 767–785 (2019). https://doi.org/10.1016/j.physa.2018.09.098
15. Néstor Diego, R.C., Molina Campoverde, P.A., Quirola Novillo, G.P., Naula Bermeo, A.K.: Development of an algorithm capable of classifying the starting, gear change and engine brake variables of a vehicle by analyzing OBD II signals. In: Botto-Tobar, M., Zamora, W., Larrea Plúa, J., Bazurto Roldan, J., Santamaría Philco, A. (eds.) ICCIS 2020. AISC, vol. 1273, pp. 121–133. Springer, Cham (2021). https://doi.org/10.1007/978-3-030-59194-6_11

16. Molina Campoverde, P.A., Rivera Campoverde, N.D., Novillo Quirola, G.P., Bermeo Naula, A.K.: Characterization of braking and clutching events of a vehicle through OBD II signals. In: Botto-Tobar, M., Zamora, W., Larrea Plúa, J., Bazurto Roldan, J., Santamaría Philco, A. (eds.) ICCIS 2020. AISC, vol. 1273, pp. 134–143. Springer, Cham (2021). https://doi.org/10.1007/978-3-030-59194-6_12
17. Huang, Y., Ng, E.C.Y., Zhou, J.L., Surawski, N.C., Chan, E.F.C.: Eco-driving technology for sustainable road transport: a review. Renew. Sustain. Energy Rev. **93**(August 2017), 596–609 (2018). https://doi.org/10.1016/j.rser.2018.05.030
18. Molina Campoverde, J.J.: Driving mode estimation model based in machine learning through PID's signals analysis obtained from OBD II. In: Botto-Tobar, M., Zambrano Vizuete, M., Torres-Carrión, P., Montes León, S., Pizarro Vásquez, G., Durakovic, B. (eds.) ICAT 2019. CCIS, vol. 1194, pp. 80–91. Springer, Cham (2020). https://doi.org/10.1007/978-3-030-42520-3_7
19. di Mascio, P., di Vito, M., Loprencipe, G., Ragnoli, A.: Procedure to determine the geometry of road alignment using GPS data. Proc. Soc. Behav. Sci. **53**, 1202–1215 (2012). https://doi.org/10.1016/j.sbspro.2012.09.969

Check for updates

Efficiency Increase of Supercharged Engines

Paúl Molina Campoverde(✉), Juan Molina Campoverde, Karina Bermeo Naula, and Gina Novillo

Universidad Politécnica Salesiana, Cuenca, Ecuador
{pmolinac1,abermeon}@ups.edu.ec

Abstract. This work studies the power reduction of a compression ignition engine during its operation in high altitude cities. In order to compensate for the loss produced by low air density, the basic operating parameters of the engine are modified in a non-intrusive way without making physical modifications, but rather with the engine bases maps variation to determine the most influential variables of the study, a design of experiments (DOE) is carried out, finding that the most important variables are fuel injection, injection rail pressure and turbocharger pressure. The information is collected by running the vehicle on a dynamometric bench to check the modifications efficiency, at the same time a gas analyzer measures the concentrations of CO, CO_2, NO_X and HC also, an opacimeter measures the particulate matter (PM) at the exhaust outlet. With the carried out modifications, a recovery of 10 kW in power is obtained without a substantial increase of engine pollutant emissions.

Keywords: Turbo charged · DOE · Statistical analysis · Design of experiments

1 Introduction

In the world it is common to find cities located thousands of meters above sea level (masl), under these conditions compression ignition engines present fuel consumption increase from 9 to 14%, a reduction in power from 8 to 13% per 1000 m and a consequent rise in polluting emissions production as a result of the decrease in air density in these geographic locations [1–3]. This type of internal combustion engines present an environmental problem as they are large producers of greenhouse gases, which is why they are required to be efficient in fuel consumption and produce reduced polluting emissions [4]. To solve this problem, engines with a high compression ratio have been implemented, but they generate high temperatures in the combustion chamber facilitating NOX production, the use of fuels enriched with alcohols reduces the amount of particulate material (PM), carbon dioxide and soot, on the other hand they also increase the NOX number. Another method consists in the controlled increase of pressure in the engine intake line using turbochargers. International standards such as Euro 6 have been implemented to control emissions from mobile sources produced by internal combustion engines. (VI standars) [5] have been implemented and to measure them, driving cycles such as the Real Driving Emission (RDE) are incorporated, which covers altitudes that vary between 0 to 700 m [6] but it does not contemplate Latin American realities such

V. Robles-Bykbaev et al. (Eds.): CITIS 2022, LNNS 607, pp. 346–356, 2023.
https://doi.org/10.1007/978-3-031-24327-1_30

as the case of the city of Cuenca Ecuador, which is located at 2,500 masl, for this reason, improve the efficiency of engines in high-altitude cities and reducing the number of polluting emissions is a priority task.

Several authors propose different solutions for this problem. Ghazikhani et al. [7], carry out an experiment using the ECE-R49 test and find that with a height variation from 390 m to 975 m, soot emissions increase by 40% due to the reduction in air density. Peter L et al. [8], use a mono cylindrical engine to simulate oxygen enrichment at the intake manifold inlet and simulate altitude conditions up to 5200 m with the use of a valve that restricts the entry of air into the engine, in this experiment they found that the Engine power delivery is primarily dependent on engine load and oxygen enrichment does not affect significantly until 2600 m. Yang et al. [9], propose a strategy to compensate the power loss of a diesel compression ignition engine in high altitude cities through the use of a two-stage turbo compression system, the high-pressure turbine is used to compensate the energy loss produced by the lift while the low pressure turbine optimizes the overall performance of the turbo compression system. Chao et al. [10], use a height simulation system to analyze the emissions produced by a diesel engine at sea level and simulated altitudes of 1000 and 2000 m, they determine that as the height increases, the emissions of HC, CO and NO_X increase, so will the emissions of NO_X vary with height but are dependent on the type of engine and working conditions, below 2000 m there is a progressive increase in HC, CO and NO_X emissions by 30%, 35%, and 14% respectively and with an increase of 1000 m more they can reach values of 34%.

In this work, with the use of design of experiments, various parameters of the vehicle's control unit are modified, including the turbocharger, which is designed to operate at sea level and at low altitudes [11] without any type of physical modification. The variation of the height and its influence will allow to improve the efficiency of a engine and to recover its power.

The following sections show the data acquisition from the dynamometer bench in addition with the exhaust gas analyzer. A model for power optimization and recovery is obtained by comparing the paramcter values underestimated versus simulation conditions through incidence factors from DOE when modifying injection maps.

2 Materials and Methods

In high-altitude cities, the air density is lower when compared to sea level, so the power of the engines is reduced and the number of pollutant emissions increases. This study tries to compensate for the loss of power from the engine and the reduction in the number of polluting emissions by modifying the electronic management of the engine and its auxiliary systems without the installation of any equipment. The experiment is developed in the city of Cuenca, Ecuador located at 2500 masl using a Hyundai Getz 1.5 CRDI vehicle with the specifications that can be seen in Table 1.

2.1 Methodology

For the model characterization the variables with the greatest influence during driving are described and are classified as: response, study, blockage and noise. The response

Table 1. Vehicle specifications

Parameters	Value
Engine capacity	1493 cm^3
Transmission gearbox	5 speed manual
Fuel	Diesel
Compression Ratio	17.8:1
Supercharger	Turbocharger
Maximum power	110 hp @ 4000 rpm
Maximum torque	215 Nm @ 1900–2750 rpm

variables are those that are at the output of the system and allow the continuous improvement of the process, the study variables directly influence the response variables, the blocking variables can be set in an operating range and not they must be evaluated and, the noise variables are all those that can modify the output and are very difficult to control during the experiment, as can be seen in Table 2 [12].

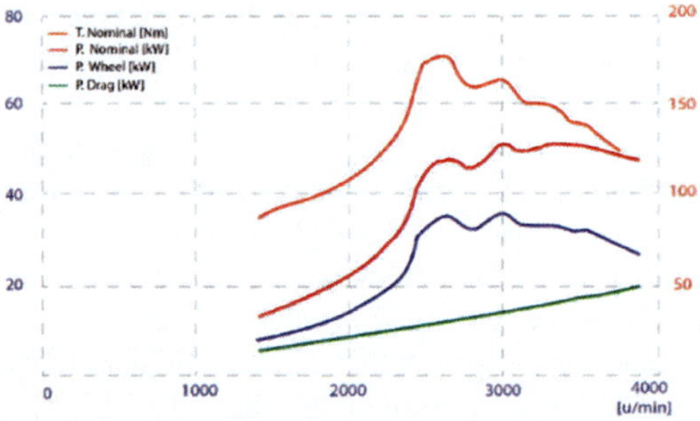

Fig. 1. Hyundai Getz vehicle characteristic curves obtained on the dynamometric bench without any modification

The power has been obtained on the wheels of the vehicle as a result of the experimentation through the dynamometric bench. Figure 1 describes the values extracted from the vehicle.

Once the influencing variables in the system have been determined, the experimental measurements are carried out. The vehicle is located on the MAHA LP 300 dynamometric bench based on the ISO 17359 2018 standard, being able to obtain power values in engine, wheel, drag and torque as shown in Fig. 1 [13]. With the vehicle running, the gas analyzer (Brain Bee AGS-688) measures the concentration of the following gases: HC, CO, CO_2, O_2 and NO_X [14]. With the MAHA MET 6.2 opacimeter it is possible to determine the particulate matter (PM) emitted by the engine during the tests. In addition, the information obtained through the OBD II port of the vehicle is stored in a computer and through a design of experiments (DOE) it is determined which parameters should

Table 2. Incidence factors

Variables	Factors	Symbol	Unit
Answer	Power	P	[kW]
	Engine torque	T	[N·m]
	Carbon monoxide	CO	[%]
	Nitrous Oxides	NO_X	[ppm]
	Particulate matter	PM	[mg·m^3]
Blocking	Coolant temperature	T Coolant	[°C]
	Rolling temperature	T Rolling	[°C]
Noise	Temperature	T_{amb}	[°C]
	Fuel temperature	T_f	[°C]
	Relative air humidity	H Rel	[%]
Study	Turbocharger pressure	Pt	[%]
	Injection rail pressure	Pr	[%]
	Injection time	T_{inj}	[ms]

be modified. Finally, with the reprogrammer module (KESS V2), the engine base maps are read and the parameters of the vehicle's control unit are varied to recover the power of the internal combustion engine and reduce the number of polluting emissions (Fig. 2).

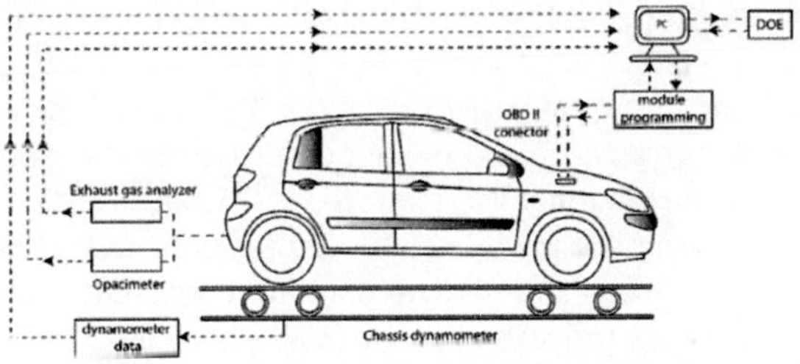

Fig. 2. Experimental samples description

2.2 Experimental Design

In this section, the fuel injection maps based on TPS, injection rail pressure and turbocharger pressure are modified using a Box-Behnken model, assigning each variable a high and low level that will subsequently allow a percentage increase over each one of the maps as seen in Table 3. Considering the value of zero as the model without modifications while the value of 10 identifies a percentage increase of that value, it should be noted that these are first approximations to describe the model.

Table 3. Design table

Factor	Name	Low	High
A	Injection time	0	10
B	Rail pressure	0	10
C	Turbo pressure	0	10

Once the planning has been carried out, 15 experimental runs are carried out, maintaining the principle of randomness of the samples and placing the highest value of the design, the intermediate and the lowest, for each experiment.

Table 4. Experimental Setup

Modification	Injection by TPS.	Rail Pressure	Turbo Pressure
1	0	5	0
...	...	...	...
15	5	0	10

3 Results and Discussion

3.1 Power and Load Ratio

In this section, using the design of experiments (DOE), each of the response variables and the influence on the process of improving power and reducing polluting emissions are analyzed. The experimentation it was realized at 80% because the max engine power is obtained around this value meanwhile at 100% its produced highest pollution emissions of the engine, Fig. 3 shows the Pareto diagram describing the variables that are statistically significant for the process, in this case all variables below this threshold (red line) are not important in the study.

The factor that directly influences power is the fuel injection according to the Pareto chart, as it is above the significance threshold in all cases, while at 100% load and 2000 rpm the three factors influence, existing a relationship between them. For the case of 80% load at 3000 rpm it shows that the turbo has significance in the process.

With factorial graphs, it is determined how much the percentage increase should be in each one of the zones. For power, it is sought that its factor is maximum, which is why it is increased by 10%, the injection pressure is set at 0%, while the turbo pressure would remain at 3% for the case of power at 80% of load at 2000 rpm. The rest of the optimization values are obtained from the factorial graphs that are observed in Fig. 4.

With the power surface graph, it can be identified that the modified parameters in the engine maps are interacting in all the percentages placed in the programming (Fig. 5).

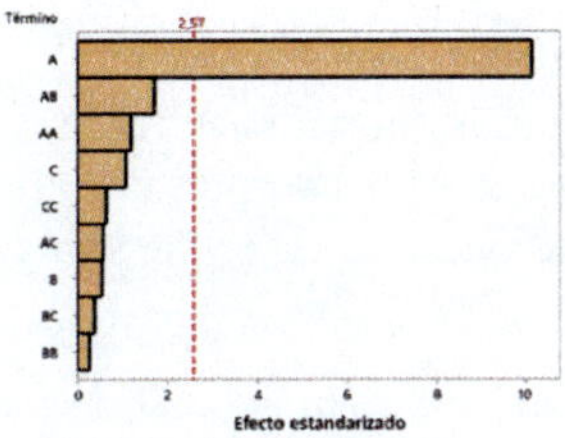

a) Power 80 % load at 2000 rpm

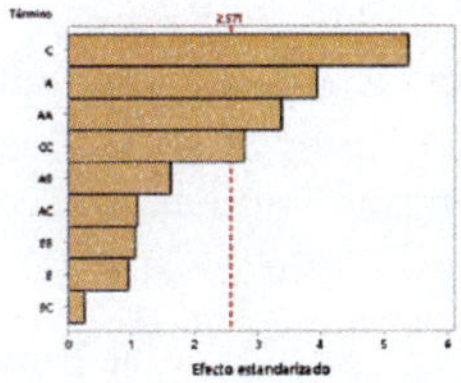

b) Power 100 % load at 2000 rpm

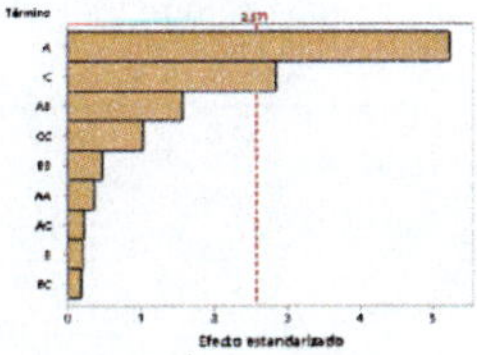

c) Power 80 % load at 3000 rpm

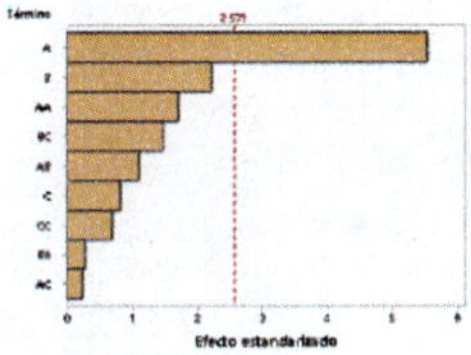

d) Power 100 % load at 3000 rpm

Fig. 3. Pareto chart (Power-Load)

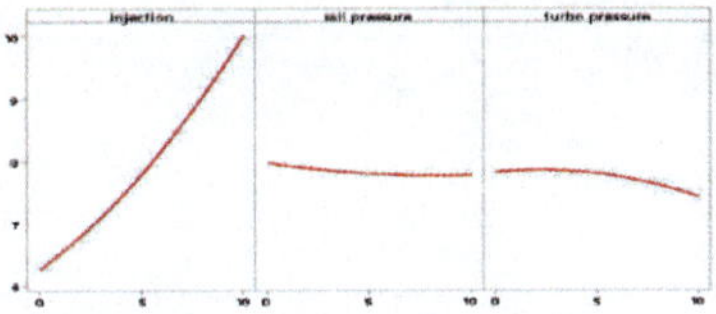

a) Power 80% load at 2000 rpm

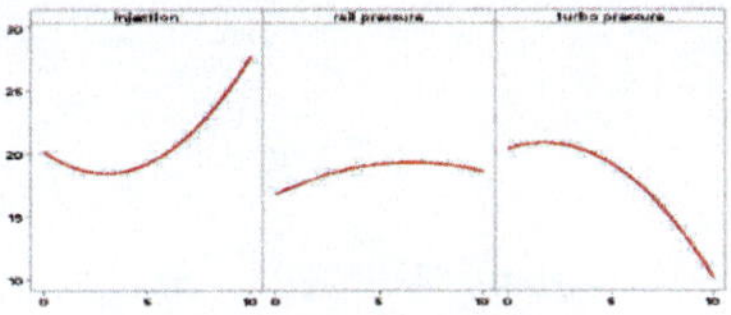

b) Power 100% load at 2000 rpm

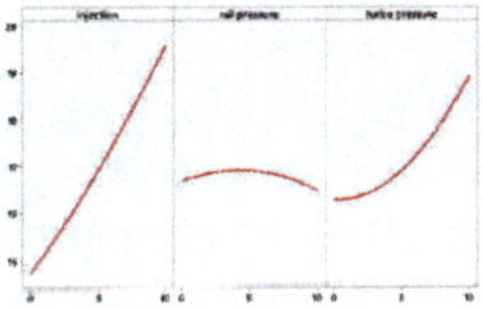

c) Power 80% load at 3000 rpm

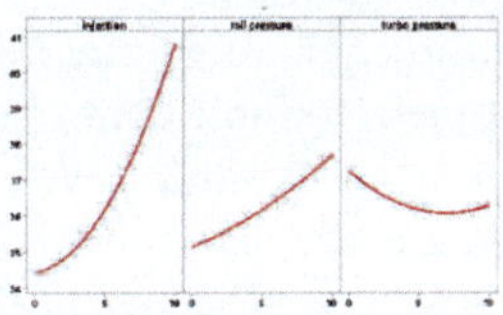

d) Power 100% load at 3000 rpm

Fig. 4. Factorial graphs of power at a given engine speed

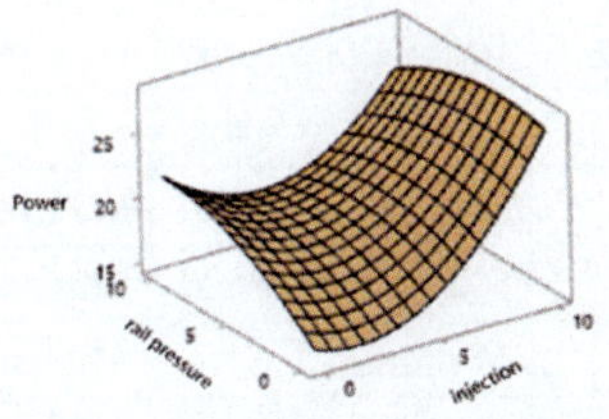

Fig. 5. Power surface graph 100% at 2000 RPM

3.2 Study Variables

For the rest of the variables, a summary table is shown (see Table 4) that shows fuel injection as the most influential factor followed by turbo pressure. It should be noted that factorial and surface studies were carried out on each of the study variables.

Table 5. Results of the Pareto chart for each of the analysis cases.

Variable	80% load at 2000 rpm			100% load at 2000 rpm			80% load at 3000 rpm			100% load at 3000 rpm		
	Influencing factors											
CO	A	–	–	A	–	C	A	AB	C	C	AC	–
Engine torque	AC	–	–	AA	–	C	–	V	C	–	–	–
HC	–	–	C	–	–	–	AA	AC	C	–	–	C
NO_X	–	–	–	–	–	–	–	–	-	–	–	–
Fuel quantity	–	AB	–	AA	–	C	–	–	–	v	–	–
CO_2	A		–	A	–	CC	A	–	–	A	–	–

3.3 Results Optimization

With the previously obtained results, the following model is developed to improve the performance of the engine that can be seen in Table 5. The obtained data are loaded into the maps of the electronic control unit of the vehicle, reaching the results shown in Table 6. The optimization values have been obtained from the percentage increase of the DOE experimental model by modifying the base maps of the reprogrammable module to the study parameters as shown in Table 7.

Table 6. Results optimization objective

Regime				
	80%–2000 rpm	100%–2000 rpm	80%–3000 rpm	100%–3000 rpm
Parameter	Objective			
Torque	Maximum	Maximum	Maximum	Maximum
PM	Minimum	Minimum	Minimum	Minimum
HC	Minimum	Minimum	Minimum	Minimum
CO	Minimum	Minimum	Minimum	Minimum
Tractive force	Maximum	Maximum	Maximum	Maximum
Power	Maximum	Maximum	Maximum	Maximum

Table 7. Values loaded into the control unit.

Optimization				
	80%–2000 rpm	100%–2000 rpm	80%–3000 rpm	100%–3000 rpm
Parameter	Value %			
Injection	10	10	10	10
Rail pressure	4	10	6.26	7.87
Turbo pressure	4.24	10	10	10

3.4 Comparison of the Estimated Values with Respect to the Simulated Ones

Some of the results show a disparity in the theoretical results with respect to those measured, this is due to the fact that the combustion process when diesel is injected at high pressures inside the combustion chamber does not correspond to a linear process. In addition, there are variables that cannot be controlled during experimentation. Modified the engine maps as can be seen in Fig. 6, an increase of 10HP and 30 Nm is achieved, with a minimum increase in emissions of (NO_X, HC and PM) while others such as (CO, CO_2 and O_2) were stable with respect to the standard values of the vehicle.

The theoretical values are those indicated in the manufacturer's data sheets (Table 8).

Table 8. Results obtained after modifying the engine maps.

Regimen				
	100% load @ 2000 rpm		80% load @ 2000 rpm	
Parameter	Theoric valuc	Real value	Theoric value	Real value
Tractive force	943 N	1375 N	438	538
Power	17.69 kW	26.5 kW	9.51	10.4
CO	0.01%	0.02%	0.002	0.02
CO_2	12.02%	12.3%	9.8	9.9
HC	7.66 ppm	5 ppm	9.35	5
NO_X	426 ppm	570 ppm	426	473
PM	7.4 mg/cm^3	39.42 mg/cm^3	9.42	21.2
Regimen				
	100% load @ 3000 rpm		80% load @ 3000 rpm	
Parameter	Theoric value	Real value	Theoric value	Real value
Tractive force	1439	1438	772.8	821

(*continued*)

Table 8. (*continued*)

Regimen				
	100% load @ 2000 rpm		80% load @ 2000 rpm	
Parameter	Theoric value	Real value	Theoric value	Real value
Power	40.31	41.6	21.4	23.6
CO	0.005	0.02	0.004	0.02
CO_2	11.61	11.6	9.2	9.1
HC	8.49	6	6.04	7
NO_X	477.5	542	356.1	373
PM	49.5	184.4	30.6	100.4

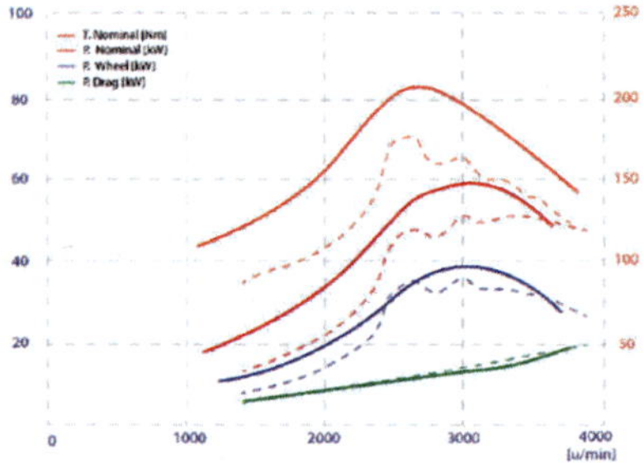

Fig. 6. Improved engine characteristic curves.

4 Conclusions and Future Works

This work shows that the performance of vehicles that operate in high-altitude cities decrease. In the car used for this study, the power measured on the chassis dynamometer was 49.1 kW @ 3280 rpm and the torque of 177.7 Nm @ 2580 rpm, while the power set by the manufacturer is 82 kW and 215 Nm of torque. Factors such as fuel consumption and polluting emissions increase due to the low density of the air, with this work various parameter of the engine are modified to compensate this reduction using a design of experiments (DOE), obtaining power values of 58.1 kW in this work. To compensate for the losses produced by oxygen deficiency, the injection rail pressure, turbocharger pressure and fuel injection are acted upon, the most influential factor for the increase in power followed by the increase in pressure in the intake manifold being the which must be controlled because an exaggerated increase in pressure can permanently compromise the internal components of the engine.

The predictions found by the formulated model were not accurate for all the variables, partly because factors such as nitrous oxides were discarded from the analysis as they did not have direct participation in the engine power. Furthermore, the combustion processes, as they are fluid-dynamic phenomena, are non-linear and are very difficult to predict by these methods.

Future works would aim to improve engine performance using unsupervised learning algorithms, since they do not require the label of variables and would allow a better understanding of the factors that influence engine operation under these atmospheric conditions.

References

1. Wang, X., Ge, Y., Yu, L., Feng, X.: Effects of altitude on the thermal efficiency of a heavy-duty diesel engine. Energy **59**, 543–548 (2013). https://doi.org/10.1016/j.energy.2013.06.050
2. Zhang, H., Shi, L., Deng, K., Liu, S., Yang, Z.: Experiment investigation on the performance and regulation rule of two-stage turbocharged diesel engine for various altitudes operation. Energy **192**, 116653 (2020). https://doi.org/10.1016/j.energy.2019.116653
3. Ramos, Á., García-Contreras, R., Armas, O.: Performance, combustion timing and emissions from a light duty vehicle at different altitudes fueled with animal fat biodiesel, GTL and diesel fuels. Appl. Energy **182**, 507–517 (2016). https://doi.org/10.1016/j.apenergy.2016.08.159
4. Fang, L., Lou, D., Hu, Z., Tan, P.: The emission characteristics of a diesel engine during start-up process at different altitudes. Energies **12**(18), 3556 (2019). https://doi.org/10.3390/en12183556
5. Rivera, N.D., Molina, P.A., Bermeo, A.K., Bermeo, O.E., Figueroa, J.L.: Driving style analysis by studying PID's signals for determination of its influence on pollutant emissions. In: Rocha, Á., López-López, P.C., Salgado-Guerrero, J.P. (eds.) Communication, Smart Technologies and Innovation for Society. SIST, vol. 252, pp. 321–331. Springer, Singapore (2022). https://doi.org/10.1007/978-981-16-4126-8_30
6. Serrano, J.R., Piqueras, P., Sanchis, E.J., Diesel, B.: A modelling tool for engine and exhaust after treatment performance analysis in altitude operation. Results Eng. **4**(August), 100054 (2019). https://doi.org/10.1016/j.rineng.2019.100054
7. Ghazikhani, M., Ebrahim Feyz, M., Mahian, O., Sabazadeh, A.: Effects of altitude on the soot emission and fuel consumption of a light-duty diesel engine. Transport **28**(2), 130–139 (2013). https://doi.org/10.3846/16484142.2013.798743
8. Perez, P.L., Boehman, A.L.: Performance of a single-cylinder diesel engine using oxygen-enriched intake air at simulated high-altitude conditions. Aerosp. Sci. Technol. **14**(2), 83–94 (2010). https://doi.org/10.1016/j.ast.2009.08.001
9. Yang, M., Gu, Y., Deng, K., Yang, Z., Liu, S.: Influence of altitude on two-stage turbocharging system in a heavy-duty diesel engine based on analysis of available flow energy. Appl. Therm. Eng. **129**, 12–21 (2018). https://doi.org/10.1016/j.applthermaleng.2017.09.138
10. He, C., et al.: Science of the total environment emission characteristics of a heavy-duty diesel engine at simulated high altitudes. Sci. Total Environ. **409**(17), 3138–3143 (2011). https://doi.org/10.1016/j.scitotenv.2011.01.029
11. Zhao, W., Yan, J., Gao, S., Lee, T.H., Li, X.: The combustion and emission characteristics of a common-rail diesel engine fueled with diesel, propanol, and pentanol blends under low intake pressures. Fuel **307**(April 2021), 121692 (2022). https://doi.org/10.1016/j.fuel.2021.121692
12. Molina Campoverde, P.A., Rivera Campoverde, N.D., Novillo Quirola, G.P., Bermeo Naula, A.K.: Characterization of braking and clutching events of a vehicle through OBD II signals. In: Botto-Tobar, M., Zamora, W., Larrea Plúa, J., Bazurto Roldan, J., Santamaría Philco, A. (eds.) ICCIS 2020. AISC, vol. 1273, pp. 134–143. Springer, Cham (2021). https://doi.org/10.1007/978-3-030-59194-6_12

13. Molina Campoverde, J.J.: Driving mode estimation model based in machine learning through PID's signals analysis obtained from OBD II. In: Botto-Tobar, M., Zambrano Vizuete, M., Torres-Carrión, P., Montes León, S., Pizarro Vásquez, G., Durakovic, B. (eds.) ICAT 2019. CCIS, vol. 1194, pp. 80–91. Springer, Cham (2020). https://doi.org/10.1007/978-3-030-42520-3_7
14. Molina Campoverde, P.A., Rivera Campoverde, N.D., Morales Espinoza, J.E., Rodriguez Fernandez, G.M., Novillo, G.P.: Influence of the road slope on NOx emissions during start up. Mater. Today Proc. (x) (2021). https://doi.org/10.1016/j.matpr.2021.07.434

FOC Control Design of VSI-SiC Power Converter for an Electric Traction System

David Galindo[1] and Efrén Fernández Palomeque[2(✉)]

[1] University of Azuay, Cuenca, Ecuador
davidgalindo@uazuay.edu.ec
[2] ERGON Center Research University of Azuay, Cuenca, Ecuador
efernandez@uazuay.edu.ec

Abstract. This study presents the design by simulation, the improvement in efficiency of a voltage source inverter (VSI), using silicon carbide (SiC) components, in a permanent magnet synchronous motor (PMSM). Using field-oriented control (FOC) at a switching frequency of 100 kHz. An analysis of efficiency and energy losses is presented, through simulations, analytical methods, device data sheets, to compare them with conventional silicon (Si) devices. For the development of simulations and control design, tools such as MATLAB and Psim are used.

Keywords: SiC devices · VSI topology · FOC controlled

1 Introduction

The rapid depletion of the earth's oil resources demands the development of highly efficient electric vehicles (EV) and hybrid electric vehicles (HEV), the latter combining the technology of internal combustion engines with electric machines and their converters. Electronics [1]. In the search for greater efficiency in electric vehicle propulsion systems, research has focused on the motor control system. The design of these systems consists of the development of DC/AC inverters used in the control of electric motors, the most popular type is PMSM. In the design of these systems the contribution of semiconductors developed with SiC play an important role.

The use of SiC devices presents many benefits explain in [1] which are interesting for the use in power converters for its high efficiency, high switching frequency, high temperature operation and also the reduction of power losses compared with the SiC devices. As shown in the Fig. 1, SiC elements have advantages over Si [2, 3].

The size, weight and power losses of the voltage converters are undoubtedly the most common problems in EV systems. Besides, the problems that can be generated when using a VSI topology based in Silicon IGBT the elements that work at a temperature range not higher than 120 °C and its activation is below 10 kHz. This involves the use of a very robust liquid cooling system that adds more weight to the system [3, 4].

There are several types of topologies of electronic power converters for electric traction systems, the main ones being:

V. Robles-Bykbaev et al. (Eds.): CITIS 2022, LNNS 607, pp. 357–367, 2023.
https://doi.org/10.1007/978-3-031-24327-1_31

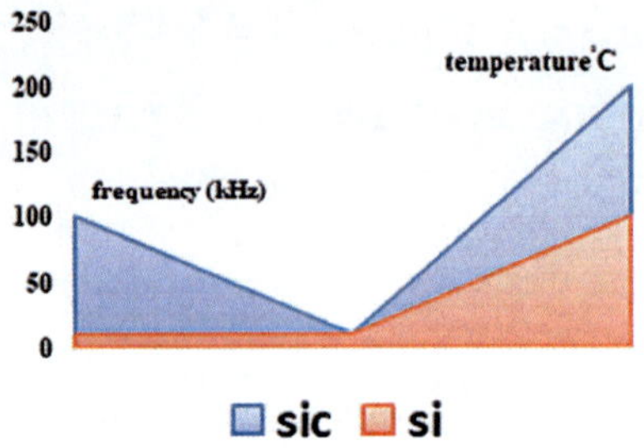

Fig. 1. Comparison of SiC vs Si devices.

VSI: it is the most used topology in hybrids and electric traction systems due to the type of voltage storage in them. The VSI uses a capacitor on the DC bus in order to filter the input current, maintaining a constant voltage level [2].

CSI: It has an inductor component that is like energy storage, it has the ability to handle high voltage, the CSI have a short circuit protection and sinusoidal output voltage by effect of AC output filter capacitors [4].

ZSI: These topologies are efficiency and of low cost, in comparison with VSI and CSI topologies, uses inductors and capacitors when working at high frequencies and has high reliability with respect to lower EMI emissions [5].

This paper contributes to the search for solutions presented in VSI topologies with conventional silicon devices, the use of SiC devices allows improving the operating conditions of this type of converters.

In this work, the implementation in simulation of a FOC control in a PMSM is presented, using a VSI topology based in SiC elements at high frequency. The work is organized as follows: Sect. 2 describes the design of the VSI topology with the high frequency silicon carbide devices proposed for the study and the design of the modulation algorithm, FOC control and characteristics of PMSM. Section 3 presents the results of the simulations. The Sect. 4 presents the discussion and finally the Sect. 5 present the conclusions of the work.

2 Method

2.1 VSI Topology

These topologies provide a three-phase voltage source, where the amplitude, phase and frequency of the voltages are controlled.

The three-phase VSI topology is shown in Fig. 2. Transistors at any position of the inverter (T1 and T2, T3 and T4, or T5 and T6) cannot be connected simultaneously, because can generate in a short circuit to the inverter.

2.2 PWM Modulation Technique

The PWM modulation technique is used to generate trigger signals for the control of power converters, these signals were originally generated using electronic circuits. Currently, microprocessors and microcontrollers have dominated the design of control signal generators.

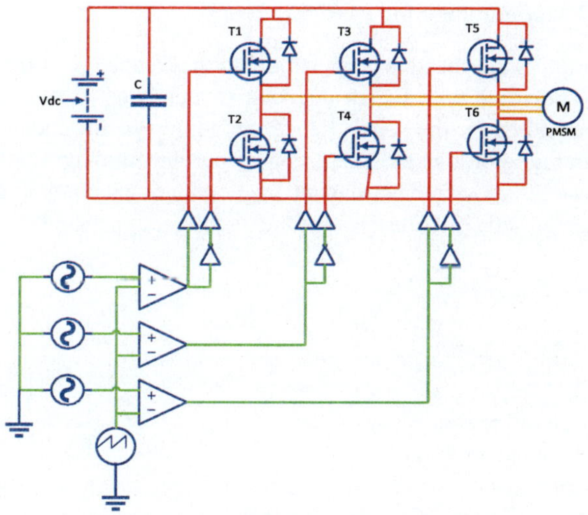

Fig. 2. VSI topology with SiC devices.

This technique is used due to its ability to control voltage, frequency and harmonic content in a single power stage. The most efficient method of controlling the gain (hence the output voltage) is to incorporate (PWM) control. Commonly used techniques are:

- Sinusoidal PWM Modulation (SPWM).
- Space Vector Based PWM Modulation (SVPWM)

The modulation technique used in this work is SPWM. The main objective of this modulation is to obtain current waveforms, this characteristic also allows the reduction of the line voltage and low harmonics distortion [7]. The SPWM modulation technique (Fig. 3) consists of comparing a high frequency wave or carrier signal to a low frequency signal known as the reference modulation signal [8] determined by the width of the inverter output voltage pulses.

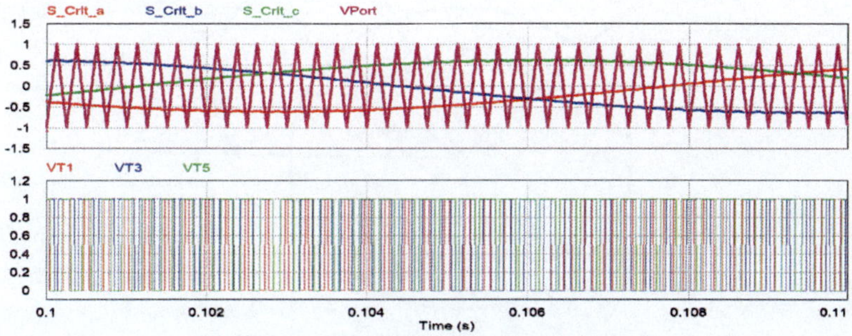

Fig. 3. SPWM modulation technique.

2.3 Design and Methodology for FOC Control

In synchronous motors the speed is proportional to the frequency of the voltage applied to its stator [8]. The speed with which the rotor rotates is equal to that of the rotating magnetic field generated by the stator, hence the name synchronous. It does not have excitation coils in the rotor, these are replaced by permanent magnets [9].

The three-phase AC-powered permanent magnet motor generates a rotating magnetic field of constant magnitude [9]. The speed of the rotating magnetic field is given by Eq. (1) [9].

$$n_s = \frac{120 * f}{P} \tag{1}$$

where:

n_s = Magnetic field speed (rpm)
f = Electrical frequency (Hz)
P = Number of poles.

The speed depends on the frequency of the voltage applied to the stator windings and the number of poles, and not on the mechanical load [9]. As the number of poles is fixed, if the same frequency of the armature voltage is maintained, the speed of the motor will remain constant with any load up to the maximum power [9].

2.4 Modeling of the Permanent Magnet Synchronous Motor

It is complicated to work with the model of a referenced PMSM in a three-phase system, it is simplified by applying the Clark transformation to a two-axis reference system [10] obtaining Eq. 2. The two transformations used are α and β, being these orthogonal each. Then a Park transformation associated with the rotor is applied, it is called dq and it is a reference system of two axes that rotate at the speed of the rotor, and Eq. 3 [10] is obtained. Figure 4 shows the relationship of axes in dq.

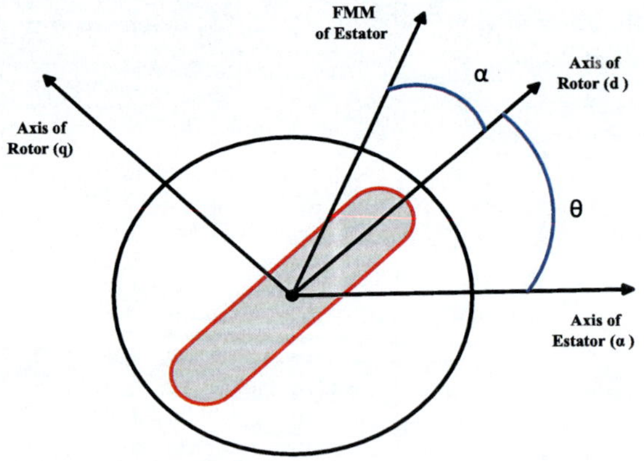

Fig. 4. Representation of the PMSM motor in the dq axes.

$$\begin{bmatrix} V\alpha \\ \overline{V\beta} \end{bmatrix} = \frac{2}{3}\begin{bmatrix} 1 & -\frac{1}{2} & -\frac{1}{2} \\ 0 & \frac{\sqrt{3}}{2} & -\frac{\sqrt{3}}{2} \end{bmatrix}\begin{bmatrix} Va \\ Vb \\ Vc \end{bmatrix} \tag{2}$$

$$\begin{bmatrix} Vd \\ \overline{Vq} \end{bmatrix} = \begin{bmatrix} \cos(\theta) & sin(\theta) \\ -sin(\theta) & cos(\theta) \end{bmatrix}\begin{bmatrix} V\alpha \\ V\beta \end{bmatrix} \tag{3}$$

where Vd and Vq are voltages on the reference axis dq. id and iq are currents in the reference axis dq. Va, Vb, and Vc are three-phase line voltages. θ angular displacement of the stator and rotor axes.

The equations of SMPMSM. In the stator-fixed β α, reference frame, the voltage equation of a PMSM is given by (4):

$$\begin{bmatrix} u_\alpha \\ u\beta \end{bmatrix} = R\begin{bmatrix} i_\alpha \\ i_\beta \end{bmatrix} + \begin{bmatrix} \psi_\alpha \\ \psi_\beta \end{bmatrix} \tag{4}$$

The voltage equations in the rotor-fixed reference frame are given by:

$$U_d = L_d i_d + Ri_d - \omega_{RS} L_d i_q \tag{5}$$

$$U_q = L_q i_q + Ri_q + \omega_{RS} L_q i_d + \omega_{RS} \psi_{PM} \tag{6}$$

The electromagnetic torque is given by:

$$T_e = \frac{3}{2}\frac{P}{2}\left(\lambda_d i_q - \lambda_q i_d\right) = \frac{3}{2}\frac{P}{2}\left(\lambda_m i_q + \left(L_d - L_q\right).i_q i_d\right) \tag{7}$$

$$P_{em} = \omega_{rm} T_e = \frac{3}{2}\omega_r\left(\lambda_d i_q - \lambda_q i_d\right) \tag{8}$$

The electromechanical power:

$$\omega_r = \frac{P}{2}\omega_{rm} \tag{9}$$

where P: Poles No; ω_{rm}: Rotor Mechanical velocity. The general mechanical equation for the motor is:

$$T_e = T_1 + T_d + B\omega_{rm} + J\rho\omega_{rm} \tag{10}$$

B: Viscous frictions coefficient; J: Inertia of shaft and load system; T_d: Dry friction; T_l: Load torque.

2.5 FOC Control Design Methodology

The Fig. 5 shows the scheme and method for the design of the FOC controller, the modules that make it up are: Three PI controllers, the inverter, the Clarke and Park transforms and their inverses, the position and speed meter and the three phase inverter.

The methodology consists of setting a speed reference value, if the control responds to the desired speed, it is determined that the control is valid and generates the signal outputs for the inverter drive; otherwise it enters the speed control block. Then it is verified if the currents id and iq are equal to the references, if they are valid, they are transformed to the alpha and beta current axes and generate the signal outputs for the inverter activation. But if they are not, they enter the current control block to later be transformed into alpha and beta.

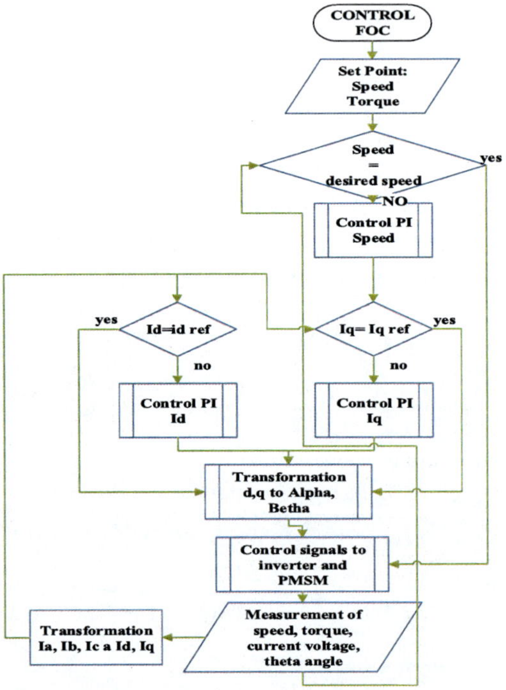

Fig. 5. Design method for FOC controller.

The FOC controller uses 2 Proportional Integral (PI) type controllers and a PID, the PID converts the speed variations to the desired current iq (IQREF), the other controller controls the current iq and its output is the voltage Vq that is given to it. Apply to the engine. The other PI controller treats the current id = 0, its output controls the voltage Vd [12] (Fig. 6).

2.6 Obtaining the Transfer Functions and Constants KP and KI

To obtain the current and speed control transfer equations, the motor values were placed and the transfer function of each of the plant stages was obtained, the Linear Analysis tool of the Control Design application, from Simulink, was used.

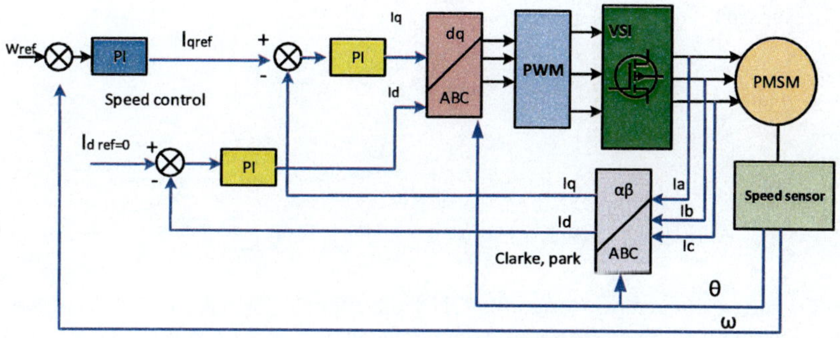

Fig. 6. Speed control system, Iq, Id, for PMSM motor.

W speed transfer function:

$$y1 = \frac{7335\,s^2 + 4.362e06\,s + 3.101e06}{s^4 + 338.4\,s^3 + 1.155e05\,s^2 + 4.362e06\,s + 3.101e06} \tag{11}$$

Iq transfer function.

$$y2 = \frac{0.1176\,s^3 + 69.88\,s^2}{s^4 + 338.4\,s^3 + 1.155e05\,s^2 + 4.362e06\,s + 3.101e06} \tag{12}$$

Id transfer function.

$$y3 = \frac{4555\,s + 2.273e06}{s^2 + 4893\,s + 2.273e06} \tag{13}$$

Once the transfer functions have been obtained, and the curves have been smoothed using the Simulink PID Tuning tool, the values of the constants KP and KI are obtained for the speed controls and currents id, iq, which are shown in Table 1. These transfer functions represent the relationship of the response of the modeled system with respect to the input signal. In this case, the input signals are speed, currents iq and id.

Table 1. Controller values Kp, Ki.

Control	Kp	Ki
Speed	2	2.7
Id	5.34	1214
Iq	0.008	3.27

2.7 Power Losses

The power losses in the VSI power converter can be analyzed in terms of conduction and switching losses. The conduction losses for the Mosfet and the diode for either SiC or Si devices can be determined by the following equations.

Mosfet losses:

$$Pcon = \left(R_{on} * i^2 (\frac{1}{8} + \frac{mcos\Phi}{3\pi} \right) \tag{14}$$

Diode losses

$$Pcon = \frac{1}{2}\left(V_D * \frac{i}{\pi} * R_{on}\frac{i^2}{4} \right) - mcos\varphi V_D \frac{i}{8} + R_{on}\frac{i}{\frac{2}{3}\pi} \tag{15}$$

where: i is the peak current, Ron is the ON resistance, V_D is the diode drain voltage, m is the modulation index and cosφ is the power factor.

The switching losses are determined by the following equations.

$$P_{sw} = f_{sw}(E_{ON} + E_{OFF})\frac{Vi}{V_{Nom^i Nom}} \tag{16}$$

$$P_{sw} = f_{sw}Err\frac{Vi}{V_{Nom^i Nom}} \tag{17}$$

where:

E_{ON} are the switching losses for activation of the Mosfet and E_{OFF} are the switching losses for turning off the Mosfet.

3 Results

In the first part, the results of VSI topology with SiC devices are shown, conduction losses and commutation losses are presented. In the second part, the results of the simulations with Si devices are shown. Finally, a comparison of the use of SiC and Si elements is presented.

The parameters for the high frequency simulation and the characteristics of the devices are presented in Table 2.

Figure 7 shows the results obtained in the PSIM simulations of the VSI and the FOC control at high switching frequency.

The Fig. 8 shows the total power losses in PSIM of SiC and Si devices used in the VSI topology at high and low frequency, respectively.

4 Discussion

In summary, the use of this methodology allows to work optimally for the use of SiC devices in VSI inverters. The design of the controllers and obtaining the transfer functions is very important for a correct operation of the electric motor. The results obtained show the effectiveness of the design of the controller if the values of the constants are not well adjusted, the controller does not work.

Table 2. Simulation parameters

Parameters	
V battery	250 v
Switching Frequency	100 kHz
Speed of reference	500 rpm
Rs stator	0.2425 Ω
Ls stator	0.00139 H
Rr rotor	0.855 Ω
Lr rotor	0.00074 H
Lm (magnetization)	0.041 H
No. Polos	4
Moment of inertia	0.002 $kg.m^2$
Torque	5 Nm

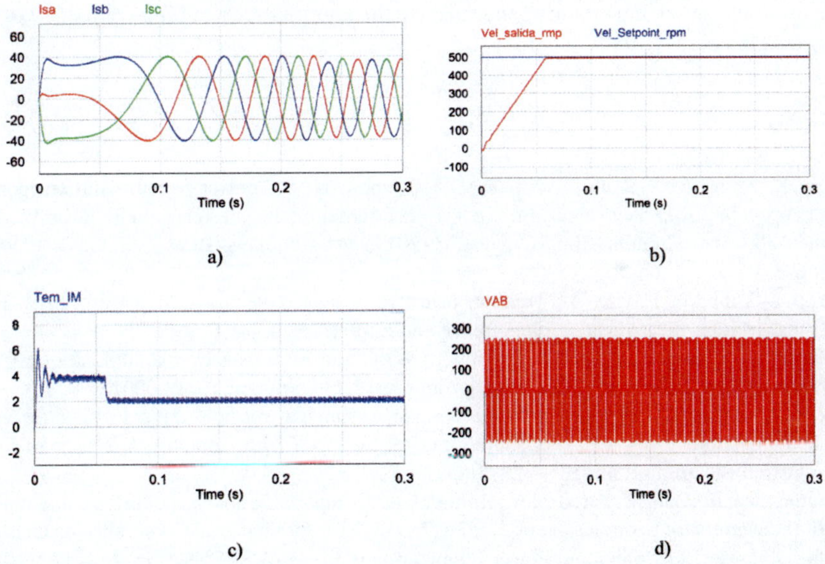

Fig. 7. Results in PSIM. a) Current of stator, b) speed control response, c) torque. d) phase voltage.

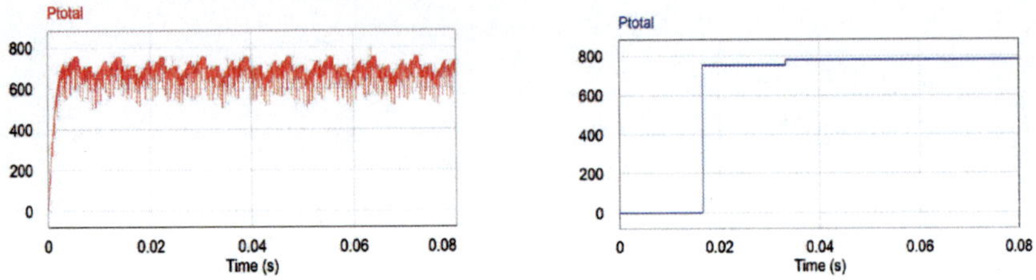

a) b)

Fig. 8. Total power losses. a) VSI SiC topology, b) VSI Si topology.

5 Conclusions

The increase in the switching frequency with the use of SiC devices reduces the volume and size of the passive elements of the system, since less current consumption is required as there are fewer losses in the switching of SiC devices.

The FOC control allows to have a stable management of the reference speed of the motor at 500 rpm, maintaining the set torque of 2 Nm with a stabilization time of 0.1 s, while eliminating the harmonics generated by the commutation at high frequency of the converter elements.

References

1. Ino, K., Miura, M., Nakano, Y., Aketa, M., Kawamoto, N.: SiC power device evolution opening a new era in power electronics. In: 2019 IEEE International Conference on Electron Devices and Solid-State Circuits (EDSSC), pp. 1–3 (2019). https://doi.org/10.1109/EDSSC.2019.8754464
2. Ye, H., Yang, Y., Emadi, A.: Traction inverters in hybrid electric vehicles. In: 2012 IEEE Transportation Electrification Conference and Expo (ITEC), pp. 1–6 (2012)
3. Zhang, L., Zheng, Z., Lou, X.: A review of WBG and Si devices hybrid applications. Chin. J. Electr. Eng. **7**(2), 1–20 (2021). https://doi.org/10.23919/CJEE.2021.000012
4. Wu, Z., Su, G.-J.: High-performance permanent magnet machine drive for electric vehicle applications using a current source inverter. In: 2008 34th Annual Conference of IEEE Industrial Electronics, pp. 2812–2817 (2008)
5. Fernández, E., Sala, V., Paredes, A., Romeral, L.: Comparative analysis of impedance source-SiC converters for traction systems. In: 2017 11th IEEE International Conference on Compatibility, Power Electronics and Power Engineering (CPE-POWERENG), pp. 464–470 (2017). https://doi.org/10.1109/CPE.2017.7915216
6. Hegazy, O., Barrero, R., Van Mierlo, J., Lataire, P., Omar, N., Coosemans, T.: An advanced power electronics interface for electric vehicles applications. IEEE Trans. Power Electron. **28**(12), 5508–5521 (2013)
7. Rashid, M.H.: Power Electronics Handbook. Academic Press, Cambridge (2001)
8. Da Silva, E.R.C., Dos Santos, E.C., Jacobina, C.B.: Pulsewidth modulation strategies. IEEE Ind. Electron. Mag. **5**(2), 37–45 (2011)
9. Patel, M., Patel, A.: Use of PWM techniques for power quality improvement. Int. J. Recent Trends Eng. **1**(4), 99–102 (2009)

10. Fitzgerald, A.E., Kingsley Jr., C., Umans, S.D.: Electric Machinery. Mcgraw-Hill Series in Electrical Engineering. Power and Energy. McGraw Hill Higher Education (2002)
11. Durán, M., Gerardo, G., Caratozzolo, P.: Diseño de controladores no lineales basados en pasividad de motores síncronos de imanes permanentes y con devanado de campo, pp. 371–376 (2004)
12. Hernández, B.L., Javier, J., San, V., De, R.M.J., Leyva, H.R.: Simulación en simulink-PSIM del control por pasividad de un motor síncrono de imanes permanentes (PMSM), vol. 1, no. October 2013, pp. 2–7 (2016)
13. Gómez, A.A., Mejía, A.E., Torres, C.A.: Control vectorial de la máquina de inducción. Sci. Tech. 3(43), 55–60 (2009)
14. Ramírez Leyva, H., Peralta Sanchez, E.: Implementación del Control de Campo Orientado de un Motor de Imanes Permanentes con base en un Controlador Digital de Señales, pp. 1–10 (2013)

A Toyota Corolla Hybrid GUI Based on Simulink and LabVIEW for Live-Data Acquisition and Monitoring System

Ángel López[1](✉), Diego Rojas[1], and Efrén Fernández[2]

[1] University of Azuay, Cuenca, Ecuador
alopezm@es.uazuay.edu.ec, drojas@uazuay.edu.ec
[2] ERGON Center Research University of Azuay, Cuenca, Ecuador
efernandez@uazuay.edu.ec

Abstract. Hybrid and electric vehicles make for an environment friendly alternative to conventional vehicles, following worldwide trends of ecological transportation, less energy-consuming systems and pollution reduction overall. From universities and academy in general a new imperative requirement for learning and training of students and technical staff about new electric drivetrain systems has become the most efficient tool to adapt these systems to our context, understanding its inner workings and developing new methods to exploit and improve performance, therefore creating a quicker way for new users to adopt and transition to an alternative-fuel vehicle. This research shows a new approach for a LabVIEW-based easy-to-use data acquisition system with a built-in GUI for live monitoring and data storage for further simulation system via Simulink, giving teachers and students the capability of observing and analyzing the actual behavior of the different physical magnitudes from a vehicle and setting a benchmark for advanced research efforts of electric and hybrid drivetrains in Ecuador.

Keywords: CAN bus · Frame ID · PIDs · GUI · Simulator · Hybrid vehicle

1 Introduction

1.1 Overview

In recent years, Europe, US, and vehicle manufactures such as Tesla and Toyota have summed efforts in great lengths to start the transition from conventional fossil fuel vehicles to electric and hybrid vehicles respectively, and more countries around the world seem to follow this more friendly-environment initiative [1–3].

Universities and academia are fully aware of the imperative demand from the new generation of students and technical staff for understanding this technology to adapt it for the reality of every country [4]. We know so far that electric and hybrid vehicles are more efficient than conventional vehicles, however we lack a proper assessment strategy to estimate the real efficiency applied in the cities of any country [5, 6].

V. Robles-Bykbaev et al. (Eds.): CITIS 2022, LNNS 607, pp. 368–380, 2023.
https://doi.org/10.1007/978-3-031-24327-1_32

Hybrid and electric vehicles employ a variety of electronic units for controlling and monitor vital performance roles such as: internal combustion engine and transmission management, body control modules, safety features (ABS, SRS) in addition to the electric motor unit, inverter, battery monitoring system, voltage converters and so on, all of them work with CAN Bus, handling different data transmission rates [7, 8].

Academic research efforts require for an advanced and robust system capable of live-data monitoring where students can observe the different physical magnitudes from the inner workings of the hybrid vehicle in motion. In addition to be capable of saving this data in any removable device for further analysis to make new estimations and deepen the scope of the research [9].

Most of the devices in the market are focused only in diagnostics to solve failure problems in-vehicle control units, which only show the parameter identification or PID and its magnitude [10].

This paper outlines a graphical user interface GUI for live data acquisition and monitoring system plugged in to the data link connector DLC via CAN Bus to examine and explore how data transmission takes place in the different electronic units of a hybrid vehicle and the inherent PIDs involving the different physical magnitudes at play, making use of a National Instruments NI 9862 data acquisition card with LabVIEW and Simulink.

The main target of this system and proposed methodology is to increase the academic level of training and teaching for new students and technicians focused on advanced monitoring and diagnostic systems for hybrid vehicle technology. All this effort to aid in the long-term goal of expanding the transition from conventional gasoline vehicles to alternative, eco-efficient drivetrains in the region. The paper presented the next sections. Section 2 describe the CAN bus parameters. Section 3 present methodology used. Section 4 presents the results. Finally, in the Sect. 5 and 6 presents the discussion and conclusion of the paper.

2 CAN Bus in Hybrid Vehicles

2.1 PID – Parameter Identification

A PID or Parameter Identification is a query for the values of indicators of different sensors and actuators of the vehicle, therefore each PID will show a corresponding physical magnitude measurement necessary for the proper working of the vehicle. For instance, the engine coolant temperature sensor data is stored in the PID #001, this means that whenever a request on the network from a scan tool with this respective address and meets the expected response node, the PID #001 shows the stored data on the engine coolant temperature sensor.

Therefore, all relevant data for physical magnitudes such as, pressure, temperature, speed, voltage, current and so on are stored in a distinct and unique PID.

2.2 Reverse Engineering

CAN Bus is the main standard for electronic module communication and it has been proven how efficient and reliable it is to ensure the operability of any embedded system, even with the ever growing addition of electronic units to fulfill specific tasks [11]. CAN Bus speeds have increased since its first introduction in the 80s and the subsequent ISO 11898 standard for CAN 2.0 A in 1995 and the extended CAN 2.0 B, now with the emerging prevalence of CAN FD [12].

Hybrid vehicle topology comprises of two main subsystems a CAN 2.0 A or B network with speeds from 100 kbps up to 1 Mbps for electric motor control and high-power applications and also a CAN 2.0 A network with speeds up to 500 kbps for auxiliary functions, such as gasoline engine control, transmission, security systems and others. Data rate speed is different between manufacturers and vehicle models [13, 14].

Access to the CAN Bus data can be obtained following the reverse engineering process proposed in [7], where in summary the authors say live data acquisition from CAN is relatively simple but rather time consuming and extensive. It is fundamental for the students to understand how CAN bus communication works and the features of the vehicle used for this purpose, rather than type of CAN, speed and standard employs the network. Figure 1 summarizes the process of reverse engineering.

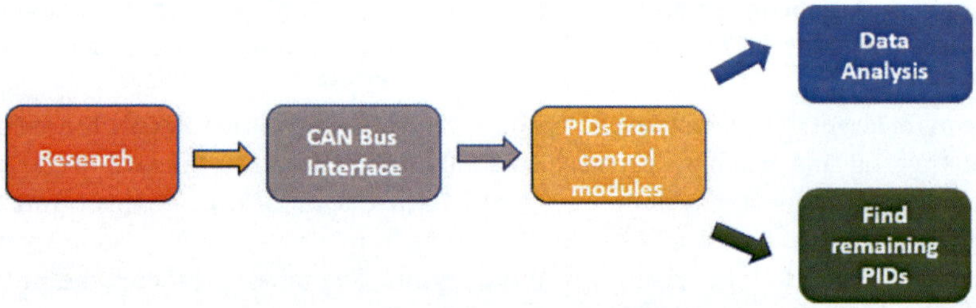

Fig. 1. Reverse engineering workflow [7]

Research: Find specialized technical information and specifications of the electronic control modules of the vehicle to use, therefore relevant information for all subsystems.

CAN Bus interface: Connection to the CAN Bus through DLC and protocol support to establish communication between the interface and the vehicle, a quick and useful reference is the standard ISO 1575-4.

PIDs from control modules: By data interpretation and filtering, find out how PIDs show info in the CAN Bus and temporarily assign it to a relevant control module, ECU, TCM, BCM and others.

Data analysis: Find and crossmatch the relationship between the PID and how it varies with time and what magnitude matches the expected physical behavior, for instance voltage, temperature, pressure. This is the most time-consuming part of the process that requires maximum skills for pattern recognition and previous knowledge of how all different sensors and actuators work in the vehicle. After this step, most of the PIDs are properly identified.

Find remaining PIDs: After the initial identification, it is necessary to relate one or two more bytes to find any missing PID and making mathematical adjustments with simple equations to match the expected magnitude [7].

3 Methodology

3.1 Hardware

The National Instruments NI 9862 flexible data rate CAN bus card up to 1 Mbps is connected to the vehicle DLC, to set first communications, a 500 kbaud data transmission rate is selected. It is evident at this point that all functionality of the software will be managed with LabVIEW and features as the next list shows: NI Manager, NI XNET, NI Max and NI Bus monitor to show CAN bus data as shown in Fig. 2:

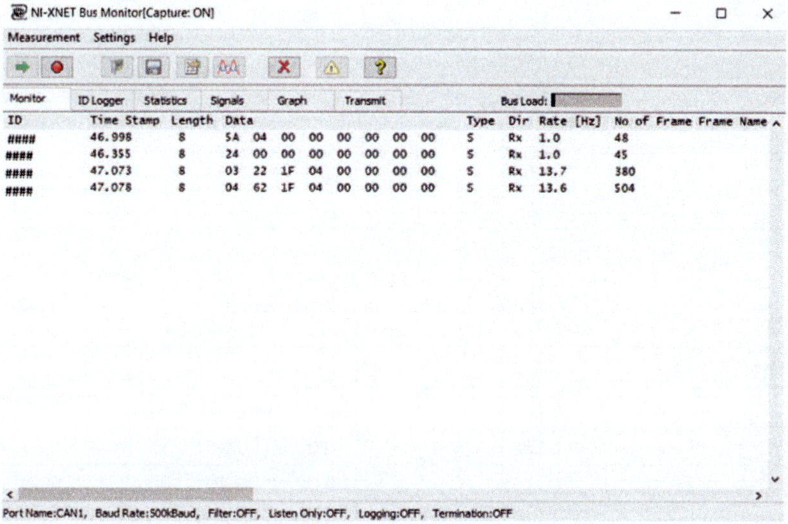

Fig. 2. NI Bus Monitor acquiring data.

NI Bus Monitor saves all data captured in the network with all relevant Frame IDs and information bytes in a.txt file, which can be imported to Microsoft Excel and handled easily. Using Microsoft Excel as the primary data editor is one the most simplistic and easy-to-use methods to explain students how CAN Bus works and how PIDs are stored in the different transmission bytes.

3.2 Vehicle

Relevant information for the vehicle used for this research is shown in Table 1.

3.3 Comparison and Magnitude Adjustment

After the initial filtering and sorting process in Microsoft Excel, the most extensive and time-consuming part begins with the identification process to find PIDs, Table 2 shows the data layout:

Table 1. Vehicle specifications and relevant info

	Details
Combustion engine	
Model	2ZR-FXE
Type	DOHC, 4 in-line cylinder, 16 valves per cylinder
Fuel	Unleaded 91 RON
Capacity	43 lt
Electric motor	
Type	AC synchronous motor
Power	53 kW
Torque	163 Nm
HV battery	
Voltage	201.6 V
Capacity	6.5 Ah

Table 2. Text file imported to Excel

Timestamp	ID	Payload
0	001	00 00 54 FF FF 06 96
0.000246	002	40 00 00 00 00 00 5E 4A
0.000731	003	03 00 10 00 70 00 00 47
0.001956	004	00 00 00 00 00 00 00 3A
0.003111	005	00 00 10 00 00 00 00 41
0.003317	006	3D 02 00 80 00 04
0.003565	007	00 02 00 00 00 00 00 55
0.003809	008	00 00 5E 40 00 00 00 4C
0.004437	009	00 00 00 00 00 00 32
0.005133	010	00 00 00 00 00 00 2D
0.005373	011	04 62 1F 5B 6E 00 00 00

The Timestamp column shows data stream every unit of time, available in microseconds, the ID column is the Frame ID which in a few words is the arbitration ID, part of the CAN message frame, when invoked, it will show the data bytes from 0 to 7 where a PID value is stored, this key explanation clears a common misunderstanding to fully comprehend what the researcher is seeing in the data and avoid confusion between what a Frame ID and PID really are. The third and last column is the Payload showing data bytes from 0 to 7 (8 bytes).

The next step is to spread the Payload column in each separate byte as Fig. 3 shows:

	A	B	C	D	E	F	G	H	I	J
1	Timestan	ID	Payload	Payload	Payload	Payload	Payload	Payload	Payload	Payload
2	0	001	19	4	2	CC	2	41	0	9
3	0.00019	009	0	0	0	0	4C			
4	0.000422	005	4	0	0	0	0	0	0	
5	0.00067	004	0	0	0	0	0	0	0	3A
6	0.00103	006	8	68	0	0	0	0	0	4C
7	0.002029	004	0	10	0	8	0	32	B2	2C
8	0.002405	009	51	2	0	80	1	19		
9	0.002649	111	40	0	0	0	0	0	34	20
10	0.003104	006	0	0	0	0	0	0	32	
11	0.003376	010	4	0	10	0	50	0	0	28
12	0.005578	056	F2	32	F2	94	F2	94	FF	65
13	0.005788	042	F2	94	F2	94	F2	94	C8	
14	0.006026	056	FF	7F	FF	0	0	F2	94	3B
15	0.006274	074	0	0	0	0	0	0	0	1

Fig. 3. Payload column showing 8 bytes of data.

At this stage in the process, it is very important to state and make clear that the Frame ID is shown in decimal form whereas the Payload Bytes are shown in hexadecimal, the next step is to select a specific Frame ID to observe how it changes byte values over time as shown in Fig. 3. It is necessary to convert these 8 columns worth of data to decimal to create a graph for each byte column, this means that for instance, a network with 100 Frame IDs may store around 800 bytes to be studied. This graph can now be compared with another graph obtained in parallel to the data acquisition and card and set a benchmark for every PID to be found in the 8 data columns from each Frame ID.

	A	B	C	D	E	F	G	H	I	J
1	Timestan	ID	Payload	Payload	Payload	Payload	Payload	Payload	Payload	Payload
2	0	009	51	2	0	80	1	19	0	49
3	0.130019	009	20	A4	0	0B	84	4E	83	B4
4	0.230422	009	40	0	0	0	0	0	34	20
5	0.33067	009	4	0	10	0	50	0	0	28
6	0.43103	009	0	0	54	0	0	6	98	5
7	0.532029	009	11	0	0	0	0	1	0	60
8	0.632405	009	0	0	0	0	0	0	0	3A
9	0.732649	009	0	0	0	0	8D	0	0	49
10	0.833104	009	A7	FF	0A	0	ED	5	0	49
11	0.933376	009	0	0	0	0	19	17	5	5
12	1.035578	009	0	0	34	40	0	0	0	22
13	1.135788	009	0	10	0	8	0	32	B2	2C
14	1.236026	009	51	2	0	80	1	19	0	49
15	1.336274	009	0	0	0	0	0	0	32	5

Fig. 4. Selection of specific Frame ID.

Another relevant aspect to be considered is the Frame ID time rate variation, which is obtained subtracting a time point from the one before it, as Fig. 4 shows a 0.10 s variation, this is an indicator of how quickly or delayed a Frame ID is detected in the network and therefore PIDs of high priority from drivetrain control units vary faster than PIDs that do not require such fast responses and may take longer time given the necessary communication request for instance temperature readings.

Whenever a PID graph matches the base reference graph, it is determined that the selected Frame ID along with its byte column may be the expected PID, at this instance as Fig. 5 shows, the only difference is the amplitude scale, where a simple mathematical equation solves this problem. For instance, in Fig. 5 the first left graph shows the reference for Hybrid Battery Voltage and the B0 and B1 graphs from the Frame ID # 001 match the curve pattern, with a simple sum of both bytes the PID is finally adjusted to physical units for the observer to interpret and analyze.

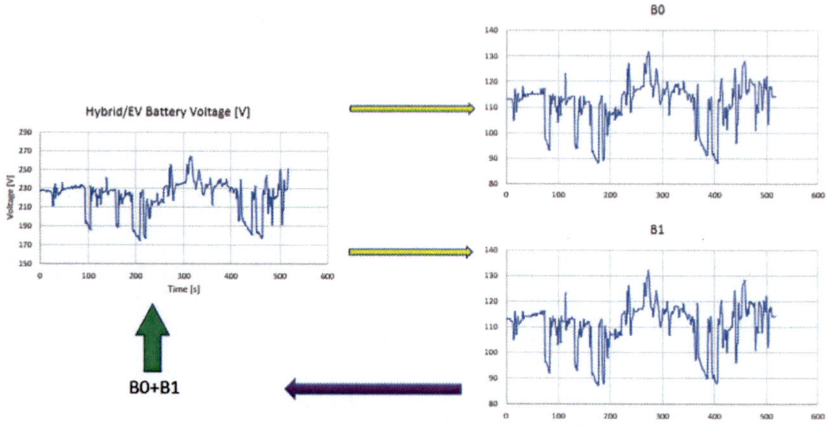

Fig. 5. Graph comparison and mathematical adjustment.

3.4 GUI in LabVIEW

The versatile nature of LabVIEW comes as the most attractive feature of the setup for a system of this nature, once the researcher has a list with identified Frame IDs and PIDs with the mathematical adjustment for magnitudes, Fig. 6 shows the block diagram.

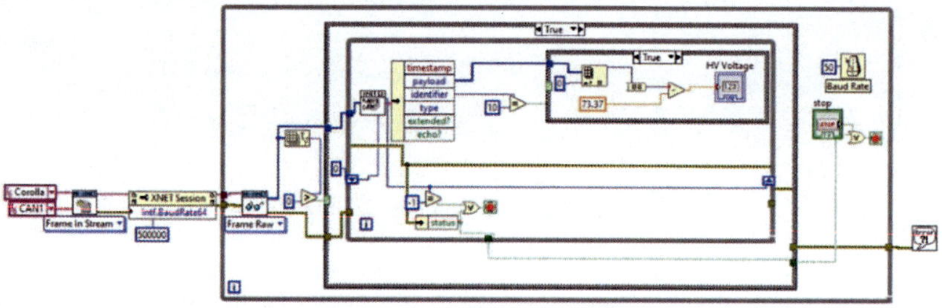

Fig. 6. LabVIEW block diagram.

A brief description for this program, shows data acquisition card setup outside the while loop, to acknowledge the presence of the card in a USB port in a laptop, CAN connection to use and data transmission speed set to 500 kbaud. Then, a new NI – XNET session is created in order to enable data reading with Frame Raw block to arrange data to be designated exactly as the layout in the Excel file, where Payload contains 8 bytes of data from B0 to B7 and for last, the identifier which is the corresponding Frame ID. In a simple logic comparator the programs sends a request to acknowledge the specified Frame ID and the byte position, right afterwards a quick conversion to hexadecimal with the mathematical adjustment formula to finally show the PID magnitude in a numeric gauge or indicator using a sample rate at the moment of 50 ms. This base program can be extended to any number of PIDs required to make a dashboard for user observation, in this case, for the students and researcher using this method. Figure 7 shows the final dashboard:

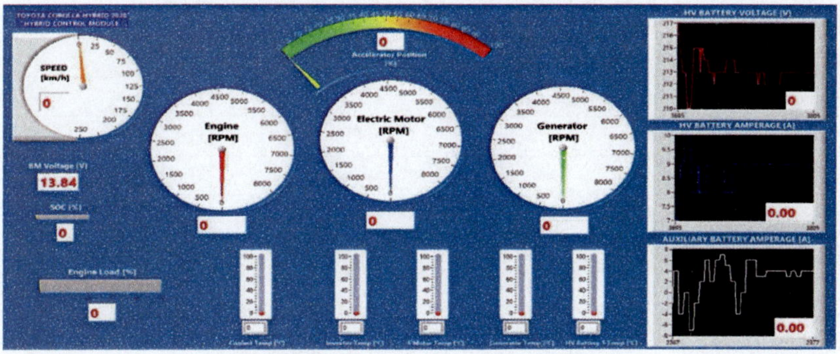

Fig. 7. GUI dashboard showing PIDs

This pragmatic approach shows the versatility of the system, with the ample variety of dashboard configurations and numeric controls, the underline and base program setup is the key to ensure the student and researcher now fully comprehend how data in CAN Bus holds the expected PIDs. Table 3 shows the PIDs list found in this example.

3.5 Simulink

The principal advantage of Simulink is that allows the investigator to recreate the conditions of a live road test drive from the database of the.txt file saved in Bus Monitor, this solves the need to retry a live-test with the vehicle and rethink study parameters or research pinpoints when availability of the vehicle is troublesome or non-existent. Another essential benefit is the ability to reenact the working conditions inside the classroom and point out certain PIDs relevant to the class or topic in study.

CAN Replay function sets the Simulink layout for initial configuration of the control panel along with the database of identified PIDs, Fig. 8 shows the block setup and Fig. 9 also shows the CAN receive block setup where a virtual channel is used to simulate CAN data stream from the PIDs database and the ID filter requests only the PID set in the respective channel, along with the sample time interval, denoting how important it is for the system to find out how PIDs are invoked in the CAN data stream.

The final dashboard follows the same arrangement as the one in LabVIEW as shown in Fig. 10:

Table 3. PIDs found and displayed in dashboard.

Frame ID and corresponding PID	Unit	Time interval
#001 SOC	[%]	0.03
#002 Battery voltage	[V]	0.49
#003 HV current	[A]	0.49
#004 Auxiliary battery current	[A]	1
#005 Calculated engine load	[%]	0.49
#006 Coolant temperature (gas engine)	[°C]	1
#007 Generator temperature	[°C]	0.1
#008 Electric motor temperature	[°C]	1
#009 Inverter temperature	[°C]	0.015
#010 HV battery block #1	[°C]	0.1
#011 Gasoline engine RPM	[rpm]	0.03
#012 Generator RPM	[rpm]	0.52
#013 Electric motor RPM	[rpm]	0.49

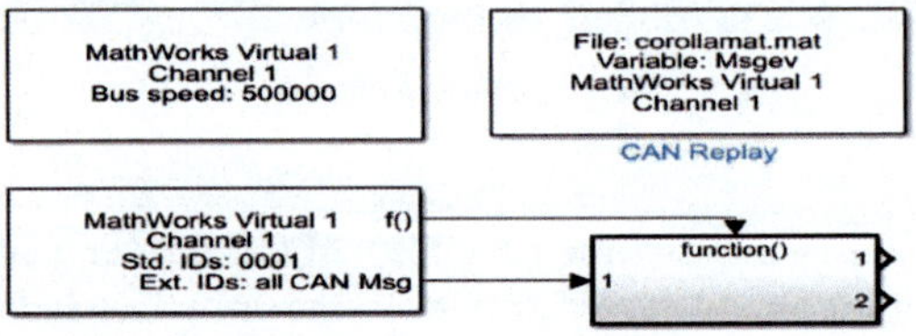

Fig. 8. Can Replay block and channel settings.

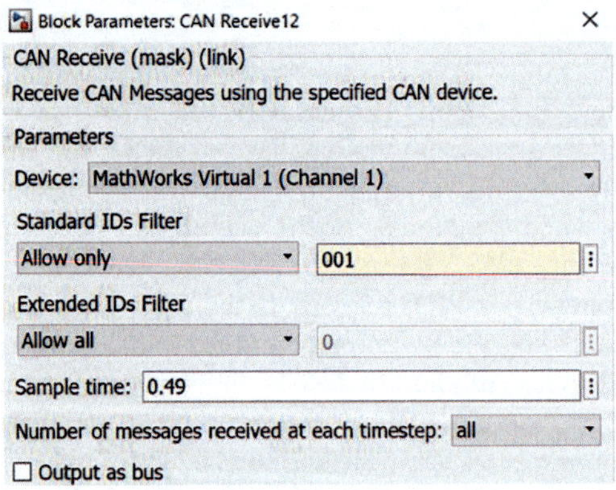

Fig. 9. Can receive setup.

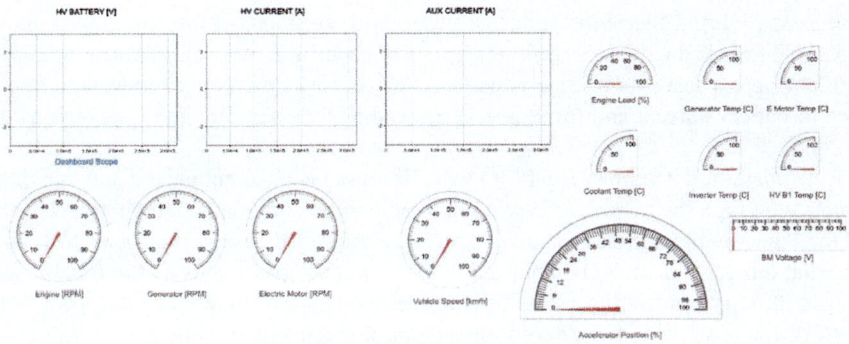

Fig. 10. Simulink PIDs dashboard.

4 Results

4.1 Validation

For the purpose to ensure reliability of the system after the mathematical adjustments with corresponding bytes, it's important to make another test in the vehicle to capture data and use the arithmetic formula again with the new data, at this juncture, if the adjusted PID matches the reference graph again, it is determined that the equation is reliable. Figure 11 shows the comparison between the reference and adjusted PID for Electric motor speed. This process must take place with every PID expected to work in the GUI LabVIEW dashboard.

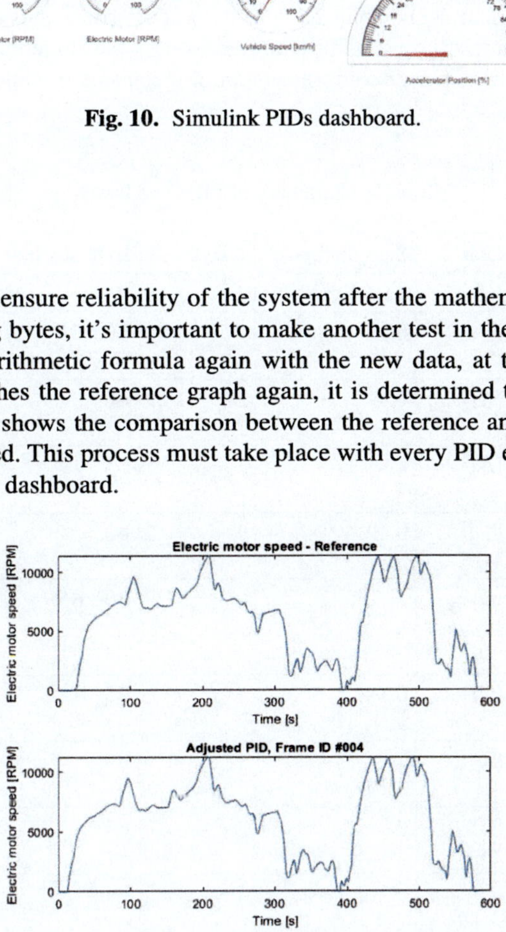

Fig. 11. Comparison between reference and adjusted graph for equation validation

5 Discussion

In summary, all identified PIDs with this method are shown in Table 4, where sample time, and byte position are the most important values for the proper working of the

LabVIEW GUI and therefore Simulink dashboard, as stated in the paragraphs above, the sample time is the data transmission speed of each PID, most of literature states that in PIDs where a fast rate change is needed, the sample time is very low, as in engine speed or battery current and for temperature readings the sample time is as high as one second.

Byte position where the actual PID value is stored is also recognized and one of the elements to graph the magnitude variation to able to be compared and identified. At the end the adjustment formula is shown, for future reference is very important to keep in mind that this formula must be adjusted with margin of error to ensure the reliability of the system. This end result must be fully understood by the student and this knowledge would be the base for a deeper comprehension of electronic module data transmission in hybrid vehicles.

Table 4. Summary of PIDs and bytes.

PID	Unit	Sample time	# Bytes	Byte position	Formula
#001 SOC	[%]	0.03	1	B4	B4 – 73.37
#002 Battery voltage	[V]	0.49	2	B0&B1	B0 + B1
#003 HV current	[A]	0.49	1	B0	(B0* – 1) + 115
#004 Auxiliary battery current	[A]	1	1	B5	B5 – 36
#005 Calculated engine load	[%]	0.49	1	B2	B2/2.55
#006 Coolant temperature (gas engine)	[C]	1	2	B0&B5	B0 – B5
#007 Generator temperature	[C]	0.1	1	B5	B5 + 20
#008 Electric motor temperature	[C]	1	1	B4	B4 – 4
#009 Inverter temperature	[C]	0.015	1	B4	B4
#010 HV battery block #1	[C]	0.1	1	B4	B4
#011 Gasoline engine RPM	[rpm]	0.03	1	B2	B2 * 25
#012 Generator RPM	[rpm]	0.52	2	B0&B1	((256 * B0 + B1)/4) * 8.5
#013 Electric motor RPM	[rpm]	0.49	1	B2	B2 * 105

6 Conclusions

CAN bus data stream transformation and analysis is one of the key factors to increase the understanding of new students on how electronic module communication takes place and how Frame ID – PID cross-reference aids in the final goal to find out how physical magnitudes and phenomena involve the inner workings of a hybrid drivetrain.

One of the minor setbacks for such system is the time-consuming process inherent to the extensive observation and comparison of reference and byte graphs, nonetheless, this opens a new branch derivative from this study where a pattern curve recognition algorithm may emerge as a solution.

LabVIEW GUI allows teachers and students to watch how live events unfold in the internal combustion engine and hybrid control module in the vehicle, most of this relevant information is not available from the manufacturer, given the circumstance that any potential user may use only primary details to avoid overwhelming the driver.

Simulink recreates the driving conditions and PIDs variation without the need to take a live road data collection session, solving logistical problems from car unavailability.

Mathematical adjustment equations sometimes match the SAE J1979 standard in a few cases, however, it is a useful data reference on how bytes work to deliver the expected PID and it is almost mandatory to remake another test to ensure the reliability of the equation for amplitude adjustment.

References

1. Christidis, P., Focas, C.: Factors affecting the uptake of hybrid and electric. Energies (2019)
2. Slowik, P., Lutsey, N.: The continued transition to electric vehicles in US cities. ICCT White Paper (2018)
3. Teece, D.J.: Tesla and the reshaping of the auto industry. Manag. Organ. Rev. **14**, 501–512 (2018). https://doi.org/10.1017/mor.2018.33
4. Aravena, C., Denny, E.: The impact of learning and short-term experience on preferences for electric vehicles. Renew. Sustain. Energy Rev. **152**, 111656 (2021). https://doi.org/10.1016/j.rser.2021.111656
5. de Souza, L.L.P., Lora, E.E.S., Palacio, J.C.E., et al.: Comparative environmental life cycle assessment of conventional vehicles with different fuel options, plug-in hybrid and electric vehicles for a sustainable transportation system in Brazil. J. Clean. Prod. **203**, 444–468 (2018). https://doi.org/10.1016/j.jclepro.2018.08.236
6. Rezvani, Z., Jansson, J., Bodin, J.: Advances in consumer electric vehicle adoption research: a review and research agenda. Transp. Res. Part D Transp. Environ. **34**, 122–136 (2015). https://doi.org/10.1016/j.trd.2014.10.010
7. Xhafa, F., Caballé, S., Barolli, L. (eds.): Advances on P2P, Parallel, Grid, Cloud and Internet Computing. Lecture Notes on Data Engineering and Communications Technologies, vol. 13. Springer, Cham (2018). https://doi.org/10.1007/978-3-319-69835-9
8. Çağatay Bayindir, K., Gözüküçük, M.A., Teke, A.: A comprehensive overview of hybrid electric vehicle: Powertrain configurations, powertrain control techniques and electronic control units. Energy Convers. Manag. **52**, 1305–1313 (2011). https://doi.org/10.1016/j.enconman.2010.09.028
9. Baghli, L, Moussaoui, A., Benmansour, K., et al.: Optimal hybrid vehicle, embedded data acquisition and tracking. In: International Symposium on Environment-Friendly Energies and Applications (EFEA), pp. 68–73 (2012). https://doi.org/10.1109/EFEA.2012.6294082

10. Sim, A.X.A., Sitohang, B.: OBD-II standard car engine diagnostic software development. In: Proceedings of 2014 International Conference on Data and Software Engineering (ICoDSE), p. 14230 (2014). https://doi.org/10.1109/ICODSE.2014.7062704
11. Navet, N., Song, Y., Simonot-Lion, F., Wilwert, C.: Trends in automotive communication systems. Proc IEEE **93**, 1204–1222 (2005). https://doi.org/10.1109/JPROC.2005.849725
12. Cook, J.A., Freudenberg, J.S.: EECS461: Controller Area Network (CAN), pp. 1–8 (2008)
13. Woo, S., Jo, H.J., Kim, I.S., Lee, D.H.: A practical security architecture for in-vehicle CAN-FD. IEEE Trans. Intell. Transp. Syst. **17**, 2248–2261 (2016). https://doi.org/10.1109/TITS.2016.2519464
14. Rojas, D., Fernandez, E.: Validation of dynamic model for communication networks in electric vehicles. In: Botto-Tobar, M., Zambrano Vizuete, M., Díaz Cadena, A. (eds.) CI3 2020. AISC, vol. 1277, pp. 41–52. Springer, Cham (2021). https://doi.org/10.1007/978-3-030-60467-7_4

Multiagent Algorithm for the Generation of Path Matrix of the Linear System Y = AX, Used in the Estimation of the Origin-Destination Matrix

Lina Patricia Zapata[1](✉), Francisco Castro Carrasco[2], and José Luis Aguayo[3]

[1] Grupo de Investigación en Desarrollo de Aplicaciones y Software Basado en las TICs (DASTIC), Universidad Politécnica Salesiana, Quito, Ecuador
lzapata@ups.edu.ec

[2] Escuela Politécnica Nacional, Quito, Ecuador

[3] Grupo de Investigación DASTIC, Universidad Politécnica Salesiana, Quito, Ecuador

Abstract. Multiagent systems try to take advantage of the concurrency available in current computer architectures, which allows for to execution of several operations simultaneously, processing several tasks concurrently, or execute at the same time different applications on a single computer. This paper covers the implementation in the Java programming language of a multi-agent algorithm for generating and loading elements in the path matrix A that is part of the linear model Y = AX, used for the estimation of the Origin. The algorithm was implemented with multiagent systems on the JADE platform (Java Agent Development Framework). In addition, an Object-Oriented algorithm was implemented to compare its level of efficiency with the multi-agent algorithm, and the results arrived in tables and graphs explaining the concepts of efficiency were presented.

Keywords: Route matrix · Object-oriented programming · Multiagent systems · Jade · FIPA · ACL

1 Introduction

Vardi, addressed inference problems in counting intensive traffic in direct link networks between pairs of nodes [1]. The archetype application context is a communication network, applicable a estudios de redes de transporte. The interest lies, at least in part, in estimating the source, or origin, to destination counts; that is, for each pair of communicating nodes i and j in the network, estimate the number of packets transmitted from i to j based on the observed link counts. Actualmente, in this area, there is long-standing interest in estimation of so-called "OD matrices" that represent the actual or expected ows or counts of traffic between all possible pairs of origin ("O") and destination ("D") nodes or zones in urban road networks during a specific time Interval [2]. Pitombeira. Loureiro, & Carvalho [3], in his work, point out that the estimation of origin-destination flows (DO) in transport networks is an important step in transport planning and proposes

V. Robles-Bykbaev et al. (Eds.): CITIS 2022, LNNS 607, pp. 381–391, 2023.
https://doi.org/10.1007/978-3-031-24327-1_33

a dynamic hierarchical Bayesian model for the estimation of DO flows day by day. On the other hand, Vardi [1] proposes the model for estimation of the Origen-Destino Matrix, starting from the $Y = AX$ model, where the Y vector of the direct links, X the OD matrix and the routing matrix A. Y and X are related through the r x c routing matrix $A = \{A_{s,a}\}$ where $A_{s,a} = 1$ if the directed link s belongs to the directed route through the network between origin-destination pair a; $A_{s,a} = 0$ otherwise. Note that A is typically singular, with a number of columns larger than the number of rows. With this definition, and with the links implicitly ordered 1, … r. Para resolver el sistema, Vardi diseña el criterio de máxima verosimilitud mediante, el Método de los Momentos. Por su parte, Tebaldi y West [4], resuelven el problema de la estimación de matrices OD a partir, del modelo propuesto por Vardi mediante la inferencia Bayesiana con el método de Monte Carlo en cadenas de Markov. The restrictive that is exhibited in these algorithms at the time of building the $Y = AX$ system is the representation of data from large transport networks where the route matrix A can become large, pretending to fill with 0 and 1 manually demands hard work, time and effort.

The main objective of this work is to design and develop an algorithm framework, with multiagent programming, which provide a typical case of concurrent behavior. This algorithm facilitates the generation of the route matrix A, , making efficient use of hardware resources and allowing greater efficiency in the final response. The Multiagent System (SMA) is constituted by a set of reactive rational agents that coordinate their skills for the resolution of the problem posed [5]. The algorithm was implemented with multiagent systems on the JADE platform (Java Agent Development Framework) [6], and performed the following functions or tasks: i) definition of the structure of matrix A and initialization of the same, ii) generation and assignment of the values corresponding to matrix A, depending on the feasibility or otherwise of the route to be followed. iii) coordinates the incorporation of the securities issued by the other agents in the matrix A resultante. In addition, three main concepts of JADE were handled, the communication of the agents based on FIPA-ACL, the behaviors of the agents and the life cycle of the agents.

The organization of this article is presented below. Section 2 shows the preliminaries. The framework for agent-based systems are located in Sect. 3. Section 4 describes the testing and analysis of the results of the multiagent algorithm design. The conclusions are shown in Sect. 5.

2 Preliminaries

Figure 1 shows the outline of a small hypothetical transport network, where the arrows represent the flow of passengers between nodes, and the circles or nodes are the stations or passenger stops.

The variables considered in the example scheme, considering only one direction of circulation are:

$x_j, j = 1, \ldots, c,$, number of passengers traveling from an origin node to a destination node (origin-destination pairs, OD). Given n nodes in the network, there are $c = n(n - 1)$ source-destination pairs.

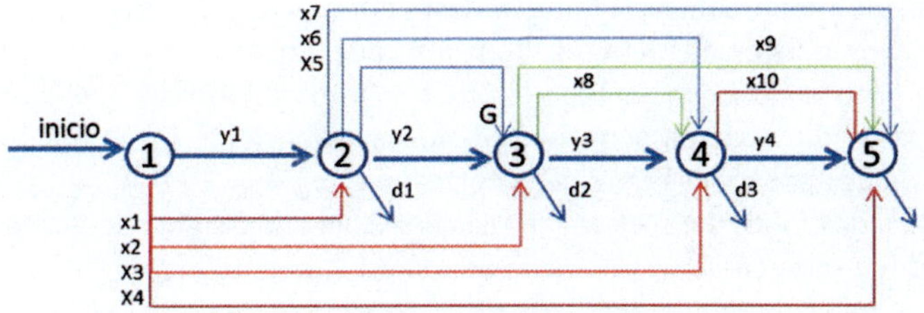

Fig. 1. Outline of a public transport network

y_i, $i = 1, \ldots, r$, flows in the links, i.e. passengers traveling from one node to another node without nodes or intermediate stops (direct links). Let r be the total number of directed links on the network. Let s = (i, j) be the link directed from node i to node j; and type Ys for traffic count in this link.

Matriz A, Y and X are related through the $r\ x\ c$ routing matrix $A = \{A_{s,a}\}$ where $A_{s,a} = 1$ if the directed link s belongs to the directed route through the network between origin-destination pair a; $A_{s,a} = 0$ otherwise. Note that A is typically singular, with a number of columns larger than the number of rows. With this definition, and with the links implicitly ordered 1, ..., *r*.

With this definition, and with the links implicitly ordered 1, ..., r; we have the defining identity:

$$\mathbf{Y} = \mathbf{AX} \tag{1}$$

Taking into account the network scheme of the example considered in Fig. 1, the resulting system of equations is:

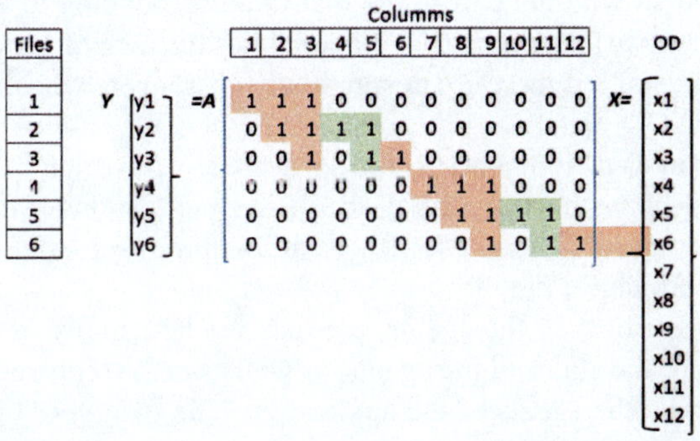

Fig. 2. The link and route structure for the network in Figure

A realistic problem will have a routing matrix that has no duplicate columns and in which each column has at least one non-zero entry. A duplicate column means a redundancy in origin-destination pair specification. A zero column is disallowed, as it would imply a non-communicating origin-destination pair. Similarly, a zero row would imply a link lying outside the network, so is also disallowed.

The n = 5 node network, in Fig. 1, has r = 4 directed links and c = 20 origin-destination pairs; the ordered sequence of nodes comprising these links and origin-destination routes appears in Fig. 2. The corresponding 4 × 20 routing matrix appears in Fig. 2, it also describes the sections of matrix A that are assigned to the agents for the generation of the same.

3 Multiagent System Framework

A multiagent system (AMS) is called a system consisting of a group of agents, which relate to each other, and the subfield of principles and designs is called distributed AI [5].

In this paper, we outline an framework for building agent-based systems, it is targeted to the JADE agent environment for deploying a network of agents that focus on collaboratively analyzing one or more streams of data. Our framework provides an integrated process to support the modeling and the configuration of agents in a JADE platform, the information flow and content in the system, and ultimately the deployment of agents from those models. We illustrate the use of this framework in the construction of a complex multi-agent system that supports the configuration and deployment de la matriz de ruta A.

The framework for building agent-based systems presented in Fig. 1 has three agent: i) the managing agent is responsible for the creation of the object instantiating in the class, and through this object makes the call to the other agents sending as an argument: the number of nodes the number of rows and columns of routing matrix A; ii) Agent matrix_triangularS1, is responsible for loading the array A, assigning the value of 1 (in the form of triangles) if there are links between the OD pairs otherwise 0 is assigned; iii) agent matrix_triangularS2 performs a task similar to agent matrix_triangularS1, with the difference that it works with the part of the matrix corresponding to the other direction (end stop-stop start); y iii) Agent matriz_diagonal has the task of generating the matrix A, assigning the value of 1 if there are destinations (Passengers who descend to the stop or station) about the OD pairs.

JADE (Java Agent Development) [7] is a software structure that simplifies the development of multi-agent systems according to FIPA specifications [8] for the interoperability of intelligent agent systems [9, 10]. With this premise, JADE provides us with the following base structure, Fig. 3:

Where the management of the agents, consists fundamentally in the identification, registration, location, and status of the agents, as well as in the registration of the specific services carried out by the agents of the application. The two types of agents in charge of performing these functions are:

- AMS (Agent Management System) is the agent that exercises supervisory control over the access and use of the agent platform. It provides, among others, the white page service and agent lifecycle, maintaining the agent identifier directory and agent status.

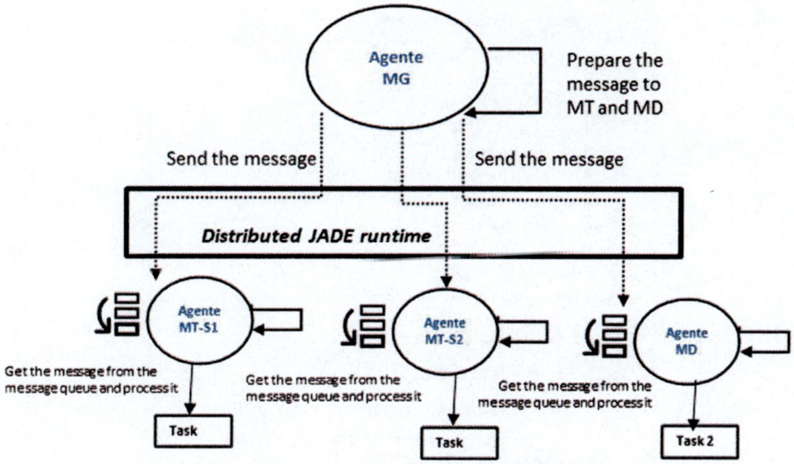

Fig. 3. Basic structure of JADE

- DF (Directory Facilitator) is the agent that provides the yellow pages service to other agents.

The Message Transport System, also called ACC (Agent Communication Channel), is the software component that controls all message exchange within the platform, including messages to and from remote platforms. The Agent class contains the basic structures for agents to perform essential interactions with the platform (logging, configuration, administration, etc.) and a basic set of methods for customizing agent tasks (sending and receiving messages, standard interaction protocols for building agent-to-agent conversations, and so on.

With all this, the definition of the specific tasks of each agent is made to develop an agent as part of the base class Agent, see Fig. 4. Each JADE agent is composed of a single execution thread, and each task is implemented using one or more behaviors. JADE is automatically and transparently used by the programmer, manages the behaviors, supporting concurrent and parallel execution.

The agents required for the generation of the path A matrix followed a summary process for its coding in Java, which is detailed below:

- Create a class to represent the agent, and inherit from the jade. Core.Agent class, with the setup() method which initializes the agent.
- Add a behavior that specifies the tasks that an agent performs, to meet its objectives, and with the action() method that executes the tasks.

Each agent has a different behavior or task that runs simultaneously. For the communication of the agents, the FIPA ACL communication mechanism is used. Figure 8 shows the sequence diagram of the Multiagent System, which details the communication that exists between them and the behavior that each one has in the process of generating the matrix A.

```
Task: Agent triangular_matrix
method TM_S1 ( val, nod, matrix)
var: acum, r
 for fil to nod - 1
   for col=0 to val
     matrix [ fil] [ col] = 1
      if sentido es igual a dos
        matrix [ fil] [ col] = 1
      endIf
    endFor
    val = nod -1
   increases r en uno
  fin para
 val= fil + 1
 acum =acum + val
```

a) Pseudocode: triangular matrix method.

```
Task: Agent diagonal_matrix
method DM (valn, nodsd, nod) hacer
  variables: val=nod
   af = nod-1
 for r1=1  to nod - 1
   i =  af
   for r2=1 to nod - 1
      matrix [ r1] [r2] = 1
       si s==2
          matrix [ r1] [r2] = 0
       endif
        nod=nod + 1
    EndFor
         af = nodos-1
         Ac= ac-1
         valorn = nod-1
  fin para
```

b) Pseudocode: diagonal matrix method.

```
Task: Agent gestor matrix
var: nod, s, matrix
   read  nod
   read s
   if s==1
       c=1
       r=1
   endif
   if s = 2
       c=n(n-1) /2
       r=n-1
    endif
    MatrixI [ f] [c]
    object of class Gestor_matrix
    method Triangular_MatrixS1
    method Triangular_MatrixS2
```

c) Pseudocódigo clase principal.

Fig. 4. Multiagent algorithms

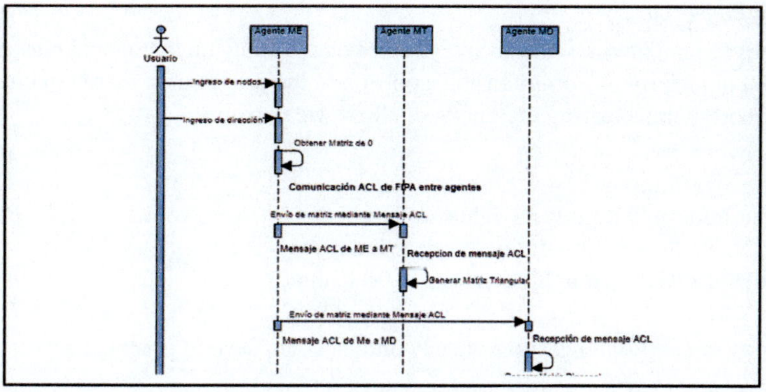

Fig. 5. Program sequence diagram.

4 Results and Discussion

Performance Measures

The most commonly used measures when determining the performance of an algorithm [11] are given two measures:

- ***T(n)*** - execution time depending on the size n of the problem. This function can be measured physically (running the program), or calculated on the code by counting the instructions to be executed and multiplying by the time required for each instruction. Thus, a simple piece of a program requires:

$$T(n) = t1 + t2 * N \tag{2}$$

- ***E(n)*** – space (RAM) depending on the size n of the problem.

Methodology

To focus on the comparison of algorithms, an object-oriented algorithm was designed with the same function that the multiagent algorithm performs, generating the matrix A, [12]. With these two algorithms defined the following steps were followed:

1. Determine the function f(n) that characterizes the resources required by each algorithm as a function of size n of the data to be processed.
2. Given two functions f(n) and g(n) an order relationship was defined between them that we will call complexity and that will tell us when one function is more complex than the other or when they are equivalent.
3. Select a series of reference functions for typical situations.
4. Define sets of functions that we will call complexity orders.

Experimental Analysis
Sometimes it is not obvious how to analyze the code to apply mathematical rules and calculate the complexity of a program and a portion of the program. In these circumstances we can resort to the experimental method, where we perform:

1. Code programming
2. We put it to run with several values of N
3. We tabulate the measured times in function n
4. We analyze which complexity group is best suited

In order to carry out measures we take into account several practical aspects:

- perform the operation a large number of times, with random data, to stay with an average value or trend
- [in the case of java] run the program a series of times before measuring to make sure that the compiler has generated optimized code.

The two algorithms were tested to know, which has better, performance at runtime for which an average of execution times was made. N = 10 tests with an equal number of nodes (60, 90, 120, 150, 180, 210) were performed on the two algorithms, Table 1 shows the results obtained from the execution of the object-oriented algorithm, and Table 2 describes the values obtained from the multiagent algorithm. And Table 3 shows the averages obtained, for each test, and it is observed that the algorithm with multi-agents is the most efficient in execution time.

Table 1. Object-oriented algorithm execution times

Object oriented algorithms						
Tests/time (sec)	60 nodes	90 nodes	120 nodes	150 nodes	180 nodes	210 nodes
Test 1	5,38	7,16	13,22	17,26	29,39	42,95
Test 2	4,53	7,07	12,32	24,50	29,30	41,61
Test 3	4,73	7,20	16,85	21,80	29,13	42,49
Test 4	4,97	6,25	13,36	19,59	29,13	43,12
Test 5	4,85	6,65	13,59	20,48	28,17	42,75
Test 6	5,16	7,66	13,41	20,87	28,25	42,75
Test 7	4,49	7,74	13,41	20,29	28,30	42,13
Test 8	4,99	8,13	12,56	19,59	28,89	41,64
Test 9	4,71	8,25	13,93	20,98	27,46	40,10
Test10	4,17	8,66	12,57	20,12	28,90	41,06

(continued)

Table 1. (*continued*)

Object oriented algorithms						
Tests/time (sec)	60 nodes	90 nodes	120 nodes	150 nodes	180 nodes	210 nodes
Total	47,97	74,76	135,22	205,47	286,91	420,60
Average	4,80	7,48	13,52	20,55	28,69	42,06

Table 2. Multiagent algorithm execution times

Multiagent algorithm						
Tests/time (sec)	60 nodes	90 nodes	120 nodes	150 nodes	180 nodes	210 nodes
Test 1	4,12	7,94	10,85	17,63	26,98	39,02
Test 2	4,02	7,05	8,99	19,12	25,76	37,83
Test 3	3,52	8,20	8,93	16,28	24,70	38,17
Test 4	3,87	7,11	9,58	15,12	24,27	39,15
Test 5	3,65	8,18	9,46	15,17	31,22	38,12
Test 6	3,73	7,97	9,49	16,31	24,98	40,22
Test 7	3,87	8,26	9,01	15,85	24,39	40,77
Test 8	3,78	8,18	8,77	16,79	26,83	39,58
Test 9	3,92	7,56	8,55	17,73	25,88	40,46
Test 10	3,84	7,82	8,78	18,01	25,42	40,08
Total	38,32	78,28	92,41	168,01	260,41	393,37
Average	3,83	7,83	9,24	16,80	26,04	39,34

Table 3. Average execution time of each algorithm

Nodes/average (sec)	Object O	% Object O	Multiagent	% Multiagent
60	4,80	4%	3,83	4%
90	7,48	6%	7,83	8%
120	13,52	12%	9,24	9%
150	20,55	18%	16,80	16%
180	28,69	25%	26,04	25%
210	42,06	36%	39,34	38%
Total	117,10	100%	103,08	100%
Average	19,52	20%	17,18	17%

Figure 5 shows the comparison between the two algorithms, showing that the multi-agent algorithm is more efficient, at run time, in the process of generating the path matrix *A* (Fig. 6).

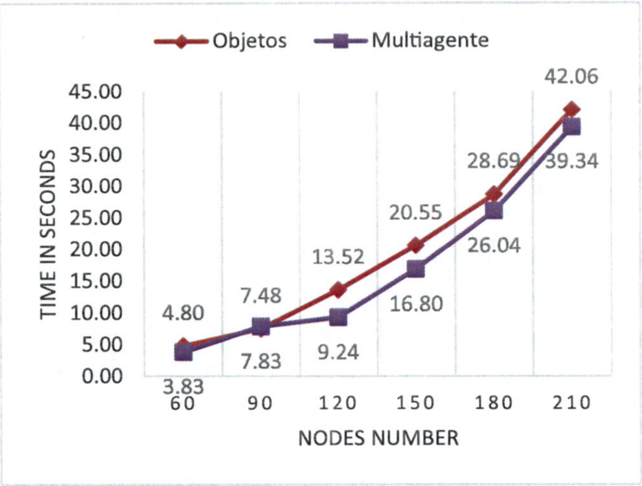

Fig. 6. Execution times of algorithms

5 Conclusions

These values were analyzed and represented graphically. For the hypothetical case analyzed, an increase in efficiency is obtained as the number of nodes in the transport network increases. The multi-agent algorithm performs better than the object-oriented sequential algorithm and increases efficiency in the solution as the number of agents increases. It can then be concluded that a multi-agent algorithm presents a good performance that influences the size of the problem to be solved. It becomes more efficient as the size of the transmission network increases, that is, as we increase the volume of work, being convenient to use it precisely to reduce calculation times concerning a purely sequential calculation.

References

1. Vardi, Y.: Network tomography: estimating source-destination traffic intensities from link data. J. Am. Stat. Assoc. **91**(433), 365–377 (1996)
2. Zapata, L.P., Flores, M., Larios, V., Maciel, R., Antunez, E.A.: Estimation of people flow in public transportation network through the origin-destination problem for the South-Eastern corridor of Quito city in the smart cities context. In: IEEE International Smart Cities Conference (ISC2), Casablanca, Marruecos (2019)

3. Pitombeira-Neto, A.R., Loureiro, C.F.G., Carvalho, L.E.: A dynamic hierarchical Bayesian model for the estimation of day-to-day origin-destination flows in transportation networks. Netw. Spatial Econ. **20**(2), 499–527 (2020)
4. Tebaldi, C., West, M.: Bayesian inference on network traffic using link count data. J. Am. Stat. Assoc. **86**(442), 442–455 (1998)
5. de Oliveira, A.S., Asato, O.L., Kaneshiro, P.J.I., Nakamoto, F.Y.: Proposta de arquitetura virtual utilizando sistemas multiagentes. In: 14th IEEE International Conference on Industry Applications (INDUSCON), no. 14, pp. 502–508 (2021)
6. Chebout, M.S.: AOM4JADE: an aspect-oriented monitoring approach for JADE-based multi-agent applications. In: International Conference on Networking and Advanced Systems (ICNAS), pp. 1–6 (2021)
7. Bellifemine, F.L., Caire, G., Greenwood, D.: Developing Multi-agent Systems with JADE (2007). ISBN 978-0-470-05747-6
8. Kadera, P., Novák, P.: Performance modeling extension of directory facilitator for enhancing communication in FIPA-compliant multiagent systems. IEEE Trans. Industr. Inf. **13**(2), 688–695 (2016)
9. Bellifemine, F., Poggi, A., Rimassa, G., Turci, P.: An object-oriented framework to realize agent systems. In: WOA, pp. 52–57 (2000)
10. García-Magariño, I., Nasralla, M.M., Lloret, J.: A repository of method fragments for agent-oriented development of learning-based edge computing systems. IEEE Netw. **35**(1), 156–162 (2021)
11. Cruz, P.S.: Universidad La Plata, 13 April 2014. http://www.frlp.utn.edu.ar/materias/algoritmos/GUIACOMPLEJIDADDEALGORITMOS.pdf
12. Banda Madrid, A.: Evaluación y análisis en la calidad en el diseño de sistemas de software orientados a objetos (2021)

Public Lighting System Management Adapted to Remote Management: A Case Study

Oscar Álvarez(✉), Leslie Mendia, and Jorge Rojas

Universidad Politécnica Salesiana, Cuenca, Ecuador
oalvarezm@est.ups.edu.ec

Abstract. In the following paper a proposal was made for a remote management system for public lighting belonging to Empresa Eléctrica C.A., analyzing the different technologies for remote control and monitoring of equipment, with the purpose of defining the most appropriate one in terms of its advantages, installation, and cost of the necessary equipment. The company has currently opted for the installation of LED lamps on Av. de los Alcaldes and Av. Aurelio Jaramillo, improving the photometric parameters with respect to the old technologies, but they still have a high energy consumption due to the operating time of the luminaires. An analysis and comparison of MINOS UMPI, OWLET, ELO and SimplySnap technologies was carried out to define the most appropriate one in consideration with the requirements of the equipment already installed, therefore a field study was carried out to verify the state of the public lighting and the transformer stations. In addition to comparing the cost of current energy consumption with the cost of energy consumption of the system with remote management. In turn, a technical-economic study of the proposal and a simulation of the lighting aspects under the current ARCONEL-CONELEC 006-20 regulations. Design a lighting system with telemanagement and radiofrequency wave communication between the operation teams through a Wi-Fi network with the monitoring equipment. Analyze the photometric effect through simulation with the Dialux program, verifying if the values obtained are in accordance with the regulations in force in the country, in addition, the technical study in the transformation stations determining their chargeability with the current and proposed system. The data obtained in the comparison of equipment allowed to determine the VAN and the IRR of the proposal obtained the following values (988.71$ and 8.29%) in a period of 10 years.

Keywords: Energy consumption · Energy efficiency · Led luminaire · Public lighting · Remote management

1 Introduction

Street lighting in society is of paramount importance as it is an infrastructure that helps with the appearance, development and atmosphere of the city's public spaces by providing safety and comfort for nighttime traffic [1]. Resource management provides energy efficiency by reducing maintenance and energy consumption costs. Currently, Ecuador

V. Robles-Bykbaev et al. (Eds.): CITIS 2022, LNNS 607, pp. 392–406, 2023.
https://doi.org/10.1007/978-3-031-24327-1_34

has an approximate consumption of 1500 GWh and it is estimated that within 5 years it could reach a total of 1700 GWh [2]. In this perspective, there is a need for proposals for public lighting management technologies that allow the reduction of power and load profiles to analyze the behavior and adequate lighting levels to enable the use of lower power lamps or alternatives to meet the growth in demand [3]. The need arises for proposals for technologies in public lighting management to reduce the power and load profile to analyze the behavior and appropriate lighting levels that enable the use of lower power lamps or alternatives to meet the growth in demand. The city of Azogues had an installed power of 2141 kW as of 2017 for the lighting system [4, 5]. Communication between devices requires a WLAN, WSN or Zigbee wireless local area network. The use of authentication protocols provides security to systems that are connected in an internet network [6, 7]. Limited resources in small networks handle cryptographic methods that are concatenated in the SimplySnap technology database and thus allow access to the platform and data of the installed network, and can also have data traffic with extensive information connected to the servers of the company or municipalities avoiding the poisoning of relevant information [8, 9]. The use of LED technology is a method to reduce energy consumption of street lighting, however, the time they remain in operation without flow of people or vehicles means a non-efficient power [10]. At present, there are several suppliers for remote management systems in the country. This is due to the fact that in other countries in Europe and Australia it has been implemented on a large scale, optimizing equipment and simplifying communication protocols [11]. SimplySnap is most economical for small-scale systems with equipment that adapts to LED luminaires with dimmable power and can take consumption measurements to store luminaire control data through modified on/off power events, which should be based on the lighting needs of the street and crosswalks [12]. The technology can also provide visualization of the area through maps that can alert the malfunctioning of the luminaires allowing decision making to solve the problems, the interface is not complex providing easy access for administrators also allows viewing energy consumption values and programmable schedules desired by the user and select the type of operation of the manual or automatic network [13, 14]. In the global positioning system, new technologies are also emerging for locating luminaires and identifying faults more adequately [15]. Power consumption occurs during critical hours or so-called peak hours that cause transformer overloading [16]. The system that supports street lighting starts from the power supply and distribution network connected through a transformer station; these elements can be improved using the power flow method to determine the load ability of the transformers and the effect they cause. For a three-phase system in an overhead network, the harmonic distortion is a value of 10% up to the luminaire power supply [17]. The LED luminaires on the market have the advantage of being programmable at various levels which can help to generate scenarios according to the needs of lighting [18]. The simulation will confirm whether the established designs can generate the energy savings without affecting the outdoor lighting environment; the Dialux program contains rules and regulations that will limit the designs [19]. Lighting energy efficiency is about reducing cost without looking at the benefits of the system or contradicting regulatory control parameters in lighting systems [20]. The design of the system is given by the road and crosswalk lighting class based on the Ecuadorian ARCERNNR regulations [21, 22]. The decision to invest in public projects involves socioeconomic factors, one of which is the cost-benefit analysis, in relation to installation costs, equipment operation and the savings generated by the proposed system [23]. The cost of the equipment has now been reduced, giving greater openness to remote management systems [24]. The capital for public investment projects is analyzed by the Central Bank of Ecuador giving an interest rate for each of the projects to be executed, which is the fundamental factor in an economic analysis [25]. At the ISCTE-IUL university in Portugal, in a telemanagement applied to the university building, it was possible to reduce energy consumption by 40% based on its climatic conditions [16]. In the streets of the Federal University of Juiz de Fora in Brazil, it was shown that by reducing the luminous flux by 40%, the standards established in the country are respected, this reduction extends the useful life of the luminaire [26].

2 Methodology and Resources

This paper presents a proposal for the management of public lighting in a group of streets in the city of Azogues with a remote management system.

First, the state of the art is analyzed, where a comparison of technological advances and concepts, principles and standards is made with reference to public lighting, LED luminaires, remote management systems and communication protocols in accordance with the ARCERNNR 006/20 regulation.

The case study is developed with SimplySnap technology in Aurelio Jaramillo Street and Av. de los Alcaldes in the city of Azogues, proceeding with the field study, load survey using ArcGIS software, field visit and with the information that will be provided by Empresa Eléctrica Azogues C.A.

The technical-economic analysis of the project is carried out under the parameters of NPV, IRR and RBC for the equipment to be used and the remote management system defined. In addition, the simulation in the DIALUX EVO program to demonstrate the power reduction does not affect the photometric parameters of the ARCERNNR regulation.

Finally, the conclusions and criteria obtained from the analysis of the proposal for the energy saving of the telemanagement system are presented.

2.1 Proposed System

Remote management is a new way to implement systems that can be controlled from any smart device, these technologies will have a scope depending on the equipment or communication protocols [2]. Table 1 shows the four telemanagement systems to be analyzed for the selection of the most suitable technology for the case study proposal, describing their characteristics, advantages, and benefits of each one.

Table 1. Characteristics of Telemanagement Systems.

	MINOS UMPI	ELO	Owlet Schreder	SimplySnap
Energy conservation	45% savings	50% in operation	Up to 85% savings	Power reduction between 30% and 40%
Maintenance	Real-time monitoring	Faults are identified automatically	Real-time monitoring	Real-time monitoring of energy consumption profiles
Reliability and safety	Notifications in case of equipment failure	Intelligent network, GSM, or Wi-Fi connections, sensors, etc.	Transmission is 10 times higher	Prevents access to unauthenticated users
Operation and services	Programming by means of astronomical clocks. Import data in XML	Operation through a control station	Implement, adapt, and reproduce lighting profiles for AP networks	Calendars with dates and on/off programming for each zone. Generates a CSV database
Environmental impact	Reduction of light pollution	Reduction of light pollution	Reduces glare	Do not generate light pollution
Software	MINOS X software connected to the server with the internet	Streetlight. Vision is 100% web and Cloud-based	Computers connected to the Internet, by means of user and password	No software is required, the application can be accessed via Wi-Fi connection or IP address

The proposed system consists of controllers compatible with the Led luminaires connected to the wired network, these communicate via radio frequency with the Gateway. The site controller receives and stores the information and through a user interface presents the information, the user can access the data through a local wifi network, it can also connect to the company's server and access through TCP/IP protocol, the connection diagram is shown in the Fig. 1.

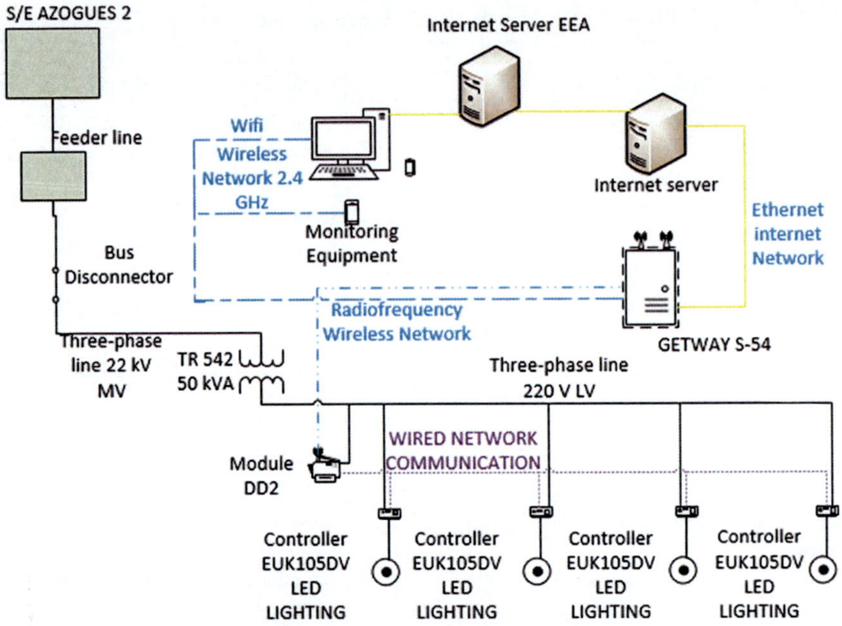

Fig. 1. System connection diagram

2.2 Communication

Wireless communication standards are the basis for remote management based on IEEE 108.11.4 formats in many versions with different frequency range and transmission speeds. The basic model works from 2.4 to 5 GHz with speeds of 2 Mbps, for radio frequency transmission a speed of 1 Mbps. The highest data transmission speed is in a wide band of 5 Ghz and speeds from 6 to 48 Mbps. The communication and information transmission in the proposal will be done through ZIGBEE topologies and radio frequency with a bandwidth of 2.4 GHz at 20 dBm.

2.3 Gateway Modules Controllers

EUK-150xxxDV (TV) series drivers are used in various applications such as lighting, tunnels, outdoor etc. Having a high efficiency up to 93% and being compact allows it to operate with lower temperatures in addition to prolonging the life of the product. In the area of operation, depending on the family of controllers, it allows the reduction of power depending on the programming. Its operation is guaranteed by the incorporation of protections against input and output overvoltage, short circuit and over temperature. Its dimensions fit Ledex lamps with inputs for alternating current of various voltages 120–220 V. The model of the Ledex luminaire, specifically HYDD-LED is shown in the Table 2, their respective values of programmable power, luminous flux, and luminous efficiency, which are installed in the areas of analysis.

The CNV-SNAP-DD2 and CNV-SNAP-AN0 stand-alone interface modules, which help control one or more Inventronics LED drivers, employ wireless lighting control

Table 2. HYDD-LED luminaire technical specifications.

Technical specifications						
Rated power (W)	60	100	150	200	240	280
Programmable power (W)	60	61–100	101–150	151–200	201–240	241–280
Voltage/frequency	90–305 VAC/ 60 Hz					
Power factor	0.95					
Harmonic distortion	10%					
Overvoltage protection	10 kV					
Overload protection	10 KA					
Luminous flux (LUMEN)	6853	12068	17769	27500	28935	35535
Luminous efficacy (LUM/W)	117	125	210	137	117	119
Shelf-life	100000 h					

networks, the latter is used for both indoor and outdoor lighting systems. Its DC power supply has inputs for two sensors that may or may not be implemented, compatible with SimplySNAP Gateway equipment allowing monitoring of either high-end or traditional controllers. The site controller is the most important element of the SimplySnap remote management system, it controls locally the luminaires without internet connection, however it has an RJ45 input in case it requires the device to be part of a LAN network. It stores energy consumption data, zones and alarms with critical behaviors and events for maintenance or troubleshooting of lighting operation, also tolerates power dimming in multi levels through programmable schedules and easy configuration. The SimplySnap Remote Access System (SSRA) has two site controllers SS420 and the SS450 model, which share most of the features, but their major difference is the communication through a Verizon 4G LTE network that the SS450 element has.

2.4 Consumption of Energy

The power consumption (kWh) maintains a harmonic distortion of 10%, due to the configuration of the power supply system to the luminaire, 220 V–60 Hz. The energy is a function of the 12 h of operation of the lighting and the average loss expansion factor that occurs in the transmission and distribution lines, being its average value (FEX = 1.0721).

$$\text{Energy Consumed} = (\text{Power} + \text{Losses}) \cdot \textit{Time} \cdot \textit{No. of lamps} \cdot \textit{FEX} \quad (1)$$

where:

Energy Consumed - (kWh).
Power - (W).
Time - (Horas).
Losses - 10%.
Fex - 1.0721.

2.5 Technical Analysis

The specifications of the transformer stations feeding the luminaires under study, which have a transformation ratio of 22/220 kV three-phase. The configuration for all transformers is delta connection for the medium voltage side and star connection on the low voltage side. During peak hours, the need arises to manage an adequate schedule and forecast of the street lighting demand with respect to the operating hours and to avoid transformer overloading, shown in Table 3.

Table 3. Transformer station data.

Case study	Transformer station	Power (kVA)	Load ability without APG (%)
Av. De los Alcaldes	542	50	17.42
	2295	75	19.3
Av. Aurelio Jaramillo	1928	30	19.13
	1083	50	50.17
	2493	75	23.15

2.6 Economic Analysis

The cost of energy for public lighting is subject to the regulation of the tariff schedule established for each electric company, in the case of EE Azogues, it takes into consideration the value of 0.04 USD and with this item the total value of the cost of public lighting is obtained. The interest rate is the price that each financial institution charges for providing credit. According to the Central Bank of Ecuador in January 2021, it was 8.28%. The net present value or net present value is calculated based on the flows of future income and expenses, analyzing whether discounting the investment made at the beginning will result in a profit on the project to be executed.

$$VAN = -I_n + \sum\nolimits_{t=1}^{n} \frac{V_t}{(1+k)^t} \tag{2}$$

where:

I_n - The initial investment when (n = 0).
n - Number of periods.
t - It is time.
V_t - Money flow in each time N.
k - Interest rate required for the investment.

The internal rate of return determines the economic feasibility of the investment, it depends on the effective flows of the project, consequently if it is higher than the reference rate the proposal can be carried out.

$$TIR = \frac{-I + \sum_{i=1}^{n} F_i}{\sum_{i=1}^{n} i \cdot F_i} \quad (3)$$

where:

F_i - Cash flow in period t.
n - Number of periods.
I - Initial value of the investment.

The benefit-cost ratio establishes the profitability of the investment decision to execute the project.

$$RBC = \frac{VA_{ofbenefits}}{VA_{ofcosts}} \quad (4)$$

where:

$VA_{delosbeneficios}$ - Annual value of benefits.
$VA_{deloscostos}$ - Annual value of costs.

2.7 Regulations and Simulation

The aspects to be complied with for road lighting according to Ecuadorian regulations are based on the vehicular flow per hour, being between 150 to 500 vehicles per hour. In the case of pedestrian traffic, nighttime use by pedestrians and cyclists is low. The photometric parameters of longitudinal uniformity and uniformity factor, as well as the average luminance, are obtained from the lighting class, which is class M3 for roads and P4 for pedestrians.

The Dialux program allows us to verify if the photometric aspects are fulfilled under the conditions established in the design. The simulation should contain the lower and upper limits of the selected lighting class and determine if the proposal is within the optimal values to avoid glare or optical problems to users.

3 Results

Based on the bibliographic research it is determined that SimplySnap technology has greater benefits and compatibility with the luminaires implemented in the company, in addition, it generates a database that is obtained from the user interface for possible lighting analysis or load projections and to determine if it is feasible to use lower power luminaires. The power flow demonstrates the load ability effect arising from the effect of street lighting shown in Table 4.

Table 4. Transformer load ability analysis.

Transformer chargeability status					
Case study	Transformer station	Load ability without APG	Load ability with AGP	Load ability with AGP-telegestion	kVA
Av Alcaldes	542	17.42	17.79	17.5	1323
	2295	19.3	19.97	19.79	0.341
	1928	19.13	21.06	19.31	0.11
Av Aurelio Jaramillo	1083	50.17	54.77	54.64	0.05
	2493	23.15	28.96	28.44	0.3901

The economic evaluation Table 5 determines that the proposed system is feasible due to the low installation and operating costs of the equipment.

Table 5. Present value of equipment.

Heading level	Example	Font size and style
	CNV-SNAP-DD2/SS420	CNV-SNAP-AN0/SS450
Interest (%)	8.28%	8.28%
Periods(years)	10	10
Investment (USD)	1939.43	2294.54
Annual savings (USD)	167.69	219.6
Annuity factor	6.39	6.39
Present value	1676.94	21.96
		VPN
	113.73	−98.54
	Select the option CNV-SNAP-DD2	

The economic evaluation Table 6 determines that the proposed system is feasible due to the low installation and operating costs of the equipment.

Table 6. Present value of the project.

Heading level	Example		Font size and style
	Actual case		SimplySnap alternative
Interest (%)	8.3%		8.28%
Periods(years)	0		10
Inversion (USD)	0		1939.43
Cost U. electricity (USD kWh)	0.04		0.04
Annual electricity cost (USD kWh)	3189.47		3021.78
Annual savings	0.00		292.81
Annuity factor	0.00	6.63	
Present value	0.00		2928.14
VPN			988.71
	It is convenient to execute the project		

In consideration of the public investment obtained by the central bank of Ecuador, the rate of return is higher than the interest rate Table 7, but the country's inflation risk must be considered.

Table 7. Project payback rate.

Rate of return	
TMRA (%)	8.28
TIR (%)	8.29

The cost-benefit analysis shows Table 8, the project is feasible and that an economic benefit will be obtained at the end of the project's useful life.

Table 8. Cost and profit cash flow.

RBC analysis			
Years	Total cost of inspection	Benefits	Costs
0		$ 0	$ 1943.43
1	$ 231.92	$ 356.89	$21.36
2	$ 231.92	$ 356.89	$21.36

(*continued*)

Table 8. (*continued*)

RBC analysis			
Years	Total cost of inspection	Benefits	Costs
3	$ 231.92	$ 356.89	$21.36
4	$ 231.92	$ 356.89	$21.36
5	$ 231.92	$ 356.89	$21.36
6	$ 231.92	$ 356.89	$21.36
7	$ 231.92	$ 356.89	$21.36
8	$ 231.92	$ 356.89	$21.36
9	$ 231.92	$ 356.89	$21.36
10	$ 231.92	$ 356.89	$21.36

The RBC parameter is greater than one, so the project is feasible to execute, shown Table 9.

Table 9. Benefit - cost relation.

VNA benefits	$ 2184.01
VNA cost	$ 1925.53
RBC	1.13

The lighting has a flux reduction due to power attenuation Fig. 2, giving a value of 178.5 W from the 210 W of the original design.

.

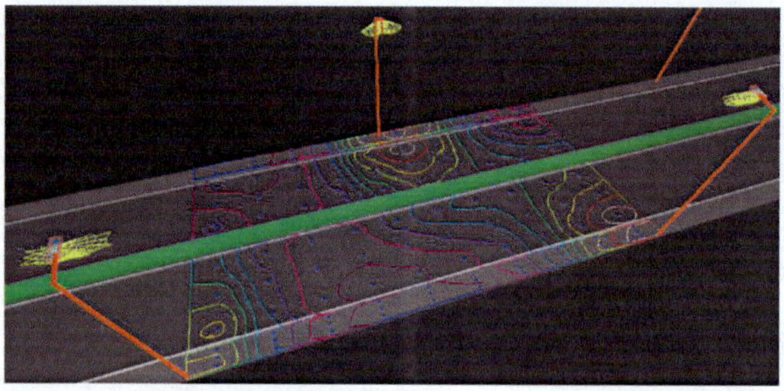

Fig. 2. Isolux curves on a surface

In the Fig. 3, the limits of illumination and other photometric factors are reduced, however, since there is no high vehicular or pedestrian flow, the power reduction can be executed.

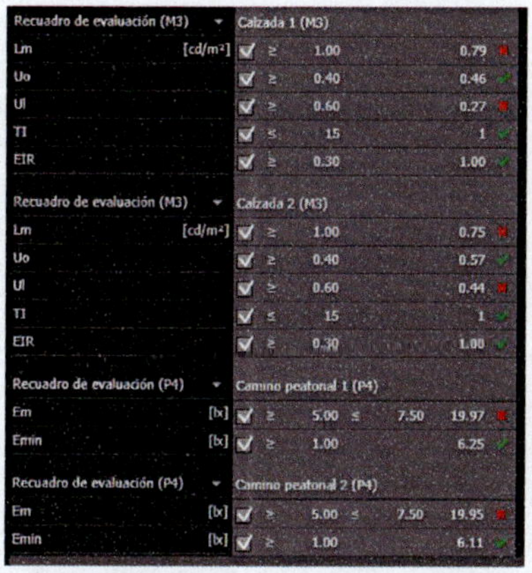

Fig. 3. Photometric aspects of regulation

The energy efficiency of power reduction is determined by the scale of lumen and power usage on a surface, shown Fig. 4.

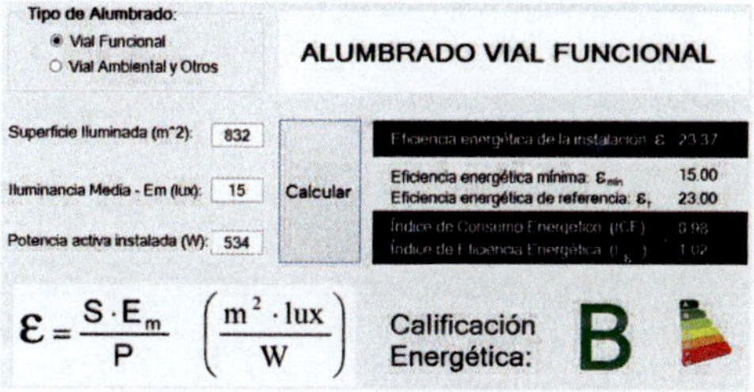

Fig. 4. Lighting energy efficiency range

4 Discussion

The state of the art showed that communication with the IEEE 108.11.4 standard predominates in the market along with the Zigbee protocol. The available equipment is low cost and contains the communication protocols, which facilitates the design of remote management systems without the need to build models and program them for operation with greater guarantee against possible failures. By working in the 2.4 GHz bandwidth and in extreme operations at 4 GHz in Verizon technology, there is no major interference in radio frequency communication or wireless wifi network. The power reduction implies reducing the established photometric parameters, however, the vehicular flow in the area is not high, which is conducive to carry out the energy efficiency proposal. The proposal can be applied to a lighting system where the influx of pedestrians and vehicles is not high and has a dimmable LED type luminaire with voltage controllers, this will facilitate the SimplySnap mesh network for data connection and interface, on the other on the other hand, in intervened or private lighting, savings increase since the cost of the service is higher and the investment is reduced by the type of labor and project contract.

5 Conclusions

It can be concluded that the SimplySnap technology is the most suitable for the public lighting conditions of the case studies, since it is compatible with the luminaires purchased by the EEA C.A.

According to the studies carried out in the technical part, the load ability of the transformers can be improved by reducing the apparent power of the machine. The data obtained in the equipment comparison allowed determining the NPV and IRR of the telemanagement proposal, obtaining the following values ($ 988.71 and 8.29%).

In the RBC calculation, it is worth mentioning that the proposal can be made considering the interest rate risk, since the economic parameters show that a total of $ 258.47 could be recovered.

In fact, reducing the power and luminous flux of the luminaires also minimizes the luminance of the roadway, however, the reduction of these factors is done at a time when there is no high vehicular or pedestrian traffic, which would not cause any inconvenience. Meanwhile, pedestrian lighting does not meet the minimum value of the standard, which indicates a reduction in power.

References

1. Langner, A.L., et al.: A pilot project of street lighting Telemanagement in a Smart Grid environment. In: 2015 IEEE PES Innovative Smart Grid Technologies Latin America, ISGT LATAM 2015, pp. 304–309 (2016). https://doi.org/10.1109/ISGT-LA.2015.7381172
2. Pallo, J.P., Manzano, S., Chicaiza, D., Nunez, C., Placencia, F., Nunez, F.: Wireless system for control, monitoring and preventive maintenance of public street lighting. In: Iberian Conference on Information Systems and Technologies (CISTI), vol. June 2018, pp. 1–6 (2018). https://doi.org/10.23919/CISTI.2018.8399382
3. Garcés-Jiménez, A., Castillo-Sequera, J.L., Del Corte-Valiente, A., Gómez-Pulido, J.M., González-Seco, E.P.D.: Analysis of artificial neural network architectures to model smart lighting systems for saving energy. IEEE Access **7**, 1–10 (2019)
4. Carli, R., Dotoli, M.: A dynamic programming approach for the decentralized control of energy retrofit in large-scale street lighting systems. IEEE Trans. Autom. Sci. Eng. **17**(3), 1140–1157 (2020). https://doi.org/10.1109/TASE.2020.2966738
5. Araujo, J., Silva, L., Oliveira, L., Fortes, M., Borba, B., Colombini, A.: Assessment of the technological update of public lighting in Brazil. IEEE Lat. Am. Trans. **18**(6), 985–991 (2020). https://doi.org/10.1109/TLA.2020.9099674
6. Daely, P.T., Reda, H.T., Satrya, G.B., Kim, J.W., Shin, S.Y.: Design of smart LED streetlight system for smart city with web-based management system. IEEE Sens. J. **17**(18), 6100–6110 (2017). https://doi.org/10.1109/JSEN.2017.2734101
7. Chen, Y., Xu, W., Peng, L., Zhang, H.: Light-weight and privacy-preserving authentication protocol for mobile payments in the context of IoT. IEEE Access **7**, 15210–15221 (2019). https://doi.org/10.1109/ACCESS.2019.2894062

8. Kim, T.H., Kumar, G., Saha, R., Buchanan, W.J., Devgun, T., Thomas, R.: LiSP-XK: extended light-weight signcryption for IoT in resource-constrained environments. IEEE Access **9**, 100972–100980 (2021). https://doi.org/10.1109/ACCESS.2021.3097267
9. Hu, X., Gong, J., Cheng, G., Zhang, G., Fan, C.: Mitigating content poisoning with name-key based forwarding and multipath forwarding based inband probe for energy management in smart cities. IEEE Access **6**, 39692–39704 (2018). https://doi.org/10.1109/ACCESS.2018.2854928
10. Berlanga-llavori, R.: Central Management System, no. 1995, pp. 58–68 (2014)
11. Novak, T., Pollhammer, K.: Traffic-adaptive control of LED-based streetlights [industry forum] public lighting in urban environments accounts for about 50% of the electricity con. IEEE Ind. Electron. Mag. **9**(2), 48–50 (2015). https://doi.org/10.1109/mie.2015.2413455
12. Poza-Lujan, J.-L., Saenz-Penafiel, J.-J., Posadas-Yague, J.-L., Conejero, J.A., Cano, J.-C.: Use of receiver operating characteristic curve to evaluate a street lighting control system. IEEE Access **9**, 144660–144675 (2021). https://doi.org/10.1109/access.2021.3121669
13. Pedone, G.: Wireless multi-sensor networks for smart cities **17**(23), 7667–7676 (2017)
14. Gagliardi, G., et al.: Advanced adaptive street lighting systems for smart cities. Smart Cities **3**(4), 1495–1512 (2020). https://doi.org/10.3390/smartcities3040071
15. Ishak, Z., Ahmad, W.S.H.M.W., Radzi, N.A.M., Sulaiman, S., Ramli, N.E.: Placement accuracy algorithm for smart street lights. Turkish J. Electr. Eng. Comput. Sci. **29**(2), 845–859 (2021). https://doi.org/10.3906/ELK-2003-7
16. Ferreira, J.C., Afonso, J.A., Monteiro, V., Afonso, J.L.: An energy management platform for public buildings. Electronics **7**(11), 1–13 (2018). https://doi.org/10.3390/electronics7110294
17. Cheng, C.A., Wu, H.C.: A cost-effective electronic lighting driver with high-power-factor and soft-switching for supplying an LED street-light lamp. Int. Trans. Electr. Energy Syst. **31**(3), 1–16 (2021). https://doi.org/10.1002/2050-7038.12801
18. Mahoor, M., Salmasi, F.R., Najafabadi, T.A.: A hierarchical smart street lighting system with brute-force energy optimization. IEEE Sens. J. **17**(9), 2871–2879 (2017). https://doi.org/10.1109/JSEN.2017.2684240
19. Perko, J., Topić, D., Šljivac, D.: Providing power supply to other use cases integrated in the system of public lighting. Int. J. Electr. Comput. Eng. Syst. **8**(1), 33–39 (2017). https://doi.org/10.32985/ijeces.8.1.5
20. Pellegrino, A., Lo Verso, V.R.M., Blaso, L., Acquaviva, A., Patti, E., Osello, A.: Lighting control and monitoring for energy efficiency: a case study focused on the interoperability of building management systems. IEEE Trans. Ind. Appl. **52**(3), 2627–2637 (2016). https://doi.org/10.1109/TIA.2016.2526969
21. Shahzad, G., Yang, H., Ahmad, A.W., Lee, C.: Energy efficient intelligent street lighting system using traffic-adaptive control. IEEE Sens. J. **16**(13), 5397–5405 (2016). https://doi.org/10.1109/JSEN.2016.2557345
22. Casagrande, C., Nogueira, F., Salmento, M., Braga, H.: Efficiency in street lighting projects by employing LED luminaires and mesopic photometry. IEEE Lat. Am. Trans. **17**(6), 921–929 (2019). https://doi.org/10.1109/TLA.2019.8896814
23. Anguelov, K., Kavaldzhieva, K.: Methodology for determining the socio-economic factors in the performance of cost-benefit analysis for the production of electricity from biomass. In: Proceedings of 2021 17th Conference on Electrical Machines, Drives and Power Systems, ELMA 2021, no. July, pp. 1–4 (2021). https://doi.org/10.1109/ELMA52514.2021.9502978
24. Sánchez Sutil, F., Cano-Ortega, A.: Smart public lighting control and measurement system using lora network. Electronics **9**(1), (2020). https://doi.org/10.3390/electronics9010124

25. Banco Central del Ecuador: Instructivo Tasas de Interés (2019). https://contenido.bce.fin.ec/docs.php?path=/documentos/Estadisticas/SectorMonFin/TasasInteres/Indice.htm
26. Nogueira, F.J., et al.: Street lighting LED luminaires using telemanagement systems: study of case. In: 2014 11th IEEE/IAS International Conference on Industry Applications, IEEE INDUSCON 2014 - Electronic Proceedings (2014). https://doi.org/10.1109/INDUSCON.2014.7059464

Education

Monitoring and Adaptation of Assessment Activities in a VLE Supported by Learning Analytic

Joe Llerena-Izquierdo[1,2](✉), M. Elena Rodriguez[1], and Ana-Elena Guerrero-Roldán[1]

[1] Universitat Oberta Catalunya, Rambla Poblenou, Barcelona, Spain
{jllerenai,mrodriguezgo,aguerreror}@uoc.edu

[2] Universidad Politécnica Salesiana, Guayaquil, Ecuador

Abstract. The complex task of developing an assessment activity and its structure requires determining the appropriate set of elements to link the learning objectives with the expected results. The analysis of learning outcomes through the use of technological techniques are challenges that are still posed in the various research works, when using them in the educational field. This paper presents a model for monitoring and adapting assessment activities in a virtual learning environment. The objective is to determine the characteristics and conditions of assessment activities that allow the teacher to guide the behavior of students in a course according to the monitoring and adaptation of assessment activities in a virtual learning environment. Therefore, this paper presents a model for monitoring and adapting evaluation activities in a virtual learning environment. The results of this study reflect those students achieve an average pass rate of 87% in four courses that actively participate in this work and an attrition rate of 13%. The limitations found suggest external factors such as connectivity and the pandemic, which affect students but can be dealt with in a preventive sense, in the face of a possible academic risk through the learning analytics exhibited to students in the virtual learning environment. The results obtained show a high commitment of the students in achieving the challenges proposed by the teacher, reaching above-average performance values. For future work, this model will be replicable to several engineering subjects in first-year courses. This work contributes to the development of a method for teacher recommendations in VLE environments.

Keywords: Learning design · Logs of learners · Learning analytics · Follow-up and adaptation of activities

1 Introduction

The emergence of new areas of research within computer science, such as Learning analytics (LA) [1], Educational Data Mining (EDM) [2, 3], the development of intelligent tutoring system (ITS) [4], applied in the field of Computer Supported Collaborative Learning (CSCL) [5, 6], have been taking directions that address studies in interdisciplinary fields such as Collaborative Interactions (IC) [7].

V. Robles-Bykbaev et al. (Eds.): CITIS 2022, LNNS 607, pp. 409–419, 2023.
https://doi.org/10.1007/978-3-031-24327-1_35

Assessment activities are part of a scaffolding within the student learning process in a framework called Learning Design (LD) that requires the teacher to take several actions [8]. Among them, the dedication of time for its development, the use of meaningful learning resources that promote learning, the knowledge of technology for development and its familiarity for application in the activity, and the use of the technology for development [9, 10].

The knowledge of what happens to students before, during and after an evaluation activity is a latent challenge for the designers of the programmatic contents of a course [11]. Designing an effective evaluation model, regardless of the learning model [12], depends on the achievement of students within a curricular content and the satisfaction of the knowledge attained by them from their perspective [13].

At the Universidad Politécnica Salesiana in Ecuador, students' intermediate and final grades obtained in a course come from various learning activities developed within a teaching plan [14]. These activities determine those achievements and learning outcomes in accordance with the objectives of the programmatic contents in courses of different specialties [15], (see Fig. 1).

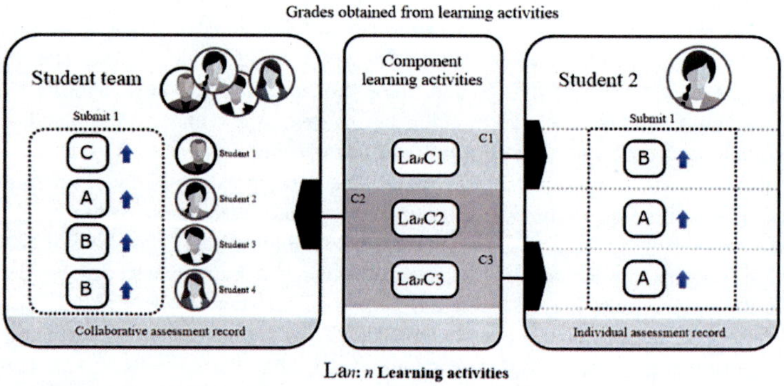

Fig. 1. Outline of learning activities and grades achieved (A, B, C, D, N) that a student achieves individually and as a group in a course.

Planned assessment activities are part of the student learning process, designed by the understanding of a teacher or teachers who develop the resources and provide appropriate feedback to students [16, 17]. No immediate feedback regarding mistakes made in a given element of the activity or caused by erroneous learning, in a given type of student [18], time to consolidate learning is reduced, especially in that activity in which the student has failed [19, 20]. The qualification is a definitive evaluation of a part of the learning process achieved during a defined time in six months, corresponding to a period or university academic level, (see Fig. 2).

The grade only evidences a weighting, where performance depends on future activities. Achieving a strategy where the teacher or group of teachers can follow up the students' course, as well as in an individualized way, would allow reducing a high number of students at risk of not passing, as well as the capacity to carry out actions of adaptations in the curriculum to avoid student desertion.

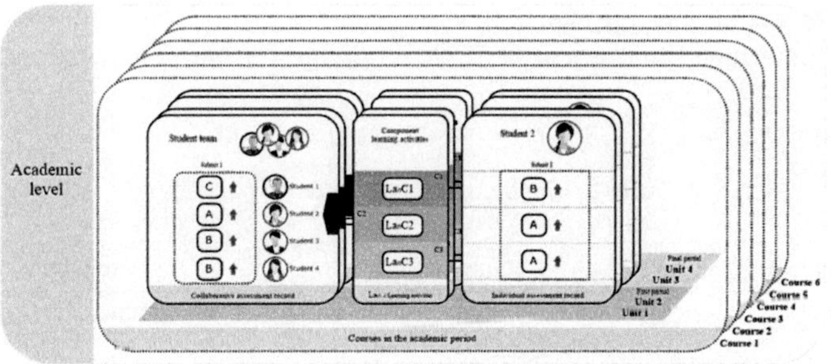

Fig. 2. Outline of a student's learning activities in different courses within an academic semester.

In addition, the results of a student's evaluation activities allow the teacher to continuously improve his evaluation model according to the set of characteristics and conditions that consolidate a follow-up supported by learning analytics [21].

This work focuses on taking advantage of the results achieved by the students, for their due collection, analysis, treatment, and presentation in a way that allows the teacher to effectively follow up and adapt the activities in a virtual learning environment (VLE) supported by learning analytics.

The objective of this study is to determine the characteristics and conditions of assessment activities that allow the teacher to guide the behavior of students in a course according to the monitoring and adaptation of assessment activities in a (VLE). The research question addressed in this work is, what are the characteristics and conditions of assessment activities in basic engineering subjects at the undergraduate level that allow the adaptation of the curriculum of a subject according to the results of students during its development?

2 Methods and Materials

The methodology followed is empirical-analytical with a quantitative approach. The method of research work is quasi-experimental using the technique of the survey to a group of 38 students who passed their subjects in the academic period of studies number 59 (October 2020 to March 2021), of the engineering careers of the Universidad Politécnica Salesiana in Ecuador.

Four courses are initially considered, Programming I (A1), Free Software (A2), Computer Networks (A3) and Linear Algebra (A4), a total of 113 students participating in a classroom monitoring method designed by one of the authors. [22]. A Moodle learning environment is used applying Kolb's experiential learning methodology [23] with an approach to Zimmerman's self-regulated learning [24, 25].

During the academic period of studies from October 2020 to March 2021, a total of 99 students, 88%, passed. A dropout or non-completion rate of 12% is evidenced, promising results to be replicated in other courses in the following periods. The students voluntarily accepted the hybrid modality of the course and knowledge of the follow-up model presented by the professor. The demographic data are presented in Table 1, showing the number of participants by gender [26]. The participation of male students maintains a higher percentage than female students. There is also a significant percentage of students older than the average of 18 years of age for the first year of studies, with 39%.

Table 1. Engineering student demographics.

Demographic categories	Sub-categories	Percentage
Gender	Male	79%
	Female	21%
Age range	18–19 years old	61%
	20–21 years old	13%
	22–23 years old	3%
	24–25 years old	5%
	26–27 years old	5%
	Greater than 27	13%

2.1 Learning Activities as a Focus for Evaluation

To carry out this study, a set of significant activities were developed, according to three components [14]: collaborative learning, experimentation practices and autonomous work. It involves a learning design that translates into an appropriate instructional design for each course [22], due to the characteristics of the students, according to Zimmerman's model [24], as competent self-regulated, cognitively oriented self-regulated, reflectively oriented self-regulated, minimally self-regulated types [27]. Classrooms identify learning conditions for students using self-regulation strategies [9] conducted in previous studies.

In this model, three dimensions are constituted for the characteristics of the activities and their adaptation: the communicative resources to support the activity, the student's performance, and the teacher's motivation (see Table 2).

3 Results

The results obtained are presented in two sections. In the first section, the results are obtained from the VLE, compiled for analysis and presentation to the students and are reflected, (see Fig. 3).

The set of students who have achieved improvement (blue line) and no improvement (orange line) as well as its tendency are shown in percentages, during the first part of the period (known as intercycle, visible on the samples from 1 to 4 on the abscissa axis), with the challenge of reaching a minimum of 25 points (out of 50), which involves the teacher's follow-up executed in a process of communication of the activities, revision of

Table 2. Characteristics of the activities to be considered according to their dimension.

Dimension	Communication resources for the activity	Student performance	Teacher motivation
External features	Access to physical documents	Access to the activity	Activity feedback
	Reading of physical document	Availability of the activity	Motivation prior to the activity
	Format of physical document	Activity enhances learning	Motivation during the activity
	Utility of the multimedia resource		Post-activity motivation
	Multimedia resources as complements		

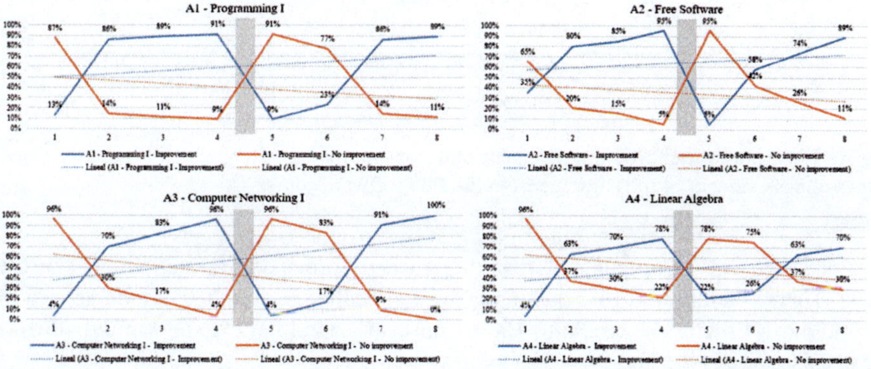

Fig. 3. Linear trends of improvement and non-improvement processes according to two stages of achievement of participants in the subjects A1 - Programming I, A2 - Free Software, A3 - Computer Networks I and A4 - Linear Algebra in Engineering degrees.

them, personalized feedback to the student's needs in an immediate way and permanent motivation.

The second part of the period (known as the *end of the cycle*), the challenge of reaching the minimum score to finish the course is 70 points (adding the 50 of the inter-cycle with the final, giving 100 points, showing from 5 to 8 on the abscissa axis), (see Fig. 3).

Graphs A1 - Programming I and A4 - Linear Algebra, correspond to first year students. Two situations are shown according to the scope of the subject, in the first subject (A1), 89% pass the course, a value that exceeds the average pass rate of 70%, and 11% fail the course. The failure rate in some courses is due to several external factors that affect the follow-up model, among them, economic situations, effects of the pandemic, health problems, as well as those who have failed the course. In the second subject (A4) an average semester pass rate of 70% is achieved.

The trend lines in graphs A1 and A4 show a continuous increase in their distance, with the trend in A1 being higher than the trend in A4, but not reaching the maximum achievable distance as in A3, the subject that is the ideal reference, since all the students finished the course.

For graph A2 - Free Software, they correspond to students in the fourth cycle. There is evidence of a rapid growth in the distances of the trend lines for the two parts of the cycle, sample 2 and sample 6 on the abscissa axis, and this responds to the diligent motivation of the students to overcome the challenges suggested in the subject, (see Fig. 4), abscissa axis.

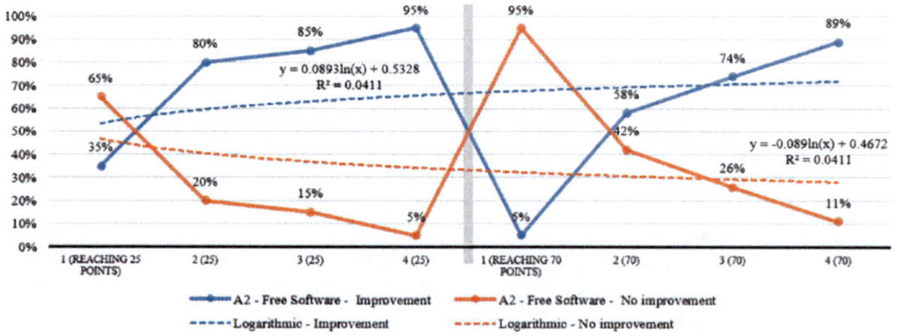

Fig. 4. Logarithmic trend of improvement and non-improvement processes according to two stages of achievement of the participants in the free software course.

For graph A3 - Computer Networking I, they correspond to students in the fifth cycle. There is a late start for reaching the initial challenge, but they overcome it in the second part of the cycle, reaching 100% approval, sample 4(25) and sample 4(70) on the abscissa axis, (see Fig. 5).

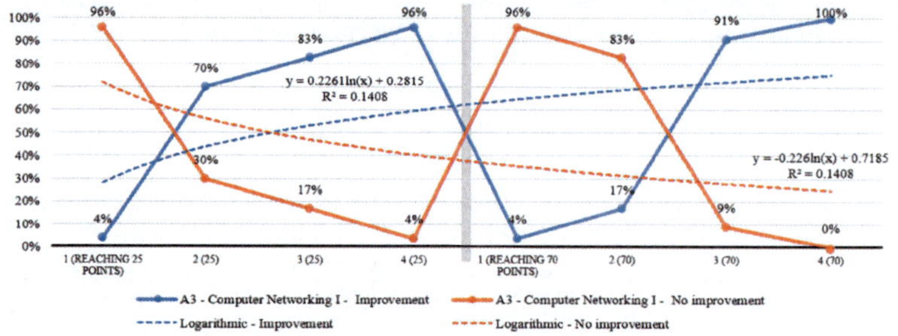

Fig. 5. Logarithmic trend of improvement and non-improvement processes according to two stages of achievement of participants in the Computer Networking I course.

The second section of the results are obtained from 38 students who voluntarily agreed to respond to a 34-item survey covering the information in Table 1 and Table 2 regarding the assessment activities, demographic data and the three dimensions determined.

Distributed in four blocks of questions, the first block corresponds to the participant's profile, the second block corresponds to the dimension, communicative resources of support for the activity, the third block corresponds to the dimension of performance, and the fourth block corresponds to the dimension of motivation carried out by the teacher. This study prioritizes the dimension of student performance according to the monitoring and adaptation of the activities in the VLE.

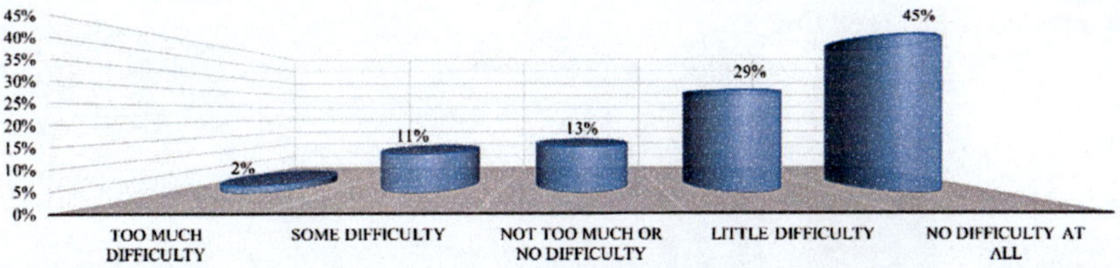

Fig. 6. Percentage of participants who determine the difficulty with access to the activity to be carried out.

With respect to the question, "Have you had difficulty accessing the activity?", 45% indicated "No difficulty at all", 29% indicated "Little difficulty", 13%, "Not too much or no difficulty", 11%, "Some difficulty" and 2%, "Too much difficulty", (see Fig. 6). There are external factors such as access to the Internet, bandwidth, difficulty with the device and office tools, which attribute 12% of having some difficulty with access to the activities in the VLE.

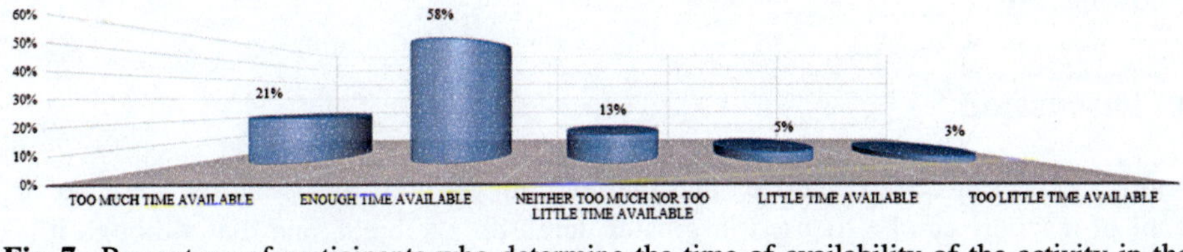

Fig. 7. Percentage of participants who determine the time of availability of the activity in the virtual environment for the learning objective.

With respect to the question, "Has the time available in the virtual environment been sufficient for the learning objective of the activity?", 21% indicated that "Too much time available", 58%, "Enough time available", 13%, "Neither too much nor too little time available", 5%, "Little time available" and 3% "Too little time available". This shows that 79% and even 92%, the time availability factor has not been a difficulty for the development of the activities, (see Fig. 7).

With respect to the question, "Do you agree that the activity has helped you to learn the topic, unit or section of the contents studied?", 50% of the students indicated that they

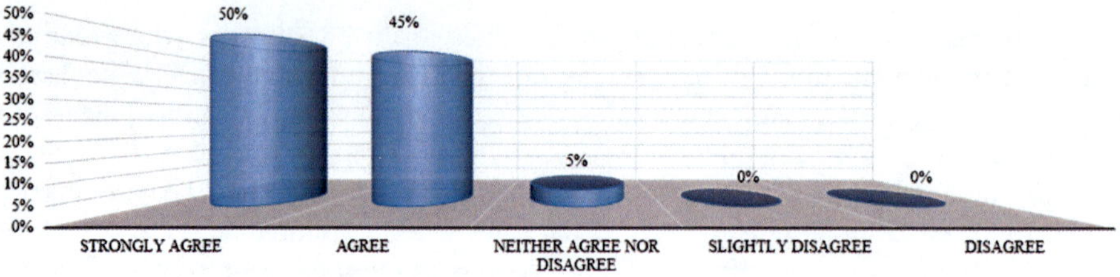

Fig. 8. Percentage of participants who agree that the activity developed helps the learning of the topic, unit, or section according to the contents studied.

"Strongly agree", 45% indicated that they "Agree" and 5% indicated that they "Neither agree nor disagree", (see Fig. 8).

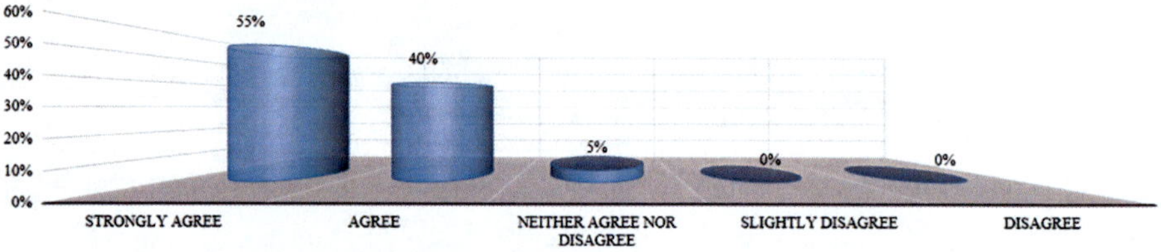

Fig. 9. Percentage of participants who agree with the feedback given by the teacher helping to reduce a low grade in the following activity.

Finally, they were asked if you agree that the feedback helps the following activity not to get a low grade, with 55% agreeing "Strongly agree", 40% agreeing "Agree" and 5% "Neither agree nor disagree", (see Fig. 9). This shows that 95% of the students agree with the feedback provided by the teacher and that it helps to reduce a low grade in the following activity.

4 Discussion

In this work, it has been analyzed that the learning activities designed by a teacher or team of teachers can be focused in greater depth on a subject, and that although it is part of the contents, the minimum contents must be fulfilled according to a level of significance that exerts the greatest effect on the learning of the studies or to a greater number of them, according to studies [16].

Characteristics of the learning activities are determined, related to three dimensions, from the communication of the learning activity, the student's performance and the motivation given by the teacher [14, 28]. Likewise, the conditions for the effectiveness of evaluation activities are related to the design of learning activities in accordance with the communicative, didactic-pedagogical and technological scope that affect students' cognitive aspects [29, 30].

In addition, external influences on the development of an activity, for example, the resources available outside the VLE or that the student has or obtains, become a limitation in the future if the design of the activities does not consider adaptations that improve the learning process according to other authors [9, 15].

5 Conclusions

This work contributes with an experience of follow-up and adaptation of evaluation activities in a VLE supported by learning analytics based on a design of learning activities for engineering courses at the Salesian Polytechnic University in the city of Guayaquil. The results reach promising values that allow replicating the proposed model to other subjects for future works, especially in subjects of initial courses where the percentage of desertion or repetition are high.

A set of characteristics of the learning activities, classified in three dimensions, and a way of monitoring students and conditions for the adaptation of these activities in four subjects of engineering careers using learning analytics in a VLE are presented.

A high commitment of the students in achieving the challenges proposed by the teacher is evidenced, reaching performance values above the average. It is also found that for courses of different specialties, the model of monitoring and adaptation of activities allows the teacher to have a complete vision of the course and to make decisions that avoid a high number of students dropping out or being at academic risk. In addition, the data are presented to the students so that they can see the development of their performance in relation to the other members of the course, generating in them a sense of prevention, awareness, and self-motivation. For future work, the experience will be replicated in different courses.

References

1. Larrabee Sønderlund, A., Hughes, E., Smith, J.: The efficacy of learning analytics interventions in higher education: a systematic review. Br. J. Educ. Technol. **50**, 2594–2618 (2019)
2. Hantoobi, S., Wahdan, A., Al-Emran, M., Shaalan, K.: A review of learning analytics studies. In: Al-Emran, M., Shaalan, K. (eds.) Recent Advances in Technology Acceptance Models and Theories. SSDC, vol. 335, pp. 119–134. Springer, Cham (2021). https://doi.org/10.1007/978-3-030-64987-6_8
3. Romero, C., Ventura, S.: Educational data mining and learning analytics: an updated survey. Wiley Interdiscip. Rev. Data Min. Knowl. Discov. **10**, e1355 (2020). https://doi.org/10.1002/WIDM.1355
4. Guo, L., Wang, D., Gu, F., Li, Y., Wang, Y., Zhou, R.: Evolution and trends in intelligent tutoring systems research: a multidisciplinary and scientometric view. Asia Pacific Educ. Rev. **22**, 441–461 (2021). https://doi.org/10.1007/S12564-021-09697-7/TABLES/6
5. Silva, L., Mendes, A.J., Gomes, A.: Computer-supported collaborative learning in programming education: a systematic literature review. In: IEEE Global Engineering Education Conference (EDUCON), 2020-April, pp. 1086–1095 (2020). https://doi.org/10.1109/EDUCON45650.2020.9125237
6. Wise, A.F., Knight, S., Shum, S.B.: Collaborative learning analytics. In: Cress, U., Rosé, C., Wise, A.F., Oshima, J. (eds.) International Handbook of Computer-Supported Collaborative Learning. CCLS, vol. 19, pp. 425–443. Springer, Cham (2021). https://doi.org/10.1007/978-3-030-65291-3_23

7. Bao, H., Li, Y., Su, Y., Xing, S., Chen, N.S., Rosé, C.P.: The effects of a learning analytics dashboard on teachers' diagnosis and intervention in computer-supported collaborative learning. Technol. Pedagog. Educ. **30**, 287–303 (2021). https://doi.org/10.1080/1475939X.2021.1902383
8. Utterberg Modén, M., Tallvid, M., Lundin, J., Lindström, B.: Intelligent tutoring systems: why teachers abandoned a technology aimed at automating teaching processes. In: Proceedings of the 54th Hawaii International Conference on System Sciences, p. 1538 (2021)
9. Llerena-Izquierdo, J., Ayala-Carabajo, R.: Integración de medios educativos digitales para la enseñanza-aprendizaje interactiva de asignaturas básicas de carreras de Ingeniería. In: La educación en Red: realidades diversas, horizontes comunes, XVII Congreso Nacional y IX Iberoamericano de Pedagogía, pp. 1173–1174. Universidad de Santiago de Compostela, Servicio de Publicaciones e Intercambio Científico, Santiago de Compostela (2021). https://doi.org/10.15304/cc.2021.1393
10. Engeness, I.: Developing teachers' digital identity: towards the pedagogic design principles of digital environments to enhance students' learning in the 21st century. Eur. J. Teach. Educ. **44**, 96–114 (2021). https://doi.org/10.1080/02619768.2020.1849129
11. Karaoglan Yilmaz, F.G.: Utilizing learning analytics to support students' academic self-efficacy and problem-solving skills. Asia Pac. Educ. Res. **31**(2), 175–191 (2021). https://doi.org/10.1007/s40299-020-00548-4
12. Jivet, I., Scheffel, M., Drachsler, H., Specht, M.: Awareness is not enough: pitfalls of learning analytics dashboards in the educational practice. In: Lavoué, É., Drachsler, H., Verbert, K., Broisin, J., Pérez-Sanagustín, M. (eds.) EC-TEL 2017. LNCS, vol. 10474, pp. 82–96. Springer, Cham (2017). https://doi.org/10.1007/978-3-319-66610-5_7
13. Kaliisa, R., Rienties, B., Mørch, A.I., Kluge, A.: Social learning analytics in computer-supported collaborative learning environments: a systematic review of empirical studies. Comput. Educ. Open. **3**, 100073 (2022). https://doi.org/10.1016/J.CAEO.2022.100073
14. Llerena, J., Ayala-Carabajo, R.: Significant learning activities (ASA) in the modality of face-to-face studies with integration of virtual educational media in engineering careers. In: 2020 XV Conferencia Latinoamericana de Tecnologias de Aprendizaje (LACLO), pp. 1–9 (2020). https://doi.org/10.1109/LACLO50806.2020.9381134
15. Nguyen, Q., Rienties, B., Toetenel, L.: Unravelling the dynamics of instructional practice: a longitudinal study on learning design and VLE activities. In: Proceedings of the 7th International Learning Analytics \& Knowledge Conference, pp. 168–177 (2017)
16. Toro-Troconis, M., et al.: Learning design in higher education: building communities of practice. In: Measurement Methodologies to Assess the Effectiveness of Global Online Learning, pp. 259–288. IGI Global (2022)
17. Laurillard, D., Kennedy, E., Charlton, P., Wild, J., Dimakopoulos, D.: Using technology to develop teachers as designers of TEL: evaluating the learning designer. Br. J. Educ. Technol. **49**, 1044–1058 (2018)
18. Rahayu, N.W., Ferdiana, R., Kusumawardani, S.S.: A systematic review of ontology use in E-Learning recommender system. Comput. Educ. Artif. Intell. **3**, 100047 (2022)
19. Lopez-Chila, R., Llerena-Izquierdo, J., Sumba-Nacipucha, N.: Using examview to create questionnaires for online evaluation in VLEs. In: Proceedings - 2021 2nd International Conference on Information Systems and Software Technologies (ICI2ST) 2021, pp. 3–9 (2021). https://doi.org/10.1109/ICI2ST51859.2021.00009
20. Alves, P., Miranda, L., Morais, C.: The influence of virtual learning environments in students' performance. Univers. J. Educ. Res. **5**, 517–527 (2017)
21. Llerena, J., Alava-Moran, N., Zamora-Galindo, J.: Learning analytics for student academic tracking, a comparison between Analytics Graphs and Edwiser Reports. In: 2021 2nd International Conference on Information Systems and Software Technologies (ICI2ST), pp. 101–107. IEEE (2021). https://doi.org/10.1109/ICI2ST51859.2021.00022

22. Llerena-Izquierdo, J.: Virtual classroom design model and its relation to student motivation and performance in a moodle learning environment during the emergency of COVID-19. In: Berrezueta, S., Abad, K. (eds.) Doctoral Symposium on Information and Communication Technologies - DSICT. LNEE, vol. 846. Springer, Cham (2022). https://doi.org/10.1007/978-3-030-93718-8_3
23. Kolb, D.A.: Experiential Learning: Experience as the Source of Learning and Development. FT Press, New Jersey (2014)
24. Zimmerman, B.J.: Becoming a self-regulated learner: an overview. Theory Pract. **41**, 64–70 (2002)
25. Saint, J., Fan, Y., Gašević, D., Pardo, A.: Temporally-focused analytics of self-regulated learning: A systematic review of literature. Comput. Educ. Artif. Intell. **3**, 100060 (2022). https://doi.org/10.1016/J.CAEAI.2022.100060
26. de las Cuevas, P., García-Arenas, M., Rico, N.: Why Not STEM? a study case on the influence of gender factors on students' higher education choice. Mathematics, **10**, 239 (2022)
27. Zheng, J., Xing, W., Zhu, G., Chen, G., Zhao, H., Xie, C.: Profiling self-regulation behaviors in STEM learning of engineering design. Comput. Educ. **143**, 103669 (2020)
28. Gomez, A., Chamba Eras, L.A., Aguilar, J.: Multi-agent systems for the management of resources and activities in a smart classroom. IEEE Lat. Am. Trans. **19**, 1511–1519 (2021). https://doi.org/10.1109/TLA.2021.9468444
29. Merchán Basabe, C.A., Merchán Basabe, C.A.: Modelamiento pedagógico de Ambientes Virtuales de Aprendizaje (AVA). Tecné, Episteme y Didaxis TED, 51–70 (2018)
30. Lucumi Useda, P., González Castañeda, M.A.: El ambiente digital en la comunicación, la actitud y las estrategias pedagógicas utilizadas por docentes. Tecné, Episteme y Didaxis TED. 1, 109–129 (2015). https://doi.org/10.17227/01213814.37TED109.129

Simulation of Synchronous and Asynchronous Generators Models for a Virtual Electrical Machines Laboratory

Omar Carrillo, Kevin Mendoza, Hugo Arcos, and Patricia Otero(✉)

Escuela Politécnica Nacional, Quito, Ecuador
patricia.otero@epn.edu.ec

Abstract. This paper presents the mathematical models for synchronous and asynchronous generators developed in MATLAB App Designer, with the purpose of implementing a virtual laboratory of electrical machines as a teaching tool for engineering students.

This tool was needed because of the SARS-CoV-2 virus pandemic, and the parameters of existing machines were used to emulate the results expected in the real laboratory. The dynamic events simulation of both the synchronous machine and the asynchronous machine use models implemented in Simulink to get transient responses and the results of each practice are displayed in graphical interfaces.

Finally, the main advantages offered by the software and the benefits given to the user in each module are detailed, focused on supporting learning, and reaching the objectives established in each laboratory practice.

Keywords: Synchronous generator · Asynchronous generator · Steady-state Model · Dynamic responses

1 Introduction

With the increasing popularity of online teaching due to the SARS-CoV-2 virus pandemic, new technologies and software teaching methods have been developed to facilitate learning [1]. This software is not the exception and focuses on solving the problem that Escuela Politécnica Nacional students of the electrical machines laboratory went through since they were not able to access the physical laboratory and carry out their practices.

In the synchronous generator simulation modules, two different power machine models have been incorporated: a model of a generation unit of the Coca Codo Sinclair Hydroelectric Plant and an existing machine in the electrical machines laboratory. On the other hand, the simulation module of the asynchronous generator focuses only on the model of the induction machine existing in the laboratory.

The objective is to develop an interactive software in a MATLAB environment that allows users to carry out laboratory practices simulating the handling of asynchronous and synchronous machines, using Simulink models defined by [2] and [3].

V. Robles-Bykbaev et al. (Eds.): CITIS 2022, LNNS 607, pp. 420–429, 2023.
https://doi.org/10.1007/978-3-031-24327-1_36

Several previous works have reviewed the saturation model of the synchronous machine, such as [4] which uses the saturation model and analyzes the cross-coupling effect between the d- and q-axis to use the model in a real generation system.

Furthermore, [5] defines and uses 'generalized flux' and 'generalized inductance' concept, which was successfully developed and applied in the derivation of numerous models of saturated smooth air-gap A.C. machines.

Those models have been reviewed to choose the best saturation model, looking forward to achieving the smallest program's execution time.

Finally, this paper shows the methods implemented to program each module for each laboratory practical. Sequentially, this document brings to readers the final product, showing the principal advantages of each module.

2 Methods

The software development is based on the implementation of mathematical steady-state models, as well as mathematical models for the dynamic events simulation.

Each module has simulation windows, where users can introduce variations in the control variables of the machines to observe and analyze their behavior over time.

Furthermore, mathematical steady-state models are used to obtain the RMS value of variables such as active power, reactive power, voltage, current, and so forth, to be presented to users as measurements. Below are the steady-state mathematical algorithms for each of the studies performed.

2.1 Tests on Synchronous Generators

The software includes the interfaces for the performance of the open-circuit test, short-circuit test, and loading test.

Open-Circuit Test. This test purpose is to obtain the open-circuit characteristic (OCC) of the synchronous machine and analyze the behavior of the armature voltage induced in open-circuit conditions. Consequently, the induced voltage is calculated by the Eq. 1 [4].

$$E_f = \mathrm{x}_{md} \cdot \left(\frac{1}{N}\right) \cdot \left(\frac{2}{3}\right) \cdot \left(\frac{v_f}{r_f + r_{var}}\right) \tag{1}$$

where x_{md} is the direct axis mutual reactance, N is the number of turns, v_f is the field circuit supply voltage, r_f is the rotor winding resistance, r_{var} is the external variable resistance connected to the rotor winding.

The values of vf and rvar are known input parameters that are entered by user-manipulated sliders through the graphical interface. The value of the reactance xmd could be represented by two regions: the linear region represented by the gap line and the non-linear region due to the presence of magnetic saturation. The value of xmd in the non-linear region is shown in Eq. 2 [2]:

$$x_{md}^{sat} = \frac{\psi_{md}^{sat}}{\psi_{md}^{unsat}} \cdot x_{md}^{unsat} \tag{2}$$

where: x_{md}^{sat} is the direct axis stator magnetizing reactance under saturation conditions, x_{md}^{unsat} is the direct axis stator magnetizing reactance under unsaturation conditions, ψ_{md}^{sat} is the direct axis mutual flow concatenations in volts, ψ_{md}^{unsat} is the direct axis mutual flow concatenations in volts.

The value obtained in Eq. 2 is entered into Eq. 1; thus, the induced voltage E_f is a value that corresponds to both unsaturated and saturated operating conditions.

The calculations presented in Eqs. 1 and 2 are performed each time the user manipulates the value of the field current by changing the sliders that control the variables v_f and $r_{var.}$

Short-Circuit Test. Obtaining steady-state responses for this study consists of the resolution of the short circuit equivalent, obtaining the armature current, which is the interest variable for the study. Equation 3 describes the generator behavior in a short-circuit steady-state condition [6].

$$E_f\angle\delta - j\left(x_d - x_q\right)\cdot I_d\angle(\delta - 90) = \left(r_s + jx_q\right)\cdot I_a\angle - \theta \tag{3}$$

where E_f is the internal voltage, δ is the rotor angle, x_d is the direct reactance, x_q is the quadrature reactance, I_d is the direct current, r_s is the stator resistance, I_a is the armature current and θ is the armature current angle.

By establishing Ef as a phasor reference and reducing the expression by separating the real part and imaginary part, Eqs. 3 and 4 are obtained.

$$\theta = tan^{-1}\frac{r_s}{x_q} \tag{4}$$

$$I_a = \frac{E_f}{r_s cos\theta + x_d \sin\theta} \tag{5}$$

Loading Test. The proposed methodology for this study is based on the resolution of the steady-state model of the synchronous machine, with three different types of loads: resistive, inductive, and capacitive. Thus, there is a system of nonlinear equations resulting from this process, whose output variables are the power angle and the armature current. These values are used to obtain the remaining variables, such as active power (P), reactive power (Q), apparent power (S), and terminal voltage (Vt).

2.2 V Curves

When a synchronous machine operates in parallel with an electrical grid and the armature resistance value is considered, reactive power, and active power can be obtained by Eqs. 6 and 7 [7].

$$P = \frac{V_t}{X_d X_q + R_a^2}\left[E_f\left(X_q \sin(\delta) + R_a \cos(\delta)\right) - R_a V_t - \frac{V_t}{2}\left(\frac{R_a^2}{X_q} + X_q\right)\sin(2\delta)\right] + \frac{V_t^2}{2X_q} sen(2\delta) \tag{6}$$

$$Q = \frac{V_t}{X_d X_q + R_a^2}\left[E_f\left(X_q \cos(\delta) - R_a \sin(\delta)\right) + \frac{V_t}{2}\left(\frac{R_a^2}{X_q} - X_q\right) - \frac{V_t}{2}\left(\frac{R_a^2}{X_q} + X_q\right)\cos(2\delta)\right] + \frac{V_t^2}{2X_q}(cos(2\delta) - 1) \tag{7}$$

Table 1. Steps to obtain V Curve.

Steps	Known variables	Equations used	Variables obtained
1	I_f	1	E_f
2	P, E_f	6, 7	Q, δ
3	P, Q, V_{fase}	8	I_a

The software uses the steps in Table 1 to establish the relation between armature current and field current.

The next expression is used to obtain the armature current.

$$\frac{\sqrt{P^2 + Q^2}}{\sqrt{3}V_f} = I_a \tag{8}$$

This process is repeated several times to get every point until the V Curve is completed.

2.3 Capability Curves

This module let students draw all the safe operating zone inside the capability curves, whose equations are defined by [8]. Users can see the operation point at the P-Q plane. If the student goes under the limit, within the limit, or goes further than the limit, a warning is shown up in the panel. Table 2 shows the signs color code for the panel lights.

Table 2. Signaling panel of operation limits.

Warning	Signs
Under the limit	Blue light
In the limit	Greenlight
Overstep limit	Red light

This module allows students to draw a line between the real-time operation point and the last operation point saved.

2.4 Dynamic Behavior of the Synchronous Machine

The program gets initial conditions using If and T0 as input parameters. Equations 6 and 7 use Ef and active power to get the reactive power. Sequentially, active power and reactive power are used to calculate initial conditions for the variables. Finally, the Simulink model [2] process the simulation. To sum up, Table 3 shows the input and output data.

Table 3. Input and output data to the synchronous machine model.

Input data		Output data	
E_{fo}	Initial induced voltage	V_t	Termina voltage
E_{ff}	Final induce voltage	i_f	Field current
T_o	Initial torque	i_a	Armature current
T_f	Final torque	P	Active power generated
t_1	Time when the perturbation begins	Q	Reactive power generated
t_2	Time when the perturbation ends	δ	Power angle
t_3	Time of the simulation during		

2.5 The Induction Generator

The objective is to operate the induction machine properly so that it works as an asynchronous generator. The conditions are:

- The prime mover linked to the induction machine is a direct current (DC) machine.
- When energizing, the induction machine operates as a motor, while the DC machine operates as a generator with an independent excitation connection.

The DC machine's connection changes to a shunt motor connection using a circuit breaker. Once the DC machine operates as a motor, it controls the system speed by varying the field current. The new rotation speed is obtained using Eq. 9, which is based on the proportionality of the change in induced voltage and speed [6].

$$\omega_f = \frac{E_{a_f}}{E_{a_o}} \cdot \omega_o \tag{9}$$

Concerning the induction machine, its variables are obtained by solving the equivalent steady-state circuit, in which the slip is an input parameter entered by users through sliders.

The software also includes an interactive electromagnetic torque-speed graph, which shows a graphical representation of the amount of torque that the machine can produce, according to the speed value and supply voltage that users assign.

3 Results

This section shows the implementation of mathematical algorithms, as well as the main advantages that the usage of the software provides.

3.1 Tests on Synchronous Generators

Below are the graphical interface results developed for the open-circuit test, short-circuit test, and loading test.

Open-Circuit Test. The open-circuit test is carried out sequentially, and the resulting graphical interface is shown in Fig. 1b. This module has multiple advantages for learning electrical machines. Among the main ones are:

- Elaboration of the OCC. The software allows obtaining the machine OCC by storing eight successive measurements of open-circuit terminal voltage (calculated by Eq. 1) and field current.
- Interactive graph of direct axis reactance. This module presents the ease of showing interactively how magnetic saturation affects the direct axis reactance, reducing its value as the field current increases.

Short-Circuit Test. This graphical interface performance is very similar to the one presented in the open-circuit section. The short-circuit test graphical interface is shown in Fig. 1c. The main advantages of the short-circuit test module are:

- Elaboration of the SCC. The software has the ease of showing the SCC once the entire procedure is finished by plotting the stored armature current and field current.
- Dynamic events simulation. This module allows to graph the dynamic behavior of the most representative machine variables when operating in short-circuit conditions.

Loading Test. This study allows analyzing the variable's behavior of the synchronous machine when changes the excitation condition or the connected load value. Sliders are shown in Fig. 1a handle the load value in ohms, both for the resistive value and the reactive value.

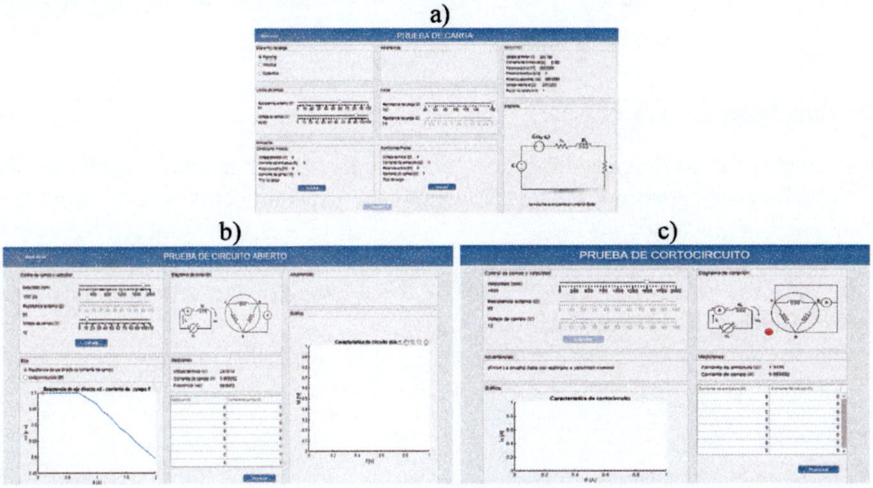

Fig. 1. Graphical interfaces: a) Loading test, b) Open circuit test, c) Short circuit test

The main advantages provided by the software are:

- Voltage drop analysis. This module allows the user to demonstrate the load connection impact on the synchronous generator variables. Users can see this effect in the "Measurements" panel.
- Selection of the type of load to connect. The software allows users to select the type of load with which they want to work and update the measurement values instantly.

3.2 V Curves

This interface has components that make it easier to study the V Curves students. These allow students to interactively draw the V Curve, so they could have better comprehension.

An advantage is the analysis of three V curves in the same plot. Likewise, users can analyze every zone of each graphic in one plot and make a comparison. An example is shown in Fig. 2, which shows the minimum point of each curve.

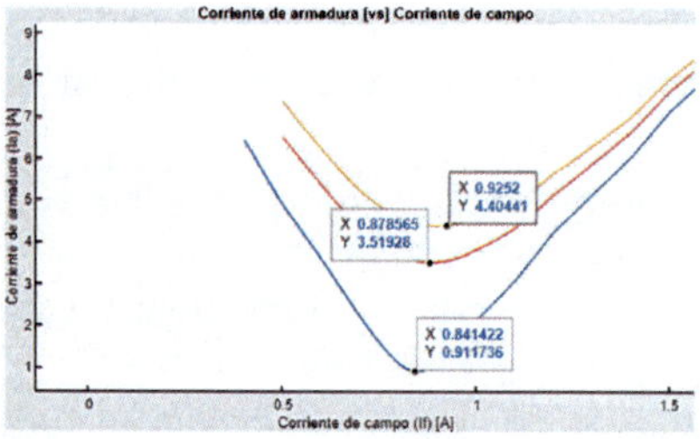

Fig. 2. Analysis of V curves built by users.

3.3 Capability Curves

This module presents the capability curves of two types of machines with which users can see the difference between both. Figure 3 shows the capability curves of the laboratory machine.

One of the benefits is that the software allows users to move through the operation limits and draw them. Section 2.3 explains the drawing of the safe operating zone. Figure 3 shows a drawing operation limit.

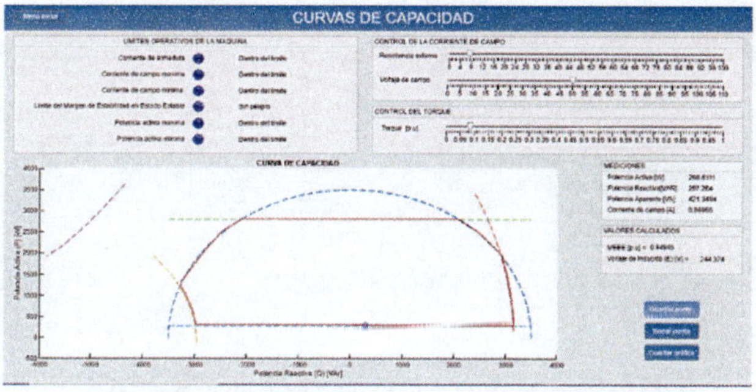

Fig. 3. Operation limit zone drawing of a unit of the laboratory machine.

3.4 Dynamic Behavior of the Synchronous Machine

This module allows users to make a sensitivity analysis, so they can understand which variables are affected by a different variation of field current or torque, especially on the transition. Figure 4 shows how much is affected the output variables when torque has changed.

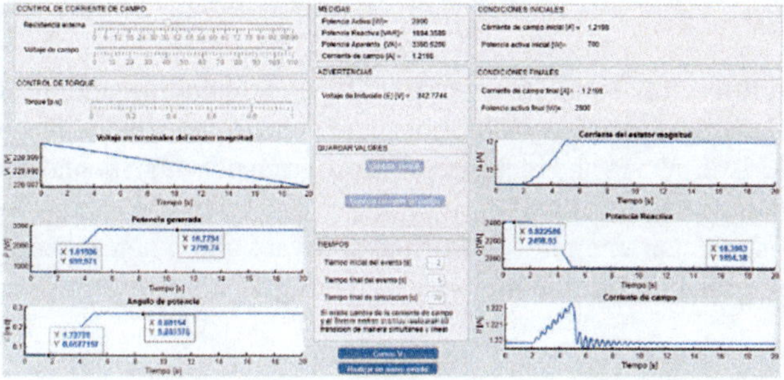

Fig. 4. Sensitive analysis when torque has increased, laboratory machine.

3.5 The Asynchronous Generator

This graphical interface has the advantage of sequentially showing users the procedure to make an induction machine works as an asynchronous generator. The graphical interface is shown in Fig. 5.

The advantages of using this module are:

- Interactive graph electromagnetic torque – speed. Using the software, users can visualize the behavior of the electromagnetic torque of the induction machine interactively, allowing them to deduce and analyze the maximum torque conditions and the effect that the induction machine armature voltage has.

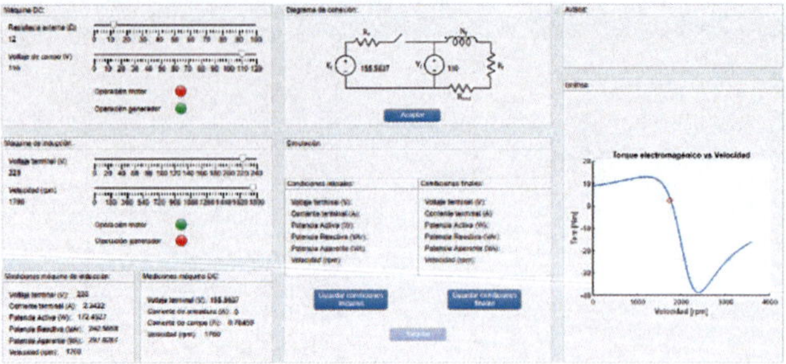

Fig. 5. Asynchronous generator graphical interface.

- Visual aid to identify the machine's operation mode. The software interactively displays the instant in which users change the machine operation mode using warning lights. The green light means that the machine operates in that state and the red light indicates that the machine works in the opposite condition, as shown in Fig. 5.
- Simulation of dynamic events. This module has the easy to show the dynamic behavior of the induction machine's main variables, mainly during the transition from the operation as a motor to its operation as a generator.

4 Discussion

As an academic tool, this software increases the understanding and allows professors to improve explanations about asynchronous and synchronous machines in every topic. Having characteristic curves and measures, the program gets high visual learning to clear up doubts.

A group of students ran a pilot test of the software and when asked if it reinforced their knowledge in the topics, the majority of them give a positive answer, as shown in Fig. 6.

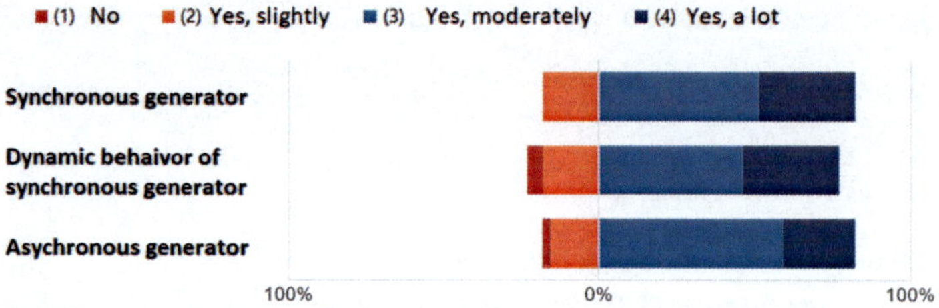

Fig. 6. Students answer to the question ¿Did the software reinforced your knowledge in the topics?

The results in Fig. 6 suggest that the tool proved to be useful as a teaching tool despite its limitations in the graphical interface.

As a software, another of its primary limitations in processing speed is that it was developed in MATLAB. For that reason, the program has considerable low-speed processing compared with programs developed in C++ or Fortran.

5 Conclusions

Using the App Designer tool facilitates to development of interactive modules, as it has a friendly programming environment that allows graphical interface customization.

The use of graphical interfaces helps to illustrate the operation of synchronous and asynchronous generators used in laboratories, allowing users to familiarize themselves with the behavior of the generators.

The illustrations of the elements used in the laboratory exemplify their utilization during the studies developed in the physical laboratory; thus, the user has the perception of equipment utilization.

The MATLAB tool facilitates software development by having internal functions like "solve". The use of Simulink allows obtaining the necessary dynamic results for the user to carry out a specific analysis.

References

1. Fegert, J.M., Vitiello, B., Plener, P.L., Clemens, V.: Challenges and burden of the coronavirus. Child and Adolescent Psychiatry (2020)
2. Ong, C.-M.: Dynamic Simulation of Electric Machinery using Matlab/Simulink, 1st edn. Prentice Hall, Hoboken (1998)
3. Martínez, H.A.: Compendio Máquinas Eléctricas. Escuela Politécnica Nacional, Quito (2019)
4. Guru, B., Huseyin, H.: Electric Machinery and Transformers. Oxford University Press, New York (2001)
5. Fitzgerald, A., Kingsley, C., Umans, S., James, B.: Electric Machinery, vol. 5. McGraw-Hill, New York (2003)
6. Játiva, J.: Incorporación de Límites por Restricciones de Servicios Auxiliares en las Curvas de Capacidad de Generadores de Centrales Eléctricas. Revista Politécnica (2013)
7. Chen, Z., Blaabjerg, F., Iov, F.: A Study of Synchronous Machine Model Implementations in Matlab/Simulink Simulations for New and Renewable Energy Systems. Institute of Energy Technology - Aalborg University, Denmark (2005)
8. Fuchs, E.F., Masoum, M.A.: Power Quality in Power Systems and Electrical Machines. Academic Press (2011)
9. Levi, E.: Saturation Modelling in D-Q Axis Models of Salient Pole Synchronous Machines, Liverpool L3 3AF, UK (1999)
10. Srivastava, K., Berggren, B.: Simulation of Synchronous Machines in Phase Coordinates Including Magnetic Saturation. Elsevier Science S.A. (1999)

Virtual Laboratory for Induction Motor Simulation

Esteban Chiguano, Luis Hidalgo, Patricia Otero(✉), and Nelson Granda

National Polytechnic School, Quito, Ecuador
{patricia.otero,nelson.granda}@epn.edu.ec

Abstract. This document presents the design of a simulation module that serves as a complement to the teaching of the practices of the physical laboratory of Electrical Machines of the Escuela Politécnica Nacional, since it allows to simulate the real behavior of polyphase induction motors at the time of their startup, as well as allows obtaining the necessary data for the calculations of its parameters, all this complying with the standards established by the IEEE. The tests established in the simulation module have been taken from the guide sheets currently used in the physical laboratory of Electrical Machines, so the module intends for the user to simulate the interaction with real machines at any time that they need it. A brief analysis of the system of equations as a function of the state variables of the three-phase and single-phase motor is presented, as well as a summary of the data collection and methodology used for the development of the simulation module with a brief explanation of the different screens it has. Finally, the results obtained both in the physical laboratory practice and in the module are presented.

Keywords: Induction motor · MatLab simulation · Induction motor test · Induction motor model

1 Introduction

Given the problems raised at the end of 2019 and the beginning of 2020 due to the Covid-19 pandemic, the suspension of academic activities in the world has pointed out that it is necessary to introduce a teaching method adapted to technological progress for the field of engineering. For this reason, a simulation module was developed in order to analyze, experiment and understand the operation of polyphase and single-phase induction motors through different experiments. The simulation process was based on the mathematical model of single-phase and polyphase induction motors, to approach the real behavior of induction motors available in the Electrical Machines laboratory of the Escuela Politécnica Nacional. The objective of this work is that the software developed can be used as a complement to the physical laboratory of Electrical Machines and a teaching tool. It was necessary to obtain the parameters of the induction motors available in the laboratory by submitting them to the different tests necessary to be able to capture their behavior in different operating scenarios such as: stable state, transitory state and for the user's own experimentation conditions.

V. Robles-Bykbaev et al. (Eds.): CITIS 2022, LNNS 607, pp. 430–438, 2023.
https://doi.org/10.1007/978-3-031-24327-1_37

2 AC Induction Motor

A motor is a machine that transforms electrical energy into mechanical energy to do work. For this reason, to properly choose an engine, should be considered the power and work it is going to have [1].

An induction motor consists of a fixed part (stator) and a moving part (rotor), which are separated by a smaller space called the air gap and does not have a commutator, slip rings or brushes, which eliminates the losses that these produce and result in a robust motor, of high reliability, high efficiency and with a minimum maintenance period [1].

2.1 Operating Principle

When a motor is excited by a balanced polyphase source, a constant but rotating magnetic field is produced in the stator in the air gap, said field cuts the rotor conductors, inducing voltages that in turn generate currents in the short-circuited windings of the rotor. The interaction between the field produced by the stator and the currents induced in the rotor produces a force or torque that causes the motor to rotate; this force is called breakout torque [2].

The speed at which the induction motor rotates will always be fixed and lower than the speed of the rotating flow (synchronous speed), this speed will also depend on the industrial frequency at which it operates. There is also a relative movement between the conductors and the rotating flux to be able to generate an electromotive force in the rotor conductors.

By considering N_m as the speed of the rotor with a given load and N_s as the synchronous speed, the rotating field of the rotor will move with a relative speed of:

$$N_r = N_s - N_m \tag{1}$$

N_r is called the slip speed and shows the relationship between the rotor speed and the magnetic field speed. The ratio between the synchronous speed and the relative speed of the motor is called slip (s) where:

$$s = \frac{N_r}{N_s} \tag{2}$$

$$s = \frac{N_s - N_m}{N_s} \tag{3}$$

You can write the rotor speed in terms of synchronous speed and slip as:

$$N_m = (1 - s)N_s \tag{4}$$

In steady state of the rotor the slip of the machine is 1 in pu so the frequency of the rotor is the same as the field, this is different when the motor is rotating due to the relative speed so the frequency is affected by the slip of the machine.

$$f_r = sf \tag{5}$$

where f_r is the rotor frequency, s is the slip, f is the network frequency.

Since the frequency of the induced electromotive force (emf) of the induction motor is directly affected and with a generally low slip, the losses in the motor core can be neglected.

2.2 Mathematical Model of the Three-Phase Induction Motor

The mathematical model in [3] was used and it was necessary to develop the equations of state variables, voltages, and currents for input into the respective blocks of the simulation software.

The goal of the mathematical model is to bring the stator components of an induction machine to a reference system qd fixed to the rotor with the positive axis d aligned with the magnetic axis of the field winding, therefore, these demonstrations are not the subject of study of this paper, so the main equations represented in terms of the motor flux concatenations are presented:

$$
\begin{aligned}
i_{qs}^{s} &= \frac{\psi_{qs}^{s}-\psi_{mq}^{s}}{x_{ls}} \\
i_{ds}^{s} &= \frac{\psi_{ds}^{s}-\psi_{md}^{s}}{x_{ls}} \\
i_{qr}^{\prime s} &= \frac{\psi_{qr}^{\prime s}-\psi_{mq}^{s}}{x_{lr}^{\prime}} \\
i_{dr}^{\prime s} &= \frac{\psi_{dr}^{\prime s}-\psi_{md}^{s}}{x_{lr}^{\prime}} \\
\psi_{qs}^{s} &= x_{ls} i_{qs}^{s} + \psi_{mq}^{s} \\
\psi_{ds}^{s} &= x_{ls} i_{ds}^{s} + \psi_{md}^{s} \\
\psi_{qr}^{\prime s} &= x_{lr}^{\prime} i_{qr}^{\prime s} + \psi_{mq}^{s} \\
\psi_{dr}^{\prime s} &= x_{lr}^{\prime} i_{dr}^{\prime s} + \psi_{md}^{s}
\end{aligned}
\tag{6}
$$

$$
\begin{aligned}
i_{0s} &= \frac{\omega_b}{x_{ls}} \int (v_{0s} - i_{0s} r_s) dt \\
i_{0s}^{\prime} &= \frac{\omega_b}{x_{lr}^{\prime}} \int \left(v_{0s}^{\prime} - i_{0r}^{\prime} r_r^{\prime}\right) dt
\end{aligned}
\tag{7}
$$

where:

v_{0s}: voltage stator referenced on the 0-axis
$v_{0s}^{\prime}$: voltage stator referenced on the 0-axis referred to the stator side
i_{qs}^{s}: steady-state stator current referenced on the q-axis
i_{ds}^{s}: steady-state stator current referenced on the d-axis
$i_{qr}^{\prime s}$: steady-state rotor current referenced on the q-axis referred to the stator side
$i_{dr}^{\prime s}$: steady-state rotor current referenced on the d-axis referred to the stator side
i_{0s}: current stator referenced on the 0-axis
$i_{0s}^{\prime}$: current stator referenced on the 0-axis referred to the stator side
$i_{0r}^{\prime}$: current rotor referenced on the 0-axis referred to the stator side
r_s: resistance stator
$r_r^{\prime}$: resistance rotor referred to the stator side
ψ_{qs}^{s}: steady-state stator flux concatenations referenced to the q-axis
ψ_{ds}^{s}: steady-state stator flux concatenations referenced to the d-axis
$\psi_{qr}^{\prime s}$: steady-state rotor flux concatenations on the q-axis referred to the stator side
$\psi_{dr}^{\prime s}$: steady-state rotor flux concatenations on the q-axis referred to the stator side
ψ_{mq}^{s}: steady state mutual flux concatenations on the q-axis.
ψ_{md}^{s}: steady state mutual flux concatenations on the d-axis.
x_{ls}: stator reactance
$x_{lr}^{\prime}$: rotor reactance referred to the stator side
ω_b: speed.

The flow concatenation equations according to the state variables to be introduced in the simulation are as follows:

$$\begin{aligned} \psi_{qs}^{s} &= \omega_b \int \left\{ v_{qs}^{s} + \frac{r_s}{x_{ls}} \left(\psi_{mq}^{s} - \psi_{qs}^{s} \right) \right\} dt \\ \psi_{ds}^{s} &= \omega_b \int \left\{ v_{ds}^{s} + \frac{r_s}{x_{ls}} \left(\psi_{md}^{s} - \psi_{ds}^{s} \right) \right\} dt \end{aligned} \tag{8}$$

$$\begin{aligned} \psi_{qr}^{\prime s} &= \omega_b \int \left\{ v_{qr}^{\prime s} + \frac{\omega_r}{\omega_b} \psi_{dr}^{\prime s} + \frac{r_r^{\prime}}{x_{lr}^{\prime}} \left(\psi_{mq}^{s} - \psi_{qr}^{\prime s} \right) \right\} dt \\ \psi_{dr}^{\prime s} &= \omega_b \int \left\{ v_{dr}^{\prime s} - \frac{\omega_r}{\omega_b} \psi_{qr}^{\prime s} + \frac{r_r^{\prime}}{x_{lr}^{\prime}} \left(\psi_{md}^{s} - \psi_{dr}^{\prime s} \right) \right\} dt \end{aligned} \tag{9}$$

$$\begin{aligned} \psi_{mq}^{s} &= x_M \left(\frac{\psi_{qs}^{s}}{x_{ls}} - \frac{\psi_{qr}^{\prime s}}{x_{lr}^{\prime}} \right) \\ \psi_{md}^{s} &= x_M \left(\frac{\psi_{ds}^{s}}{x_{ls}} - \frac{\psi_{dr}^{\prime s}}{x_{lr}^{\prime}} \right) \end{aligned} \tag{10}$$

2.3 Mathematical Model of the Single-Phase Induction Motor

The main disadvantage of this type of motor is that, having only one phase in the stator winding, is that the magnetic field does not rotate, therefore, it has no starting torque; for this reason, an auxiliary winding is required to produce the rotating field [2].

The auxiliary winding shaft is placed approximately 90° ahead of the main winding, this winding is smaller in cross section, requires a smaller number of turns and has a higher inductance resistance ratio than the main winding.

By advancing the axis of the auxiliary winding we already have a perpendicularity between the auxiliary winding and the main winding, for this reason to simplify the analysis, the stationary *qd* axes can be aligned with the perpendicular axes of the physical windings.

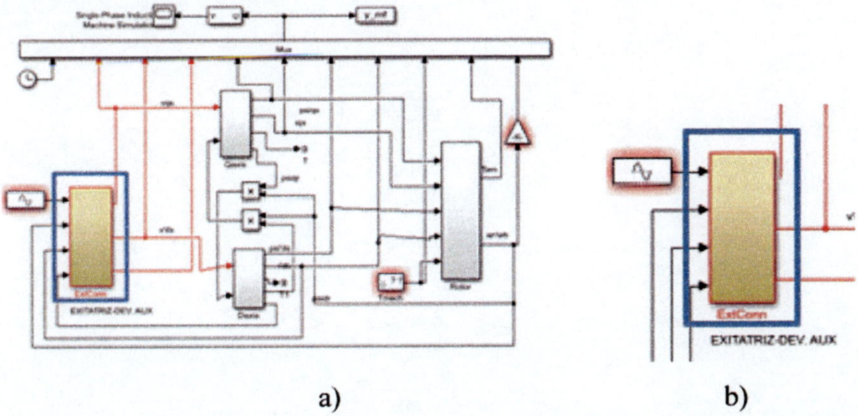

Fig. 1. a) Modeling of the single-phase induction motor, b) Modelling of the auxiliary winding [3]

The development of the single-phase mathematical model in Fig. 1a) is designed from the three-phase motor model. For the simulation, the modeling of the "qd" and rotor axes are like those of the three-phase induction motor model. A block representing the auxiliary winding of the single-phase motor is added, Fig. 1b), which has been developed in [3].

3 Methodology for the Development of the Module

In this section, the methodology and simulation tool that will be used to collect, develop, and validate the information of this degree work are described.

3.1 Compilation of Information from the Physical Laboratory of Electrical Machines

The objective is to emulate the physical laboratory in a virtual for this, the modules must have the same elements used in the practices, the information of each of the engines available in the laboratory is presented in [4].

3.2 Simulation Software

The module has been developed in MatLab software version 2020b, due to the existing compatibility between App designer and Simulink, to create an independent executable file.

4 Implementation of the Simulation Module

The Fig. 2 shows the sequence of activities in the application in a general way, considering that within each step there are subsystems of each practice that are detailed later.

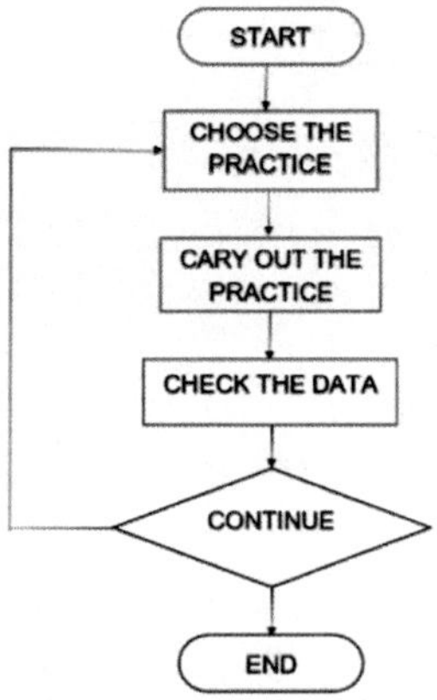

Fig. 2. General flow diagram

The following is a summary of the laboratory practices that the module allows the user to carry out for subsequent analysis and validation (Table 1):

Table 1. Practices summary

Screen	Practice description
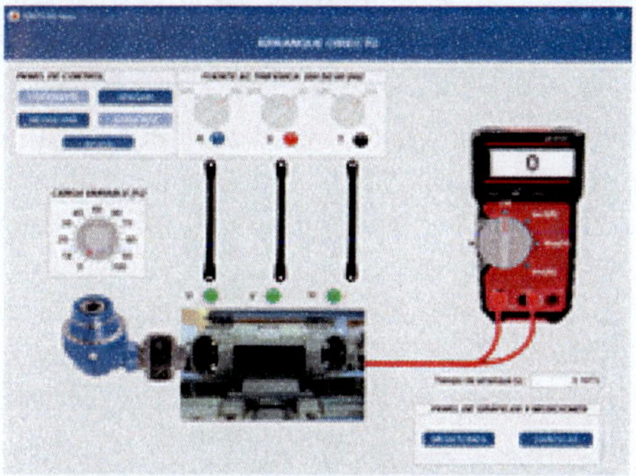	Practice 1: consists of several starting types: direct starting, Y - Δ starting, rotor resistance starting, reduced voltage starting, phase change braking, These types of starting and braking comply with the same dynamics where the simulation module has the necessary restrictions for the user to carry out the practice correctly, in addition, it has a "HELP" button in each practice that provides additional information for its correct handling.
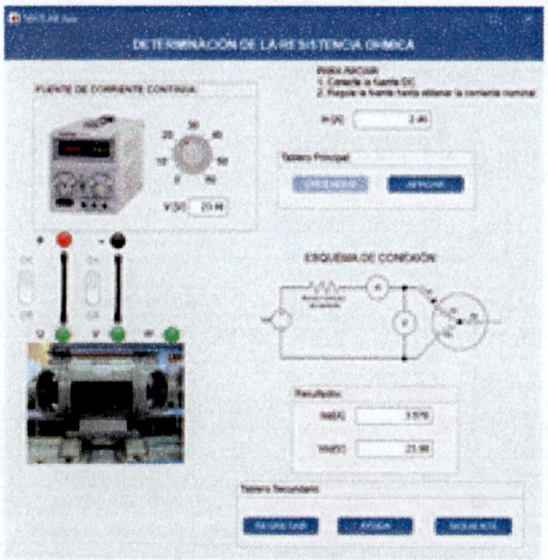	Practice 2: Starting with data entry and then moving on to the tests that must be done to the engine in the order in which it is necessary to calculate the machine parameters established in the guide sheets and as they are carried out in the IEEE standards. This practice is divided into three parts: DC test, open circuit test, locked rotor test The standards establish that the locked rotor test must be performed at different frequencies [5], however, for study purposes of this project it was developed in the way it is physically performed in the laboratory and additionally
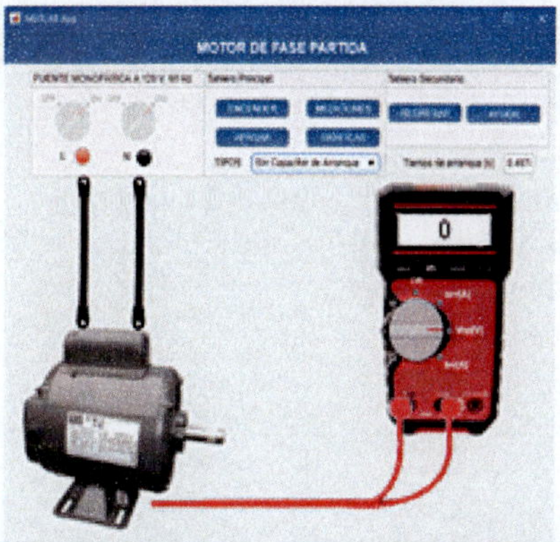	Practice 3: shows the types of starting that the single-phase split-phase induction motor has and for each start, the ignition, measurement, and shutdown are fulfilled [6]. Single-phase induction motors are divided into different types such as: normal split phase, split phase with start capacitor, split phase with start capacitor and start capacitor

5 Analysis of Results

This section details the tests that were carried out in the physical laboratory of electrical machines for their subsequent evaluation, analysis and comparison with the results obtained from the simulation software.

Once the data from the respective motor starting tests were obtained, the laboratory results were compared with those of the software to obtain the errors described in Table 2.

Which are justified because there are several factors that alter the measurement such as: phase imbalance, worn and inaccurate measurement equipment, measurement appreciation.

The measurement equipment was not able to determine with accuracy of the transient values of the starting currents and depend on the appreciation of the observer, so that at the time of the tests it does not have any of these characteristics that, unlike the software, being a program designed based on the mathematical model.

The induction motor presents results with precision and accuracy in transient processes from which, by means of programming codes and algorithms, the exact value of parameters can be extracted.

Table 2. Error calculations in motors No. 1 and 2

Start test errors												
MOTOR N°1												
	LABORATORY				SOFTWARE				RELATIV ERROR [%]			
	Direct start	Y-D start	R.R. start	V.R. stat	Direct start	Y-D start	R.R. start	V.R. stat	Direct start	Y-D start	R.R. start	V.R. stat
Vop [V]	213,2	213,2	213,2	213,1	220	220	220	220	3,1895	3,1895	3,1895	3,2379
Varr [V]	0	0	0	116,9	0	0	0	117,7				0,6843
Iarr [A]	16,9	19,6	11,5	16,2	21,7	17,2	11,26	20,21	28,4024	12,2449	2,0870	24,7531
Iee [A]	3,22	3,22	3,43	3,22	3,412	3,414	3,414	3,778	5,9627	6,0248	0,4665	17,3292
MOTOR N°2												
	LABORATORY				SOFTWARE				RELATIV ERROR [%]			
	Direct start	Y-D start	R.R. start	V.R. stat	Direct start	Y-D start	R.R. start	V.R. stat	Direct start	Y-D start	R.R. start	V.R. stat
Vop [V]	218	218	219,4	215	220	220	220	220	0,9174	0,9174	0,2735	2,3256
Varr [V]	0	0	0	102,2	0	0	0	102,4				0,1957
Iarr [A]	29,9	21,1	15,4	22,4	21,7	17,5	11,48	16,9	27,4247	17,0616	25,4545	24,5536
Iee [A]	5,91	5,7	6,19	5,8	3,411	3,411	2,786	3,708	42,2843	40,1579	54,9919	36,0690

For the second part of the analysis of results, the parameters of the machines were calculated with the values of the physical tests and the values of the simulation to show the error between the parameters directly as shown in Table 3.

Table 3. Error calculations for motors No. 1 and 2

	MOTOR 1					
	TEST 1	SOFTWARE	RELATIVE ERROR [%]	TEST 2	SOFTWARE	RELATIVE ERROR [%]
Xls [Ω]	5,7543	5,6976	0,99	5,9395	5,8672	1,22
X'lr [Ω]	5,7543	5,6976	0,99	5,9395	5,8672	1,22
Xm [Ω]	99,902	89,9058	10,01	99,6658	89,7044	9,99
R'r [Ω]	4,8983	4,9495	1,05	4,7396	4,7902	1,07
Rs [Ω]	1,6056	1,60595	0,02	1,6304	1,6306	0,01
J [kg m2]	0,0116	0,0116	0,00	0,0116	0,0116	0,00
	MOTOR 2					
	TEST 1	SOFTWARE	RELATIVE ERROR [%]	TEST 2	SOFTWARE	RELATIVE ERROR [%]
Xls [Ω]	2,5646	2,5289	1,39	2,6046	2,5664	1,47
X'lr [Ω]	2,5646	2,5289	1,39	2,6046	2,5664	1,47
Xm [Ω]	55,7541	53,7847	3,53	56,2312	54,2124	3,59
R'r [Ω]	1,6226	1,6249	0,14	1,5616	1,5645	0,19
Rs [Ω]	0,8537	0,8534	0,04	0,8025	0,8027	0,02
J [kg m2]	0,04119	0,04119	0,00	0,04119	0,04119	0,00

6 Discussion

Aside the simulation results, once the software was completed a group of 100 engineering students were volunteer to run a preliminary pilot test. When they were asked the question: ¿Did the software reinforced your knowledge about induction motors? the results are shown in Fig. 3

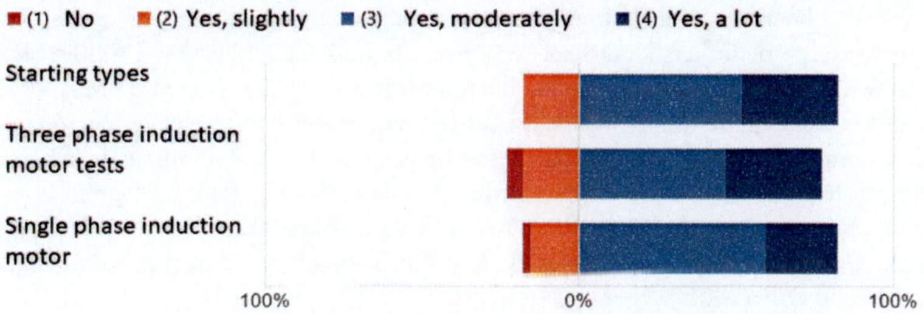

Fig. 3. Pilot test results

Most of them have a positive response which might indicate that these type of tools for teaching are valuable as a complement and specially useful in cases where the student can't assist in person to the laboratory.

7 Conclusions

In the present work, a simulation module of any type of induction motor was developed. The software shows the behavior of the three-phase motor against different types of starts, as well as the tests carried out to obtain the electrical data of the machine with an error percentage of less than 3%, which makes this module a learning tool reliable.

The mathematical models of the three-phase and single-phase induction motor are complied with, which have been adapted to the needs required to carry out the types of starts, motor tests and conditioning of variables in order that the developed model works for any type of induction motor with emphasis on the motors found in the physical laboratory of electrical machines of the EPN.

A simulation module capable of providing specific data of current, voltage and power that appear at the transient moment of the machine (within the first 10 ms) was implemented, this allows to know how convenient it is for the machine to be started using a method. Another, in addition to knowing the robustness and efficiency of the connected motor.

The mathematical model of the three-phase and single-phase induction motor provide a response consistent with the data on the nameplate of a motor, which is why it has been validated with the values of current, voltage and power in steady state, in addition to that, for the transitory period, the starting currents show curves as a function of time that meet the objective of showing the difference between the different types of starting.

References

1. Motores eléctricos y accionamientos para la industria de procesamiento | Editores (2021). https://www.editores-srl.com.ar/Motores%20el%C3%A9ctricos%20y%20accionamientos%20para%20la%20industria%20de%20procesamiento. Accessed 02 May 2021
2. Fitzegarl, A.E., Kingsles, C., Umans, S.: Máquinas Eléctricas, Sexta. Mc Graw Hill
3. Chee Mung, O.: Dynamic simulation of Electric Machinery Using MATLAB, Primera, vol. 1. Upper Saddle River, New Jersey (1998)
4. Sandoval, T., Javier, D.: Diagnóstico y actualización del tablero de control y mesas de trabajo del laboratorio de máquinas eléctricas de la Escuela Politécnica Nacional (2009) [En línea]. Disponible en: http://bibdigital.epn.edu.ec/handle/15000/1598. Accessed 02 May 2021
5. Luna, S., Graciela, K.: Estimación de parámetros de un motor trifásico de inducción a partir de mediciones en el arranque directo de corrientes transitorias del estator utilizando la herramienta de estimación de parámetros de Matlab/Simulink dic (2018). [En línea]. Disponible en: http://bibdigital.epn.edu.ec/handle/15000/19907. Accessed 02 May 2021
6. Guru, B., Huseyin, H.: Máquinas eléctricas y transformadores, Tercera. Alfaomega Grupo Editor, México D.F. (2006)

Homogeneity of Engineering Courses from an Assignment Management Model in Virtual Learning Environments

Alicia Cedeño-Tello[1,2,3] and Joe Llerena-Izquierdo[1,4](✉)

[1] Universidad Politécnica Salesiana, Guayaquil, Ecuador
jllerena@ups.edu.ec

[2] Maestría TIC para la Educación, Universidad Politécnica Salesiana, Guayaquil, Ecuador

[3] Grupo de Investigación de Educación e Información Científica, GIEDIC, Guayaquil, Ecuador

[4] Grupo de Investigación en Enseñanza-Aprendizaje de las Ciencias para La Ingeniería, Guayaquil, Ecuador

Abstract. The management models for assigning courses to professors in higher education have shown the need to involve technologies for the construction of homogeneous virtual classrooms. This work presents a proposal for the creation of pilot virtual classrooms that allow a quick start of the teaching-learning process in first year students, measuring its impact. The objective is to determine the levels of acceptance of pre-existing significant learning resources in large groups of the same subject. The results, collected through a survey of 64 participants who passed the Programming and Linear Algebra course of the Engineering Careers at Universidad Politécnica Salesiana of Guayaquil, in Ecuador, reflect satisfaction values that reach 91%. In addition, the content of the classrooms influences a 94% positivity in learning and a 94% usefulness from the students' perspective. The limitations found suggest a continuous improvement in the early year's teacher assignment management model. It is expected to carry out a following study based on the experience of the teachers responsible for the courses in a longitudinal work for the next courses to be offered. This work contributes to the development of good practices in the educational management of higher education institutions.

Keywords: Virtual classroom · Teaching · LMS platform

1 Introduction

The COVID-19 pandemic has highlighted the need for the use of Information and Communication Technologies in educational management models at all levels [1, 2]. Teaching and learning processes in this context, during and after the COVID-19 pandemic [3, 4], have been integrating methodological proposals with the use of technologies or technological services that allow an effective development, especially in the direction of higher education [5, 6].

The development of content for higher education courses that allow the incorporation of technologies according to different contexts [7, 8] will depend on two key factors for

V. Robles-Bykbaev et al. (Eds.): CITIS 2022, LNNS 607, pp. 439–447, 2023.
https://doi.org/10.1007/978-3-031-24327-1_38

effective action [9, 10], the appropriate technological infrastructure [11] and the training of teachers with digital competencies [12, 13] in the management of technologies for innovative processes [14].

In Ecuador, Universidad Politécnica Salesiana offers common and generic courses every semester to all students, especially in the engineering careers, allocating about 10 courses to cover a total of 400 students in the first year.

The investment of resources in technology has not been spared by the authorities of the institution, with the latest version of the LMS platform (Moodle) for virtual environments and a human team in the Information Technology department to support teachers [15].

The course management model for the assignment of professors is foreseen within a planning prior to a new academic period. In this process the assignment of a professor can be affected by a change, withdrawal, or new distribution of his assignment (due to an increase in the number of courses or their closure) and with this there are delays in the personalization of the classrooms. Students at the beginning of each semester access the classroom and find two situations, classrooms complete with resources for the subject or classrooms in which the assigned tutor has had to be replaced and therefore there is no visible material for the participants. The teacher's response times depend largely on the ability to use the Moodle platform (with an extremely rigid template and customized by the institution) and its adaptation to the available technological structure.

The problem detected is evident from three areas: on the part of the institution, there is evidence of classrooms with scattered resources and underutilization of the infrastructure provided to the teacher; on the part of the faculty, the short times for personalization, after being assigned to a new course, evidence of delays due to technical management in assigning the course as well as the delivery of access credentials, planning of assigned hours and their changes, also the crossing of hours affects having a virtual classroom on time; And on the student's side, there is evidence of discomfort, due to the disadvantages of not having the resources at the agreed times, as well as visualizing different instructional designs in the environments when comparing them between subjects [16]. All this affects the student's learning process, so it is necessary a preventive management of the use of technology in the teaching staff.

The objective of this work is to determine the levels of acceptance of pre-existing significant learning resources in large groups of the same subject, establishing the impact of the homogeneous use of virtual classrooms from a management model in teaching-learning processes for the optimization of human and technological resources.

1.1 Planning Management of Virtual Environments

Universidad Politécnica Salesiana belongs to the group of higher education institutions in Ecuador. The university has three campuses, the main one in the city of Cuenca, and one in the city of Quito and another in the city of Guayaquil. This work is defined in the city of Guayaquil. Its strategic lines are directed in four pillars, Teaching, Research, Linkage with Society and Salesian Management and Identity. The contribution made by this research is to Teach.

The Academic Development Department, with the support of the Basic Area, carries out the timetable distribution for two months before the beginning of each academic period, the assignment of subjects is carried out considering the affinity between the teacher's degree, his or her previous teaching evaluations, and his or her academic performance (see Fig. 1). In addition, there is a group of teachers called teaching staff or support team, which assists all teachers who teach the subject, they are responsible for updating the teaching resources and carry out the teaching-learning process according to the results and objectives of the analytical plan, is called the team designer of the learning process of the course.

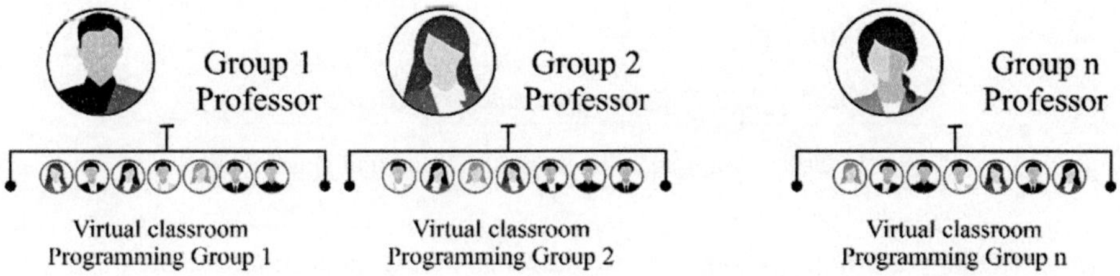

Fig. 1. Several courses are offered at the university each academic semester.

Universidad Politécnica Salesiana has an undergraduate academic offer with technical and humanistic careers that are supported by a virtual learning environment on an LMS, Moodle and its hosting in the Internet cloud on Amazon Web Service (AWS). The number of careers per city are, in the city of Cuenca 21 careers are offered, in the city of Quito 22 careers, and in the city of Guayaquil 18 careers, to these must be added 4 careers offered only online. A total of 65 university careers nationwide. In addition, it has a postgraduate academic offer, with 8 master's degrees in the city of Cuenca, 10 master's degrees in the city of Quito and 8 master's degrees in the city of Guayaquil, who also use the virtual learning platform. With a total of 962 professors, 22880 undergraduate students and 1528 graduate students, the virtual learning environment serves 25370 users directly related to teaching. A population that continuously accesses the use of virtual environment resources.

2 Material and Methods

The development of the following work uses the analytical empirical method, of a quasi-experimental type with a quantitative approach for the analysis of results. We have chosen to use the survey technique directed to a group of 64 students of the first cycle of engineering careers, who belonging to the Universidad Politécnica Salesiana, in the city of Guayaquil, Ecuador, and who participate on a voluntary basis.

In the period 2021–2022 (59, from September 2021 to February 2022) the cloning of a basic virtual classroom in the subjects of Programming and another one in Linear Algebra was carried out as a first experience, where the teaching staff, that is, the group of professors who teach a specific subject selects renewable learning objects, After a joint review, the resources are cloned in each of the subjects where the offer to students is high. With this, the groups of assigned teachers have the responsibility to review the

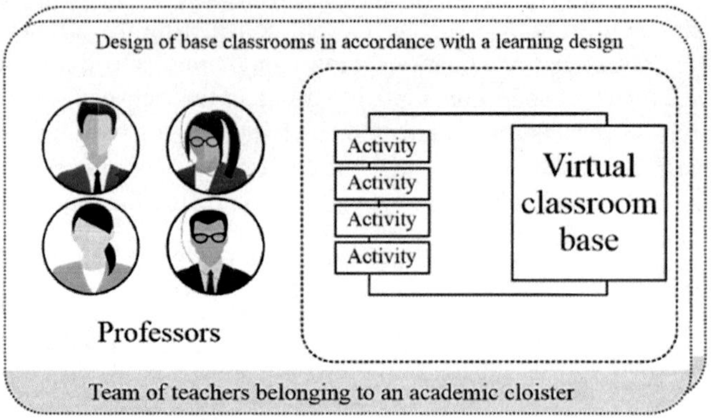

Fig. 2. The teaching team in charge of the virtual classroom.

basic contents, contribute with new ones and focus their time to improve the quality of the educational process (see Fig. 2).

The cloning of virtual classrooms comes from an allocation management model by the Academic Development Department. The number of classrooms is created according to the offer to the city's students and their acceptance. Likewise, the opening of a new course from a cloning takes no more than five minutes until the access verification by the teacher. With this, the new teacher, in addition to having the resources and activities of the virtual learning environment, is provided with the entire set of didactic educational media and significant learning activities based on an organizational and instructional design of the subject (see Fig. 3). Finally, in the continuous meetings of the virtual cloisters, the teachers share ideas and experiences that allow improving the execution of the teaching and learning process from the scope of the analytical plan of the subject.

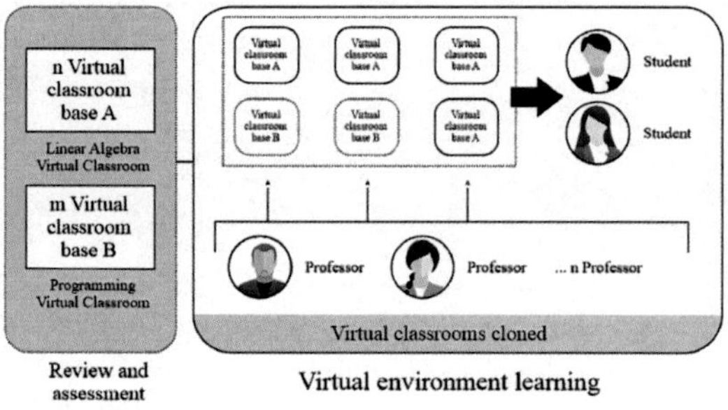

Fig. 3. Model of duplication of virtual classrooms with homogeneous base

3 Results

The results of the survey to 64 students who passed the course and participated voluntarily in this study, belong to the first courses of the Engineering Careers, 41% correspond to the area of Programming and 59% correspond to the area of Linear Algebra (see Fig. 4). With this, we have a proportional percentage of participants for each of the pilot virtual classrooms created in this first experience, which allows us to have an evaluation of the proposal.

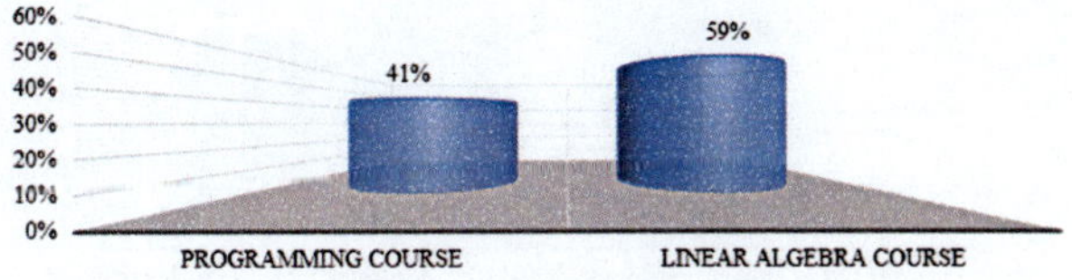

Fig. 4. Percentage of participating students by course.

Students were asked, "Do you feel satisfied with the virtual classroom that contains all the resources and activities pre-established from the beginning of the course?", according to the question, 64% are "Very satisfied", 27% are "Satisfied", 6% are "Neither satisfied nor not satisfied", 2% are "Dissatisfied" and 1% are "Very dissatisfied", (see Fig. 5), this indicates that 91% are satisfied for having found a virtual classroom with all the resources and activities ready from the beginning of the course.

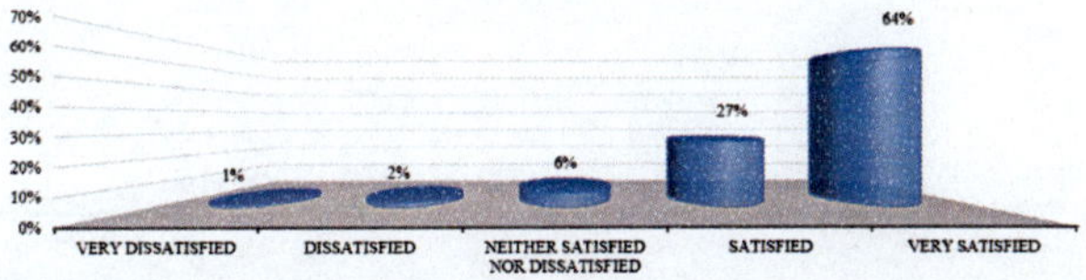

Fig. 5. Percentage of student satisfaction with the virtual classroom that contains all the resources and activities pre-established from the beginning of the course.

According to the question, "Has it been easy for you to understand the aspect of the sections of the course in the virtual classroom?", 78% said "Yes" and 22% said "No", (see Fig. 6). This result guides us to continue improving the organizational design of the sections in the virtual environment.

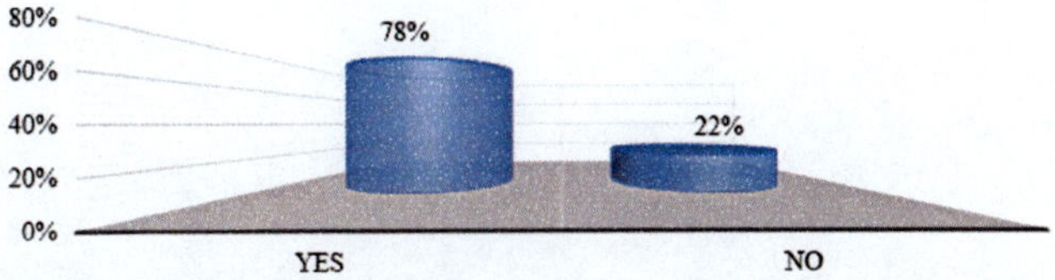

Fig. 6. Percentage of students who have found it easy to understand the look and feel of the course sections in the virtual classroom.

To the question, "Can you indicate which of the sections you found very interesting?", 16% indicated "unit 1", 20% indicated "unit 2", 14% indicated "unit 3" and 50%

indicated "all units", (see Fig. 7). This result evidence the preference of student interest in one or all the sections and that it is related to the specific contents of the topics to be covered. It was found that half of the students were interested in all the units, and a significant proportional value in the rest of them.

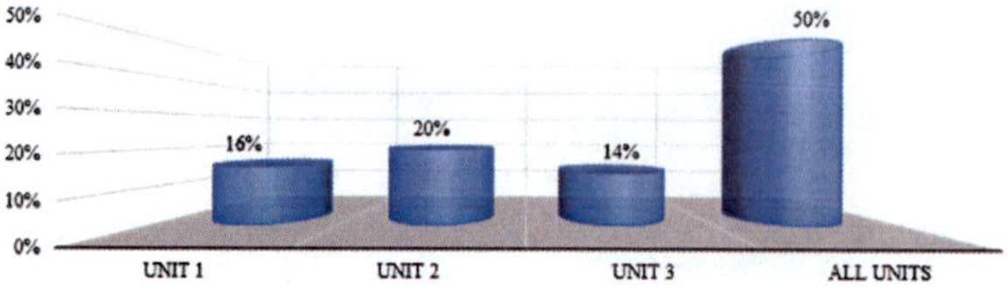

Fig. 7. Percentage of students indicating what is interesting in each unit or section.

To the question, "How do you perceive the content of the virtual classroom for your learning of the subject?", it is obtained that 61% indicate that "Very positive", 33%, "Positive", 6%, "Neither positive nor negative", (see Fig. 8). This shows that 94% of the students see the content chosen by the teacher(s) as positive for the diversity of learning in the classroom.

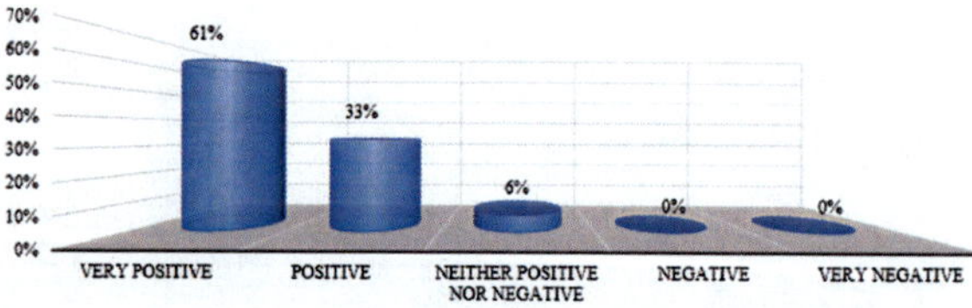

Fig. 8. Percentage of perception of the content of the virtual classroom for the learning of the subject.

According to the question, "Has the structure and design of the virtual environment been useful for your learning in this pilot classroom?", 77% found it "Totally useful", 17% "Somewhat useful" and 6% "Not very useful", (see Fig. 9). In other words, 94% find the structure and design of the virtual classroom useful for learning.

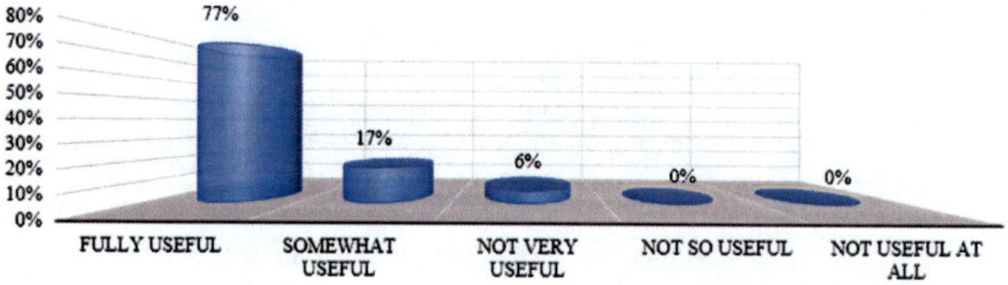

Fig. 9. Percentage of usefulness of for student learning in the pilot classroom.

4 Discussion

Since 2010, the Universidad Politécnica Salesiana has incorporated technologies to the face-to-face modality, strengthening teaching activities. After the pandemic, higher education institutions are required to take a look, once again, at the role of information technologies in the curricula [17, 18]. It is discussed that the entry of new technologies, platforms and devices generate new interaction ecosystems between teachers and students [11, 19], adding new data that represent challenges for the design of learning contents and that must be analyzed from a formative vision focused on the learner [20, 21].

The limitations encountered suggest a continuous improvement in the management model for the assignment of professors in the first years. It is expected to carry out in future works, a study from the experience of the professors responsible for the courses in a longitudinal work for the next courses to be offered. In addition to its replication in the other branches at the national level in the same university. This work contributes to the development of good practices in educational management in teaching and learning processes in higher education institutions.

5 Conclusions

The benefits of virtual environments for the operation of a course assignment management to teachers in hybrid study modalities (face-to-face or non-face-to-face) show a positive debate in front of the emerging online education situation that is still maintained with the presence of COVID-19 in our society.

This work establishes the impact of the homogeneous use of virtual classrooms from a management model in teaching-learning processes for the optimization of human and technological resources. In addition, it describes the incidence of the homogeneous use of virtual classrooms from the perspective, interest, and motivation of students to have analysis inputs that can be used by teachers for the improvement of the learning model of the subject, as well as the development of good practices in educational management involving the teaching and learning processes of higher education institutions.

Finally, the resources used, such as renewable learning objects, are in accordance with the instructional model for the creation of innovative processes in the development of hybrid education.

Acknowledgements. We would like to thank the staff of the Master in ICT for Education of the Universidad Politécnica Salesiana for the development of this work and the research groups for the research and methodological process, as well as for the accompanying process, the Research Group on Education and Scientific Information, GIEDIC (https://www.investigacion.ups.edu.ec/grupo/giedic/) and the Educational Innovation Group GIETICEA - GIEACI (https://gieaci.blog.ups.edu.ec/).

References

1. Wang, M.-J., Yang, L.-Z., Yang, W.-H.: The perceived effectiveness of the blended team-based learning (BTBL) model on promoting intercultural awareness in hospitality education. J. Hosp. Tour. Educ. 1–11 (2022). https://doi.org/10.1080/10963758.2022.2056472

2. Castro, C.N., Carvalho, L.C., Moreira, S.B.: Distance education in higher education: modern applications and management courses. In: Carvalho, L.C., Teixeira, N., and Pardal, P. (eds.) Interdisciplinary and Practical Approaches to Managerial Education and Training, pp. 25–45. IGI Global, Hershey, PA, USA (2022). https://doi.org/10.4018/978-1-7998-8239-8.ch002
3. Areşan, D., Gabriel Ţîru, L.: Students satisfaction with the online teaching process. Acad. Int. Sci. J. **25**, 184–193 (2022). https://doi.org/10.7336/academicus.2022.25.11
4. Chaveesuk, S., Chaiyasoonthorn, W.: COVID-19 in emerging countries and students' intention to use cloud classroom: evidence from Thailand. Educ. Res. Int. **2022**, 1–13 (2022). https://doi.org/10.1155/2022/6909120
5. Ferri, F., Grifoni, P., Guzzo, T.: Online learning and emergency remote teaching: opportunities and challenges in emergency situations. In: Societies, p. 86 (2020). https://doi.org/10.3390/soc10040086
6. Meristo, M., Ljalikova, A., Heero, A.: Foreign language teacher perceptions of the virtual classroom. Int. J. Comput. Lang. Learn. Teach. **12**, 1–19 (2022). https://doi.org/10.4018/IJCALLT.291112
7. Songkram, N., Songkram, N., Chootongchai, S., Samanakupt, T.: Developing students' learning and innovation skills using the virtual smart classroom. Int. J. Emerg. Technol. Learn. **16**(4), 34–51 (2021). https://doi.org/10.3991/ijet.v16i04.15221
8. Wannapiroon, N., Pimdee, P.: Thai undergraduate science, technology, engineering, arts, and math (STEAM) creative thinking and innovation skill development: a conceptual model using a digital virtual classroom learning environment. Educ. Inf. Technol. 1–28 (2022). https://doi.org/10.1007/S10639-021-10849-W/FIGURES/5
9. Ansari, F., Muzyyab Ajmal, M., Basit, A., Ali Khattak, H., Nawaz, A., Ahmed, S.: Evaluating readiness for technology integration in the classroom post-pandemic. In: 2022 Advances in Science and Engineering Technology International Conferences (ASET), pp. 1–7. IEEE (2022). https://doi.org/10.1109/ASET53988.2022.9735101
10. Jiang, S.: The new trend of EFL E-learning in the post-pandemic period from the perspective of ubiquitous learning. In: the International Conference on Educational Innovation and Teaching Methodology (2020)
11. Rehatschek, H.: Outline of possible synchronous solutions and experiences in order to supply large groups of students with learning content in classroom and mixed classroom/distance scenarios. In: International Conference on Interactive Collaborative Learning, pp. 523–534 (2022). https://doi.org/10.1007/978-3-030-93904-5_52
12. Alfoudari, A.M., Durugbo, C.M., Aldhmour, F.M.: Understanding socio-technological challenges of smart classrooms using a systematic review. Comput. Educ. **173**, 104282 (2021). https://doi.org/10.1016/J.COMPEDU.2021.104282
13. Yang, Y., Zhang, H., Chai, H., Xu, W.: Design and application of intelligent teaching space for blended teaching. Interact. Learn. Environ. 1–18 (2022). https://doi.org/10.1080/10494820.2022.2028857
14. Sumba Nacipucha, N., Cueva Estrada, J.M., Conde Lorenzo, E., Mármol Castillo, M.: Enseñanza superior en el Ecuador en tiempos de COVID 19 en el marco del modelo TPACK. Rev. San Gregor. 171–186 (2020).https://doi.org/10.36097/rsan.v1i43.1524
15. Llerena Izquierdo, J.: Uso de AVAC en la Universidad Politécnica Salesiana. In: Congreso Nacional Pedagogía de Don Bosco: Reflexiones, experiencias y desafíos "Memoria académica", pp. 335–345. Quito (2014)
16. Llerena-Izquierdo, J.: Virtual classroom design model and its relation to student motivation and performance in a moodle learning environment during the emergency of COVID-19. In: Springer (ed.) Berrezueta, S., Abad, K. (eds) Doctoral Symposium on Information and Communication Technologies - DSICT. Lecture Notes in Electrical Engineering, pp. 21–32. Springer, Cham (2022). https://doi.org/10.1007/978-3-030-93718-8_3

17. Wood-Harper, T.: Emerging EdTechs amidst the COVID-19 pandemic: cases in higher education institutions. In: Fostering Communication and Learning With Underutilized Technologies in Higher Education, pp. 93–107. IGI Global (2021)
18. Llerena-Izquierdo, J., Ayala-Carabajo, R.: University teacher training during the COVID-19 emergency: the role of online teaching-learning tools. In: Rocha, Á., Ferrás, C., López-López, P.C., Guarda, T. (eds.) ICITS 2021. AISC, vol. 1331, pp. 90–99. Springer, Cham (2021). https://doi.org/10.1007/978-3-030-68418-1_10
19. Idrissi, W.E.L.M.E.L., Chemsi, G., Kababi, K.E.L., Radid, M.: The impact of serious game on the nursing students' learning, behavioral engagement, and motivation. Int. J. Emerg. Technol. Learn. **17**, 18–35 (2022)
20. Tärning, B., Ternblad, E.-M., Haake, M., Gulz, A., Nirme, J.: Lessons learned from a study on distractions in virtual learning environments: reliability, ecological validity, and an elusive social component. Presence Teleoperators Virtual Environ. **28**, 65–85 (2019). https://doi.org/10.1162/pres_a_00342
21. Mahmood, S.: Instructional strategies for online teaching in COVID-19 pandemic. Hum. Behav. Emerg. Technol. **3**, 199–203 (2021). https://doi.org/10.1002/HBE2.218

Author Index

V. Robles-Bykbaev et al. (Eds.): CITIS 2022, LNNS 607, pp. 449–450, 2023.
https://doi.org/10.1007/978-3-031-24327-1